Applied Probability and Statist

BAILEY · The Elements … cations to the Natural Sciences

BARTHOLOMEW · Stoc… *Second Edition*

BENNETT and FRANKL… …istry and the Chemical Industry

BHAT · Elements of Applied Stochastic Processes

BLOOMFIELD · Fourier Analysis of Time Series: An Introduction

BOX and DRAPER · Evolutionary Operation: A Statistical Method for Process Improvement

BROWN and HOLLANDER · Statistics: A Biomedical Introduction

BROWNLEE · Statistical Theory and Methodology in Science and Engineering, *Second Edition*

BURY · Statistical Models in Applied Science

CHERNOFF and MOSES · Elementary Decision Theory

CHOW · Analysis and Control of Dynamic Economic Systems

CLELLAND, deCANI, BROWN, BURSK, and MURRAY · Basic Statistics with Business Applications, *Second Edition*

COCHRAN · Sampling Techniques, *Third Edition*

COCHRAN and COX · Experimental Designs, *Second Edition*

COX · Planning of Experiments

COX and MILLER · The Theory of Stochastic Processes, *Second Edition*

DANIEL · Application of Statistics to Industrial Experimentation

DANIEL and WOOD · Fitting Equations to Data

DAVID · Order Statistics

DEMING · Sample Design in Business Research

DODGE and ROMIG · Sampling Inspection Tables. *Second Edition*

DRAPER and SMITH · Applied Regression Analysis

DUNN and CLARK · Applied Statistics: Analysis of Variance and Regression

ELANDT-JOHNSON · Probability Models and Statistical Methods in Genetics

FLEISS · Statistical Methods for Rates and Proportions

GNANADESIKAN · Methods for Statistical Data Analysis of Multivariate Observations

GOLDBERGER · Econometric Theory

GROSS and CLARK · Survival Distributions

GROSS and HARRIS · Fundamentals of Queueing Theory

GUTTMAN, WILKS and HUNTER · Introductory Engineering Statistics, *Second Edition*

HAHN and SHAPIRO · Statistical Models in Engineering

HALD · Statistical Tables and Formulas

HALD · Statistical Theory with Engineering Applications

HARTIGAN · Clustering Algorithms

HILDEBRAND, LAING and ROSENTHAL · Prediction Analysis of Cross Classifications

HOEL · Elementary Statistics, *Fourth Edition*

HOLLANDER and WOLFE · Nonparametric Statistical Methods

HUANG · Regression and Econometric Methods

continued on back

Linear Regression Analysis

LINEAR REGRESSION ANALYSIS

G.A.F. SEBER

Professor of Statistics
University of Auckland
Auckland, New Zealand

John Wiley & Sons

New York · London · Sydney · Toronto

Library of Congress Cataloging in Publication Data:

Seber, George Arthur Frederick.
Linear regression analysis.

(Wiley series in probability and mathematical statistics)
Includes bibliographical references and index.
1. Regression analysis. I. Title.
QA278.2.S4 519.5′36 76-40117
ISBN 0-471-01967-4

Printed in the United States of America

10 9 8 7 6 5 4 3 2 1

Preface

Regression analysis is an often used tool in the statistician's toolbox. The theory is elegant and the computational problems intriguing, so that both "pure" and "applied" statisticians can feel at home in the subject. For example, among the theoreticians there is still an ongoing interest in least squares with all its generalizations and special cases. At the same time practitioners continue to develop a wide range of graphic methods for testing models and examining underlying assumptions. As numerical analysis and statistics have slowly intertwined, statisticians have been made to realize the difficulties associated with certain time-honored computational procedures. The development of regression computer programs that are efficient and accurate is now recognized as an important part of statistical research.

However, this continuing research interest in regression across the pure–applied spectrum creates problems for the textbook writer. As demonstrated by current books on the subject, the material can be presented at a variety of levels of mathematical difficulty ranging from the very general treatments of Seber [1966], Searle [1971], and Rao [1973], for example, to the more discursive approaches of Williams [1959] and Sprent [1969]. Clearly such a variety of books is needed because there is a wide diversity of readers. However, while teaching the subject continuously over the past 10 years I have become more and more aware of the need for a suitable university text that takes a middle road between giving no proofs of results and giving proofs in complete generality. Because regression analysis and much of analysis of variance are concerned with full rank models, it would appear that the less than full rank case tends to be overemphasized for the sake of greater generality. In particular, the generalized inverse has been rather overworked to the detriment of the simple geometric ideas that underlie least squares. Clearly the generalized inverse has its uses, but its role needs to be kept in perspective.

Regression analysis is an applied subject describing methods for handling data; ideally, any theoretical course should be backed up by practical work. This raises the question of whether a single textbook should try to deal, in detail, with both the theoretical and computational aspects of the subject. Clearly such a task is not easy and in this age of statistical package programs it would seem more appropriate to treat the two aspects quite separately. For example, procedures that have an elegant theory may be computationally unsatisfactory, whereas messy, complicated algorithms may be efficient and accurate. Because packages vary from place to place according to the local computer facilities, the ultimate solution might be to have a theoretical textbook that includes the computational aspects of regression in broad outline only and a practical "manual" giving numerical examples and details of packages; a forerunner of the latter type of book is Daniel and Wood [1971].

With the above thoughts in mind I have endeavored to write a theoretical book that is satisfying for the mathematically minded reader but that does not lose the reader in generalities. I have also endeavored to give an up-to-date account of computational methods and algorithms currently in use without getting entrenched in minor computing details. Since the research literature on regression continues to grow rapidly, I have surveyed the better-known statistical journals with the aim of producing a book that I hope will be useful as a general reference. The basic prerequisites for reading this book are a good knowledge of matrix algebra and some acquaintance with straight-line regression and simple analysis of variance models.

The first four chapters provide a fairly standard formal treatment of least squares fitting and hypothesis testing for the multiple linear regression model. In Chapter 1 expectation and covariance operators for vectors of random variables are introduced gently, and in Chapter 2 the multivariate normal distribution and certain theorems on quadratic forms are considered. Chapter 3 deals with least squares estimation and includes generalized least squares, the less than full rank case, and estimation with restrictions. Chapter 4 considers, in detail, the F-test for a linear hypothesis, and in Chapter 5 there is a discussion of confidence intervals and simultaneous inference as applied to regression models; prediction and inverse prediction (discrimination) intervals are also considered. Chapter 6 examines the assumptions underlying the least squares theory and various methods of testing these assumptions are provided. Because straight-line and polynomial fitting are important techniques, Chapters 7 and 8 are devoted to these two topics, respectively. Chapter 9 exploits the close relationship that exists between regression and analysis of variance models and provides simple procedures for carrying out an analysis of variance; attention is confined mainly to balanced (orthogonal) designs. In Chapter 10 the analysis of covariance is also considered from a regression viewpoint and, because of close ties with analysis of covariance, the topic of missing observations is considered in detail. Chapters 11 and 12 deal with the computational aspects of regression analysis: Chapter 11 is devoted to algorithms for least squares fitting and Chapter 12 considers the problem of choosing the best regression subset from a set of likely regressor (independent) variables.

Appendixes A and B contain a number of matrix results whose proofs are not always readily accessible, and Appendix C describes probability plotting. Appendixes D, E, and F give some statistical tables which are useful in simultaneous inference. Finally, there is a set of outline solutions for the exercises.

It has not been easy to find or make up theoretical problems that are relevant and yet not too difficult. It is hoped that the 200 or so problems scattered throughout the book will not only help the student but also provide some ideas for teachers.

This book is based on several courses that I have been giving at Auckland University, New Zealand, during the past 10 years and I wish to thank the many students who have stimulated my teaching interest in the subject. Special thanks are also due to Heather Lucas for reading a first draft and to Peggy Haworth for typing most of the manuscript.

ACKNOWLEDGMENTS

For permission to reproduce certain published tables and figures, thanks are due to the authors and Editors of *Biometrika* (Appendix E), *Journal of the American Statistical Association* (Tables 5.1, 5.2, and Appendix D), *Journal of the Royal Statistical Society, Series B* (Appendix F), and *Technometrics* (Table 5.3).

G. A. F. SEBER

Auckland, New Zealand
July 1976

Contents

Linear Regression Analysis

CHAPTER 1

Vectors of Random Variables

1.1 NOTATION

Matrices and vectors are denoted by boldface letters $\mathbf{A}$ and $\mathbf{a}$, respectively, and scalars by italics. Random variables are represented by capital letters, and their values by lowercase letters (for example, Y and y, respectively). This use of capitals for random variables, which seems to be widely accepted, is particularly useful in regression when distinguishing between fixed and random regressor (independent) variables. However, it does cause problems because a vector of random variables, $\mathbf{Y}$, say, then looks like a matrix. Occasionally in Chapter 11, because of a shortage of letters, a boldface lowercase letter represents a vector of random variables.

If X and Y are random variables then the symbols $E[Y]$, $\mathrm{var}[Y]$, $\mathrm{cov}[X,Y]$, and $E[X|Y=y]$ (or, more briefly, $E[X|Y]$) represent expectation, variance, covariance, and conditional expectation, respectively.

The $n\times n$ matrix with diagonal elements $d_1, d_2, \ldots, d_n$ and zeros elsewhere is denoted by $\mathrm{diag}(d_1, d_2, \ldots, d_n)$, and when all the d_i's are unity we have the identity matrix $\mathbf{I}_n$.

If $\mathbf{a}$ is an $n\times 1$ column vector with elements $a_1, a_2, \ldots, a_n$, we write $\mathbf{a}=[(a_i)]$, and the *length* or *norm* of $\mathbf{a}$ is denoted by $\|\mathbf{a}\|$. Thus

$$\|\mathbf{a}\| = \sqrt{\mathbf{a}'\mathbf{a}} = \left(a_1^2 + a_2^2 + \cdots + a_n^2\right)^{1/2},$$

The vector with elements all equal to unity is represented by $\mathbf{1}_n$.

If the $m\times n$ matrix $\mathbf{A}$ has elements a_{ij} we write $\mathbf{A}=[(a_{ij})]$, and the sum of the diagonal elements, called the trace of $\mathbf{A}$, is denoted by $\mathrm{tr}\,\mathbf{A}$ ($=a_{11}+a_{22}+\cdots+a_{kk}$, where k is the smaller of m and n). The transpose of $\mathbf{A}$ is represented by $\mathbf{A}'=[(a'_{ij})]$, where $a'_{ij}=a_{ji}$. If $\mathbf{A}$ is square its determinant is written $|\mathbf{A}|$, and if $\mathbf{A}$ is nonsingular its inverse is denoted by $\mathbf{A}^{-1}$. The space spanned by the columns of $\mathbf{A}$, called the range space of $\mathbf{A}$, is denoted by $\mathcal{R}[\mathbf{A}]$. The null space or kernel of $\mathbf{A}$ $(=\{\mathbf{x}:\mathbf{Ax}=\mathbf{0}\})$ is denoted by $\mathcal{N}[\mathbf{A}]$.

We say that $Y \sim N(\theta, \sigma^2)$ if Y is normally distributed with mean θ and variance σ^2: Y has a *standard normal* distribution if $\theta = 0$ and $\sigma^2 = 1$. The t- and chi-square distributions with k degrees of freedom are denoted by t_k and χ_k^2, respectively, and the F-distribution with m and n degrees of freedom is denoted by $F_{m,n}$.

Finally we mention the "dot" and "bar" notation representing sum and average, respectively; for example,

$$a_{i\cdot} = \sum_{j=1}^{J} a_{ij} \qquad \text{and} \qquad \bar{a}_{i\cdot} = \frac{a_{i\cdot}}{J}.$$

In the case of a single subscript we omit the dot and write $\bar{a}$.

Some knowledge of linear algebra by the reader is assumed, and for a short review course several books are available (e.g., Scheffé [1959: Appendix], Graybill [1961, 1969], Rao [1973: Chapter 1]). However, a number of matrix results are included in Appendixes A and B at the end of this book, and references to these Appendixes are denoted by, for example, *A2.3*.

1.2 LINEAR REGRESSION MODELS

A common problem in statistics is that of estimating the relationship that exists (if any) between two random variables X and Y; for instance, height and weight, income and intelligence quotient (IQ), ages of husband and wife at marriage, length and breadth of leaves, temperature and pressure of a certain volume of gas, or the length of a metal rod and its temperature. If we have n pairs of observations (x_i, y_i) $(i = 1, 2, \ldots, n)$, we can plot these points, giving the so-called *scatter diagram*, and endeavor to fit a smooth curve through the points in such a way that the points are as "close" to the curve as possible. Clearly we would not expect an exact fit because both variables in the above examples are subject to chance fluctuations owing to factors outside our control. Even if there is an exact relationship between variables like temperature and pressure, fluctuations would still show up in the scatter diagram because of errors of measurement.

Frequently the type of curve to be fitted is suggested by empirical evidence or theoretical arguments, as in the following examples.

EXAMPLE 1.1 Ohm's law states that $Y = rX$ where Y amperes is the current through a resistor of r ohms, and X volts is the voltage across the resistor. This gives us a straight line through the origin so that a linear scatter diagram will lend support to the law and r can be estimated from the slope of the fitted line.

EXAMPLE 1.2 From the laws of mechanics the force Y gram-weight necessary to prevent an object weighing w gram-weight from slipping down a smooth inclined plane at angle θ is given by the formula $Y = w\sin\theta$. If we put $X = \sin\theta$ we once again have a straight line through the origin. In this case the observed values (x_i, y_i) would deviate slightly from a linear relationship because of errors in measuring Y and θ and the presence of friction between the object and the plane.

EXAMPLE 1.3 Theoretical chemistry predicts that, for a given sample of gas kept at constant temperature, the volume V and pressure P of the gas approximately satisfy the relationship $PV = c$. Thus writing $Y = P$ and $X = 1/V$ we have $Y = cX$.

EXAMPLE 1.4 From more carefully controlled experiments it is found that the equation relating pressure and volume is of the form $PV^{\gamma} = c$ where $\gamma = 1$. However, we can still achieve linearity by taking logarithms; thus

$$\log P = \log c - \gamma \log V$$

or

$$Y = a + bX.$$

Hence $\log c$ and $-\gamma$ can be estimated from the straight line fitted to experimental data.

EXAMPLE 1.5 The inverse square law states that the force of gravity F between two bodies distance D apart is given by

$$F = \frac{c}{D^{\beta}}$$

where $\beta = 2$. Taking logarithms leads to

$$\log F = \log c - \beta \log D,$$

and from experimental data we can estimate β and test whether $\beta = 2$.

EXAMPLE 1.6 Experiments show that a metal rod expands when it is heated, and the extension is proportional to the rise in temperature. Also the increase from two identical rods placed end to end is twice that of a single rod, so that the extension is proportional to the original length.

Hence we are led to consider the straight-line model $Y_T = Y_0(1+\alpha T)$, where Y_T is the length at temperature T (T being measured with respect to a suitable origin) and α is the so-called coefficient of linear expansion. For more accurate work a quadratic model is suggested, namely,

$$Y_T = Y_0(1+\alpha T+\beta T^2).$$

When there are no underlying theoretical or experimental considerations to help us, it is sometimes difficult to decide just what type of curve to fit, as for example in Fig. 1.1. Here a straight line seems to be as good a curve to fit as any other because it involves only a few parameters, though clearly what is needed is some measure of goodness of fit for comparing different curves. Sometimes there is no tendency for points on the scatter diagram to group in such a way as to suggest some sort of trend. For example, the scatter diagram in Fig. 1.2 suggests little or no relationship between X and Y.

In many cases one of the variables, X, say, is not random but is fixed or controlled. For example, X could refer to the year of production and Y to the number of items manufactured by a firm in the given year. An example where X is controlled is an experiment in which the yield Y per unit area is measured for fixed quantities X of a certain fertilizer. In both examples we have, for each value of $X=x$, a random variable Y with mean $\varphi(x)$; that is, $Y=\varphi(x)+\epsilon$ where $E[\epsilon]=0$. Here $\varphi(x)$ is called the regression curve or regression function of Y on X.

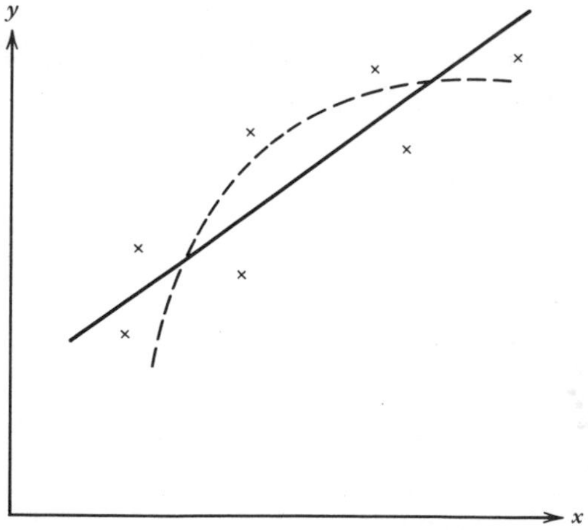

Fig. 1.1 Two different curves fitted to the same data.

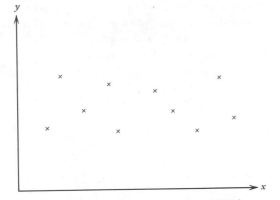

Fig. 1.2 Scatter diagram: x and y unrelated.

To describe how φ might be estimated from data pairs (x_i, y_i) we consider the simple straight-line case $\varphi(x) = \beta_0 + \beta_1 x$. Our model is then

$$Y_i = \beta_0 + \beta_1 x_i + \epsilon_i \qquad (i = 1, 2, \ldots, n).$$

A very elegant method for estimating β_0 and β_1 is the so-called method of least squares. This method of estimation, which leads to estimates with certain optimal properties, is based on the appealing idea of choosing β_0 and β_1 to minimize the squares of the vertical deviations of the data points from the fitted line (Fig. 1.3); that is, we minimize $\Sigma_i \epsilon_i^2 = \Sigma_i (Y_i - \beta_0 - \beta_1 x_i)^2$ with respect to β_0 and β_1. Clearly this least squares principle can be

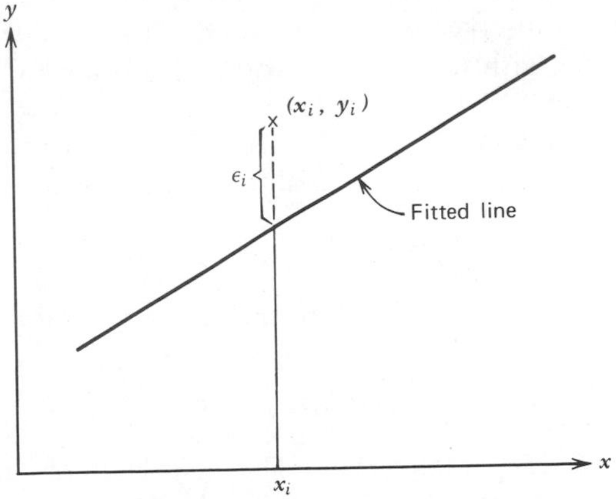

Fig. 1.3 The method of least squares consists of minimizing $\Sigma \varepsilon_i^2$.

applied to any regression curve $\varphi(x)$, though the minimization can be difficult unless φ linear in the unknown parameters. For example, $\varphi(x)=\beta_0 e^{-x\beta_1}$ is nonlinear and $\varphi(x)=\beta_0+\beta_1 x+\beta_2 x^2$ is linear in the β_j.

We have seen from the above examples that X and Y may both be random. In this case X and Y have a joint distribution and we can define two regression functions $E[Y|X=x]$ and $E[X|Y=y]$. For example, assuming a straight-line relationship, we have

$$E[Y|X=x]=\beta_0+\beta_1 x,$$

and we effectively proceed as though X is not random at all, though any inferences made are of course conditional on the observed values of X.

EXAMPLE 1.7 Suppose that a population of N animals is trapped on n consecutive occasions. It is assumed that the probability of trapping an animal on a particular occasion is constant and equal to p. Let Y_i be the number caught on the ith occasion and let $X_i=\Sigma_{j=1}^{i-1} Y_j$ $(i=2,\ldots,n; X_1=0)$ be the number of animals caught before the ith occasion. Then, using a binomial model,

$$E[Y_i|x_i]=(N-x_i)p=Np-px_i,$$

which is a straight line.

An important application of regression models is that of prediction. In this case we fit a model for the purpose of predicting Y for future values of x. Clearly we have to be very sure of our model if our predictions are to be at all reliable. For example, suppose the true model is given by Fig. 1.4. Then although we may be able to achieve a good fit to the left half of the curve, it would be inappropriate to make predictions for values of x greater than x_0 if we have observed Y only for values of x less than x_0. We are on safer ground if we wish to predict Y for x within the experimental range considered thus far.

Frequently a random variable Y depends on a number of variables $X_1, X_2, \ldots, X_k$, say, so that we may wish to find a regression surface

$$E[Y|X_1=x_1,\ X_2=x_2,\ldots,\ X_k=x_k]=\varphi(x_1,x_2,\ldots,x_k).$$

In this book we direct our attention to the important class of *linear* models, that is,

$$\varphi(x_1,x_2,\ldots,x_k)=\beta_0+\beta_1 x_1+\cdots+\beta_k x_k,$$

which is linear in the parameters β_j. This restriction to linearity is not as restrictive as one might think. For example, many functions of several

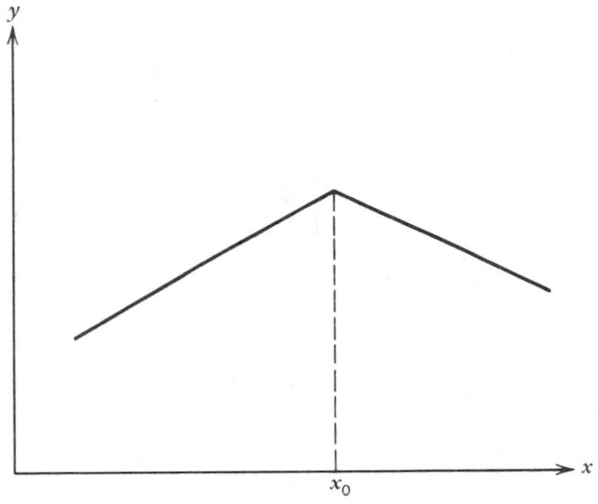

Fig. 1.4 An underlying model consisting of two straight lines.

variables are approximately linear over sufficiently small regions, or they may be made linear by a suitable transformation as we saw above. Also the x_i could be functions of other variables z, w, etc., for example, $x_1 = \sin z$, $x_2 = \log w$, $x_3 = zw$, or $x_i = x^i$ (which leads to polynomial regression). We can also include "categorical" models under our umbrella by using dummy x-variables. For example, suppose we wish to compare the means of two populations, say, $\beta_i = E[U_i]$ $(i = 1,2)$. Then we can combine the data into the single model

$$E[Y] = \beta_1 x_1 + \beta_2 x_2, \tag{1.1}$$

where $x_i = 1$ when Y is a U_i observation, and $x_i = 0$ otherwise $(i = 1,2)$. Having outlined briefly the nature of linear models we now give a brief résumé of the main topics covered in the book.

Once a model has been selected, then the unknown parameters β_j can be estimated using the method of least squares. This method is best described geometrically using matrix theory; theoretical details are given in Chapter 3 and computational algorithms are described in Chapter 11. By specifying the joint distribution of the Y_i's we can examine the statistical properties of the least squares estimates.

The next problem considered is that of hypothesis testing. For instance, we might be interested in testing $\beta = 2$ in Example 1.5, $\beta = 0$ in Example 1.6, or $\beta_1 - \beta_2 = 0$ in (1.1). These are all special cases of testing $\mathbf{a}'\boldsymbol{\beta} = \mathbf{c}$, where $\mathbf{a}'\boldsymbol{\beta}$ is a linear combination of the β_j's. In general we may wish to test a set of linear combinations, namely, $\mathbf{A}\boldsymbol{\beta} = \mathbf{c}$; this problem is discussed in Chapter 4 and applications to models like (1.1) are described in Chapters 9 and 10.

In addition to hypothesis testing we may wish to construct confidence intervals for various parameters or linear combinations of parameters, or derive prediction intervals for future observations; this is done in Chapter 5. Then an important chapter, Chapter 6, looks at the distributional assumptions underlying the theory developed thus far. Tests and graphs are described for checking up on the assumptions.

Because straight-line regression is important in its own right, it is discussed in some detail in Chapter 7. Polynomial regression, which has its own peculiar problems, follows in Chapter 8.

Thus far we have generally assumed that the underlying model had already been chosen. However, in many practical situations the problem is to decide which of the possible x_j's should be included in the model. Chapter 12 is devoted to this important topic.

The remainder of this chapter and Chapter 2, which deals with the multivariate normal distribution, are intended to provide some basic theory for the rest of the book.

1.3 EXPECTATION AND COVARIANCE OPERATORS

Let Z_{ij} $(i=1,2,\ldots,m; j=1,2,\ldots,n)$ be a set of random variables with expected values $E[Z_{ij}]$. Expressing both the random variables and their expectations in matrix form we can define the general expectation operator $\mathcal{E}$ of the matrix $\mathbf{Z}=[(Z_{ij})]$, namely:

Definition.

$$\mathcal{E}[\mathbf{Z}]=\left[\left(\mathrm{E}[Z_{ij}]\right)\right].$$

In particular, when $m=n=1$, *then* $\mathcal{E}[\mathbf{Z}]=\mathrm{E}[Z_{11}]$.

THEOREM 1.1 If $\mathbf{A}=[(a_{ij})]$, $\mathbf{B}=[(b_{ij})]$, and $\mathbf{C}=[(c_{ij})]$ are $l\times m$, $n\times p$, and $l\times p$ matrices, respectively, of constants, then

$$\mathcal{E}[\mathbf{AZB}+\mathbf{C}]=\mathbf{A}\mathcal{E}[\mathbf{Z}]\mathbf{B}+\mathbf{C}.$$

Proof. Let $\mathbf{W}=\mathbf{AZB}+\mathbf{C}$; then $W_{ij}=\sum_{r=1}^{m}\sum_{s=1}^{n}a_{ir}Z_{rs}b_{sj}+d_{ij}$ and

$$\mathcal{E}[\mathbf{AZB}+\mathbf{C}]=[(E[W_{ij}])]=\left[\left(\sum_r\sum_s a_{ir}E[Z_{rs}]b_{sj}+c_{ij}\right)\right]$$

$$=\left[\left((\mathbf{A}\mathcal{E}[\mathbf{Z}]\mathbf{B})_{ij}\right)\right]+\left[(c_{ij})\right]=\mathbf{A}\mathcal{E}[\mathbf{Z}]\mathbf{B}+\mathbf{C}. \qquad \square$$

In this proof we note that l, m, n, and p are any positive integers, and the matrices of constants can take any values. For example, we have the following.

COROLLARY If $\mathbf{X}$ is an $m\times 1$ vector of random variables, then

$$\mathcal{E}[\mathbf{AX}]=\mathbf{A}\mathcal{E}[\mathbf{X}].$$

THEOREM 1.2 If $\mathbf{A}$ and $\mathbf{B}$ are $m\times n$ matrices of constants and $\mathbf{X}$ and $\mathbf{Y}$ are $n\times 1$ vectors of random variables, then

$$\mathcal{E}[\mathbf{AX}+\mathbf{BY}]=\mathbf{A}\mathcal{E}[\mathbf{X}]+\mathbf{B}\mathcal{E}[\mathbf{Y}].$$

Proof. This is straightforward and is left as an exception. □

COROLLARY

$$\mathcal{E}[a\mathbf{X}+b\mathbf{Y}]=a\mathcal{E}[\mathbf{X}]+b\mathcal{E}[\mathbf{Y}].$$

In a similar manner we can generalize the notions of variance and covariance for vectors. If $\mathbf{X}$ and $\mathbf{Y}$ are $m\times 1$ and $n\times 1$ vectors of random variables then we define the generalized covariance operator $\mathcal{C}$ as follows:

Definition.

$$\mathcal{C}[\mathbf{X},\mathbf{Y}]=[(\text{cov}[X_i,Y_j])].$$

THEOREM 1.3

$$\mathcal{C}[\mathbf{X},\mathbf{Y}]=\mathcal{E}[(\mathbf{X}-\mathcal{E}[\mathbf{X}])(\mathbf{Y}-\mathcal{E}[\mathbf{Y}])'].$$

Proof. Let $\mathcal{E}[\mathbf{X}]=\boldsymbol{\alpha}$ and $\mathcal{E}[\mathbf{Y}]=\boldsymbol{\beta}$; then

$$\begin{aligned}\mathcal{C}[\mathbf{X},\mathbf{Y}]&=[(\text{cov}[X_i,Y_j])]\\&=[(E[(X_i-\alpha_i)(Y_j-\beta_j)])]\\&=\mathcal{E}[((X_i-\alpha_i)(Y_j-\beta_j))] \qquad \text{(Theorem 1.1)}\\&=\mathcal{E}[(\mathbf{X}-\boldsymbol{\alpha})(\mathbf{Y}-\boldsymbol{\beta})']\\&=\mathcal{E}[(\mathbf{X}-\mathcal{E}[\mathbf{X}])(\mathbf{Y}-\mathcal{E}[\mathbf{Y}])'].\end{aligned}$$

□

Definition. *When* $\mathbf{X}=\mathbf{Y}$, $\mathcal{C}[\mathbf{Y},\mathbf{Y}]$, *written as* $\mathcal{D}[\mathbf{Y}]$, *is called the dispersion (variance–covariance) matrix of* $\mathbf{Y}$. *Thus*

$$\mathcal{D}[\mathbf{Y}]=\left[\left(\operatorname{cov}[Y_i,Y_j]\right)\right]$$

$$=\begin{bmatrix} \operatorname{var}[Y_1], & \operatorname{cov}[Y_1,Y_2], & \dots, & \operatorname{cov}[Y_1,Y_n] \\ \operatorname{cov}[Y_2,Y_1], & \operatorname{var}[Y_2], & \dots, & \operatorname{cov}[Y_2,Y_n] \\ \cdots & \cdots & \cdots & \cdots \\ \operatorname{cov}[Y_n,Y_1], & \operatorname{cov}[Y_n,Y_2], & \dots, & \operatorname{var}[Y_n] \end{bmatrix} \tag{1.2}$$

Since $\operatorname{cov}[Y_i,Y_j]=\operatorname{cov}[Y_j,Y_i]$, *the above matrix is symmetric. We note that when* $\mathbf{Y}=Y_1$, $\mathcal{D}[\mathbf{Y}]=\operatorname{var}[Y_1]$.

EXAMPLE 1.8 If $\mathbf{a}$ is any $n\times 1$ vector of constants, then

$$\mathcal{D}[\mathbf{Y}-\mathbf{a}]=\mathcal{D}[\mathbf{Y}].$$

Solution. From

$$E\left[Y_i-a_i-E[Y_i-a_i]\right]=E\left[Y_i-E[Y_i]\right]$$

we have $\operatorname{cov}[Y_i-a_i,Y_j-a_j]=\operatorname{cov}[Y_i,Y_j]$, etc.

EXAMPLE 1.9 Prove that

$$\mathcal{E}[(\mathbf{Y}-\mathbf{a})(\mathbf{Y}-\mathbf{a})']=\mathcal{D}[\mathbf{Y}]+(\mathcal{E}[\mathbf{Y}]-\mathbf{a})(\mathcal{E}[\mathbf{Y}]-\mathbf{a})'.$$

Solution. Expanding $(\mathbf{X}-\mathcal{E}[\mathbf{X}])(\mathbf{X}-\mathcal{E}[\mathbf{X}])'$ and using Theorem 1.1, we can readily show that

$$\mathcal{D}[\mathbf{X}]=\mathcal{E}[\mathbf{X}\mathbf{X}']-(\mathcal{E}[\mathbf{X}])(\mathcal{E}[\mathbf{X}])'. \tag{1.3}$$

Setting $\mathbf{X}=\mathbf{Y}-\mathbf{a}$ and using Example 1.8 leads to the required result.

THEOREM 1.4 If $\mathbf{X}$ and $\mathbf{Y}$ are $m\times 1$ and $n\times 1$ vectors of random variables, and $\mathbf{A}$ and $\mathbf{B}$ are $l\times m$ and $p\times n$ matrices of constants, respectively, then

$$\mathcal{C}[\mathbf{AX},\mathbf{BY}]=\mathbf{A}\mathcal{C}[\mathbf{X},\mathbf{Y}]\mathbf{B}'.$$

Proof. Let $\mathbf{U}=\mathbf{AX}$ and $\mathbf{V}=\mathbf{BY}$. Then, by Theorems 1.3 and 1.1

$$\begin{aligned}\mathcal{C}[\mathbf{AX},\mathbf{BY}]&=\mathcal{C}[\mathbf{U},\mathbf{V}]\\&=\mathcal{E}[(\mathbf{U}-\mathcal{E}[\mathbf{U}])(\mathbf{V}-\mathcal{E}[\mathbf{V}])']\\&=\mathcal{E}[(\mathbf{AX}-\mathbf{A}\mathcal{E}[\mathbf{X}])(\mathbf{BY}-\mathbf{B}\mathcal{E}[\mathbf{Y}])']\\&=\mathcal{E}[\mathbf{A}(\mathbf{X}-\mathcal{E}[\mathbf{X}])(\mathbf{Y}-\mathcal{E}[\mathbf{Y}])'\mathbf{B}']\\&=\mathbf{A}\mathcal{E}[(\mathbf{X}-\mathcal{E}[\mathbf{X}])(\mathbf{Y}-\mathcal{E}[\mathbf{Y}])']\mathbf{B}'\\&=\mathbf{A}\mathcal{C}[\mathbf{X},\mathbf{Y}]\mathbf{B}'.\end{aligned}$$

□

COROLLARY 1

$$\mathcal{C}[\mathbf{AX},\mathbf{Y}]=\mathbf{A}\mathcal{C}[\mathbf{X},\mathbf{Y}]; \qquad \mathcal{C}[\mathbf{X},\mathbf{BY}]=\mathcal{C}[\mathbf{X},\mathbf{Y}]\mathbf{B}'.$$

COROLLARY 2

$$\begin{aligned}\mathcal{D}[\mathbf{AX}]&=\mathcal{C}[\mathbf{AX},\mathbf{AX}]\\&=\mathbf{A}\mathcal{C}[\mathbf{X},\mathbf{X}]\mathbf{A}'\\&=\mathbf{A}\mathcal{D}[\mathbf{X}]\mathbf{A}'.\end{aligned}\tag{1.4}$$

THEOREM 1.5 If $\mathbf{X}$, $\mathbf{Y}$, $\mathbf{U}$, and $\mathbf{V}$ are any (not necessarily distinct) $n\times 1$ vectors of random variables, then for all real numbers a, b, c, and d (including zero)

$$\mathcal{C}[a\mathbf{X}+b\mathbf{Y},c\mathbf{U}+d\mathbf{V}]=ac\mathcal{C}[\mathbf{X},\mathbf{U}]+ad\mathcal{C}[\mathbf{X},\mathbf{V}]+bc\mathcal{C}[\mathbf{Y},\mathbf{U}]+bd\mathcal{C}[\mathbf{Y},\mathbf{V}].$$

Proof. By Theorems 1.3 and 1.2 (corollary) we have

$$\begin{aligned}&\mathcal{C}[a\mathbf{X}+b\mathbf{Y},c\mathbf{U}+d\mathbf{V}]\\&\quad=\mathcal{E}[(a\mathbf{X}+b\mathbf{Y}-a\mathcal{E}[\mathbf{X}]-b\mathcal{E}[\mathbf{Y}])(c\mathbf{U}+d\mathbf{V}-c\mathcal{E}[\mathbf{U}]-d\mathcal{E}[\mathbf{V}])']\\&\quad=\mathcal{E}[(a(\mathbf{X}-\mathcal{E}[\mathbf{X}])+b(\mathbf{Y}-\mathcal{E}[\mathbf{Y}]))(c(\mathbf{U}-\mathcal{E}[\mathbf{U}])+d(\mathbf{V}-\mathcal{E}[\mathbf{V}]))']\\&\quad=\mathcal{E}[ac(\mathbf{X}-\mathcal{E}[\mathbf{X}])(\mathbf{U}-\mathcal{E}[\mathbf{U}])'+\cdots+bd(\mathbf{Y}-\mathcal{E}[\mathbf{Y}])(\mathbf{V}-\mathcal{E}[\mathbf{V}])']\\&\quad=ac\mathcal{C}[\mathbf{X},\mathbf{U}]+ad\mathcal{C}[\mathbf{X},\mathbf{V}]+bc\mathcal{C}[\mathbf{Y},\mathbf{U}]+bd\mathcal{C}[\mathbf{Y},\mathbf{V}].\end{aligned}$$

□

COROLLARY Setting $\mathbf{X}=\mathbf{U}$, $\mathbf{Y}=\mathbf{V}$, $a=c$, and $b=d$, we have

$$\begin{aligned}\mathcal{D}[a\mathbf{X}+b\mathbf{Y}] &= \mathcal{C}[a\mathbf{X}+b\mathbf{Y}, a\mathbf{X}+b\mathbf{Y}] \\ &= a^2\mathcal{D}[\mathbf{X}]+2ab\mathcal{C}[\mathbf{X},\mathbf{Y}]+b^2\mathcal{D}[\mathbf{Y}]. \qquad (1.5)\end{aligned}$$

THEOREM 1.6 If $\mathbf{X}$ is a vector of random variables such that no element of $\mathbf{X}$ is a linear combination of the remaining elements (that is, there do not exist $\mathbf{a}$ ($\neq \mathbf{0}$) and b such that $\mathbf{a}'\mathbf{X}=b$ for all values of $\mathbf{X}=\mathbf{x}$), then $\mathcal{D}[\mathbf{X}]$ is a positive definite matrix (see *A4*).

Proof. For any constant vector $\mathbf{c}$ we have:

$$\begin{aligned} 0 &\leqslant \operatorname{var}[\mathbf{c}'\mathbf{X}] \\ &= \mathcal{D}[\mathbf{c}'\mathbf{X}] \\ &= \mathbf{c}'\mathcal{D}[\mathbf{X}]\mathbf{c} \qquad \text{(Theorem 1.4, Corollary 2).}\end{aligned}$$

Now equality holds if and only if $\mathbf{c}'\mathbf{X}$ is a constant, that is, if and only if $\mathbf{c}'\mathbf{X}=d$ ($\mathbf{c}\neq\mathbf{0}$) or $\mathbf{c}=\mathbf{0}$. Because the former possibility is ruled out, $\mathbf{c}=\mathbf{0}$ and $\mathcal{D}[\mathbf{X}]$ is positive definite. □

EXAMPLE 1.10 If $\mathbf{X}$ and $\mathbf{Y}$ are $m\times 1$ and $n\times 1$ vectors of random variables such that no element of $\mathbf{X}$ is a linear combination of the remaining elements, prove that there exists an $n\times m$ matrix $\mathbf{M}$ such that $\mathcal{C}[\mathbf{X},\mathbf{Y}-\mathbf{MX}]=\mathbf{0}$.

Solution. Setting $a=1$, $b=0$, $c=1$, $d=-1$, $\mathbf{U}=\mathbf{Y}$, and $\mathbf{V}=\mathbf{MX}$ in Theorem 1.5, and using Theorem 1.4, we have

$$\begin{aligned}\mathcal{C}[\mathbf{X},\mathbf{Y}-\mathbf{MX}] &= \mathcal{C}[\mathbf{X},\mathbf{Y}]-\mathcal{C}[\mathbf{X},\mathbf{MX}] \\ &= \mathcal{C}[\mathbf{X},\mathbf{Y}]-\mathcal{C}[\mathbf{X},\mathbf{X}]\mathbf{M}' \\ &= \mathcal{C}[\mathbf{X},\mathbf{Y}]-\mathcal{D}[\mathbf{X}]\mathbf{M}'. \qquad (1.6)\end{aligned}$$

By the preceding theorem $\mathcal{D}[\mathbf{X}]$ is positive definite and therefore nonsingular (*A4.1*). Hence the above expression is zero for $\mathbf{M}'=(\mathcal{D}[\mathbf{X}])^{-1}\mathcal{C}[\mathbf{X},\mathbf{Y}]$, or for $M=\mathcal{C}[\mathbf{Y},\mathbf{X}](\mathcal{D}[\mathbf{X}])^{-1}$.

EXERCISES 1a

1. If $\mathbf{X}$ and $\mathbf{Y}$ are $m\times 1$ and $n\times 1$ vectors of random variables, and $\mathbf{a}$ and $\mathbf{b}$

are $m\times 1$ and $n\times 1$ vectors of constants, prove that

$$\mathcal{C}[\mathbf{X}-\mathbf{a},\mathbf{Y}-\mathbf{b}]=\mathcal{C}[\mathbf{X},\mathbf{Y}].$$

2. Prove that $\mathcal{C}[\mathbf{X},\mathbf{Y}]=\mathcal{E}[\mathbf{XY}']-(\mathcal{E}[\mathbf{X}])(\mathcal{E}[\mathbf{Y}])'$.
3. Let $\mathbf{X}=(X_1,X_2,\ldots,X_n)'$ be a vector of random variables, and let $Y_1=X_1$, $Y_i=X_i-X_{i-1}$ $(i=2,3,\ldots,n)$. If the Y_i are mutually independent random variables each with unit variance, find $\mathcal{D}[\mathbf{X}]$.
4. If $X_1,X_2,\ldots,X_n$ are random variables each with variance σ^2, and $X_{i+1}=\rho X_i+a$ $(i=1,2,\ldots,n-1)$, where a and ρ are constants, find $\mathcal{D}[\mathbf{X}]$.

1.4 MEAN AND VARIANCE OF QUADRATIC FORMS

THEOREM 1.7 Let $\mathbf{X}=[(X_i)]$ be an $n\times 1$ vector of random variables and let $\mathbf{A}$ be an $n\times n$ symmetric matrix. If $\mathcal{E}[\mathbf{X}]=\boldsymbol{\theta}$ and $\mathcal{D}[\mathbf{X}]=\boldsymbol{\Sigma}=[(\sigma_{ij})]$, then

$$E[\mathbf{X}'\mathbf{AX}]=\operatorname{tr}[\mathbf{A}\boldsymbol{\Sigma}]+\boldsymbol{\theta}'\mathbf{A}\boldsymbol{\theta}.$$

Proof. $E[\mathbf{X}'\mathbf{AX}]=E[(\mathbf{X}-\boldsymbol{\theta})'\mathbf{A}(\mathbf{X}-\boldsymbol{\theta})+\boldsymbol{\theta}'\mathbf{AX}+\mathbf{X}'\mathbf{A}\boldsymbol{\theta}-\boldsymbol{\theta}'\mathbf{A}\boldsymbol{\theta}]$. Now $\mathbf{X}'\mathbf{A}\boldsymbol{\theta}=(\mathbf{X}'\mathbf{A}\boldsymbol{\theta})'=\boldsymbol{\theta}'\mathbf{A}'\mathbf{X}=\boldsymbol{\theta}'\mathbf{AX}$ and

$$E[\boldsymbol{\theta}'\mathbf{AX}]=\mathcal{E}[\boldsymbol{\theta}'\mathbf{AX}]=\boldsymbol{\theta}'\mathbf{A}\mathcal{E}[\mathbf{X}]=\boldsymbol{\theta}'\mathbf{A}\boldsymbol{\theta}.$$

Hence

$$\begin{aligned}E[\mathbf{X}'\mathbf{AX}]&=E[(\mathbf{X}-\boldsymbol{\theta})'\mathbf{A}(\mathbf{X}-\boldsymbol{\theta})]+\boldsymbol{\theta}'\mathbf{A}\boldsymbol{\theta}\\&=\sum_i\sum_j a_{ij}E[(X_i-\theta_i)(X_j-\theta_j)]+\boldsymbol{\theta}'\mathbf{A}\boldsymbol{\theta}\\&=\sum_i\sum_j a_{ij}\sigma_{ij}+\boldsymbol{\theta}'\mathbf{A}\boldsymbol{\theta}\\&=\operatorname{tr}[\mathbf{A}\boldsymbol{\Sigma}]+\boldsymbol{\theta}'\mathbf{A}\boldsymbol{\theta}.\end{aligned}$$

□

COROLLARY 1 By setting $\mathbf{Y}=\mathbf{X}-\mathbf{b}$, and noting that $\mathcal{D}[\mathbf{Y}]=\mathcal{D}[\mathbf{X}]$ (Example 1.8), we have

$$E[(\mathbf{X}-\mathbf{b})'\mathbf{A}(\mathbf{X}-\mathbf{b})]=\operatorname{tr}[\mathbf{A}\boldsymbol{\Sigma}]+(\boldsymbol{\theta}-\mathbf{b})'\mathbf{A}(\boldsymbol{\theta}-\mathbf{b}).$$

COROLLARY 2 If $\boldsymbol{\Sigma}=\sigma^2\mathbf{I}_n$ then $\operatorname{tr}[\mathbf{A}\boldsymbol{\Sigma}]=\sigma^2\operatorname{tr}\mathbf{A}$. Thus in this case we have the following simple rule:

$$E[\mathbf{X}'\mathbf{AX}]=\sigma^2(\text{sum of coefficients of } X_i^2)+(\mathbf{X}'\mathbf{AX})_{\mathbf{X}=\boldsymbol{\theta}}.$$

EXAMPLE 1.11 If $X_1, X_2, \ldots, X_n$ are independently and identically distributed with mean θ and variance σ^2, find the expected value of $Q=(X_1-X_2)^2+(X_2-X_3)^2+\cdots+(X_{n-1}-X_n)^2$.

Solution. Here $\text{cov}[X_i, X_j]=0$ $(i \neq j)$ so that $\mathfrak{D}[\mathbf{X}]=\sigma^2\mathbf{I}_n$; Corollary 2 now applies with $Q=\mathbf{X}'\mathbf{A}\mathbf{X}$. Setting $X_i=\theta$ in Q we obtain zero for the second term of $E[\mathbf{X}'\mathbf{A}\mathbf{X}]$. Hence from

$$Q=2\sum_{i=1}^{n} X_i^2 - X_1^2 - X_n^2 - 2\sum_{i=1}^{n-1} X_i X_{i+1}$$

we have $\text{tr}\,\mathbf{A}=2n-2$ and, finally, $E[Q]=\sigma^2(2n-2)$.

EXAMPLE 1.12 If $X_1, X_2, \ldots, X_n$ have a common mean θ, and $\mathfrak{D}[\mathbf{X}]=\mathbf{\Sigma}$ where $\sigma_{ii}=\sigma^2$ and $\sigma_{ij}=\rho\sigma^2$ $(i\neq j)$, show that $Q=\Sigma(X_i-\overline{X})^2$ is an unbiased estimate of $\sigma^2(1-\rho)(n-1)$.

Solution. Now $Q=\mathbf{X}'\mathbf{A}\mathbf{X}$, where $\mathbf{A}=[(\delta_{ij}-n^{-1})]$, and

$$\mathbf{A\Sigma}=\sigma^2\begin{bmatrix} 1-\frac{1}{n} & -\frac{1}{n} & \cdots & -\frac{1}{n} \\ -\frac{1}{n} & 1-\frac{1}{n} & \cdots & -\frac{1}{n} \\ \cdots & \cdots & \cdots & \cdots \\ -\frac{1}{n} & -\frac{1}{n} & \cdots & 1-\frac{1}{n} \end{bmatrix}\begin{bmatrix} 1 & \rho & \cdots & \rho \\ \rho & 1 & \cdots & \rho \\ \cdot & \cdot & \cdots & \cdot \\ \rho & \rho & \cdots & 1 \end{bmatrix}$$

$$=\sigma^2(1-\rho)\mathbf{A}. \tag{1.7}$$

Once again the second term in $E[Q]$ is zero so that

$$E[Q]=\text{tr}[\mathbf{A\Sigma}]=\sigma^2(1-\rho)\text{tr}\,\mathbf{A}=\sigma^2(1-\rho)(n-1).$$

THEOREM 1.8 Let $X_1, X_2, \ldots, X_n$ be n independent random variables with means $\theta_1, \theta_2, \ldots, \theta_n$, common variance μ_2, and common third and fourth moments about their means, μ_3 and μ_4, respectively (that is, $\mu_r=E[(X_i-\theta_i)^r]$). If $\mathbf{A}$ is any $n\times n$ symmetric matrix, and $\mathbf{a}$ is the column vector of the diagonal elements of $\mathbf{A}$, then

$$\text{var}[\mathbf{X}'\mathbf{A}\mathbf{X}]=(\mu_4-3\mu_2^2)\mathbf{a}'\mathbf{a}+2\mu_2^2\,\text{tr}\,\mathbf{A}^2+4\mu_2\boldsymbol{\theta}'\mathbf{A}^2\boldsymbol{\theta}+4\mu_3\boldsymbol{\theta}'\mathbf{A}\mathbf{a}.$$

(This result is stated without proof in Atiqullah [1962].)

Proof. We note that $\mathcal{E}[\mathbf{X}]=\boldsymbol{\theta}$, $\mathcal{D}[\mathbf{X}]=\mu_2\mathbf{I}_n$, and

$$\operatorname{var}[\mathbf{X}'\mathbf{A}\mathbf{X}]=E\left[(\mathbf{X}'\mathbf{A}\mathbf{X})^2\right]-\left(E[\mathbf{X}'\mathbf{A}\mathbf{X}]\right)^2. \tag{1.8}$$

Now

$$\mathbf{X}'\mathbf{A}\mathbf{X}=(\mathbf{X}-\boldsymbol{\theta})'\mathbf{A}(\mathbf{X}-\boldsymbol{\theta})+2\boldsymbol{\theta}'\mathbf{A}(\mathbf{X}-\boldsymbol{\theta})+\boldsymbol{\theta}'\mathbf{A}\boldsymbol{\theta}$$

so that squaring gives

$$\begin{aligned}(\mathbf{X}'\mathbf{A}\mathbf{X})^2=&\left[(\mathbf{X}-\boldsymbol{\theta})'\mathbf{A}(\mathbf{X}-\boldsymbol{\theta})\right]^2+4\left[\boldsymbol{\theta}'\mathbf{A}(\mathbf{X}-\boldsymbol{\theta})\right]^2+(\boldsymbol{\theta}'\mathbf{A}\boldsymbol{\theta})^2\\&+2\boldsymbol{\theta}'\mathbf{A}\boldsymbol{\theta}\left[(\mathbf{X}-\boldsymbol{\theta})'\mathbf{A}(\mathbf{X}-\boldsymbol{\theta})+2\boldsymbol{\theta}'\mathbf{A}(\mathbf{X}-\boldsymbol{\theta})\right]\\&+4\boldsymbol{\theta}'\mathbf{A}(\mathbf{X}-\boldsymbol{\theta})(\mathbf{X}-\boldsymbol{\theta})'\mathbf{A}(\mathbf{X}-\boldsymbol{\theta}).\end{aligned}$$

Setting $\mathbf{Y}=\mathbf{X}-\boldsymbol{\theta}$ we have $\mathcal{E}[\mathbf{Y}]=\mathbf{0}$ and, using Theorem 1.7 (Corollary 1), we have

$$\begin{aligned}E\left[(\mathbf{X}'\mathbf{A}\mathbf{X})^2\right]=&E\left[(\mathbf{Y}'\mathbf{A}\mathbf{Y})^2\right]+4E\left[(\boldsymbol{\theta}'\mathbf{A}\mathbf{Y})^2\right]+(\boldsymbol{\theta}'\mathbf{A}\boldsymbol{\theta})^2\\&+2\boldsymbol{\theta}'\mathbf{A}\boldsymbol{\theta}(\mu_2\operatorname{tr}\mathbf{A})+4E\left[\boldsymbol{\theta}'\mathbf{A}\mathbf{Y}\mathbf{Y}'\mathbf{A}\mathbf{Y}\right].\end{aligned}$$

As a first step in evaluating the above expression we note that

$$(\mathbf{Y}'\mathbf{A}\mathbf{Y})^2=\sum_i\sum_j\sum_k\sum_l a_{ij}a_{kl}Y_iY_jY_kY_l.$$

Since the Y_i are mutually independent with the same first four moments about the origin, we have:

$$E\left[Y_iY_jY_kY_l\right]\begin{aligned}&=\mu_4 && (i=j=k=l)\\&=\mu_2^2 && (i=j,k=l;\ i=k,\ j=l;\ i=l,\ j=k)\\&=0 && \text{(otherwise).}\end{aligned}$$

Hence

$$\begin{aligned}E\left[(\mathbf{Y}'\mathbf{A}\mathbf{Y})^2\right]&=\mu_4\sum_i a_{ii}^2+\mu_2^2\left(\underset{i\neq k}{\sum\sum}a_{ii}a_{kk}+\underset{i\neq j}{\sum\sum}a_{ij}^2+\underset{i\neq j}{\sum\sum}a_{ij}a_{ji}\right)\\&=(\mu_4-3\mu_2^2)\mathbf{a}'\mathbf{a}+\mu_2^2\left[(\operatorname{tr}\mathbf{A})^2+2\operatorname{tr}\mathbf{A}^2\right],\end{aligned} \tag{1.9}$$

since $a_{ij}=a_{ji}$ and $\sum\sum_{ij}a_{ij}^2=\operatorname{tr}\mathbf{A}^2$. Also

$$(\boldsymbol{\theta}'\mathbf{A}\mathbf{Y})^2=(\mathbf{b}'\mathbf{Y})^2=\sum_i\sum_j b_ib_jY_iY_j,$$

say, and

$$(\boldsymbol{\theta}'\mathbf{AY})(\mathbf{Y}'\mathbf{AY}) = \sum_i \sum_j \sum_k b_i a_{jk} Y_i Y_j Y_k ,$$

so that

$$E\left[(\boldsymbol{\theta}'\mathbf{AY})^2\right] = \mu_2 \sum_i b_i^2 = \mu_2 \mathbf{b}'\mathbf{b} = \mu_2 \boldsymbol{\theta}'\mathbf{A}^2\boldsymbol{\theta}$$

and

$$E\left[(\boldsymbol{\theta}'\mathbf{AY})(\mathbf{Y}'\mathbf{AY})\right] = \mu_3 \sum_i b_i a_{ii} = \mu_3 \mathbf{b}'\mathbf{a} = \mu_3 \boldsymbol{\theta}'\mathbf{Aa} .$$

Finally, collecting all the terms, using

$$E\left[\mathbf{X}'\mathbf{AX}\right] = \mu_2 \operatorname{tr} \mathbf{A} + \boldsymbol{\theta}'\mathbf{A}\boldsymbol{\theta} ,$$

and substituting in Equation (1.8) leads to the desired result. □

COROLLARY 1 If $X_1, X_2, \ldots, X_n$ are also normally distributed, then $\mu_3 = 0$, $\mu_4 = 3\mu_2^2$, and

$$\operatorname{var}\left[\mathbf{X}'\mathbf{AX}\right] = 2\mu_2^2 \operatorname{tr} \mathbf{A}^2 + 4\mu_2 \boldsymbol{\theta}'\mathbf{A}^2\boldsymbol{\theta}.$$

COROLLARY 2 If $\boldsymbol{\theta} = \mathbf{0}$ and $\mu_2 = \sigma^2$ in Corollary 1, then

$$\operatorname{var}\left[\mathbf{X}'\mathbf{AX}\right] = 2\sigma^4 \operatorname{tr} \mathbf{A}^2.$$

EXERCISES 1b

1. If $\mathbf{A}$ is a symmetric matrix and $\mathbf{X}$ is a vector of random variables, prove that

$$E\left[\mathbf{X}'\mathbf{AX}\right] = \operatorname{tr}\left[\mathbf{A}\mathcal{E}\left[\mathbf{XX}'\right]\right].$$

   Hence deduce Theorem 1.7.
2. If $X_1, X_2, \ldots, X_n$ are mutually independent random variables with common mean θ, and variances $\sigma_1^2, \sigma_2^2, \ldots, \sigma_n^2$, respectively, prove that $\Sigma_i (X_i - \overline{X})^2 / [n(n-1)]$ is an unbiased estimate of var$[\overline{X}]$.
3. The random variables $X_1, X_2, \ldots, X_n$ have a common mean θ, a common variance σ^2, and the correlation between any pair of random variables is a known constant ρ.
   (a) Find var$[\overline{X}]$ and hence prove that $-1/(n-1) \leqslant \rho \leqslant 1$.

(b) If

$$Q = a\sum_{i=1}^{n} X_i^2 + b\left(\sum_{i=1}^{n} X_i\right)^2$$

is an unbiased estimate of σ^2, find a and b. Hence show that, in this case,

$$Q = \sum_{i=1}^{n} \frac{\left(X_i - \bar{X}\right)^2}{(1-\rho)(n-1)}.$$

4. If $X \sim N(\theta, \sigma^2)$, find the moment generating function of X and hence prove that $\mu_3 = 0$ and $\mu_4 = 3\mu_2^2$.
5. Let $X_1, X_2, \ldots, X_n$ be independently and identically distributed as $N(\theta, \sigma^2)$. Define

$$S^2 = \frac{1}{n-1}\sum_{i=1}^{n}\left(X_i - \bar{X}\right)^2$$

and

$$Q = \frac{1}{2(n-1)}\sum_{i=1}^{n-1}(X_{i+1} - X_i)^2.$$

(a) Prove that $\text{var}[S^2] = 2\sigma^4/(n-1)$.
(b) Show that Q is an unbiased estimate of σ^2.
(c) Find the variance of Q and hence show that, as $n \to \infty$, the efficiency of Q relative to S^2 is $\frac{2}{3}$.

1.5 INDEPENDENCE OF RANDOM VARIABLES

Two vectors of random variables $\mathbf{X}$ and $\mathbf{Y}$ are said to be (statistically) independent if the joint density function $f(\mathbf{x}, \mathbf{y})$ factorizes in the form

$$f(\mathbf{x}, \mathbf{y}) = f_1(\mathbf{x}) f_2(\mathbf{y})$$

where f_1 and f_2 are the respective marginal density functions.

THEOREM 1.9 If $\mathbf{X}$ and $\mathbf{Y}$ are independent, and the functions $a(\mathbf{X})$ and $b(\mathbf{Y})$ are random variables (that is, a and b are measurable* functions), then $a(X)$ and $b(Y)$ are independent.

*All continuous functions, and most "well-behaved" functions are measurable.

Proof. The above result follows immediately from the factorization of the joint characteristic function; namely,

$$E\left[e^{isa(\mathbf{X})+itb(\mathbf{Y})}\right]=E\left[e^{isa(\mathbf{X})}\right]E\left[e^{itb(\mathbf{Y})}\right]. \qquad \square$$

It is well-known that cov$[X,Y]=0$ does not in general imply that X and Y are independent. However, in one important case, namely, the bivariate normal distribution, X and Y are independent if and only if cov$[X,Y]=0$; a generalization of this result applied to the multivariate normal distribution is given in Theorem 2.6 (Section 2.3).

When considering more than two random variables, say, X_1, X_2, and X_3, one is often faced with the problem of proving that they are *mutually* independent, that is, proving

$$f(x_1,x_2,x_3)=f_1(x_1)f_2(x_2)f_3(x_3), \tag{1.10}$$

where f_i is the probability density function of X_i. It is tempting to try and solve this problem by proving that each pair of random variables (X_i,X_j) are independent. However, although mutual independence implies pairwise independence [this can be seen by integrating out one of the x_i's in Equation (1.10)], the converse does not hold in general. For example, suppose

$$\begin{aligned}f(x_1,x_2,x_3)=&(2\pi)^{-3/2}\exp\left[-\tfrac{1}{2}(x_1^2+x_2^2+x_3^2)\right]\\&\times\left\{1+x_1x_2x_3\exp\left[-\tfrac{1}{2}(x_1^2+x_2^2+x_3^2)\right]\right\}\\&-\infty<x_i<\infty \qquad (i=1,2,3),\end{aligned}$$

then the second term in the brace is an odd function of x_3 so that its integral over $-\infty<x_3<\infty$ is zero (see also Exercise 1 below). Hence

$$\begin{aligned}f(x_1,x_2)&=\frac{1}{2\pi}e^{-(1/2)(x_1^2+x_2^2)}\\&=f_1(x_1)f_2(x_2),\end{aligned}$$

and X_1 and X_2 are independent standard normal variables. We therefore conclude that X_1, X_2, and X_3 are pairwise independent and have marginal standard normal distributions, but they are not mutually independent because

$$f(x_1,x_2,x_3)\neq(2\pi)^{-3/2}\exp\left[-\tfrac{1}{2}(x_1^2+x_2^2+x_3^2)\right]=f_1(x_1)f_2(x_2)f_3(x_3).$$

However, in the special case when the random variables have a joint multivariate normal distribution, pairwise independence implies mutual independence (cf. Theorem 2.6, Corollary, Section 2.3).

Because of the central role of the multivariate normal distribution in regression analysis, we discuss the properties of this distribution in some detail in the next chapter.

EXERCISES 1c

1. Show that for all x_1, x_2, x_3, $1+x_1x_2x_3\exp[-\frac{1}{2}(x_1^2+x_2^2+x_3^2)]>0$.
2. If X and Y are random variables with the same variance, prove that $\text{cov}[X+Y,X-Y]=0$. Give a counterexample which shows that zero covariance does not necessarily imply independence.
3. Let X and Y be discrete random variables taking values 0 or 1 only, and let $\text{pr}[X=i, Y=j]=p_{ij}$ $(i=1,0;j=1,0)$. Prove that X and Y are independent if and only if $\text{cov}[X,Y]=0$.
4. If X is a random variable with a symmetric density function [that is, $f(x)=f(-x)$] and zero mean, prove that $\text{cov}[X,X^2]=0$.
5. If X, Y, and Z have the joint probability density function

$$f(x,y,z)=\tfrac{1}{8}(1+xyz), \qquad -1\leqslant x,y,z\leqslant 1,$$

prove that they are pairwise independent but not mutually independent.

1.6 CHI-SQUARE DISTRIBUTION

In Theorem 1.10 we prove a basic result about the difference of two chi-square random variables which can be used to show that certain quadratic forms have a chi-square distribution. But first a lemma.

LEMMA If $Y\sim\chi_k^2$ then the moment generating function of Y is

$$M(t)=(1-2t)^{-(1/2)k}.$$

Proof.

$$\begin{aligned}M(t)&=E[\exp(tY)]\\&=\int_0^\infty \frac{1}{2^k\Gamma(\frac{1}{2}k)}y^{(1/2)k-1}e^{ty-(1/2)y}\,dy\\&=(1-2t)^{-(1/2)k}\int_0^\infty\frac{1}{2^k\Gamma(\frac{1}{2}k)}z^{(1/2)k-1}e^{-(1/2)z}\,dz\\&=(1-2t)^{-(1/2)k}.\end{aligned}$$

[Here we use the substitution $z=(1-2t)y$ and assume that $|t|<\frac{1}{2}$]. □

If a moment generating function exists in some interval containing the origin ($t=0$), then it uniquely determines the distribution. Thus the chi-square distribution and, for that matter, most of the standard distributions (including the normal distribution) are uniquely determined by their moment generating functions. We can use this fact to prove the following theorem.

THEOREM 1.10 If $Q_i \sim \chi^2_{r_i}$ for $i=1,2$, $r_1 > r_2$, and $Q = Q_1 - Q_2$ is statistically independent of Q_2, then $Q \sim \chi^2_r$ where $r = r_1 - r_2$.

Proof.

$$\begin{aligned}(1-2t)^{-(1/2)r_1} &= E[\exp(tQ_1)]\\ &= E[\exp(tQ + tQ_2)]\\ &= E[\exp(tQ)]E[\exp(tQ_2)]\\ &= E[\exp(tQ)](1-2t)^{-(1/2)r_2}.\end{aligned}$$

Hence

$$E[\exp(tQ)] = (1-2t)^{-(1/2)(r_1 - r_2)},$$

which is the moment generating function of $\chi^2_{r_1 - r_2}$. □

MISCELLANEOUS EXERCISES 1

1. Let $X_1, X_2, \ldots, X_n$ be random variables with a common mean θ. Suppose $\operatorname{cov}[X_i, X_j] = 0$ for all i, j such that $j > i+1$. If

$$Q_1 = \sum_{i=1}^{n} (X_i - \bar{X})^2$$

and

$$Q_2 = (X_1 - X_2)^2 + (X_2 - X_3)^2 + \cdots + (X_{n-1} - X_n)^2 + (X_n - X_1)^2,$$

prove that

$$E\left[\frac{3Q_1 - Q_2}{n(n-3)}\right] = \operatorname{var}[\bar{X}].$$

2. If X and Y are random variables, prove that

$$\operatorname{var}[X] = \underset{Y}{E}\{\operatorname{var}[X|Y]\} + \underset{Y}{\operatorname{var}}\{E[X|Y]\}.$$

Generalize this result to vectors $\mathbf{X}$ and $\mathbf{Y}$ of random variables.

3. Let $\mathbf{X}$ be a 3×1 vector of random variables such that

$$\mathscr{D}[\mathbf{X}] = \begin{bmatrix} 5 & 2 & 3 \\ 2 & 3 & 0 \\ 3 & 0 & 2 \end{bmatrix}.$$

(a) Find the variance of $X_1 - 2X_2 + X_3$.
(b) Find the variance–covariance matrix of $\mathbf{Y} = (Y_1, Y_2)'$, where $Y_1 = X_1 + X_2$ and $Y_2 = X_1 + X_2 + X_3$.

4. Given a random sample X_1, X_2, X_3 from the distribution with density function

$$f(x) = \tfrac{1}{2}, \qquad -1 \leqslant x \leqslant 1,$$

find the variance of $(X_1 - X_2)^2 + (X_2 - X_3)^2 + (X_3 - X_1)^2$.

5. If $X_1, X_2, \ldots, X_n$ are independently and identically distributed as $N(0, \sigma^2)$, and $\mathbf{A}$ and $\mathbf{B}$ are any $n \times n$ symmetric matrices, prove that

$$\operatorname{cov}[\mathbf{X}'\mathbf{A}\mathbf{X}, \mathbf{X}'\mathbf{B}\mathbf{X}] = 2\sigma^4 \operatorname{tr}[\mathbf{AB}].$$

CHAPTER 2

Multivariate Normal Distribution

2.1 DEFINITION

By analogy with the (univariate) normal density function

$$f(y)=(2\pi\sigma^2)^{-1/2}\exp\left[-\frac{1}{2\sigma^2}(y-\theta)^2\right] \qquad (-\infty<y<\infty)$$

$$=(2\pi v)^{-1/2}\exp\left[-\tfrac{1}{2}(y-\theta)v^{-1}(y-\theta)\right] \qquad (v=\sigma^2>0)$$

we can define the multivariate density function

$$f(y_1,y_2,\ldots,y_n)=k^{-1}\exp\left[-\tfrac{1}{2}(\mathbf{y}-\boldsymbol{\theta})'\boldsymbol{\Sigma}^{-1}(\mathbf{y}-\boldsymbol{\theta})\right], \tag{2.1}$$

where $-\infty<y_i<\infty$ $(i=1,2,\ldots,n)$ and $\boldsymbol{\Sigma}$ is an $n\times n$ positive definite matrix.

THEOREM 2.1 If $\mathbf{Y}=(Y_1,Y_2,\ldots,Y_n)'$ is a vector of random variables with probability density function (2.1), then:

(i) $k=(2\pi)^{(1/2)n}|\boldsymbol{\Sigma}|^{1/2}$.
(ii) $\mathcal{E}[\mathbf{Y}]=\boldsymbol{\theta}$ and $\mathfrak{D}[\mathbf{Y}]=\boldsymbol{\Sigma}$.
(iii) $Q=(\mathbf{Y}-\boldsymbol{\theta})'\boldsymbol{\Sigma}^{-1}(\mathbf{Y}-\boldsymbol{\theta})\sim\chi_n^2$.

Proof. (i) Because $\boldsymbol{\Sigma}$ is positive definite there exists a real orthogonal matrix $\mathbf{T}$ such that

$$\mathbf{T}'\boldsymbol{\Sigma}\mathbf{T}=\mathrm{diag}(\lambda_1,\lambda_2,\ldots,\lambda_n)=\boldsymbol{\Lambda},$$

say, where $\lambda_1,\lambda_2,\ldots,\lambda_n$, the eigenvalues of $\boldsymbol{\Sigma}$, are positive (*A4.1*). Let

$(\mathbf{y}-\boldsymbol{\theta})=\mathbf{T}\mathbf{x}$; then the Jacobian of this transformation is

$$J=\left\|\left[\left(\frac{\partial y_i}{\partial x_j}\right)\right]\right\|=|\mathbf{T}|,$$

and the absolute value of J, $|J|$, say, is unity as the determinant of an orthogonal matrix is ± 1 $(1=|\mathbf{T}'\mathbf{T}|=|\mathbf{T}'||\mathbf{T}|=|\mathbf{T}|^2)$. Therefore, since $\mathbf{T}^{-1}=\mathbf{T}'$, we have

$$\operatorname{diag}(\lambda_1^{-1},\lambda_2^{-1},\ldots,\lambda_n^{-1})=(\mathbf{T}'\boldsymbol{\Sigma}\mathbf{T})^{-1}=\mathbf{T}'\boldsymbol{\Sigma}^{-1}\mathbf{T}$$

and

$$\int_{-\infty}^{\infty}\cdots\int_{-\infty}^{\infty}\exp\left[-\tfrac{1}{2}(\mathbf{y}-\boldsymbol{\theta})'\boldsymbol{\Sigma}^{-1}(\mathbf{y}-\boldsymbol{\theta})\right]dy_1\,dy_2\cdots dy_n$$

$$=\int_{-\infty}^{\infty}\cdots\int_{-\infty}^{\infty}\exp\left\{-\tfrac{1}{2}\sum_{i=1}^{n}\left(\frac{x_i^2}{\lambda_i}\right)\right\}dx_1\,dx_2\cdots dx_n$$

$$=\prod_{i=1}^{n}\int_{-\infty}^{\infty}\exp\left[-\tfrac{1}{2}\left(\frac{x_i^2}{\lambda_i}\right)\right]dx_i$$

$$=\prod_{i=1}^{n}(2\pi\lambda_i)^{1/2}. \qquad (2.2)$$

Now f represents a density function if the multiple integral of (2.1) is unity. Therefore we must have

$$k=(2\pi)^{(1/2)n}\left(\prod_i\lambda_i\right)^{1/2}$$

$$=(2\pi)^{(1/2)n}|\boldsymbol{\Sigma}|^{1/2},$$

since

$$\prod_i\lambda_i=|\boldsymbol{\Lambda}|=|\mathbf{T}'\boldsymbol{\Sigma}\mathbf{T}|=|\mathbf{T}'\mathbf{T}||\boldsymbol{\Sigma}|=|\boldsymbol{\Sigma}|. \qquad (2.3)$$

(ii) If $\mathbf{Y}-\boldsymbol{\theta}=\mathbf{T}\mathbf{X}$, then the density function of $\mathbf{X}$ is given by

$$g(x_1,x_2,\ldots,x_n)=f(\mathbf{y}(\mathbf{x}))|J|$$

$$=\prod_{i=1}^{n}(2\pi\lambda_i)^{-1/2}\exp\left(-\tfrac{1}{2}\frac{x_i^2}{\lambda_i}\right).$$

This factorization of the joint density function implies that the X_i are mutually independent and $X_i \sim N(0,\lambda_i)$; in particular $\mathcal{E}[\mathbf{X}]=\mathbf{0}$ and $\mathcal{D}[\mathbf{X}]=\mathbf{\Lambda}$. We then have $\mathcal{E}[\mathbf{Y}-\boldsymbol{\theta}]=\mathbf{0}$, that is, $\mathcal{E}[\mathbf{Y}]=\boldsymbol{\theta}$, and

$$\begin{aligned}\mathcal{D}[\mathbf{Y}] &= \mathcal{E}\left[(\mathbf{Y}-\boldsymbol{\theta})(\mathbf{Y}-\boldsymbol{\theta})'\right]\\ &= \mathcal{E}\left[\mathbf{TXX'T'}\right]\\ &= \mathbf{T}\mathcal{E}\left[\mathbf{XX'}\right]\mathbf{T'}\\ &= \mathbf{T}\mathcal{D}\left[\mathbf{X}\right]\mathbf{T'}\\ &= \mathbf{T\Lambda T'}\\ &= \mathbf{T(T'\Sigma T)T'}\\ &= \mathbf{\Sigma}.\end{aligned}$$

(iii)

$$\begin{aligned}Q &= \mathbf{X'T'\Sigma^{-1}TX}\\ &= \mathbf{X'\Lambda^{-1}X}\\ &= \sum_i \left(\frac{X_i^2}{\lambda_i}\right)\\ &= \sum_i Z_i^2, \qquad (2.4)\end{aligned}$$

where the Z_i are independently distributed as $N(0,1)$. Thus $Z_i^2 \sim \chi_1^2$, and $Q \sim \chi_n^2$ as the sum of independent chi-square random variables is also chi-square. □

Definition. *If* $\mathbf{Y}$ *has the density function* (2.1) *we say that* $\mathbf{Y}$ *has a multivariate normal distribution and write* $\mathbf{Y} \sim N_n(\boldsymbol{\theta},\mathbf{\Sigma})$. *When* $n=1$ *we drop the subscript.*

COROLLARY 1 If $\mathbf{Y} \sim N_n(\boldsymbol{\theta},\mathbf{\Sigma})$ then $\mathbf{Y}-\boldsymbol{\theta} \sim N_n(\mathbf{0},\mathbf{\Sigma})$.

COROLLARY 2 If $Y_1, Y_2, \ldots, Y_n$ are mutually independent normal random variables with means $\theta_1, \theta_2, \ldots, \theta_n$, respectively, and common variance σ^2, then $\mathbf{Y} \sim N_n(\boldsymbol{\theta}, \sigma^2\mathbf{I}_n)$.

Proof.

$$f(y_1,y_2,\ldots,y_n)=\prod_{i=1}^{n} f_i(y_i)$$

$$=(2\pi\sigma^2)^{-(1/2)n}\exp\left[-\frac{\sum_i (y_i-\theta_i)^2}{2\sigma^2}\right]$$

$$=(2\pi)^{-(1/2)n}|\sigma^2\mathbf{I}_n|^{-1/2}\exp\left\{-\tfrac{1}{2}(\mathbf{y}-\boldsymbol{\theta})'(\sigma^2\mathbf{I}_n)^{-1}(\mathbf{y}-\boldsymbol{\theta})\right\}.$$

COROLLARY 3

$$\int_{-\infty}^{\infty}\cdots\int_{-\infty}^{\infty}\exp\{-\tfrac{1}{2}(\mathbf{y}-\boldsymbol{\theta})'\boldsymbol{\Sigma}^{-1}(\mathbf{y}-\boldsymbol{\theta})\}\,dy_1\,dy_2\cdots dy_n$$

$$=(2\pi)^{(1/2)n}|\boldsymbol{\Sigma}|^{1/2} \tag{2.5}$$

$$=(2\pi)^{(1/2)n}|\boldsymbol{\Sigma}^{-1}|^{-1/2}. \tag{2.6}$$

EXAMPLE 2.1 (BIVARIATE NORMAL DISTRIBUTION) Let $\mathbf{Y}=(Y_1,Y_2)'$ have a joint density function

$$f(y_1,y_2)=\left\{2\pi\sigma_1\sigma_2(1-\rho^2)^{1/2}\right\}^{-1}$$

$$\times\exp\left[-\frac{1}{2(1-\rho^2)}\left\{\frac{(y_1-\theta_1)^2}{\sigma_1^2}-\frac{2\rho(y_1-\theta_1)(y_2-\theta_2)}{\sigma_1\sigma_2}+\frac{(y_2-\theta_2)^2}{\sigma_2^2}\right\}\right]$$

where $\sigma_i>0$, $-\infty<y_i<\infty$ $(i=1,2)$, and $|\rho|<1$. Prove that $\mathbf{Y}\sim N_2(\boldsymbol{\theta},\boldsymbol{\Sigma})$, and find the correlation between Y_1 and Y_2.

Solution.

$$f(y_1,y_2)=k^{-1}\exp\left[-\tfrac{1}{2}(\mathbf{y}-\boldsymbol{\theta})'\mathbf{V}(\mathbf{y}-\boldsymbol{\theta})\right]$$

where

$$\mathbf{V}=\frac{1}{(1-\rho^2)}\begin{pmatrix}\dfrac{1}{\sigma_1^2} & -\dfrac{\rho}{\sigma_1\sigma_2}\\[2ex] -\dfrac{\rho}{\sigma_1\sigma_2} & \dfrac{1}{\sigma_2^2}\end{pmatrix}.$$

Let

$$\mathbf{\Sigma}=\mathbf{V}^{-1}=\begin{pmatrix} \sigma_1^2 & \rho\sigma_1\sigma_2 \\ \rho\sigma_1\sigma_2 & \sigma_2^2 \end{pmatrix},$$

then $\mathbf{\Sigma}$ is positive definite as all its leading minor determinants are positive (*A4.7*), that is, $\sigma_1^2>0$ and $|\mathbf{\Sigma}|=\sigma_1^2\sigma_2^2(1-\rho^2)>0$. Therefore since $k=(2\pi)^{2/2}|\mathbf{\Sigma}|^{1/2}$ we have shown that the distribution of $\mathbf{Y}$ can be written in the form of $N_2(\boldsymbol{\theta},\mathbf{\Sigma})$.

Finally we note that $\sigma_{12}=\text{cov}[Y_1,Y_2]=\sigma_1\sigma_2\rho$ and $\sigma_{ii}^2=\sigma_i^2$ $(i=1,2)$, so that ρ is the required correlation coefficient.

EXAMPLE 2.2 (Graybill [1961: pp. 60–61]) Let $\mathbf{Y}$ have a bivariate normal distribution $f(y_1,y_2)=k^{-1}\exp(-\frac{1}{2}Q)$, where

$$Q=y_1^2+2y_2^2-y_1y_2-3y_1-2y_2+4.$$

Identify $\boldsymbol{\theta}$ and $\mathbf{\Sigma}$.

Solution. We note first that

$$Q=(\mathbf{y}-\boldsymbol{\theta})'\mathbf{\Sigma}^{-1}(\mathbf{y}-\boldsymbol{\theta})=\mathbf{y}'\mathbf{\Sigma}^{-1}\mathbf{y}-2\mathbf{y}'\mathbf{\Sigma}^{-1}\boldsymbol{\theta}+\boldsymbol{\theta}'\mathbf{\Sigma}^{-1}\boldsymbol{\theta}.$$

From *A6*, $\partial Q/\partial\mathbf{y}=2\mathbf{\Sigma}^{-1}\mathbf{y}-2\mathbf{\Sigma}^{-1}\boldsymbol{\theta}$, and $\boldsymbol{\theta}$ is the solution of $\partial Q/\partial\mathbf{y}=\mathbf{0}$. Thus

$$\frac{\partial Q}{\partial y_1}=2y_1-y_2-3=0,$$

$$\frac{\partial Q}{\partial y_2}=-y_1+4y_2-2=0,$$

and solving these two equations yields $y_1=2$ and $y_2=1$; that is, $\boldsymbol{\theta}'=(2,1)$. Also $\mathbf{Y}'\mathbf{\Sigma}^{-1}\mathbf{Y}=y_1^2+2y_2^2-y_1y_2$, and it is readily shown that

$$\mathbf{\Sigma}=\frac{1}{7}\begin{pmatrix} 8 & 2 \\ 2 & 4 \end{pmatrix}.$$

Remark. If $\mathbf{Y}\sim N_n(\boldsymbol{\theta},\mathbf{\Sigma})$ and $\mathbf{Y}'=(\mathbf{Y}_1',\mathbf{Y}_2')$, Miller [1975] derives the probability density functions for the following random variables: (1) $\|\mathbf{Y}\|$ (the so-called Rayleigh distribution), (2) $\mathbf{Y}_1'\mathbf{Y}_2$, (3) the angle between $\mathbf{Y}_1$ and $\mathbf{Y}_2$ when these two vectors are of the same dimension, (4) $\|\mathbf{Y}_1\|\times\|\mathbf{Y}_2\|$, (5) $\|\mathbf{Y}_1\|/\|\mathbf{Y}_2\|$, (6) $\|\mathbf{Y}_1\|^2+\|\mathbf{Y}_2\|^2$, and (7) $\|\mathbf{Y}_1\|^2-\|\mathbf{Y}_2\|^2$. The density functions simplify considerably when $\mathbf{Y}_1$ and $\mathbf{Y}_2$ are independent.

EXERCISES 2a

1. Evaluate

$$\int_{-\infty}^{\infty} \exp(-(y_1^2+2y_1y_2+4y_2^2))\,dy_1\,dy_2,$$

2. If $f(y_1,y_2)=k^{-1}\exp[-\frac{1}{2}(2y_1^2+y_2^2+2y_1y_2-22y_1-14y_2+65)]$ represents a bivariate normal density function, prove that $k=2\pi$.
3. If $\mathbf{Y}\sim N_2(\mathbf{0},\mathbf{\Sigma})$, where $\mathbf{\Sigma}=[(\sigma_{ij})]$, prove that

$$\left(\mathbf{Y}'\mathbf{\Sigma}^{-1}\mathbf{Y}-\frac{Y_1^2}{\sigma_{11}}\right)\sim\chi_1^2.$$

2.2 MOMENT GENERATING FUNCTION

If $\mathbf{Y}\sim N_n(\boldsymbol{\theta},\mathbf{\Sigma})$ then $M(\mathbf{t})$, the moment generating function of $\mathbf{Y}$, is derived as follows: Let $\mathbf{X}=\mathbf{Y}-\boldsymbol{\theta}$; then by Corollary 1 of Theorem 2.1, $\mathbf{X}$ is $N_n(\mathbf{0},\mathbf{\Sigma})$ and

$$\begin{aligned}
M(\mathbf{t}) &= E\left[\exp\left(\sum_i t_i Y_i\right)\right]\\
&= E\left[\exp(\mathbf{t}'\mathbf{Y})\right]\\
&= E\left[\exp\{\mathbf{t}'(\mathbf{X}+\boldsymbol{\theta})\}\right]\\
&= \int_{-\infty}^{\infty}\cdots\int_{-\infty}^{\infty} k^{-1}\exp\left[-\tfrac{1}{2}\mathbf{x}'\mathbf{\Sigma}^{-1}\mathbf{x}+\mathbf{t}'(\mathbf{x}+\boldsymbol{\theta})\right]dx_1\,dx_2\cdots dx_n\\
&= \int_{-\infty}^{\infty}\cdots\int_{-\infty}^{\infty} k^{-1}\exp\left[-\tfrac{1}{2}(\mathbf{x}-\mathbf{\Sigma t})'\mathbf{\Sigma}^{-1}(\mathbf{x}-\mathbf{\Sigma t})\right]dx_1\,dx_2\cdots dx_n\\
&\qquad\times\exp\left[\mathbf{t}'\boldsymbol{\theta}+\tfrac{1}{2}\mathbf{t}'\mathbf{\Sigma t}\right]\\
&= \exp\left[\mathbf{t}'\boldsymbol{\theta}+\tfrac{1}{2}\mathbf{t}'\mathbf{\Sigma t}\right].
\end{aligned}\tag{2.7}$$

The last step follows from the fact that the integrand represents the density $N_n(\mathbf{x}-\mathbf{\Sigma t},\mathbf{\Sigma})$ so that the multiple integral is unity. We note that (2.7) is an obvious generalization of the moment generating function for the (univariate) normal distribution $N(\theta,\sigma^2)$, namely,

$$M(t)=\exp(t\theta+\tfrac{1}{2}\sigma^2t^2). \tag{2.8}$$

It can be shown that when $\boldsymbol{\Sigma}$ is positive definite, $M(\mathbf{t})$ uniquely determines the density function of $\mathbf{Y}$, and we use this result to prove the following theorem.

THEOREM 2.2 If $\mathbf{Y}\sim N_n(\boldsymbol{\theta},\boldsymbol{\Sigma})$ and $\mathbf{C}$ is a $p\times n$ matrix of rank p, then $\mathbf{CY}\sim N_p(\mathbf{C}\boldsymbol{\theta},\mathbf{C}\boldsymbol{\Sigma}\mathbf{C}')$.

Proof. Let $\mathbf{X}=\mathbf{CY}$; then for every real $\mathbf{t}$

$$\begin{aligned}E\left[\exp(\mathbf{t}'\mathbf{X})\right]&=E\left[\exp(\mathbf{t}'\mathbf{CY})\right]\\&=E\left[\exp(\mathbf{s}'\mathbf{Y})\right],\qquad \text{say}\\&=\exp(\mathbf{s}'\boldsymbol{\theta}+\tfrac{1}{2}\mathbf{s}'\boldsymbol{\Sigma}\mathbf{s})\qquad\left[\text{by (2.7)}\right]\\&=\exp\left[\mathbf{t}'(\mathbf{C}\boldsymbol{\theta})+\tfrac{1}{2}\mathbf{t}'(\mathbf{C}\boldsymbol{\Sigma}\mathbf{C}')\mathbf{t}\right].\end{aligned}\tag{2.9}$$

Because $\mathbf{C}\boldsymbol{\Sigma}\mathbf{C}'$ is positive definite (*A4.5*), the above expression represents the moment generating function of $N_p(\mathbf{C}\boldsymbol{\theta},\mathbf{C}\boldsymbol{\Sigma}\mathbf{C}')$, and the theorem is proved. □

THEOREM 2.3 Mutually independent univariate normal variables with the same variance remain mutually independent normal variables under orthogonal transformations.*

Proof. Let $Y_1,Y_2,\ldots,Y_n$ be mutually independent normal variables with means $\theta_1,\theta_2,\ldots,\theta_n$ and common variance σ^2. Then, by Theorem 2.1 (Corollary 2), $\mathbf{Y}=(Y_1,Y_2,\ldots,Y_n)'\sim N_n(\boldsymbol{\theta},\sigma^2\mathbf{I}_n)$. Let $\mathbf{X}=\mathbf{TY}$, where $\mathbf{T}$ is an $n\times n$ orthogonal matrix; then $\mathscr{D}[\mathbf{X}]=\mathbf{T}\mathscr{D}[\mathbf{Y}]\mathbf{T}'=\sigma^2\mathbf{TT}'=\sigma^2\mathbf{I}_n$ and, by Theorem 2.2, we have $\mathbf{X}\sim N_n(\mathbf{T}\boldsymbol{\theta},\sigma^2\mathbf{I}_n)$. Hence the X_i are mutually independent normal variables with common variance σ^2. □

EXAMPLE 2.3 Let $Y_1,Y_2,\ldots,Y_n$ be independently and identically distributed as $N(\theta,\sigma^2)$. Prove that $\overline{Y}$ is statistically independent of $Q=\Sigma_i(Y_i-\overline{Y})^2/\sigma^2$, and $Q\sim\chi^2_{n-1}$.

*The above invariance property is a unique feature of the normal distribution (Lancaster [1954: p. 251]). Further characterizations of the univariate and multivariate normal distributions are given by Lukacs [1956], Laha [1957], Rao [1969, 1972a, 1973], Kingman and Graybill [1970], Patil and Boswell [1970], and Anderson [1971].

Solution. Let $U_i=(Y_i-\theta)/\sigma$; then $U_i \sim N(0,1)$ and $\mathbf{U}\sim N_n(\mathbf{0},\mathbf{I}_n)$. Let

$$X_1=\frac{U_1+U_2+\cdots+U_n}{\sqrt{n}},$$

$$X_2=\frac{U_1-U_2}{\sqrt{2}},$$

$$X_3=\frac{U_1+U_2-2U_3}{\sqrt{6}},$$

$$\cdots$$

$$X_n=\frac{U_1+U_2+\cdots+U_{n-1}-(n-1)U_n}{\sqrt{n(n-1)}}; \qquad (2.10)$$

that is, $\mathbf{X}=\mathbf{TU}$, where $\mathbf{T}$ is an $n\times n$ orthogonal matrix (the above transformation is known as Helmert's transformation). By Theorem 2.3 $\mathbf{X}\sim N_n(\mathbf{0},\mathbf{I}_n)$ so that the X_i are mutually independent standard normal random variables. Now

$$\begin{aligned}\sum_{i=1}^{n} X_i^2 &= \mathbf{X}'\mathbf{X}\\ &= \mathbf{U}'\mathbf{T}'\mathbf{T}\mathbf{U}\\ &= \mathbf{U}'\mathbf{U}\\ &= \sum_i U_i^2\\ &= n\overline{U}^2+\sum_{i=1}^{n}\left(U_i-\overline{U}\right)^2\\ &= X_1^2+\sum_{i=1}^{n}\left(U_i-\overline{U}\right)^2 \end{aligned} \qquad (2.11)$$

and $Q=\Sigma_i(U_i-\overline{U})^2=\Sigma_{i=2}^n X_i^2$. Since X_1 is independent of $(X_2,X_3,\ldots,X_n)$, X_1 is independent of Q (Theorem 1.9 in Section 1.5). Hence $\overline{Y}$ is independent of Q. Also $X_i^2\sim\chi_1^2$ so that $Q\sim\chi_{n-1}^2$. (It should be noted that any orthogonal matrix $\mathbf{T}$ with the same first row will do.)

THEOREM 2.4 If $\mathbf{Y}\sim N_n(\boldsymbol{\theta},\boldsymbol{\Sigma})$, then the marginal distribution of any subset of the elements of $\mathbf{Y}$ is also multivariate normal.

Proof. Without loss of generality we can take as our subset $\mathbf{X}'=(Y_1, Y_2, \ldots, Y_p)$. Then

$$\begin{aligned}\mathbf{X} &= (\mathbf{I}_p, \mathbf{O})\mathbf{Y}\\ &= \mathbf{CY},\end{aligned}$$

where $\mathbf{C}$ is a $p \times n$ matrix of rank p. Hence, by Theorem 2.2, $\mathbf{X} \sim N_p(\mathbf{C}\boldsymbol{\theta}, \mathbf{C}\boldsymbol{\Sigma}\mathbf{C}')$, where $\mathbf{C}\boldsymbol{\theta}=(\theta_1, \theta_2, \ldots, \theta_p)'$ and $\mathbf{C}\boldsymbol{\Sigma}\mathbf{C}'$ is the leading $p \times p$ submatrix of $\boldsymbol{\Sigma}$. □

THEOREM 2.5 $\mathbf{Y}$ has a multivariate normal distribution if and only if $\mathbf{a}'\mathbf{Y}$ is univariate normal for all real vectors $\mathbf{a}$ $(\mathbf{a} \neq \mathbf{0})$.

Proof. Suppose $\mathcal{E}[\mathbf{Y}]=\boldsymbol{\theta}$ and $\mathcal{D}[\mathbf{Y}]=\boldsymbol{\Sigma}$, where $\boldsymbol{\Sigma}$ is $n \times n$. If $X=\mathbf{a}'\mathbf{Y}$ is univariate normal, then $E[X]=\mathbf{a}'\boldsymbol{\theta}$, $\text{var}[X]=\mathcal{D}[\mathbf{a}'\mathbf{Y}]=\mathbf{a}'\boldsymbol{\Sigma}\mathbf{a}$ (Theorem 1.4, Corollary 2) and $X \sim N(\mathbf{a}'\boldsymbol{\theta}, \mathbf{a}'\boldsymbol{\Sigma}\mathbf{a})$. Hence the moment generating function of X is given by [Equation (2.8)]

$$E\left[\exp(Xt)\right]=\exp\left[(\mathbf{a}'\boldsymbol{\theta})t+\tfrac{1}{2}\mathbf{a}'\boldsymbol{\Sigma}\mathbf{a}t^2\right], \tag{2.12}$$

and this equation holds for all t. Setting $t=1$ we have, for every $\mathbf{a}$,

$$\begin{aligned}E\left[\exp(\mathbf{a}'\mathbf{Y})\right] &= \exp(\mathbf{a}'\boldsymbol{\theta}+\tfrac{1}{2}\mathbf{a}'\boldsymbol{\Sigma}\mathbf{a})\\ &= M(\mathbf{a}),\end{aligned}$$

which is the moment generating function for $N_n(\boldsymbol{\theta}, \boldsymbol{\Sigma})$. Since $\text{var}[X]>0$ for $\mathbf{a} \neq \mathbf{0}$, $\mathbf{a}'\boldsymbol{\Sigma}\mathbf{a}>0$ and $\boldsymbol{\Sigma}$ is positive definite. Hence $\mathbf{Y} \sim N_n(\boldsymbol{\theta}, \boldsymbol{\Sigma})$. Conversely, if $\mathbf{Y} \sim N_n(\boldsymbol{\theta}, \boldsymbol{\Sigma})$ then $\mathbf{a}'\mathbf{Y}$ is univariate normal (Theorem 2.2: $\text{rank}(\mathbf{a}')=1$).

COROLLARY If $\mathbf{Y}'=(Y_1, Y_2)$ and the marginal distributions of Y_1 and Y_2 are both univariate normal, then $\mathbf{Y}$ has a bivariate normal distribution if and only if $a_1Y_1+a_2Y_2$ is univariate normal for all real a_1 and a_2.

We have shown that the multivariate normal distribution is the only multivariate distribution which has the property that every linear combination $\mathbf{a}'\mathbf{Y}$ is univariate normal; this uniqueness property may in fact be used to define the multivariate normal distribution. We also note that if the marginal distributions of Y_1 and Y_2 are both normal, then the joint distribution is not necessarily bivariate normal as seen by the counterexample

$$f(y_1,y_2)=(2\pi)^{-1}\exp\left[-\tfrac{1}{2}(y_1^2+y_2^2)\right]\left\{1+y_1y_2\exp\left[-\tfrac{1}{2}(y_1^2+y_2^2)\right]\right\}.$$

Numerous counterexamples are available (cf. Pierce and Dykstra [1969], and they can be constructed using, for example, the methods of Joshi [1970] and Kowalski [1973]. We note from the above theorem that we must have $a_1Y_1+a_2Y_2$ normal for *all* real a_1 and a_2, and not just for $a_1=1$, $a_2=0$ and $a_1=0$, $a_2=1$.

EXERCISES 2b

1. Find the moment generating function of the bivariate normal distribution given in Example 2.1.
2. If $\mathbf{Y}\sim N_n(\boldsymbol{\theta},\boldsymbol{\Sigma})$, prove that $Y_i\sim N(\theta_i,\sigma_{ii})$.
3. Suppose that $\mathbf{Y}=(Y_1,Y_2,Y_3)'\sim N_3(\boldsymbol{\theta},\boldsymbol{\Sigma})$, where

$$\boldsymbol{\theta}=\begin{bmatrix}2\\1\\2\end{bmatrix}\quad\text{and}\quad\boldsymbol{\Sigma}=\begin{bmatrix}2&1&1\\1&3&0\\1&0&1\end{bmatrix}.$$

   Find the joint distribution of

$$Z_1=Y_1+Y_2+Y_3$$

   and

$$Z_2=Y_1-Y_2.$$

4. Given $\mathbf{Y}\sim N_n(\boldsymbol{\theta},\mathbf{I}_n)$, find the joint distribution of $L=\mathbf{a}'\mathbf{Y}$ and $M=\mathbf{b}'\mathbf{Y}$, where $\mathbf{a}'\mathbf{b}=0$, and hence show that L and M are statistically independent.
5. Prove Theorem 2.1 (iii) using moment generating functions.
6. Use Theorem 2.5 to prove Theorem 2.4.
7. Let $\begin{pmatrix}X_1\\Y_1\end{pmatrix},\begin{pmatrix}X_2\\Y_2\end{pmatrix},\ldots,\begin{pmatrix}X_n\\Y_n\end{pmatrix}$ be a random sample from $N_2(\boldsymbol{\theta},\boldsymbol{\Sigma})$. Find the joint density function of the sample means $\overline{X}$ and $\overline{Y}$.
8. If Y_1 and Y_2 are random variables such that Y_1+Y_2 and Y_1-Y_2 are independent standard normal variables, prove that Y_1 and Y_2 have a bivariate normal distribution.
9. Let X and Y have a joint probability density function

$$f(x,y)=\frac{1}{2\pi}\exp\left[-\tfrac{1}{2}(x^2+y^2)\right]\left(1-\frac{xy}{(1+x^2)(1+y^2)}\right),\qquad -\infty<x,y<\infty.$$

Prove that the marginal distributions of X and Y are normal.

Joshi [1970]

10. Given that $Y_1, Y_2, \dots, Y_n$ are independently distributed as $N(0,1)$, find the joint moment generating function of $\overline{Y}, Y_1 - \overline{Y}, Y_2 - \overline{Y}, \dots, Y_n - \overline{Y}$ and hence deduce that $\overline{Y}$ and $\Sigma_i (Y_i - \overline{Y})^2$ are statistically independent.

Hogg and Craig [1970]

2.3 INDEPENDENCE OF NORMAL VARIABLES

It is well known that independence implies zero covariance. In this section we give two theorems in which the converse is also true.

THEOREM 2.6 Let $\mathbf{Y} \sim N_n(\boldsymbol{\theta}, \boldsymbol{\Sigma})$ and suppose $\mathbf{Y}$ is partitioned in the form of

$$\mathbf{Y} = \begin{pmatrix} \mathbf{Y}^{(1)} \\ \mathbf{Y}^{(2)} \end{pmatrix},$$

where $\mathbf{Y}^{(1)}$ has p elements $(p < n)$. Then $\mathbf{Y}^{(1)}$ and $\mathbf{Y}^{(2)}$ are statistically independent if and only if $\mathcal{C}[\mathbf{Y}^{(1)}, \mathbf{Y}^{(2)}] = \mathbf{O}$.

Proof. If $\mathbf{Y}^{(1)}$ and $\mathbf{Y}^{(2)}$ are independent, then so are any pair of elements $Y_i^{(1)}$, $Y_j^{(2)}$. Hence

$$\begin{aligned} \operatorname{cov}\left[Y_i^{(1)}, Y_j^{(2)} \right] &= E\left[\left(Y_i^{(1)} - \theta_i^{(1)} \right)\left(Y_j^{(2)} - \theta_j^{(2)} \right) \right] \\ &= E\left[Y_i^{(1)} - \theta_i^{(1)} \right] E\left[Y_j^{(2)} - \theta_j^{(2)} \right] \\ &= 0, \end{aligned}$$

and

$$\mathcal{C}\left[\mathbf{Y}^{(1)}, \mathbf{Y}^{(2)} \right] = \mathbf{O}. \tag{2.13}$$

Conversely, if

$$\boldsymbol{\Sigma}_{12} = \mathcal{C}\left[\mathbf{Y}^{(1)}, \mathbf{Y}^{(2)} \right] = \mathbf{O},$$

then

$$\boldsymbol{\Sigma} = \begin{pmatrix} \boldsymbol{\Sigma}_{11} & \boldsymbol{\Sigma}_{12} \\ \boldsymbol{\Sigma}_{21} & \boldsymbol{\Sigma}_{22} \end{pmatrix} = \begin{pmatrix} \boldsymbol{\Sigma}_{11} & \mathbf{O} \\ \mathbf{O} & \boldsymbol{\Sigma}_{22} \end{pmatrix},$$

$$|\boldsymbol{\Sigma}| = |\boldsymbol{\Sigma}_{11}||\boldsymbol{\Sigma}_{22}|,$$

$$\boldsymbol{\Sigma}^{-1} = \begin{pmatrix} \boldsymbol{\Sigma}_{11}^{-1} & \mathbf{O} \\ \mathbf{O} & \boldsymbol{\Sigma}_{22}^{-1} \end{pmatrix},$$

and

$$(\mathbf{Y}-\boldsymbol{\theta})'\boldsymbol{\Sigma}^{-1}(\mathbf{Y}-\boldsymbol{\theta})=\sum_{i=1}^{2}(\mathbf{Y}^{(i)}-\boldsymbol{\theta}^{(i)})'\boldsymbol{\Sigma}_{ii}^{-1}(\mathbf{Y}^{(i)}-\boldsymbol{\theta}^{(i)})$$

$$=Q_1+Q_2, \qquad \text{say.}$$

Hence

$$\begin{aligned}
f(\mathbf{y})&=(2\pi)^{-(1/2)n}|\boldsymbol{\Sigma}|^{-1/2}\exp\left[-\tfrac{1}{2}(Q_1+Q_2)\right]\\
&=(2\pi)^{-(1/2)p}|\boldsymbol{\Sigma}_{11}|^{-1/2}\exp\left(-\tfrac{1}{2}Q_1\right)\\
&\quad\times(2\pi)^{-(1/2)(n-p)}|\boldsymbol{\Sigma}_{22}|^{-1/2}\exp\left(-\tfrac{1}{2}Q_2\right)\\
&=f_1(\mathbf{y}^{(1)})f_2(\mathbf{y}^{(2)}),
\end{aligned}$$

and the theorem is proved. □

COROLLARY The above theorem is readily generalized as follows: if $\mathbf{Y}'=(\mathbf{Y}^{(1)'},\mathbf{Y}^{(2)'},\ldots,\mathbf{Y}^{(k)'})$ and $\mathcal{C}[\mathbf{Y}^{(i)},\mathbf{Y}^{(j)}]=\mathbf{O}$ for each $i,j(i\neq j)$, then the $\mathbf{Y}^{(i)}$ $(i=1,2,\ldots,k)$ are mutually independent (and not just pairwise independent).

THEOREM 2.7 Suppose $\mathbf{Y}\sim N_n(\boldsymbol{\theta},\sigma^2\mathbf{I}_n)$ and let $\mathbf{U}=\mathbf{AY}$ and $\mathbf{V}=\mathbf{BY}$. Let $\mathbf{A}_1$ represent the linearly independent rows of $\mathbf{A}$ and let $\mathbf{U}_1=\mathbf{A}_1\mathbf{Y}$. Then if $\mathcal{C}[\mathbf{U},\mathbf{V}]=\mathbf{O}$ we have:

(i) $\mathbf{U}_1$ is independent of $\mathbf{V}'\mathbf{V}$.
(ii) $\mathbf{U}'\mathbf{U}$ and $\mathbf{V}'\mathbf{V}$ are independent.

Proof. Let $\mathbf{B}_1$ represent the linearly independent rows of $\mathbf{B}$. Because (by Theorem 1.4)

$$\mathbf{O}=\mathcal{C}[\mathbf{U},\mathbf{V}]=\mathbf{A}\mathcal{D}[\mathbf{Y}]\mathbf{B}'=\sigma^2\mathbf{AB}', \tag{2.14}$$

every row of $\mathbf{A}$ is orthogonal to every row of $\mathbf{B}$. Hence if

$$\mathbf{C}_1=\begin{pmatrix}\mathbf{A}_1\\ \mathbf{B}_1\end{pmatrix},$$

the rows of $\mathbf{C}_1$ are linearly independent and $\mathbf{C}_1\mathbf{Y}$ has a multivariate normal distribution (Theorem 2.2). Let $\mathbf{V}_1=\mathbf{B}_1\mathbf{Y}$; then $\mathbf{AB}'=\mathbf{O}$ implies that $\mathbf{A}_1\mathbf{B}_1'=\mathbf{O}$, $\mathcal{C}[\mathbf{U}_1,\mathbf{V}_1]=\mathbf{O}$, and $\mathbf{U}_1$ and $\mathbf{V}_1$ are statistically independent (Theorem 2.6).

Now without loss of generality we can assume that

$$\mathbf{B}=\begin{pmatrix}\mathbf{B}_1\\ \mathbf{B}_2\end{pmatrix},$$

where the rows of $\mathbf{B}_2$ are linearly dependent on the rows of $\mathbf{B}_1$; that is, there exists a matrix $\mathbf{H}$ such that $\mathbf{B}_2'=\mathbf{B}_1'\mathbf{H}'$ or $\mathbf{B}_2=\mathbf{H}\mathbf{B}_1$. Hence

$$\begin{aligned}\mathbf{V}&=\mathbf{B}\mathbf{Y}\\ &=\begin{pmatrix}\mathbf{B}_1\\ \mathbf{H}\mathbf{B}_1\end{pmatrix}\mathbf{Y}\\ &=\begin{pmatrix}\mathbf{I}\\ \mathbf{H}\end{pmatrix}\mathbf{B}_1\mathbf{Y}\\ &=\mathbf{M}\mathbf{V}_1, \qquad \text{say,}\end{aligned}$$

and, by Theorem 1.9, $\mathbf{U}_1$ is independent of $\mathbf{V}'\mathbf{V}$ $(=\mathbf{V}_1'\mathbf{M}'\mathbf{M}\mathbf{V}_1)$. Using a similar argument, $\mathbf{U}=\mathbf{L}\mathbf{U}_1$, and $\mathbf{V}'\mathbf{V}$ is independent of $\mathbf{U}'\mathbf{U}$ $(=\mathbf{U}_1'\mathbf{L}'\mathbf{L}\mathbf{U}_1)$. □

COROLLARY 1 Generalizing the above argument, and using the corollary of Theorem 2.6, we can prove the following: if $\mathbf{U}=\mathbf{A}\mathbf{Y}$, $\mathbf{V}=\mathbf{B}\mathbf{Y}$, $\mathbf{W}=\mathbf{C}\mathbf{Y},\ldots,$ and the covariance of any pair is zero, then $\mathbf{U}'\mathbf{U},\mathbf{V}'\mathbf{V},\mathbf{W}'\mathbf{W},\ldots$ are *mutually* independent.

COROLLARY 2 The above theorem still holds if $\mathbf{Y}\sim N_n(\boldsymbol{\theta},\boldsymbol{\Sigma})$.

Proof. Since $\boldsymbol{\Sigma}$ is positive definite, there exists a nonsingular matrix $\mathbf{R}$ such that $\boldsymbol{\Sigma}=\mathbf{R}\mathbf{R}'$ (*A4.2*). Let $\mathbf{Y}=\mathbf{R}\mathbf{X}$; then $\mathbf{U}=\mathbf{A}\mathbf{R}\mathbf{X}$, $\mathbf{V}=\mathbf{B}\mathbf{R}\mathbf{X}$, and $\mathbf{U}_1=\mathbf{A}_1\mathbf{R}\mathbf{X}$. Here the rows of $\mathbf{A}_1\mathbf{R}$ are linearly independent as the rank of a matrix is unchanged by postmultiplying by a nonsingular matrix (*A2.2*). Now $\mathbf{X}=\mathbf{R}^{-1}\mathbf{Y}$ so that

$$\mathcal{D}[\mathbf{X}]=\mathbf{R}^{-1}\mathcal{D}[\mathbf{Y}]\mathbf{R}^{-1\prime}=\mathbf{R}^{-1}\mathbf{R}\mathbf{R}'\mathbf{R}'^{-1}=\mathbf{I}_n \tag{2.15}$$

and, by Theorem 2.2, $\mathbf{X}\sim N_n(\mathbf{R}^{-1}\boldsymbol{\theta},\mathbf{I}_n)$. The proof is completed by applying Theorem 2.7 to $\mathbf{X}$. □

COROLLARY 3 If $\mathbf{Y}\sim N_n(\boldsymbol{\theta},\sigma^2\mathbf{I}_n)$, then $\mathbf{a}'\mathbf{Y}$ and $\mathbf{b}'\mathbf{Y}$ are independent if and only if $\mathbf{a}'\mathbf{b}=0$.

Proof. Let $U=\mathbf{a}'\mathbf{Y}$ and $V=\mathbf{b}'\mathbf{Y}$; then

$$\operatorname{cov}(U,V)=\mathcal{C}[\mathbf{a}'\mathbf{Y},\mathbf{b}'\mathbf{Y}]=\mathbf{a}'\mathfrak{D}[\mathbf{Y}]\mathbf{b}=\sigma^2\mathbf{a}'\mathbf{b}$$

and the result follows from the argument following Equation (2.14) above. □

From a teaching viewpoint I have found that the above theorem and its corollaries are generally more useful than a wide variety of theorems quoted in the literature concerning the independence of quadratic and linear forms. However, for completeness, some of these other theorems are given in Section 2.4.

EXAMPLE 2.4 Let $Y_1, Y_2, \ldots, Y_n$ be independently and identically distributed as $N(0,\sigma^2)$. Prove that $\overline{Y}$ and $Q=\Sigma_i(Y_i-\overline{Y})^2$ are independent.

Solution. Now $\mathbf{Y}\sim N_n(\mathbf{0},\sigma^2\mathbf{I}_n)$, $U=\overline{Y}=\mathbf{a}'\mathbf{Y}$, and $Q=\mathbf{V}'\mathbf{V}$, where

$$\mathbf{V}=\begin{bmatrix} Y_1-\overline{Y} \\ Y_2-\overline{Y} \\ \cdots \\ Y_n-\overline{Y} \end{bmatrix}=\mathbf{BY},$$

say. For $i=1,2,\ldots,n$ we have

$$\begin{aligned}\operatorname{cov}[U,V_i]&=\operatorname{cov}\left[\overline{Y},Y_i-\overline{Y}\right]\\ &=\operatorname{cov}\left[\overline{Y},Y_i\right]-\operatorname{cov}\left[\overline{Y},\overline{Y}\right] \qquad \text{(by Theorem 1.5)}\\ &=\frac{1}{n}\operatorname{var}[Y_i]-\operatorname{var}[\overline{Y}]\\ &=\frac{\sigma^2}{n}-\frac{\sigma^2}{n}=0.\end{aligned}$$

Hence, by Theorem 2.7, U is independent of $\mathbf{V}'\mathbf{V}$.

EXERCISES 2c

1. If $Y_1, Y_2, \ldots, Y_n$ have a multivariate normal distribution and each pair Y_i, Y_j $(i\neq j)$ are independent, prove that the Y_i are all mutually independent.

2. Suppose $\mathbf{Y} \sim N_n(\boldsymbol{\theta}, \boldsymbol{\Sigma})$ and consider the partitions

$$\mathbf{Y} = \begin{pmatrix} \mathbf{X}_1 \\ \mathbf{X}_2 \end{pmatrix}, \qquad \boldsymbol{\theta} = \begin{pmatrix} \boldsymbol{\alpha}_1 \\ \boldsymbol{\alpha}_2 \end{pmatrix}, \qquad \text{and} \qquad \boldsymbol{\Sigma} = \begin{pmatrix} \boldsymbol{\Sigma}_{11} & \boldsymbol{\Sigma}_{12} \\ \boldsymbol{\Sigma}_{21} & \boldsymbol{\Sigma}_{22} \end{pmatrix},$$

where $\mathbf{X}_1$ and $\boldsymbol{\alpha}_1$ are $p \times 1$ vectors and $\boldsymbol{\Sigma}_{11}$ is a $p \times p$ matrix $(p < n)$. If $\mathbf{W} = \mathbf{X}_1 - \boldsymbol{\Sigma}_{12}\boldsymbol{\Sigma}_{22}^{-1}\mathbf{X}_2$, prove that $\mathcal{C}[\mathbf{W}, \mathbf{X}_2] = \mathbf{O}$ and deduce that $\mathbf{W}$ and $\mathbf{X}_2$ are independent. Hence prove that the conditional distribution of $\mathbf{X}_1$, given $\mathbf{X}_2 = \mathbf{x}_2$, is

$$N_p\left(\boldsymbol{\alpha}_1 + \boldsymbol{\Sigma}_{12}\boldsymbol{\Sigma}_{22}^{-1}(\mathbf{x}_2 - \boldsymbol{\alpha}_2), \boldsymbol{\Sigma}_{11} - \boldsymbol{\Sigma}_{12}\boldsymbol{\Sigma}_{22}^{-1}\boldsymbol{\Sigma}_{21}\right)$$

3. If Y_1 and Y_2 have the bivariate normal distribution with density function

$$f(y_1, y_2) = \left[2\pi\sigma_1\sigma_2(1-\rho^2)^{1/2}\right]^{-1}$$

$$\times \exp\left[-\tfrac{1}{2}\left\{\frac{(y_1-\theta_1)^2}{\sigma_1^2} - \frac{2\rho(y_1-\theta_1)(y_2-\theta_2)}{\sigma_1\sigma_2} + \frac{(y_2-\theta_2)^2}{\sigma_2^2}\right\}\right],$$

use Exercise 2 to find the conditional distribution of Y_1 given $Y_2 = y_2$.

4. Let $\mathbf{Y} \sim N_n(\theta\mathbf{1}_n, \boldsymbol{\Sigma})$ where $\sigma_{ii} = \sigma^2$ (all i) and $\sigma_{ij} = \sigma^2(1-\rho)$ (all $i, j, i \neq j$). Prove that $\bar{Y}$ is statistically independent of $\Sigma_i(Y_i - \bar{Y})^2$.

5. Given $\mathbf{Y} \sim N_3(\boldsymbol{\theta}, \boldsymbol{\Sigma})$, where

$$\Sigma = \begin{bmatrix} 1 & \rho & 0 \\ \rho & 1 & \rho \\ 0 & \rho & 1 \end{bmatrix},$$

for what value of ρ are $Y_1 + Y_2 + Y_3$ and $Y_1 - Y_2 - Y_3$ statistically independent?

2.4 QUADRATIC FORMS IN NORMAL VARIABLES

At this point it is very tempting to give a large number of "popular" theorems on quadratic forms that have been proved over the past 20–30 years. Although these theorems have their own interest, they are not needed in the treatment that follows; Theorems 2.7 (Section 2.3) and 1.10 (Section 1.5) are generally sufficient. However, there are two very elegant results relating to idempotent matrices which are very easy to use, and these are given below.

THEOREM 2.8 Let $\mathbf{Y} \sim N_n(\boldsymbol{\theta}, \sigma^2 \mathbf{I}_n)$ and let $\mathbf{P}$ be an $n \times n$ symmetric matrix of rank r. Then $Q = (\mathbf{Y} - \boldsymbol{\theta})'\mathbf{P}(\mathbf{Y} - \boldsymbol{\theta})/\sigma^2$ is distributed as χ_r^2 if and only if $\mathbf{P}^2 = \mathbf{P}$ (that is, $\mathbf{P}$ is idempotent).

Proof. Given $\mathbf{P}^2 = \mathbf{P}$, then $\mathbf{P}$ has r eigenvalues equal to unity and $n - r$ eigenvalues equal to zero (*A5.1*). Hence there exists an orthogonal matrix $\mathbf{T}$ such that $\mathbf{T}'\mathbf{P}\mathbf{T} = \boldsymbol{\Lambda}$, where $\boldsymbol{\Lambda}$ is defined in the proof of *A5.1*. If $\mathbf{Z} = \mathbf{T}'(\mathbf{Y} - \boldsymbol{\theta})$ then, since $\mathbf{Y} - \boldsymbol{\theta} \sim N_n(\mathbf{0}, \sigma^2 \mathbf{I}_n)$, $\mathbf{Z} \sim N_n(\mathbf{0}, \sigma^2 \mathbf{I}_n)$ (Theorem 2.3) and the Z_i are independently and identically distributed as $N(0, \sigma^2)$. Therefore

$$Q = \frac{\mathbf{Z}'\mathbf{T}'\mathbf{P}\mathbf{T}\mathbf{Z}}{\sigma^2} = \frac{Z_1^2 + Z_2^2 + \cdots + Z_r^2}{\sigma^2}, \tag{2.16}$$

and $Q \sim \chi_r^2$.

Conversely; given $Q \sim \chi_r^2$, then $E[\exp(tQ)] = (1 - 2t)^{-(1/2)r}$. Now because $\mathbf{P}$ is symmetric there exists an orthogonal matrix $\mathbf{S}$ such that $\mathbf{S}'\mathbf{P}\mathbf{S} = \operatorname{diag}(\lambda_1, \lambda_2, \ldots, \lambda_n) = \boldsymbol{\Lambda}$, say, where the λ_i are the eigenvalues of $\mathbf{P}$. Let $\mathbf{Y} - \boldsymbol{\theta} = \mathbf{S}\mathbf{Z}$; then

$$E[\exp(tQ)] = E\left[\exp\left(\frac{t\mathbf{Z}'\mathbf{S}'\mathbf{P}\mathbf{S}\mathbf{Z}}{\sigma^2}\right)\right]$$

$$= E\left[\exp\left(\frac{t\sum_i \lambda_i Z_i^2}{\sigma^2}\right)\right].$$

However, $\mathbf{Z} = \mathbf{S}'(\mathbf{Y} - \boldsymbol{\theta}) \sim N_n(\mathbf{0}, \boldsymbol{\Lambda})$ so that the Z_i are independently distributed as $N(0, \lambda_i)$, and

$$E\left[\exp\left(\frac{t\lambda_i Z_i^2}{\sigma^2}\right)\right] = \int_{-\infty}^{\infty} (2\pi\sigma^2)^{-1/2} \exp\left(\frac{-\frac{1}{2}z_i^2(1 - 2\lambda_i t)}{\sigma^2}\right) dz_i$$

$$= (1 - 2\lambda_i t)^{-1/2}.$$

Therefore

$$(1 - 2t)^{-(1/2)r} = E[\exp(tQ)]$$

$$= \prod_{i=1}^{n} (1 - 2t\lambda_i)^{-1/2}$$

and

$$(1-2t)^r = \prod_{i=1}^{n} (1-2t\lambda_i) \tag{2.17}$$

identically in t for $|t|$ sufficiently small. By the uniqueness of polynomial roots we must have r λ_i's equal to unity and $n-r$ equal to zero. Thus $\mathbf{P}^2 = \mathbf{P}$ and rank $\mathbf{P} = r$ (*A5.1*). □

EXAMPLE 2.5 If $\mathbf{Y} \sim N_n(\mathbf{0}, \mathbf{I}_n)$, $\mathbf{Y}'\mathbf{Y} = \mathbf{Y}'\mathbf{A}\mathbf{Y} + \mathbf{Y}'\mathbf{B}\mathbf{Y}$, and $\mathbf{Y}'\mathbf{A}\mathbf{Y} \sim \chi_r^2$, prove that $\mathbf{Y}'\mathbf{B}\mathbf{Y} \sim \chi_{n-r}^2$.

Solution. By the above theorem $\mathbf{A}$ is a symmetric idempotent matrix of rank r. Also $\mathbf{B} = \mathbf{I}_n - \mathbf{A}$ is symmetric, and

$$\mathbf{B}^2 = (\mathbf{I}_n - \mathbf{A})^2 = \mathbf{I}_n - 2\mathbf{A} + \mathbf{A}^2 = \mathbf{I}_n - \mathbf{A} = \mathbf{B}.$$

Applying the above theorem once again we have $\mathbf{Y}'\mathbf{B}\mathbf{Y} \sim \chi_p^2$ where, by *A5.2*,

$$p = \operatorname{rank} \mathbf{B} = \operatorname{tr} \mathbf{B} = n - \operatorname{tr} \mathbf{A} = n - \operatorname{rank} \mathbf{A} = n - r.$$

EXAMPLE 2.6 Let $\mathbf{Y} \sim N_n(\mathbf{0}, \mathbf{I}_n)$ and let $Q_i = \mathbf{Y}'\mathbf{P}_i\mathbf{Y}$ $(i = 1, 2)$. Given that Q_1 and Q_2 are each distributed as chi-square, show that Q_1 and Q_2 are independent if and only if $\mathbf{P}_1\mathbf{P}_2 = \mathbf{O}$.

Solution. Since Q_i is chi-square, $\mathbf{P}_i$ is symmetric and idempotent. Given $\mathbf{P}_1\mathbf{P}_2 = \mathbf{O}$, then $\mathcal{C}[\mathbf{P}_1\mathbf{Y}, \mathbf{P}_2\mathbf{Y}] = \mathbf{P}_1\mathcal{D}[\mathbf{Y}]\mathbf{P}_2' = \mathbf{P}_1\mathbf{P}_2 = \mathbf{O}$ and Q_1 $(= \mathbf{Y}'\mathbf{P}_1\mathbf{Y} = \mathbf{Y}'\mathbf{P}_1^2\mathbf{Y} = (\mathbf{P}_1\mathbf{Y})'(\mathbf{P}_1\mathbf{Y}))$ is independent of Q_2 (cf Theorem 2.7).

Conversely, given Q_1 and Q_2 are independent chi-square variables, then $Q_1 + Q_2$ is also distributed as chi-square; that is,

$$\begin{aligned}\mathbf{P}_1 + \mathbf{P}_2 &= (\mathbf{P}_1 + \mathbf{P}_2)^2 \\ &= \mathbf{P}_1^2 + \mathbf{P}_1\mathbf{P}_2 + \mathbf{P}_2\mathbf{P}_1 + \mathbf{P}_2^2 \\ &= \mathbf{P}_1 + \mathbf{P}_1\mathbf{P}_2 + \mathbf{P}_2\mathbf{P}_1 + \mathbf{P}_2,\end{aligned}$$

or

$$\mathbf{P}_1\mathbf{P}_2 + \mathbf{P}_2\mathbf{P}_1 = \mathbf{O}. \tag{2.18}$$

Multiplying Equation (2.18) on the left (right) by $\mathbf{P}_1$ we obtain the two equations

$$\mathbf{P}_1\mathbf{P}_2 + \mathbf{P}_1\mathbf{P}_2\mathbf{P}_1 = \mathbf{O} \qquad \text{and} \qquad \mathbf{P}_1\mathbf{P}_2\mathbf{P}_1 + \mathbf{P}_2\mathbf{P}_1 = \mathbf{O}.$$

Hence $\mathbf{P}_1\mathbf{P}_2 = \mathbf{P}_2\mathbf{P}_1$ and, from Equation (2.18),

$$\mathbf{P}_2\mathbf{P}_1 = \mathbf{O}.$$

(We note that the chi-square assumption is not actually needed; see Lancaster [1969] for a proof.)

EXAMPLE 2.7 If $\mathbf{Y} \sim N_n(\mathbf{0}, \mathbf{I}_n)$, prove that

$$\sum_{i=1}^{n} \left(Y_i - \overline{Y}\right)^2 \sim \chi^2_{n-1}.$$

Solution. As in Example 1.12 in Section 1.4 we find that $\Sigma(Y_i - \overline{Y})^2 = \mathbf{Y}'\mathbf{A}\mathbf{Y}$, where $A = [(\delta_{ij} - n^{-1})]$ is idempotent. Hence rank $\mathbf{A} = \operatorname{tr} \mathbf{A} = n-1$ and the result follows from Theorem 2.8.

THEOREM 2.9 (Hogg and Craig [1958, 1970]) Let $\mathbf{Y} \sim N_n(\boldsymbol{\theta}, \sigma^2\mathbf{I}_n)$ and let $Q_i = (\mathbf{Y} - \boldsymbol{\theta})'\mathbf{P}_i(\mathbf{Y} - \boldsymbol{\theta})/\sigma^2$ $(i = 1, 2)$. If $Q_i \sim \chi^2_{r_i}$ and $Q_1 - Q_2 \geqslant 0$, then $Q_1 - Q_2$ and Q_2 are independently distributed as $\chi^2_{r_1 - r_2}$ and χ_{r_2}, respectively.

Proof. If $Q_i \sim \chi^2_{r_i}$, then $\mathbf{P}_i^2 = \mathbf{P}_i$ (Theorem 2.8). Also $Q_1 - Q_2 \geqslant 0$ implies that $\mathbf{P}_1 - \mathbf{P}_2$ is positive semidefinite, and therefore idempotent (*A5.5*). Hence, by Theorem 2.8, $Q_1 - Q_2 \sim \chi^2_r$ where

$$\begin{aligned} r &= \operatorname{rank}\left[\mathbf{P}_1 - \mathbf{P}_2\right] \\ &= \operatorname{tr}\left[\mathbf{P}_1 - \mathbf{P}_2\right] \\ &= \operatorname{tr}\mathbf{P}_1 - \operatorname{tr}\mathbf{P}_2 \\ &= \operatorname{rank}\mathbf{P}_1 - \operatorname{rank}\mathbf{P}_2 \\ &= r_1 - r_2. \end{aligned} \tag{2.19}$$

Also by *A5.5* $\mathbf{P}_1\mathbf{P}_2 = \mathbf{P}_2\mathbf{P}_1 = \mathbf{P}_2$, and $(\mathbf{P}_1 - \mathbf{P}_2)\mathbf{P}_2 = \mathbf{O}$. Therefore, since $\mathbf{Z} = (\mathbf{Y} - \boldsymbol{\theta})/\sigma^2 \sim N_n(\mathbf{0}, \mathbf{I}_n)$, we have, by Example 2.6, that $Q_1 - Q_2$ $(= \mathbf{Z}'(\mathbf{P}_1 - \mathbf{P}_2)\mathbf{Z})$ is independent of Q_2 $(= \mathbf{Z}'\mathbf{P}_2\mathbf{Z})$. □

There is an extensive literature on the independence of quadratic forms in normal variables and the reader is referred to Styan [1970] for further references. A number of apparently shorter proofs of certain well-known results are given by Searle [1971: Section 2.5, theorems 3 and 4]. However, these proofs, which utilize Theorem 2.6 in this chapter, should be examined more closely because they involve complex valued linear combinations of normal variables (for example, $\mathbf{L}$ in his theorem 3 may be complex valued).

More general versions of Theorems 2.8 and 2.9 are available and the interested reader should consult Searle [1971: Section 2.5].

EXERCISES 2d

1. If $\mathbf{Y} \sim N_n(\mathbf{0}, \boldsymbol{\Sigma})$ and $\mathbf{A}$ is any given $n \times n$ symmetric matrix, show that the moment generating function of $Q = \mathbf{Y}'\mathbf{A}\mathbf{Y}$ is
$$M(t) = |\mathbf{I}_n - 2t\mathbf{A}\boldsymbol{\Sigma}|^{-1/2}.$$
Given $\boldsymbol{\Sigma} = \mathbf{I}_n$ and $\mathbf{A}$ is idempotent of rank r, deduce that
$$M(t) = (1-2t)^{-(1/2)r}.$$
2. Given $\mathbf{Y} \sim N_n(\mathbf{0}, \mathbf{I}_n)$, find the joint moment generating function of $Q_1 = \mathbf{Y}'\mathbf{A}\mathbf{Y}$ and $Q_2 = \mathbf{Y}'\mathbf{B}\mathbf{Y}$. If $\mathbf{A}\mathbf{B} = \mathbf{O}$ prove that Q_1 and Q_2 are statistically independent. Conversely, if Q_1 and Q_2 are independent and positive semidefinite, show, by considering the variance of $Q_1 + Q_2$, that $\mathbf{A}\mathbf{B} = \mathbf{O}$.
3. Let $\mathbf{Y} \sim N_n(\mathbf{0}, \mathbf{I}_n)$ and let $Q_i = \mathbf{Y}'\mathbf{A}_i\mathbf{Y}$ $(i = 1, 2, \ldots, k)$ be positive semidefinite. If the Q_i are pairwise independent, prove that they are mutually independent.
4. Given $\mathbf{Y} \sim N_n(\boldsymbol{\theta}, \boldsymbol{\Sigma})$ prove that the third moments of the elements of $\mathbf{Y} - \boldsymbol{\theta}$ are zero, and hence deduce that
$$\mathcal{C}[\mathbf{Y}, \mathbf{Y}'\mathbf{A}\mathbf{Y}] = 2\boldsymbol{\Sigma}\mathbf{A}\boldsymbol{\theta}.$$
5. Let $\mathbf{Y} \sim N_n(\mathbf{0}, \boldsymbol{\Sigma})$ and let $\mathbf{A}$ be any $n \times n$ symmetric matrix of rank r. Prove that $\mathbf{Y}'\mathbf{A}\mathbf{Y} \sim \chi_r^2$ if and only if $\mathbf{A}\boldsymbol{\Sigma}\mathbf{A} = \mathbf{A}$.
6. Let $\mathbf{Y} \sim N_n(\mathbf{0}, \mathbf{I}_n)$ and let $Q = Q_1 + Q_2 + Q_3$, where $Q_i = \mathbf{Y}'\mathbf{A}_i\mathbf{Y}$ $(i = 1, 2, 3)$. If $Q \sim \chi_r^2$, $Q_i \sim \chi_{r_i}^2$ $(i = 1, 2)$, and $Q_3 \geqslant 0$, show that Q_1, Q_2, and Q_3 are mutually independent and that $Q_3 \sim \chi_{r_3}^2$, where $r_3 = r - r_1 - r_2$.
Hogg and Craig [1970]

MISCELLANEOUS EXERCISES 2

1. Evaluate the integral
$$\int_{-\infty}^{\infty}\int_{-\infty}^{\infty}(x^2 + xy + 3y^2)\exp\left[-(x^2 + 2xy + 2y^2)\right].$$
2. Let $\mathbf{Y} \sim N_n(\mathbf{0}, \boldsymbol{\Sigma})$, where $\boldsymbol{\Sigma} = \sigma^2[(1-\rho)\mathbf{I}_n + \rho\mathbf{1}\mathbf{1}']$ $(0 \leqslant \rho < 1)$. Using Helmert's transformation given in Example 2.3 (Section 2.2), prove that $\overline{Y}$ is independent of $Q = \Sigma_i(Y_i - \overline{Y})^2/\sigma^2(1-\rho)$, and that $Q \sim \chi_{n-1}^2$.

3. If $\mathbf{X}$ and $\mathbf{Y}$ are n-dimensional vectors with independent multivariate normal distributions, prove that $a\mathbf{X}+b\mathbf{Y}$ is also multivariate normal.
4. Let $\mathbf{Y}_1,\mathbf{Y}_2,\ldots,\mathbf{Y}_n$ be a random sample of n observations from $N_n(\boldsymbol{\theta},\boldsymbol{\Sigma})$ and let $\overline{\mathbf{Y}}=\Sigma_{i=1}^n\mathbf{Y}_i/n$.
   (a) Find the distribution of $\overline{\mathbf{Y}}$.
   (b) Prove that $\mathcal{C}[\overline{\mathbf{Y}},\mathbf{Y}_i-\overline{\mathbf{Y}}]=\mathbf{O}$.
   (c) Show that

$$\mathcal{E}\left[\frac{1}{n-1}\sum_{i=1}^{n}(\mathbf{Y}_i-\overline{\mathbf{Y}})(\mathbf{Y}_i-\overline{\mathbf{Y}})'\right]=\boldsymbol{\Sigma}$$

5. If $Y_1,Y_2,\ldots,Y_n$ is a random sample from $N(0,\sigma^2)$, prove that $\overline{Y}$ is independent of $\Sigma_{i=1}^{n-1}(Y_i-Y_{i+1})^2$.
6. Let $\mathbf{Y}=(Y_1,Y_2,\ldots,Y_n)'$ be a vector of n random variables ($n\geqslant 3$) with probability density function

$$f(\mathbf{y})=(2\pi)^{-(1/2)n}\left[\exp\left(-\tfrac{1}{2}\sum_i y_i^2\right)\right]\left\{1+\prod_{i=1}^{n}\left[y_i\exp\left(-\tfrac{1}{2}y_i^2\right)\right]\right\},$$

$$-\infty<y_i<\infty.$$

   Prove that any subset of $n-1$ random variables are mutually independent $N(0,1)$ variables. Pierce and Dykstra [1969]
7. If $\mathbf{Y}\sim N_n(\mathbf{0},\mathbf{I}_n)$ find the variance of

$$(Y_1-Y_2)^2+(Y_2-Y_3)^2+\cdots+(Y_{n-1}-Y_n)^2.$$

8. Let $\mathbf{Y}\sim N_n(\mathbf{0},\mathbf{I}_n)$, and let $\mathbf{X}=\mathbf{AY}$, $\mathbf{U}=\mathbf{BY}$, and $\mathbf{V}=\mathbf{CY}$, where $\mathbf{A}$, $\mathbf{B}$, and $\mathbf{C}$ are all $r\times n$ matrices of rank $r(r<n)$. If $\mathcal{C}[\mathbf{X},\mathbf{U}]=\mathbf{O}$ and $\mathcal{C}[\mathbf{X},\mathbf{V}]=\mathbf{O}$, prove that $\mathbf{X}$ is independent of $\mathbf{U}+\mathbf{V}$.
9. Suppose that $\mathbf{Y}=(Y_1,Y_2,Y_3,Y_4)'\sim N(\mathbf{0},\mathbf{I}_4)$ and let $Q=Y_1Y_2-Y_3Y_4$.
   (a) Prove that Q does not have a chi-square distribution.
   (b) Find the moment generating function of Q.
10. If $\mathbf{Y}\sim N_n(\mathbf{0},\mathbf{I}_n)$ prove that the conditional distribution of $\mathbf{Y}'\mathbf{Y}$ given $\mathbf{a}'\mathbf{Y}=0$ is χ^2_{n-1}.
11. If $\mathbf{Y}\sim N_n(\theta\mathbf{1}_n,\boldsymbol{\Sigma})$, where $\boldsymbol{\Sigma}=[(\sigma_{ij})]$ with $\sigma_{ii}=1$ $(i=1,2,\ldots,n)$ and $\sigma_{ij}=\rho$ $(i\neq j)$, find the moment generating function of $Q=\Sigma_i(Y_i-\overline{Y})^2/(1-\rho)$ and hence deduce that $Q\sim\chi^2_{n-1}$.
12. Given $\mathbf{Y}\sim N_n(\boldsymbol{\theta},\boldsymbol{\Sigma})$, prove that

$$\text{var}[\mathbf{Y}'\mathbf{A}\mathbf{Y}]=2\,\text{tr}[\mathbf{A}\boldsymbol{\Sigma}\mathbf{A}\boldsymbol{\Sigma}]+4\boldsymbol{\theta}'\mathbf{A}\boldsymbol{\Sigma}\mathbf{A}\boldsymbol{\theta}.$$

   [Hint: let $\mathbf{X}=\mathbf{R}^{-1}\mathbf{Y}$ where $\boldsymbol{\Sigma}=\mathbf{R}\mathbf{R}'$ (*A4.2*).]

CHAPTER 3

Linear Regression: Estimation and Distribution Theory

3.1 LEAST SQUARES ESTIMATION

Let Y be a random variable which fluctuates about an unknown parameter η; that is, $Y = \eta + \varepsilon$ where ε is the fluctuation or "error." For example, ε may be a "natural" fluctuation inherent in the experiment which gives rise to η, or it may represent the error in measuring η so that η is the true response and Y is the observed response.

Suppose now that η can be expressed in the form

$$\eta = \beta_0 + \beta_1 x_1 + \cdots + \beta_{p-1} x_{p-1}$$

where $x_1, x_2, \ldots, x_{p-1}$ are known constants (for example, experimental variables which are controlled by the experimenter and which are measured with negligible error), and the β_j $(j = 0, 1, \ldots, p-1)$ are unknown parameters to be estimated. If the x_j are varied and n values, $Y_1, Y_2, \ldots, Y_n$, of Y are observed, then

$$Y_i = \beta_0 + \beta_1 x_{i1} + \cdots + \beta_{p-1} x_{i,p-1} + \varepsilon_i \qquad (i = 1, 2, \ldots, n), \qquad (3.1)$$

where x_{ij} is the ith value of x_j. Writing these n equations in matrix form we have

$$\begin{bmatrix} Y_1 \\ Y_2 \\ \cdots \\ Y_n \end{bmatrix} = \begin{bmatrix} x_{10} & x_{11} & x_{12} & \cdots & x_{1,p-1} \\ x_{20} & x_{21} & x_{22} & \cdots & x_{2,p-1} \\ \cdots & \cdots & \cdots & \cdots & \cdots \\ x_{n0} & x_{n1} & x_{n2} & \cdots & x_{n,p-1} \end{bmatrix} \begin{bmatrix} \beta_0 \\ \beta_1 \\ \cdots \\ \beta_{p-1} \end{bmatrix} + \begin{bmatrix} \varepsilon_1 \\ \varepsilon_2 \\ \cdots \\ \varepsilon_n \end{bmatrix},$$

or

$$\mathbf{Y} = \mathbf{X}\boldsymbol{\beta} + \boldsymbol{\varepsilon} \qquad (3.2)$$

where $x_{10}=x_{20}=\cdots=x_{n0}=1$. The $n\times p$ matrix $\mathbf{X}$ will be called the *regression matrix*, and the x_{ij}'s are generally chosen so that the columns of $\mathbf{X}$ are linearly independent; that is, $\mathbf{X}$ has rank p. However, in some experimental design situations the elements of $\mathbf{X}$ are chosen to be 0 or 1, and the columns of $\mathbf{X}$ may be linearly dependent. In this case $\mathbf{X}$ is commonly called the *design matrix*.

It has been the custom in the past to call the x_j's the independent variables and Y the dependent variable. However, this terminology is confusing, so we follow the more contemporary usage and refer to x_j as a *regressor* or *predictor* variable and Y as the *response* variable.

We note that the model (3.1) is a very general one. For example, setting $x_{ij}=x_i^j$ and $k=p-1$, we have the polynomial model

$$Y_i=\beta_0+\beta_1 x_i+\beta_2 x_i^2+\cdots+\beta_k x_i^k+\varepsilon_i.$$

Again,

$$Y_i=\beta_0+\beta_1 e^{w_{i1}}+\beta_2 w_{i1}w_{i2}+\beta_3\sin w_{i3}+\varepsilon_i$$

is also a special case. The essential aspect of (3.1) is that it is linear in the unknown parameters β_j; for this reason it is called a *linear model*. In contrast,

$$Y_i=\beta_0+\beta_1 e^{-\beta_2 x_i}+\varepsilon_i \tag{3.3}$$

is a nonlinear model, being nonlinear in β_2.

Before considering the problem of estimating $\boldsymbol{\beta}$, we note that all the theory in this and subsequent chapters is developed for the model (3.2), where x_{i0} is not necessarily constrained to be unity. In the case when $x_{i0}\neq 1$ the reader may question the use of a notation in which i runs from 0 to $p-1$ rather than 1 to p. However, since the major application of the theory is to the case $x_{i0}\equiv 1$, it is convenient to "separate" β_0 from the other β_j's right from the outset.

One method of obtaining an estimate of $\boldsymbol{\beta}$ is the so-called method of least squares. This method consists of minimizing $\sum_i \varepsilon_i^2$ with respect to $\boldsymbol{\beta}$; that is, setting $\boldsymbol{\theta}=\mathbf{X}\boldsymbol{\beta}$, we minimize $\boldsymbol{\varepsilon}'\boldsymbol{\varepsilon}=\|\mathbf{Y}-\boldsymbol{\theta}\|^2$ subject to $\boldsymbol{\theta}\in\mathcal{R}[\mathbf{X}]=\Omega$, where Ω is the range space of $\mathbf{X}$ $(=\{\mathbf{y}:\mathbf{y}=\mathbf{X}\mathbf{x}$ for any $\mathbf{x}\})$. If we let $\boldsymbol{\theta}$ vary in Ω, $\|\mathbf{Y}-\boldsymbol{\theta}\|^2$ (the square of the length of $\mathbf{Y}-\boldsymbol{\theta}$) will be a minimum for $\boldsymbol{\theta}=\hat{\boldsymbol{\theta}}$ when $(\mathbf{Y}-\hat{\boldsymbol{\theta}})\perp\Omega$ (cf. Fig. 3.1). Thus

$$\mathbf{X}'(\mathbf{Y}-\hat{\boldsymbol{\theta}})=\mathbf{0}$$

or

$$\mathbf{X}'\hat{\boldsymbol{\theta}}=\mathbf{X}'\mathbf{Y}. \tag{3.4}$$

Here $\hat{\boldsymbol{\theta}}$ is uniquely determined, being the *unique* orthogonal projection of $\mathbf{Y}$ onto Ω (see Appendix B). Now given the columns of $\mathbf{X}$ are linearly

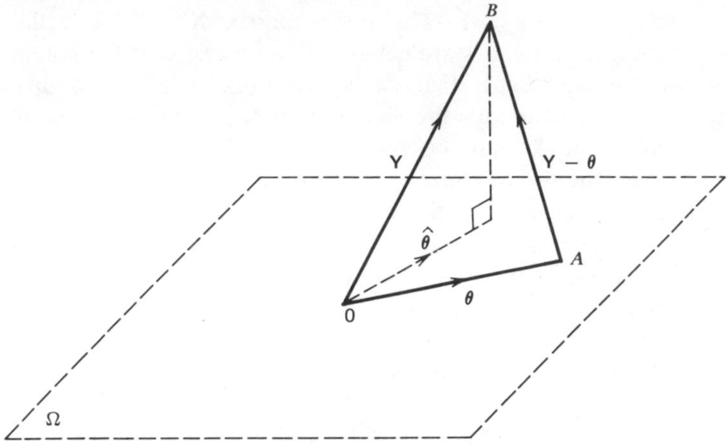

Fig. 3.1 The method of least squares consists of finding A such that AB is a minimum.

independent there exists a unique vector $\hat{\boldsymbol{\beta}}$ such that $\hat{\boldsymbol{\theta}}=\mathbf{X}\hat{\boldsymbol{\beta}}$. Therefore substituting in (3.4) we have

$$\mathbf{X}'\mathbf{X}\hat{\boldsymbol{\beta}}=\mathbf{X}'\mathbf{Y}, \tag{3.5}$$

the so-called *normal equation(s)*. As we assume $\mathbf{X}$ has rank p, $\mathbf{X}'\mathbf{X}$ is positive definite (*A4.6*) and therefore nonsingular. Hence (3.5) has a unique solution, namely,

$$\hat{\boldsymbol{\beta}}=(\mathbf{X}'\mathbf{X})^{-1}\mathbf{X}'\mathbf{Y}.$$

Here $\hat{\boldsymbol{\beta}}$ is called the (ordinary) least squares estimate of $\boldsymbol{\beta}$, and computational methods for actually calculating the estimate are given in Chapter 11.

We note that $\hat{\boldsymbol{\beta}}$ can also be obtained by writing

$$\begin{aligned}\boldsymbol{\varepsilon}'\boldsymbol{\varepsilon}&=(\mathbf{Y}-\mathbf{X}\boldsymbol{\beta})'(\mathbf{Y}-\mathbf{X}\boldsymbol{\beta})\\&=\mathbf{Y}'\mathbf{Y}-2\boldsymbol{\beta}'\mathbf{X}'\mathbf{Y}+\boldsymbol{\beta}'\mathbf{X}'\mathbf{X}\boldsymbol{\beta}\end{aligned}$$

(using the fact that $\boldsymbol{\beta}'\mathbf{X}'\mathbf{Y}=(\boldsymbol{\beta}'\mathbf{X}'\mathbf{Y})'=\mathbf{Y}'\mathbf{X}\boldsymbol{\beta}$), and differentiating $\boldsymbol{\varepsilon}'\boldsymbol{\varepsilon}$ with respect to $\boldsymbol{\beta}$. Thus from $\partial\boldsymbol{\varepsilon}'\boldsymbol{\varepsilon}/\partial\boldsymbol{\beta}=\mathbf{0}$ we have (*A6*)

$$-2\mathbf{X}'\mathbf{Y}+2\mathbf{X}'\mathbf{X}\boldsymbol{\beta}=\mathbf{0} \tag{3.6}$$

or

$$\mathbf{X}'\mathbf{X}\boldsymbol{\beta}=\mathbf{X}'\mathbf{Y}.$$

This solution for $\boldsymbol{\beta}$ gives us a stationary value of $\boldsymbol{\varepsilon}'\boldsymbol{\varepsilon}$, and a simple algebraic identity (see Exercise 1 of Exercises 3a) confirms that $\hat{\boldsymbol{\beta}}$ is a minimum.

In addition to the method of least squares there are several other methods used for estimating $\boldsymbol{\beta}$. These are described in Section 3.10.

The fitted regression $\mathbf{X}\hat{\boldsymbol{\beta}}$ is denoted by $\hat{\mathbf{Y}}$ $(=[(\hat{Y}_i)]$, and the elements of

$$\begin{aligned}\mathbf{e} &= \mathbf{Y} - \hat{\mathbf{Y}} \\ &= \mathbf{Y} - \mathbf{X}\hat{\boldsymbol{\beta}} \\ &= \left(\mathbf{I}_n - \mathbf{X}(\mathbf{X}'\mathbf{X})^{-1}\mathbf{X}'\right)\mathbf{Y} \\ &= (\mathbf{I}_n - \mathbf{P})\mathbf{Y}, \quad \text{say,} \end{aligned} \tag{3.7}$$

are called the *residuals*. The minimum value of $\boldsymbol{\varepsilon}'\boldsymbol{\varepsilon}$, namely,

$$\begin{aligned}\mathbf{e}'\mathbf{e} &= (\mathbf{Y} - \mathbf{X}\hat{\boldsymbol{\beta}})'(\mathbf{Y} - \mathbf{X}\hat{\boldsymbol{\beta}}) \\ &= \mathbf{Y}'\mathbf{Y} - 2\hat{\boldsymbol{\beta}}'\mathbf{X}'\mathbf{Y} + \hat{\boldsymbol{\beta}}'\mathbf{X}'\mathbf{X}\hat{\boldsymbol{\beta}} \\ &= \mathbf{Y}'\mathbf{Y} - \hat{\boldsymbol{\beta}}'\mathbf{X}'\mathbf{Y} + \hat{\boldsymbol{\beta}}'\left[\mathbf{X}'\mathbf{X}\hat{\boldsymbol{\beta}} - \mathbf{X}'\mathbf{Y}\right] \\ &= \mathbf{Y}'\mathbf{Y} - \hat{\boldsymbol{\beta}}'\mathbf{X}'\mathbf{Y} \qquad [\text{by (3.5)}], & (3.8) \\ &= \mathbf{Y}'\mathbf{Y} - \hat{\boldsymbol{\beta}}'\mathbf{X}'\mathbf{X}\hat{\boldsymbol{\beta}}, & (3.9)\end{aligned}$$

is called the *residual sum of squares* (RSS).

We note that $\hat{\mathbf{Y}}$ and $\mathbf{e}$ are unique.

EXAMPLE 3.1 Let Y_1 and Y_2 be independent random variables with means θ and 2θ, respectively. Find the least squares estimate of θ and the residual sum of squares.

Solution. Writing

$$\begin{pmatrix} Y_1 \\ Y_2 \end{pmatrix} = \begin{pmatrix} 1 \\ 2 \end{pmatrix}\theta + \begin{pmatrix} \varepsilon_1 \\ \varepsilon_2 \end{pmatrix},$$

we have $\mathbf{Y} = \mathbf{X}\boldsymbol{\beta} + \boldsymbol{\varepsilon}$ where $\mathbf{X} = \begin{pmatrix} 1 \\ 2 \end{pmatrix}$ and $\boldsymbol{\beta} = \theta$. Hence, by the above theory,

$$\begin{aligned}\hat{\theta} &= (\mathbf{X}'\mathbf{X})^{-1}\mathbf{X}'\mathbf{Y} \\ &= \left\{(1,2)\begin{pmatrix} 1 \\ 2 \end{pmatrix}\right\}^{-1}(1,2)\mathbf{Y} \\ &= \tfrac{1}{5}(1,2)\begin{pmatrix} Y_1 \\ Y_2 \end{pmatrix} \\ &= \tfrac{1}{5}(Y_1 + 2Y_2),\end{aligned}$$

and

$$\begin{aligned}\mathbf{e}'\mathbf{e} &= \mathbf{Y}'\mathbf{Y} - \hat{\boldsymbol{\beta}}'\mathbf{X}'\mathbf{Y}\\ &= \mathbf{Y}'\mathbf{Y} - \hat{\theta}\,(Y_1 + 2Y_2)\\ &= Y_1^2 + Y_2^2 - \tfrac{1}{5}(Y_1 + 2Y_2)^2.\end{aligned}$$

The problem can also be solved by first principles as follows: $\boldsymbol{\varepsilon}'\boldsymbol{\varepsilon} = (Y_1 - \theta)^2 + (Y_2 - 2\theta)^2$ and $\partial\boldsymbol{\varepsilon}'\boldsymbol{\varepsilon}/\partial\theta = 0$ implies that $\hat{\theta} = \frac{1}{5}(Y_1 + 2Y_2)$. Further,

$$\begin{aligned}\mathbf{e}'\mathbf{e} &= (Y_1 - \hat{\theta}\,)^2 + (Y_2 - 2\hat{\theta}\,)^2\\ &= Y_1^2 + Y_2^2 - \hat{\theta}\,(2Y_1 + 4Y_2) + 5\hat{\theta}^2\\ &= Y_1^2 + Y_2^2 - \tfrac{1}{5}(Y_1 + 2Y_2)^2.\end{aligned}$$

In practice both approaches are used.

Since $\hat{\boldsymbol{\theta}} = \mathbf{X}\hat{\boldsymbol{\beta}} = \mathbf{X}(\mathbf{X}'\mathbf{X})^{-1}\mathbf{X}'\mathbf{Y} = \mathbf{P}\mathbf{Y}$, $\mathbf{P}$ is the linear transformation representing the orthogonal projection of n-dimensional Euclidean space, E_n, onto Ω. Similarly, $\mathbf{I}_n - \mathbf{P}$ represents the orthogonal projection of E_n onto the orthogonal complement, $\Omega^{\perp}$, of Ω. Thus $\mathbf{Y} = \mathbf{P}\mathbf{Y} + (\mathbf{I}_n - \mathbf{P})\mathbf{Y}$ represents a unique orthogonal decomposition of $\mathbf{Y}$ into two components, one in Ω and the other in $\Omega^{\perp}$. Some basic properties of $\mathbf{P}$ and $(\mathbf{I}_n - \mathbf{P})$ are proved in Theorem 3.1, though these properties follow directly from the more general results concerning orthogonal projections stated in Appendix B.

THEOREM 3.1

(i) $\mathbf{P}$ and $\mathbf{I}_n - \mathbf{P}$ are symmetric and idempotent.
(ii) $\text{rank}[\mathbf{I}_n - \mathbf{P} = \text{tr}[\mathbf{I}_n - \mathbf{P}] = n - p$.
(iii) $(\mathbf{I}_n - \mathbf{P})\mathbf{X} = \mathbf{O}$.

Proof. (i) $\mathbf{P}$ is obviously symmetric and $(\mathbf{I}_n - \mathbf{P})' = \mathbf{I}_n - \mathbf{P}' = \mathbf{I}_n - \mathbf{P}$. Also

$$\begin{aligned}\mathbf{P}^2 &= \mathbf{X}(\mathbf{X}'\mathbf{X})^{-1}\mathbf{X}'\mathbf{X}(\mathbf{X}'\mathbf{X})^{-1}\mathbf{X}'\\ &= \mathbf{X}\mathbf{I}_p(\mathbf{X}'\mathbf{X})^{-1}\mathbf{X}' = \mathbf{P},\end{aligned}$$

and $(\mathbf{I}_n - \mathbf{P})^2 = \mathbf{I}_n - 2\mathbf{P} + \mathbf{P}^2 = \mathbf{I}_n - \mathbf{P}$.

(ii) Since $\mathbf{I}_n - \mathbf{P}$ is symmetric and idempotent we have, by *A5.2*,

$$\begin{aligned}\text{rank}[\mathbf{I}_n - \mathbf{P}] &= \text{tr}[\mathbf{I}_n - \mathbf{P}]\\ &= n - \text{tr}\,\mathbf{P},\end{aligned}$$

where

$$\begin{aligned}\operatorname{tr}\mathbf{P} &= \operatorname{tr}\left[\mathbf{X}(\mathbf{X}'\mathbf{X})^{-1}\mathbf{X}'\right] \\ &= \operatorname{tr}\left[\mathbf{X}'\mathbf{X}(\mathbf{X}'\mathbf{X})^{-1}\right] \qquad (\text{by } A1.2) \\ &= \operatorname{tr}\mathbf{I}_p \\ &= p.\end{aligned}$$

(iii) $(\mathbf{I}_n - \mathbf{P})\mathbf{X} = \mathbf{X} - \mathbf{X}(\mathbf{X}'\mathbf{X})^{-1}\mathbf{X}'\mathbf{X} = \mathbf{X} - \mathbf{X} = \mathbf{O}$. □

If $\mathbf{X}$ has rank r ($r<p$), then $\hat{\boldsymbol{\theta}} = \mathbf{P}\mathbf{Y}$ still defines a unique projection matrix $\mathbf{P}$ (Appendix B). To find an expression for $\mathbf{P}$ let the $n\times r$ matrix $\mathbf{X}_1$ consist of r linearly independent columns of $\mathbf{X}$. Then, from (3.4), $\mathbf{X}_1'(\mathbf{Y}-\hat{\boldsymbol{\theta}}) = \mathbf{0}$, $\hat{\boldsymbol{\theta}} = \mathbf{X}_1\hat{\boldsymbol{\alpha}}$ for some $\hat{\boldsymbol{\alpha}}$, and $\mathbf{P} = \mathbf{X}_1(\mathbf{X}_1'\mathbf{X}_1)^{-1}\mathbf{X}_1'$. In this case Theorem 3.1 still holds, but with p replaced by r. We note that $\mathbf{P}$ can also be expressed in the form $\mathbf{X}(\mathbf{X}'\mathbf{X})^{-}\mathbf{X}'$, where $(\mathbf{X}'\mathbf{X})^{-}$ is a generalized inverse of $\mathbf{X}'\mathbf{X}$ (cf. *B1.8*).

The above geometrical approach can be extended in a number of ways, for example, Seber [1966], Drygas [1970], Seely and Zyskind [1971], and Watson [1972]. Rao [1974] gives a general theory based on "oblique" rather than orthogonal projections; his theory allows $\mathbf{X}$ to have less than full rank and $\mathscr{D}[\mathbf{Y}]$ is possibly singular.

EXERCISES 3a

1. Show that

$$(\mathbf{Y}-\mathbf{X}\boldsymbol{\beta})'(\mathbf{Y}-\mathbf{X}\boldsymbol{\beta}) = (\mathbf{Y}-\mathbf{X}\hat{\boldsymbol{\beta}})'(\mathbf{Y}-\mathbf{X}\hat{\boldsymbol{\beta}}) + (\hat{\boldsymbol{\beta}}-\boldsymbol{\beta})'\mathbf{X}'\mathbf{X}(\hat{\boldsymbol{\beta}}-\boldsymbol{\beta}),$$

and hence deduce that the left side is minimized when $\boldsymbol{\beta} = \hat{\boldsymbol{\beta}}$.

2. Prove that $\Sigma_{i=1}^{n}(Y_i - \hat{Y}_i) = 0$.
3. Let

$$\begin{aligned}Y_1 &= \theta + \varepsilon_1 \\ Y_2 &= 2\theta - \phi + \varepsilon_2 \\ Y_3 &= \theta + 2\phi + \varepsilon_3,\end{aligned}$$

where $E[\varepsilon_i] = 0$ ($i = 1,2,3$). Find the least squares estimates of θ and ϕ.

4. Consider the regression model

$$E[Y_i] = \beta_0 + \beta_1 x_i + \beta_2(3x_i^2 - 2) \qquad (i = 1,2,3),$$

where $x_1 = -1$, $x_2 = 0$, and $x_3 = +1$. Find the least squares estimates of β_0, β_1, and β_2. Show that the least squares estimates of β_0 and β_1 are unchanged if $\beta_2 = 0$.

5. The observed tension T in a nonextensible string required to maintain a body of unknown weight w in equilibrium on a smooth inclined plane of angle θ $(0 < \theta < \pi/2)$ is a random variable with mean $E[T] = w \sin\theta$. If for $\theta = \theta_i$ $(i = 1, 2, \ldots, n)$ the corresponding values of T are T_i $(i = 1, 2, \ldots, n)$, find the least squares estimate of w.
6. If $\mathbf{P} = \mathbf{X}(\mathbf{X}'\mathbf{X})^{-1}\mathbf{X}'$, show that $\mathcal{R}[\mathbf{P}] = \mathcal{R}[\mathbf{X}]$.
7. For a general full rank regression model show that

$$\sum_{i=1}^{n} \hat{Y}_i (Y_i - \hat{Y}_i) = 0.$$

8. Suppose that we scale the regressor variables so that $x_{ij} = k_j w_{ij}$ for all i, j. By expressing $\mathbf{X}$ in terms of a new matrix $\mathbf{W}$, prove that $\hat{\mathbf{Y}}$ remains unchanged under this change of scale.

3.2 PROPERTIES OF LEAST SQUARES ESTIMATES

If we assume that the "errors" are unbiased, that is, $\mathcal{E}[\boldsymbol{\varepsilon}] = \mathbf{0}$, then

$$\begin{aligned} \mathcal{E}[\hat{\boldsymbol{\beta}}] &= (\mathbf{X}'\mathbf{X})^{-1}\mathbf{X}'\mathcal{E}[\mathbf{Y}] \\ &= (\mathbf{X}'\mathbf{X})^{-1}\mathbf{X}'\mathbf{X}\boldsymbol{\beta} \\ &= \boldsymbol{\beta}, \end{aligned} \tag{3.10}$$

and $\hat{\boldsymbol{\beta}}$ is an unbiased estimate of $\boldsymbol{\beta}$. If we further assume that the ε_i are uncorrelated and have the same variance, that is, $\text{cov}[\varepsilon_i, \varepsilon_j] = \delta_{ij}\sigma^2$, then $\mathcal{D}[\boldsymbol{\varepsilon}] = \sigma^2 \mathbf{I}_n$ and

$$\mathcal{D}[\mathbf{Y}] = \mathcal{D}[\mathbf{Y} - \mathbf{X}\boldsymbol{\beta}] = \mathcal{D}[\boldsymbol{\varepsilon}].$$

Hence, by Theorem 1.4 (Corollary 2),

$$\begin{aligned} \mathcal{D}[\hat{\boldsymbol{\beta}}] &= \mathcal{D}[(\mathbf{X}'\mathbf{X})^{-1}\mathbf{X}'\mathbf{Y}] \\ &= (\mathbf{X}'\mathbf{X})^{-1}\mathbf{X}'\mathcal{D}[\mathbf{Y}]\mathbf{X}(\mathbf{X}'\mathbf{X})^{-1} \\ &= \sigma^2(\mathbf{X}'\mathbf{X})^{-1}(\mathbf{X}'\mathbf{X})(\mathbf{X}'\mathbf{X})^{-1} \\ &= \sigma^2(\mathbf{X}'\mathbf{X})^{-1}. \end{aligned} \tag{3.11}$$

The question now arises as to why we chose $\hat{\boldsymbol{\beta}}$ as our estimate of $\boldsymbol{\beta}$ and not some other estimate. We show below that, for a reasonable class of estimates, $\hat{\beta}_j$ is the estimate of β_j with the smallest variance. Here $\hat{\beta}_j$ can be readily "extracted" from $\hat{\boldsymbol{\beta}}=(\hat{\beta}_0,\hat{\beta}_1,\ldots,\hat{\beta}_{p-1})'$ by simply premultiplying by the row vector $\mathbf{c}'$ which contains unity in the $(j+1)$th position and zeros elsewhere. It transpires that this special property of $\hat{\beta}_j$ can be generalized to the case of any linear combination $\mathbf{a}'\hat{\boldsymbol{\beta}}$ using the following theorem.

THEOREM 3.2 Let $\hat{\boldsymbol{\theta}}$ be the least squares estimate of $\boldsymbol{\theta}=\mathbf{X}\boldsymbol{\beta}$. Then among the class of linear unbiased estimates of $\mathbf{c}'\boldsymbol{\theta}$, $\mathbf{c}'\hat{\boldsymbol{\theta}}$ is the unique estimate with minimum variance. (We say that $\mathbf{c}'\hat{\boldsymbol{\theta}}$ is the Best Linear Unbiased Estimate, or BLUE, of $\mathbf{c}'\boldsymbol{\theta}$.)

Proof. Now from Section 3.1 $\hat{\boldsymbol{\theta}}=\mathbf{PY}$, where $\mathbf{PX}=\mathbf{X}$. Hence $E[\mathbf{c}'\hat{\boldsymbol{\theta}}]=\mathbf{c}'\mathbf{P}\boldsymbol{\theta}=\mathbf{c}'\boldsymbol{\theta}$ for all $\boldsymbol{\theta}\in\Omega=\mathcal{R}[\mathbf{X}]$, and $\mathbf{c}'\hat{\boldsymbol{\theta}}$ $[=(\mathbf{Pc})'\mathbf{Y}]$ is a linear unbiased estimate of $\mathbf{c}'\boldsymbol{\theta}$. Let $\mathbf{d}'\mathbf{Y}$ be any other linear unbiased estimate of $\mathbf{c}'\boldsymbol{\theta}$. Then $\mathbf{c}'\boldsymbol{\theta}=E[\mathbf{d}'\mathbf{Y}]=\mathbf{d}'\boldsymbol{\theta}$ or $(\mathbf{c}-\mathbf{d})'\boldsymbol{\theta}=0$ so that $(\mathbf{c}-\mathbf{d})\perp\Omega$. Therefore $\mathbf{P}(\mathbf{c}-\mathbf{d})=\mathbf{0}$ and $\mathbf{Pc}=\mathbf{Pd}$.

Now

$$\begin{aligned}\operatorname{var}[(\mathbf{Pd})'\mathbf{Y}]&=\mathcal{D}[(\mathbf{Pd})'\mathbf{Y}]\\&=\sigma^2\mathbf{d}'\mathbf{P}'\mathbf{P}\mathbf{d}\\&=\sigma^2\mathbf{d}'\mathbf{P}^2\mathbf{d}\\&=\sigma^2\mathbf{d}'\mathbf{P}\mathbf{d}\qquad\text{(Theorem 3.1)}\end{aligned}$$

so that

$$\begin{aligned}\operatorname{var}[\mathbf{d}'\mathbf{Y}]-\operatorname{var}[\mathbf{c}'\hat{\boldsymbol{\theta}}]&=\operatorname{var}[\mathbf{d}'\mathbf{Y}]-\operatorname{var}[(\mathbf{Pd})'\mathbf{Y}]\\&=\sigma^2(\mathbf{d}'\mathbf{d}-\mathbf{d}'\mathbf{Pd})\\&=\sigma^2\mathbf{d}'(\mathbf{I}_n-\mathbf{P})\mathbf{d}\\&=\sigma^2\mathbf{d}'(\mathbf{I}_n-\mathbf{P})'(\mathbf{I}_n-\mathbf{P})\mathbf{d}\\&=\sigma^2\mathbf{d}_1'\mathbf{d}_1,\qquad\text{say,}\\&\geqslant 0\end{aligned}$$

with equality only if $(\mathbf{I}_n-\mathbf{P})\mathbf{d}=\mathbf{0}$ or $\mathbf{d}=\mathbf{Pd}=\mathbf{Pc}$. Hence $\mathbf{c}'\hat{\boldsymbol{\theta}}$ has minimum variance and is unique. □

In this section we have assumed that $\mathbf{X}$ has full rank so that $\mathbf{P}=\mathbf{X}(\mathbf{X}'\mathbf{X})^{-1}\mathbf{X}'$ and $\boldsymbol{\theta}=\mathbf{X}\boldsymbol{\beta}$ implies that $\boldsymbol{\beta}=(\mathbf{X}'\mathbf{X})^{-1}\mathbf{X}'\boldsymbol{\theta}$. Hence setting $\mathbf{c}'=\mathbf{a}'(\mathbf{X}'\mathbf{X})^{-1}\mathbf{X}'$ we have that $\mathbf{a}'\hat{\boldsymbol{\beta}}$ $(=\mathbf{c}'\hat{\boldsymbol{\theta}})$ is the BLUE of $\mathbf{a}'\boldsymbol{\beta}$ for every $\mathbf{a}$.

We note that the above theorem still holds when $\mathbf{X}$ has less than full rank: $\mathbf{P}$ becomes the projection matrix with the properties described in *B1*.

Thus far we have not made any assumptions about the distribution of the ε_i. However, when the ε_i are independently and identically distributed as $N(0,\sigma^2)$, that is, $\boldsymbol{\varepsilon}\sim N(\mathbf{0},\sigma^2\mathbf{I}_n)$ or equivalently, $\mathbf{Y}\sim N_n(\mathbf{X}\boldsymbol{\beta},\sigma^2\mathbf{I}_n)$, then $\mathbf{a}'\hat{\boldsymbol{\beta}}$ has minimum variance for the entire class of unbiased estimates, and not just for linear estimates (cf. Rao [1973: p. 319] for proof). In particular $\hat{\beta}_i$, which is also the maximum likelihood estimate of β_i (Section 4.1.2), is the most efficient estimate of β_i.

When the common underlying distribution of the ε_i is not normal, then the least squares estimate of β_i is not the same as the asymptotically most efficient maximum likelihood estimate. The asymptotic efficiency of the least squares estimate is, for this case, derived by Cox and Hinkley [1968].

Eicker [1963] has discussed the question of the consistency and asymptotic normality of $\hat{\boldsymbol{\beta}}$ as $n\to\infty$. Under weak restrictions he shows that $\hat{\boldsymbol{\beta}}$ is a consistent estimate of $\boldsymbol{\beta}$ if and only if the smallest eigenvalue of $\mathbf{X}'\mathbf{X}$ tends to infinity. This condition on the smallest eigenvalue is a mild one so that the result has wide applicability. Eicker also proves a theorem giving necessary and sufficient conditions for the asymptotic normality of each $\hat{\beta}_j$ (see Anderson [1971: pp. 23–27].

EXERCISES 3b

1. Let $Y_i=\beta_0+\beta_1 x_i+\varepsilon_i$ $(i=1,2,\ldots,n)$, where $\mathcal{E}[\boldsymbol{\varepsilon}]=\mathbf{0}$ and $\mathcal{D}[\boldsymbol{\varepsilon}]=\sigma^2\mathbf{I}_n$. Find the least squares estimates of β_0 and β_1. Prove that they are uncorrelated if and only if $\bar{x}=0$.
2. In order to estimate two parameters θ and ϕ it is possible to make observations of three types:
   (a) the first type have expectation θ,
   (b) the second type have expectation $\theta+\phi$, and
   (c) the third type have expectation $\theta-2\phi$.
   All observations are subject to uncorrelated errors of mean zero and constant variance.

   If m observations of type (a), m observations of type (b), and n observations of type (c) are made, find the least squares estimates $\hat{\theta}$ and $\hat{\phi}$. Prove that these estimates are uncorrelated if $m=2n$.
3. Let $Y_1, Y_2,\ldots, Y_n$ be a random sample from $N(\theta,\sigma^2)$. Find the linear unbiased estimate of θ with minimum variance.

4. Let

$$Y_i = \beta_0 + \beta_1 (x_{i1} - \bar{x}_1) + \beta_2 (x_{i2} - \bar{x}_2) + \varepsilon_i \qquad (i = 1, 2, \ldots, n),$$

where $\bar{x}_j = \sum_{i=1}^n x_{ij}/n$, $\mathscr{E}[\boldsymbol{\varepsilon}] = \mathbf{0}$, and $\mathscr{D}[\boldsymbol{\varepsilon}] = \sigma^2 \mathbf{I}_n$. If $\hat{\beta}_1$ is the least squares estimate of β_1, show that

$$\operatorname{var}[\hat{\beta}_1] = \frac{\sigma^2}{\sum_i (x_{i1} - \bar{x}_1)^2 (1 - r_{12}^2)},$$

where r_{12} is the correlation coefficient of the pairs (x_{i1}, x_{i2}).

3.3 ESTIMATION OF σ^2

We now focus our attention on σ^2 $(= \operatorname{var} \varepsilon_i)$. An unbiased estimate is described in the following theorem.

THEOREM 3.3 If $\mathscr{E}[\mathbf{Y}] = \mathbf{X}\boldsymbol{\beta}$, where $\mathbf{X}$ is an $n \times p$ matrix of rank p, and $\mathscr{D}[\mathbf{Y}] = \sigma^2 \mathbf{I}_n$, then

$$S^2 = \frac{(\mathbf{Y} - \mathbf{X}\hat{\boldsymbol{\beta}})'(\mathbf{Y} - \mathbf{X}\hat{\boldsymbol{\beta}})}{n-p} = \frac{\text{RSS}}{n-p}$$

is an unbiased estimate of σ^2.

Proof. By Equation (3.7),

$$\mathbf{Y} - \mathbf{X}\hat{\boldsymbol{\beta}} = (\mathbf{I}_n - \mathbf{P})\mathbf{Y}$$

and, from Theorem 3.1,

$$\begin{aligned}(n-p)S^2 &= \mathbf{Y}'(\mathbf{I}_n - \mathbf{P})'(\mathbf{I}_n - \mathbf{P})\mathbf{Y} \\ &= \mathbf{Y}'(\mathbf{I}_n - \mathbf{P})^2\mathbf{Y} \\ &= \mathbf{Y}'(\mathbf{I}_n - \mathbf{P})\mathbf{Y}. \qquad (3.12)\end{aligned}$$

Hence by Theorems 1.7 (Corollary 2) and 3.1 (iii),

$$\begin{aligned}E[\mathbf{Y}'(\mathbf{I}_n - \mathbf{P})\mathbf{Y}] &= \sigma^2 \operatorname{tr}(\mathbf{I}_n - \mathbf{P}) + \boldsymbol{\beta}'\mathbf{X}'(\mathbf{I}_n - \mathbf{P})\mathbf{X}\boldsymbol{\beta} \\ &= \sigma^2 (n-p)\end{aligned}$$

and $E[S^2] = \sigma^2$. □

It transpires that S^2, like $\hat{\boldsymbol{\beta}}$, has certain minimum properties which are partly summarized in the following theorem.

THEOREM 3.4 (Atiqullah [1962]) Let $Y_1, Y_2, \ldots, Y_n$ be n independent random variables with common variance σ^2 and common third and fourth moments, μ_3 and μ_4, respectively, about their means. If $\mathscr{E}[\mathbf{Y}] = \mathbf{X}\boldsymbol{\beta}$, where $\mathbf{X}$ is $n \times p$ of rank p, then $(n-p)S^2$ is the unique nonnegative quadratic unbiased estimate of $(n-p)\sigma^2$ with minimum variance when $\mu_4 = 3\sigma^4$ or when the diagonal elements of $\mathbf{P}$ are all equal.

Proof. Since $\sigma^2 \geqslant 0$ it is not unreasonable to follow Rao [1952] and consider estimates which are nonnegative. Let $\mathbf{Y}'\mathbf{A}\mathbf{Y}$ be a member of the class $\mathcal{C}$ of nonnegative quadratic unbiased estimates of $(n-p)\sigma^2$. Then, by Theorem 1.7,

$$(n-p)\sigma^2 = E[\mathbf{Y}'\mathbf{A}\mathbf{Y}] = \sigma^2 \operatorname{tr} \mathbf{A} + \boldsymbol{\beta}'\mathbf{X}'\mathbf{A}\mathbf{X}\boldsymbol{\beta}$$

for all $\boldsymbol{\beta}$, so that $\operatorname{tr}\mathbf{A} = n-p$ (setting $\boldsymbol{\beta} = \mathbf{0}$) and $\boldsymbol{\beta}'\mathbf{X}'\mathbf{A}\mathbf{X}\boldsymbol{\beta} = 0$ for all $\boldsymbol{\beta}$. Thus $\mathbf{X}'\mathbf{A}\mathbf{X} = \mathbf{O}$ (*A9.2*) or, since $\mathbf{A}$ is positive semidefinite, $\mathbf{A}\mathbf{X} = \mathbf{O}$ (*A3.5*). Hence if $\mathbf{a}$ is the vector of diagonal elements of $\mathbf{A}$, and $\gamma_2 = (\mu_4 - 3\sigma^4)/\sigma^4$, then from Theorem 1.8:

$$\begin{aligned} \operatorname{var}[\mathbf{Y}'\mathbf{A}\mathbf{Y}] &= \sigma^4\gamma_2\mathbf{a}'\mathbf{a} + 2\sigma^4 \operatorname{tr}\mathbf{A}^2 + 4\sigma^2\boldsymbol{\beta}'\mathbf{X}'\mathbf{A}^2\mathbf{X}\boldsymbol{\beta} + 4\mu_3\boldsymbol{\beta}'\mathbf{X}'\mathbf{A}\mathbf{a} \\ &= \sigma^4\gamma_2\mathbf{a}'\mathbf{a} + 2\sigma^4 \operatorname{tr}\mathbf{A}^2. \end{aligned} \tag{3.13}$$

Now, by Theorem 3.3, $(n-p)S^2$ $(=\mathbf{Y}'(\mathbf{I}_n - \mathbf{P})\mathbf{Y} = \mathbf{Y}'\mathbf{R}\mathbf{Y}$, say) is a member of the class $\mathcal{C}$. Also, by Theorem 3.1,

$$\operatorname{tr}\mathbf{R}^2 = \operatorname{tr}\mathbf{R} = n-p$$

so that, if we substitute in (3.13),

$$\operatorname{var}[\mathbf{Y}'\mathbf{R}\mathbf{Y}] = \sigma^4\gamma_2\mathbf{r}'\mathbf{r} + 2\sigma^4(n-p). \tag{3.14}$$

To find sufficient conditions for $\mathbf{Y}'\mathbf{R}\mathbf{Y}$ to have minimum variance for class $\mathcal{C}$, let $\mathbf{A} = \mathbf{R} + \mathbf{D}$. Then $\mathbf{D}$ is symmetric, and $\operatorname{tr}\mathbf{A} = \operatorname{tr}\mathbf{R} + \operatorname{tr}\mathbf{D}$; thus $\operatorname{tr}\mathbf{D} = 0$. Since $\mathbf{A}\mathbf{X} = \mathbf{O}$ we have $\mathbf{A}\mathbf{P} = \mathbf{A}\mathbf{X}(\mathbf{X}'\mathbf{X})^{-1}\mathbf{X}' = \mathbf{O}$, and combining this equation with $\mathbf{P}^2 = \mathbf{P}$, that is, $\mathbf{R}\mathbf{P} = \mathbf{O}$, leads to

$$\mathbf{O} = \mathbf{A}\mathbf{P} = \mathbf{R}\mathbf{P} + \mathbf{D}\mathbf{P} = \mathbf{D}\mathbf{P}$$

and

$$\mathbf{D}\mathbf{R} = \mathbf{D} \qquad (= \mathbf{D}' = \mathbf{R}\mathbf{D}).$$

Hence

$$\begin{aligned}\mathbf{A}^2 &= \mathbf{R}^2 + \mathbf{DR} + \mathbf{RD} + \mathbf{D}^2\\ &= \mathbf{R} + 2\mathbf{D} + \mathbf{D}^2\end{aligned}$$

and

$$\begin{aligned}\operatorname{tr}\mathbf{A}^2 &= \operatorname{tr}\mathbf{R} + 2\operatorname{tr}\mathbf{D} + \operatorname{tr}\mathbf{D}^2\\ &= (n-p) + \operatorname{tr}\mathbf{D}^2.\end{aligned}$$

Substituting in (3.13), setting $\mathbf{a}=\mathbf{r}+\mathbf{d}$, and using (3.14), we have

$$\begin{aligned}\operatorname{var}[\mathbf{Y}'\mathbf{AY}] &= \sigma^4\gamma_2\mathbf{a}'\mathbf{a} + 2\sigma^4\left[(n-p) + \operatorname{tr}\mathbf{D}^2\right]\\ &= \sigma^4\gamma_2(\mathbf{r}'\mathbf{r} + 2\mathbf{r}'\mathbf{d} + \mathbf{d}'\mathbf{d}) + 2\sigma^4\left[(n-p) + \operatorname{tr}\mathbf{D}^2\right]\\ &= \sigma^4\gamma_2\mathbf{r}'\mathbf{r} + 2\sigma^4(n-p) + 2\sigma^4\left[\gamma_2(\mathbf{r}'\mathbf{d} + \tfrac{1}{2}\mathbf{d}'\mathbf{d}) + \operatorname{tr}\mathbf{D}^2\right]\\ &= \operatorname{var}[\mathbf{Y}'\mathbf{RY}] + 2\sigma^4\left[\gamma_2\left(\sum_i r_{ii}d_{ii} + \tfrac{1}{2}\sum_i d_{ii}^2\right) + \sum_i\sum_j d_{ij}^2\right].\end{aligned}$$

To find the estimate with minimum variance we must minimize var[$\mathbf{Y}'\mathbf{AY}$] subject to $\operatorname{tr}\mathbf{D}=\mathbf{0}$ and $\mathbf{DR}=\mathbf{D}$. The minimization in general is difficult (cf Hsu [1938]), but can be done readily in two important special cases. First, if $\gamma_2=0$, then

$$\operatorname{var}[\mathbf{Y}'\mathbf{AY}] = \operatorname{var}[\mathbf{Y}'\mathbf{RY}] + 2\sigma^4\sum_i\sum_j d_{ij}^2,$$

which is mimimized when $d_{ij}=0$ for all i, j, that is, when $\mathbf{D}=\mathbf{O}$ and $\mathbf{A}=\mathbf{R}$. Second, if the diagonal elements of $\mathbf{P}$ are all equal, then they are equal to p/n (since, by Theorem 3.1 (ii), $\operatorname{tr}\mathbf{P}=p$). Hence $r_{ii}=(n-p)/n$ for each i and

$$\begin{aligned}\operatorname{var}[\mathbf{Y}'\mathbf{AY}] &= \operatorname{var}[\mathbf{Y}'\mathbf{RY}] + 2\sigma^4\left[\gamma_2\left(0 + \tfrac{1}{2}\sum_i d_{ii}^2\right) + \sum\sum d_{ij}^2\right]\\ &= \operatorname{var}[Y'RY] + 2\sigma^4\left[(\tfrac{1}{2}\gamma_2 + 1)\sum_i d_{ii}^2 + \sum_{i\neq j}\sum d_{ij}^2\right]\end{aligned}$$

as $\sum_i r_{ii}d_{ii} = [(n-p)/n]\operatorname{tr}\mathbf{D}=0$. Now $\gamma_2 > -2$ (*A11.1*) so that var[$\mathbf{Y}'\mathbf{AY}$] is minimized when $d_{ij}=0$ for all i, j. Thus in both cases we have minimum variance if and only if $\mathbf{A}=\mathbf{R}$. □

This theorem highlights the fact that a uniformly minimum variance quadratic unbiased estimator of σ^2 exists only under certain restrictive conditions like those stated in the enunciation of the theorem. If normality can be assumed ($\gamma_2=0$) then it transpires that (Rao [1973: p. 319]) S^2 is the minimum variance unbiased estimator of σ^2 in the entire class of unbiased estimators (and not just the class of quadratic estimators).

Rao [1970, 1972b] has also introduced another criterion for choosing the estimator of σ^2 called MINQUE, Minimum Norm Quadratic Unbiased Estimation. Irrespective of whether we assume normality or not, this criterion also leads to S^2 (cf. Rao [1970, 1974: p. 448]).

EXERCISES 3c

1. Suppose $\mathbf{Y}\sim N_n(\mathbf{X}\boldsymbol{\beta},\sigma^2\mathbf{I}_n)$, where $\mathbf{X}$ is $n\times p$ of rank p.
   (a) Find var$[S^2]$.
   (b) Evaluate $E[(\mathbf{Y}'\mathbf{A}_1\mathbf{Y}-\sigma^2)^2]$ for

$$\mathbf{A}_1=\frac{1}{n-p+2}\left(\mathbf{I}_n-\mathbf{X}(\mathbf{X}'\mathbf{X})^{-1}\mathbf{X}'\right).$$

   (c) Prove that $\mathbf{Y}'\mathbf{A}_1\mathbf{Y}$ is an estimator of σ^2 with a smaller mean square error than S^2. Theil and Schweitzer [1961]

2. Let $Y_1, Y_2, \ldots, Y_n$ be independently and identically distributed with mean θ and variance σ^2. Find the nonnegative quadratic unbiased estimate of σ^2 with the minimum variance.

3.4 DISTRIBUTION THEORY

Up till now the only assumptions we have made about the ε_i are that $\mathcal{E}[\boldsymbol{\varepsilon}]=\mathbf{0}$ and $\mathcal{D}[\boldsymbol{\varepsilon}]=\sigma^2\mathbf{I}_n$. If we assume that the ε_i are also normally distributed, then $\boldsymbol{\varepsilon}\sim N_n(\mathbf{0},\sigma^2\mathbf{I}_n)$ and hence $\mathbf{Y}\sim N_n(\mathbf{X}\boldsymbol{\beta},\sigma^2\mathbf{I}_n)$. A number of distributional results then follow.

THEOREM 3.5 If $\mathbf{Y}\sim N_n(\mathbf{X}\boldsymbol{\beta},\sigma^2\mathbf{I}_n)$, where $\mathbf{X}$ is $n\times p$ of rank p, then:
(i) $\hat{\boldsymbol{\beta}}\sim N_p(\boldsymbol{\beta},\sigma^2(\mathbf{X}'\mathbf{X})^{-1})$.
(ii) $(\hat{\boldsymbol{\beta}}-\boldsymbol{\beta})'\mathbf{X}'\mathbf{X}(\hat{\boldsymbol{\beta}}-\boldsymbol{\beta})/\sigma^2\sim\chi_p^2$.
(iii) $\hat{\boldsymbol{\beta}}$ is independent of S^2.
(iv) $\text{RSS}/\sigma^2=(n-p)S^2/\sigma^2\sim\chi_{n-p}^2$.

Proof. (i) Since $\hat{\boldsymbol{\beta}}=(\mathbf{X}'\mathbf{X})^{-1}\mathbf{X}'\mathbf{Y}=\mathbf{C}\mathbf{Y}$, say, where $\mathbf{C}$ is a $p\times n$ matrix such that rank $\mathbf{C}=$ rank $\mathbf{X}'=$ rank $\mathbf{X}=p$ (by *A2.4*), $\hat{\boldsymbol{\beta}}$ has a multivariate normal distribution (Theorem 2.2 in Section 2.2). In particular, from Equations (3.10) and (3.11), we have $\hat{\boldsymbol{\beta}}\sim N_p(\boldsymbol{\beta},\sigma^2(\mathbf{X}'\mathbf{X})^{-1})$.

(ii) $(\hat{\boldsymbol{\beta}}-\boldsymbol{\beta})'\mathbf{X}'\mathbf{X}(\hat{\boldsymbol{\beta}}-\boldsymbol{\beta})/\sigma^2=(\hat{\boldsymbol{\beta}}-\boldsymbol{\beta})'(\mathscr{D}[\hat{\boldsymbol{\beta}}])^{-1}(\hat{\boldsymbol{\beta}}-\boldsymbol{\beta})$ which, by (i) and Theorem 2.1 (iii), is distributed as χ_p^2.

(iii)

$$\begin{aligned}\mathscr{C}[\hat{\boldsymbol{\beta}},\mathbf{Y}-\mathbf{X}\hat{\boldsymbol{\beta}}]&=\mathscr{C}\left[(\mathbf{X}'\mathbf{X})^{-1}\mathbf{X}'\mathbf{Y},(\mathbf{I}_n-\mathbf{P})\mathbf{Y}\right]\\&=(\mathbf{X}'\mathbf{X})^{-1}\mathbf{X}'\mathscr{D}[\mathbf{Y}](\mathbf{I}_n-\mathbf{P})'\\&=\sigma^2(\mathbf{X}'\mathbf{X})^{-1}\mathbf{X}'(\mathbf{I}_n-\mathbf{P})\\&=\mathbf{O}\qquad[\text{Theorem 3.1 (iii)}].\end{aligned}$$

If $\mathbf{U}_1=\hat{\boldsymbol{\beta}}$ and $\mathbf{V}=\mathbf{Y}-\mathbf{X}\hat{\boldsymbol{\beta}}$ in Theorem 2.7 (Section 2.3), $\hat{\boldsymbol{\beta}}$ is independent of $(\mathbf{Y}-\mathbf{X}\hat{\boldsymbol{\beta}})'(\mathbf{Y}-\mathbf{X}\hat{\boldsymbol{\beta}})$ and therefore of S^2.

(iv) This result can be proved in various ways depending on which theorems in Chapter 2 we are prepared to invoke. It is instructive to examine three methods of proof.

Method 1

$$\begin{aligned}Q_1&=(\mathbf{Y}-\mathbf{X}\boldsymbol{\beta})'(\mathbf{Y}-\mathbf{X}\boldsymbol{\beta})\\&=(\mathbf{Y}-\mathbf{X}\hat{\boldsymbol{\beta}}+\mathbf{X}(\hat{\boldsymbol{\beta}}-\boldsymbol{\beta}))'(\mathbf{Y}-\mathbf{X}\hat{\boldsymbol{\beta}}+\mathbf{X}(\hat{\boldsymbol{\beta}}-\boldsymbol{\beta}))\\&=(\mathbf{Y}-\mathbf{X}\hat{\boldsymbol{\beta}})'(\mathbf{Y}-\mathbf{X}\hat{\boldsymbol{\beta}})+2(\hat{\boldsymbol{\beta}}-\boldsymbol{\beta})'\mathbf{X}'(\mathbf{Y}-\mathbf{X}\hat{\boldsymbol{\beta}})+(\hat{\boldsymbol{\beta}}-\boldsymbol{\beta})'\mathbf{X}'\mathbf{X}(\hat{\boldsymbol{\beta}}-\boldsymbol{\beta})\\&=(\mathbf{Y}-\mathbf{X}\hat{\boldsymbol{\beta}})'(\mathbf{Y}-\mathbf{X}\hat{\boldsymbol{\beta}})+(\hat{\boldsymbol{\beta}}-\boldsymbol{\beta})'\mathbf{X}'\mathbf{X}(\hat{\boldsymbol{\beta}}-\boldsymbol{\beta})\\&=Q+Q_2,\text{ say,}\end{aligned}\tag{3.15}$$

since

$$(\hat{\boldsymbol{\beta}}-\boldsymbol{\beta})'\mathbf{X}'(\mathbf{Y}-\mathbf{X}\hat{\boldsymbol{\beta}})=(\hat{\boldsymbol{\beta}}-\boldsymbol{\beta})'(\mathbf{X}'\mathbf{Y}-\mathbf{X}'\mathbf{X}\hat{\boldsymbol{\beta}})=0.\tag{3.16}$$

Now Q_1/σ^2 $(=\Sigma_i\varepsilon_i^2/\sigma^2)$ is χ_n^2, and $Q_2/\sigma^2\sim\chi_p^2$ [by (ii)]. Also Q_2 is a continuous function of $\hat{\boldsymbol{\beta}}$ so that by Theorem 1.9 (Section 1.5) and (iii), Q is independent of Q_2. Hence $Q/\sigma^2\sim\chi_{n-p}^2$ (Theorem 1.10, Section 1.6).

Method 2

Using Theorem 3.3 we have

$$\begin{aligned}\text{RSS} &= \mathbf{Y}'(\mathbf{I}_n - \mathbf{P})\mathbf{Y} \\ &= (\mathbf{Y} - \mathbf{X}\boldsymbol{\beta})'(\mathbf{I}_n - \mathbf{P})(\mathbf{Y} - \mathbf{X}\boldsymbol{\beta}) \qquad [\text{Theorem 3.1 (iii)}] \\ &= \boldsymbol{\varepsilon}'(\mathbf{I}_n - \mathbf{P})\boldsymbol{\varepsilon}, \end{aligned} \tag{3.17}$$

where $\mathbf{I}_n - \mathbf{P}$ is symmetric and idempotent of rank $n-p$. Since $\boldsymbol{\varepsilon} \sim N_n(\mathbf{0}, \sigma^2\mathbf{I}_n)$, $\text{RSS}/\sigma^2 \sim \chi^2_{n-p}$ (Theorem 2.8, in Section 2.4).

Method 3

Using the decomposition (3.15) we have $Q_1/\sigma^2 \sim \chi^2_n$, $Q_2/\sigma^2 \sim \chi^2_p$ [by (ii)] and $Q = Q_1 - Q_2 \geqslant 0$. Hence by Theorem 2.9 in Section 2.4, $Q/\sigma^2 \sim \chi^2_{n-p}$. □

EXERCISES 3d

1. Given $Y_1, Y_2, \ldots, Y_n$ independently distributed as $N(\theta, \sigma^2)$, use Theorem 3.5 to prove that:
   (a) $\bar{Y}$ is statistically independent of $Q = \Sigma_i (Y_i - \bar{Y})^2$, and
   (b) $Q/\sigma^2 \sim \chi^2_{n-p}$.
2. Use Theorem 2.7 to prove that for the general regression model RSS is independent of $(\hat{\boldsymbol{\beta}} - \boldsymbol{\beta})'\mathbf{X}'\mathbf{X}(\hat{\boldsymbol{\beta}} - \boldsymbol{\beta})$.

3.5 ORTHOGONAL STRUCTURE IN THE DESIGN MATRIX

Suppose in the model $\mathcal{E}[\mathbf{Y}] = \mathbf{X}\boldsymbol{\beta}$ we can divide the matrix $\mathbf{X}$ into $k+1$ sets of columns denoted in matrix form by

$$\mathbf{X} = (\mathbf{X}_0, \mathbf{X}_1, \ldots, \mathbf{X}_k).$$

A corresponding division can be made in $\boldsymbol{\beta}$ so that

$$\boldsymbol{\beta} = \begin{bmatrix} \boldsymbol{\beta}_0 \\ \boldsymbol{\beta}_1 \\ \vdots \\ \boldsymbol{\beta}_k \end{bmatrix},$$

where the number of elements in $\boldsymbol{\beta}_i$ is equal to the number of columns in $\mathbf{X}_i$ $(i = 0, 1, \ldots, k)$. Then the model can be written in the form

$$\mathcal{E}[\mathbf{Y}] = \mathbf{X}_0\boldsymbol{\beta}_0 + \mathbf{X}_1\boldsymbol{\beta}_1 + \cdots + \mathbf{X}_k\boldsymbol{\beta}_k.$$

Suppose now that the columns of $\mathbf{X}_i$ are mutually orthogonal to $\mathbf{X}_j$ for all i, j $(i \neq j)$, namely, $\mathbf{X}_i'\mathbf{X}_j = \mathbf{O}$. Then

$$\hat{\boldsymbol{\beta}} = (\mathbf{X}'\mathbf{X})^{-1}\mathbf{X}'\mathbf{Y}$$

$$= \begin{bmatrix} \mathbf{X}_0'\mathbf{X}_0 & \mathbf{O} & \cdots & \mathbf{O} \\ \mathbf{O} & \mathbf{X}_1'\mathbf{X}_1 & \cdots & \mathbf{O} \\ \cdots & \cdots & \cdots & \mathbf{O} \\ \mathbf{O} & \mathbf{O} & \cdots & \mathbf{X}_k'\mathbf{X}_k \end{bmatrix}^{-1} \begin{bmatrix} \mathbf{X}_0'\mathbf{Y} \\ \mathbf{X}_1'\mathbf{Y} \\ \cdots \\ \mathbf{X}_k'\mathbf{Y} \end{bmatrix}$$

$$= \begin{bmatrix} (\mathbf{X}_0'\mathbf{X}_0)^{-1}\mathbf{X}_0'\mathbf{Y} \\ (\mathbf{X}_1'\mathbf{X}_1)^{-1}\mathbf{X}_1'\mathbf{Y} \\ \cdots \\ (\mathbf{X}_k'\mathbf{X}_k)^{-1}\mathbf{X}_k'\mathbf{Y} \end{bmatrix}$$

$$= \begin{bmatrix} \hat{\boldsymbol{\beta}}_0 \\ \hat{\boldsymbol{\beta}}_1 \\ \vdots \\ \hat{\boldsymbol{\beta}}_k \end{bmatrix},$$

say, where $\hat{\boldsymbol{\beta}}_i$ turns out to be the least squares estimate of $\boldsymbol{\beta}_i$ for the model $\mathcal{E}[\mathbf{Y}] = \mathbf{X}_i\boldsymbol{\beta}_i$. This means that the least squares estimate of $\boldsymbol{\beta}_i$ is unchanged if any of the other $\boldsymbol{\beta}_j$ $(j \neq i)$ are put equal to zero. Also, from Equation (3.8), the residual sum of squares takes the form

$$\begin{aligned} \text{RSS} &= \mathbf{Y}'\mathbf{Y} - \hat{\boldsymbol{\beta}}'\mathbf{X}'\mathbf{Y} \\ &= \mathbf{Y}'\mathbf{Y} - \sum_{r=0}^{k} \hat{\boldsymbol{\beta}}_r'\mathbf{X}_r'\mathbf{Y} \end{aligned}$$

so that if we put $\boldsymbol{\beta}_i = \mathbf{0}$ in the model, the only change in the residual sum of squares is the addition of the term $\hat{\boldsymbol{\beta}}_i'\mathbf{X}_i'\mathbf{Y}$; that is, we now have

$$\mathbf{Y}'\mathbf{Y} - \sum_{\substack{r=0 \\ r \neq i}}^{k} \hat{\boldsymbol{\beta}}_r'\mathbf{X}_r'\mathbf{Y}. \qquad (3.18)$$

In the simplest case where each $\mathbf{X}_i$ consists of one column only, say, $\mathbf{x}_i$, then

$$\hat{\beta}_i = \frac{\mathbf{x}_i'\mathbf{Y}}{\mathbf{x}_i'\mathbf{x}_i},$$

and

$$\begin{aligned}\text{RSS} &= \mathbf{Y}'\mathbf{Y} - \sum_{r=0}^{k} \hat{\beta}_r \mathbf{x}_r'\mathbf{Y} \\ &= \mathbf{Y}'\mathbf{Y} - \sum_{r=0}^{k} \hat{\beta}_r^2 (\mathbf{x}_r'\mathbf{x}_r). \end{aligned} \tag{3.19}$$

Two applications of this model are discussed in Sections 8.2.1 and 8.5.2.

An interesting property of regression models, proved by Hotelling (see Exercise 3 below), is the following: given any design matrix $\mathbf{X}$ such that $\mathbf{x}_i'\mathbf{x}_i = c_i^2$, then

$$\text{var}[\hat{\beta}_i] \geqslant \frac{\sigma^2}{c_i^2},$$

and the minimum is attained when $\mathbf{x}_i'\mathbf{x}_j = 0$ (all j, $j \neq i$). Hence given $\mathbf{x}_i'\mathbf{x}_i = c_i^2$ $(i = 0, 1, 2, \ldots, k)$, the "optimum" choice of $\mathbf{X}$ is the design matrix with mutually orthogonal columns. The following proof of this result is instructive.

LEMMA Consider the model

$$Y_i = \beta_0 + \beta_1 x_{i1} + \cdots + \beta_k x_{ik} + \varepsilon_i \qquad (i = 1, 2, \ldots, n)$$

where the x_{ij} are standardized so that, for $j = 1, 2, \ldots, k$, $\Sigma_i x_{ij} = 0$ and $\Sigma_i x_{ij}^2 = c$. Then

$$\frac{1}{k+1} \sum_{j=0}^{k} \text{var}[\hat{\beta}_j] \tag{3.20}$$

is minimized when the columns of $\mathbf{X}$ are mutually orthogonal.

Proof.

$$\mathbf{X}'\mathbf{X} = \begin{pmatrix} n & \mathbf{0}' \\ \mathbf{0} & \mathbf{C} \end{pmatrix}$$

and

$$\begin{aligned}\sum_{j=0}^{k} \operatorname{var}[\hat{\beta}_j] &= \operatorname{tr} \mathfrak{D}[\hat{\boldsymbol{\beta}}] \\ &= \sigma^2\left(\operatorname{tr}\mathbf{C}^{-1} + \frac{1}{n}\right) \\ &= \sigma^2 \sum_{j=0}^{k} \lambda_j^{-1}, \qquad (3.21)\end{aligned}$$

where $\lambda_0 = n$ and λ_j $(j=1,2,\ldots,k)$ are the eigenvalues of $\mathbf{C}$ (*A1.5*). Now the minimum of (3.21) subject to the condition $\operatorname{tr}[\mathbf{X}'\mathbf{X}] = n + kc$, or $\operatorname{tr}\mathbf{C} = kc$, is given by λ_j = constant, that is, $\lambda_j = c$ $(j=1,2,\ldots,k)$. Hence there exists an orthogonal matrix $\mathbf{T}$ such that $\mathbf{T}'\mathbf{C}\mathbf{T} = c\mathbf{I}_k$, or $\mathbf{C} = c\mathbf{I}_k$, so that the columns of $\mathbf{X}$ must be mutually orthogonal.

EXERCISES 3e

1. Prove the above statement that the minimum is given by $\lambda_j = c$ $(j = 1,2,\ldots,k)$.
2. It is required to fit a regression model of the form

$$E[Y_i] = \beta_0 + \beta_1 x_i + \beta_2 \phi(x_i), \qquad i=1,2,3.$$

where $\phi(x)$ is a second-degree polynomial. If $x_1 = -1$, $x_2 = 0$, and $x_3 = 1$, find ϕ such that the design matrix $\mathbf{X}$ has mutually orthogonal columns.

3. Suppose $\mathbf{X} = [\mathbf{x}_0, \mathbf{x}_1, \ldots, \mathbf{x}_{k-1}, \mathbf{x}_k] = [\mathbf{W}, \mathbf{x}_k]$ has linearly independent columns.

(a) Using *A7* prove that

$$|\mathbf{X}'\mathbf{X}| = |\mathbf{W}'\mathbf{W}|\left(\mathbf{x}_k'\mathbf{x}_k - \mathbf{x}_k'\mathbf{W}(\mathbf{W}'\mathbf{W})^{-1}\mathbf{W}'\mathbf{x}_k\right).$$

(b) Deduce that

$$\frac{|\mathbf{W}'\mathbf{W}|}{|\mathbf{X}'\mathbf{X}|} \geqslant \frac{1}{\mathbf{x}_k'\mathbf{x}_k},$$

and hence show that $\operatorname{var}[\hat{\beta}_k] \geqslant \sigma^2(\mathbf{x}_k'\mathbf{x}_k)^{-1}$ with equality if and only if $\mathbf{x}_k'\mathbf{x}_j = 0$ $\quad (j=0,1,\ldots,k-1)$. Rao[1973: p. 236]

4. What modifications in the statement of the lemma proved above can be made if the term β_0 is omitted?
5. Show that the full rank model

$$Y_i = \beta_0 + \beta_1 x_{i1} + \cdots + \beta_k x_{ik} + \varepsilon_i$$

can always, by reparametrization, be expressed in the form

$$Y_i = \gamma_0 + \gamma_1 z_{i1} + \cdots + \gamma_k z_{ik} + \varepsilon_i,$$

where the design matrix now has orthogonal columns and $\gamma_r = \gamma_{r+1} = \cdots = \gamma_k = 0$ if and only if $\beta_r = \beta_{r+1} = \cdots = \beta_k = 0$ $(r = 0, 1, \ldots, k)$.
6. Suppose we wish to find the weights β_i $(i = 1, 2, \ldots, k)$ of k objects. One method is to weigh each object r times and take the average; this requires a total of kr weighings, and the variance of each average is σ^2/r (σ^2 being the variance of the weighing error). Another method is to weigh the objects in combinations; some of the objects are distributed between the two pans and weights are placed in one pan to achieve equilibrium. The regression model for such a scheme is

$$Y = \beta_1 x_1 + \beta_2 x_2 + \cdots + \beta_k x_k + \varepsilon$$

where $x_i = 0$, 1, or -1 according as the ith object is not used, placed in the left pan or in the right pan, ε is the weighing error (assumed to be the same for all weighings), and Y is the weight required for equilibrium (Y is regarded as negative if placed in the left pan). After n such weighing operations we can find the least squares estimates $\hat{\beta}_i$ of the weights,
   (a) Show that the estimates of the weights have maximum precision (i.e., minimum variance) when each entry in the design matrix $\mathbf{X}$ is ± 1 and the columns of $\mathbf{X}$ are mutually orthogonal.
   (b) If the objects are weighed individually, show that kn weighings are required to achieve the same precision as that given by the optimal design with n weighings. Rao [1973: p. 309]

3.6 GENERALIZED LEAST SQUARES

Having developed a least squares theory for the model $\mathbf{Y} = \mathbf{X}\boldsymbol{\beta} + \boldsymbol{\varepsilon}$, where $\mathscr{E}[\boldsymbol{\varepsilon}] = \mathbf{0}$ and $\mathscr{D}[\boldsymbol{\varepsilon}] = \sigma^2 \mathbf{I}_n$, we now consider what modifications are necessary if we allow the ε_i to be correlated. In particular, we assume that $\mathscr{D}[\boldsymbol{\varepsilon}] = \sigma^2 \mathbf{V}$, where $\mathbf{V}$ is a *known* $n \times n$ positive definite matrix.

Since $\mathbf{V}$ is positive definite, there exists an $n \times n$ nonsingular matrix $\mathbf{K}$

such that $\mathbf{V}=\mathbf{K}\mathbf{K}'$ (*A4.2*). Therefore setting $\mathbf{Z}=\mathbf{K}^{-1}\mathbf{Y}$, $\mathbf{B}=\mathbf{K}^{-1}\mathbf{X}$, and $\boldsymbol{\eta}=\mathbf{K}^{-1}\boldsymbol{\varepsilon}$, we have the model $\mathbf{Z}=\mathbf{B}\boldsymbol{\beta}+\boldsymbol{\eta}$, where $\mathbf{B}$ is $n\times p$ of rank p (*A2.2*), $\mathcal{E}[\boldsymbol{\eta}]=\mathbf{0}$ and $\mathcal{D}[\boldsymbol{\eta}]=\sigma^2\mathbf{I}_n$ [by Equation (2.15) in Section 2.3]. Minimizing $\boldsymbol{\eta}'\boldsymbol{\eta}$ with respect to $\boldsymbol{\beta}$, and using the theory of Section 3.1, the least squares estimate of $\boldsymbol{\beta}$ for this transformed model is

$$\begin{aligned}\boldsymbol{\beta}^* &= (\mathbf{B}'\mathbf{B})^{-1}\mathbf{B}'\mathbf{Z}\\ &= \left(\mathbf{X}'(\mathbf{K}\mathbf{K}')^{-1}\mathbf{X}\right)^{-1}\mathbf{X}'(\mathbf{K}\mathbf{K}')^{-1}\mathbf{Y}\\ &= (\mathbf{X}'\mathbf{V}^{-1}\mathbf{X})^{-1}\mathbf{X}'\mathbf{V}^{-1}\mathbf{Y},\end{aligned}$$

with expected value

$$\mathcal{E}[\boldsymbol{\beta}^*]=(\mathbf{X}'\mathbf{V}^{-1}\mathbf{X})^{-1}\mathbf{X}'\mathbf{V}^{-1}\mathbf{X}\boldsymbol{\beta}=\boldsymbol{\beta},$$

dispersion matrix

$$\begin{aligned}\mathcal{D}[\boldsymbol{\beta}^*] &= \sigma^2(\mathbf{B}'\mathbf{B})^{-1}\\ &= \sigma^2(\mathbf{X}'\mathbf{V}^{-1}\mathbf{X})^{-1}\end{aligned} \tag{3.22}$$

and the residual sum of squares

$$\begin{aligned}\mathbf{f}'\mathbf{f} &= (\mathbf{Z}-\mathbf{B}\boldsymbol{\beta}^*)'(\mathbf{Z}-\mathbf{B}\boldsymbol{\beta}^*)\\ &= (\mathbf{Y}-\mathbf{X}\boldsymbol{\beta}^*)'(\mathbf{K}\mathbf{K}')^{-1}(\mathbf{Y}-\mathbf{X}\boldsymbol{\beta}^*)\\ &= (\mathbf{Y}-\mathbf{X}\boldsymbol{\beta}^*)'\mathbf{V}^{-1}(\mathbf{Y}-\mathbf{X}\boldsymbol{\beta}^*).\end{aligned}$$

Alternatively, we can obtain $\boldsymbol{\beta}^*$ by simply differentiating

$$\begin{aligned}\boldsymbol{\eta}'\boldsymbol{\eta} &= \boldsymbol{\varepsilon}'\mathbf{V}^{-1}\boldsymbol{\varepsilon}\\ &= (\mathbf{Y}-\mathbf{X}\boldsymbol{\beta})'\mathbf{V}^{-1}(\mathbf{Y}-\mathbf{X}\boldsymbol{\beta})\\ &= \mathbf{Y}'\mathbf{Y}-2\boldsymbol{\beta}'\mathbf{X}'\mathbf{V}^{-1}\mathbf{Y}+\boldsymbol{\beta}'\mathbf{X}'\mathbf{V}^{-1}\mathbf{X}\boldsymbol{\beta}\end{aligned}$$

with respect to $\boldsymbol{\beta}$. Thus, by *A6*,

$$\frac{\partial\boldsymbol{\eta}'\boldsymbol{\eta}}{\partial\boldsymbol{\beta}}=-2\mathbf{X}'\mathbf{V}^{-1}\mathbf{Y}+2\mathbf{X}'\mathbf{V}^{-1}\mathbf{X}\boldsymbol{\beta} \tag{3.23}$$

and setting this equal to zero leads once again to $\boldsymbol{\beta}^*$: $\mathbf{X}'\mathbf{V}^{-1}\mathbf{X}$ has an inverse as it is positive definite, by *A4.5*. We note that the coefficient of $2\boldsymbol{\beta}$ in (3.23) gives us the inverse of $\mathcal{D}[\boldsymbol{\beta}^*]/\sigma^2$.

There is some variation in terminology among books dealing with the above model: some texts call $\boldsymbol{\beta}^*$ the weighted least squares estimate. However, we call $\boldsymbol{\beta}^*$ the *generalized least squares estimate* and reserve the expression *weighted least squares* for the case when $\mathbf{V}$ is a diagonal matrix: the diagonal case is discussed in various places throughout this book (see Index).

EXAMPLE 3.2 Let $\mathbf{Y}=\mathbf{x}\beta+\boldsymbol{\varepsilon}$, where $\mathbf{Y}=[(Y_i)]$ and $\mathbf{x}=[(x_i)]$ are $n\times 1$ vectors, $\mathcal{E}[\boldsymbol{\varepsilon}]=\mathbf{0}$ and $\mathcal{D}[\boldsymbol{\varepsilon}]=\sigma^2\mathbf{V}$. If $\mathbf{V}=\text{diag}(w_1^{-1},w_2^{-1},\ldots,w_n^{-1})$ $(w_i>0)$, find the weighted least squares estimate of β and its variance.

Solution. It is simpler in this case to differentiate $\boldsymbol{\eta}'\boldsymbol{\eta}$ directly rather than use the general matrix theory. Thus, since $\mathbf{V}^{-1}=\text{diag}(w_1,w_2,\ldots,w_n)$,

$$\boldsymbol{\eta}'\boldsymbol{\eta}=\sum_i (Y_i-x_i\beta)^2 w_i$$

and

$$\frac{\partial\boldsymbol{\eta}'\boldsymbol{\eta}}{\partial\beta}=-2\sum_i x_i(Y_i-x_i\beta)w_i. \tag{3.24}$$

Setting the right side of (3.24) equal to zero leads to

$$\beta^*=\frac{\sum_i w_iY_ix_i}{\sum_i w_ix_i^2}$$

and, from the coefficient of 2β,

$$\text{var}[\beta^*]=\sigma^2\left(\sum_i w_ix_i^2\right)^{-1}.$$

We can also find the variance directly from

$$(\mathbf{X}'\mathbf{V}^{-1}\mathbf{X})^{-1}=(\mathbf{x}'\mathbf{V}^{-1}\mathbf{x})^{-1}=\left(\sum w_ix_i^2\right)^{-1}.$$

Since the generalized least squares estimate is simply the ordinary least squares estimate (OLSE) for a transformed model we would expect $\boldsymbol{\beta}^*$ to have the same optimal properties, namely, that $\mathbf{a}'\boldsymbol{\beta}^*$ is the best linear unbiased estimate (BLUE) of $\mathbf{a}'\boldsymbol{\beta}$. To see this we note first of all that

$$\mathbf{a}'\boldsymbol{\beta}^*=\mathbf{a}'(\mathbf{X}'\mathbf{V}^{-1}\mathbf{X})^{-1}\mathbf{X}'\mathbf{V}^{-1}\mathbf{Y}=\mathbf{b}'\mathbf{Y},$$

say, is linear and unbiased. Let $\mathbf{b}_1'\mathbf{Y}$ be any other linear unbiased estimate of $\mathbf{a}'\boldsymbol{\beta}$. Then, using the transformed model, $\mathbf{a}'\boldsymbol{\beta}^* = \mathbf{a}'(\mathbf{B}'\mathbf{B})^{-1}\mathbf{B}'\mathbf{Z}$ and $\mathbf{b}_1'\mathbf{Y} = \mathbf{b}_1'\mathbf{K}\mathbf{K}^{-1}\mathbf{Y} = (\mathbf{K}'\mathbf{b}_1)'\mathbf{Z}$. By Theorem 3.2 (Section 3.2) and the ensuing argument

$$\operatorname{var}[\mathbf{a}'\boldsymbol{\beta}^*] \leqslant \operatorname{var}[(\mathbf{K}'\mathbf{b}_1)'\mathbf{Z}] = \operatorname{var}[\mathbf{b}_1'\mathbf{Y}].$$

Equality occurs if and only if $(\mathbf{K}'\mathbf{b}_1)' = \mathbf{a}'(\mathbf{B}'\mathbf{B})^{-1}\mathbf{B}'$, or

$$\mathbf{b}_1' = \mathbf{a}'(\mathbf{B}'\mathbf{B})^{-1}\mathbf{B}'\mathbf{K}^{-1} = \mathbf{a}'(\mathbf{X}'\mathbf{V}^{-1}\mathbf{X})^{-1}\mathbf{X}'\mathbf{V}^{-1} = \mathbf{b}'.$$

Thus $\mathbf{a}'\boldsymbol{\beta}^*$ is the unique BLUE estimate of $\mathbf{a}'\boldsymbol{\beta}$.

Having derived $\boldsymbol{\beta}^*$, we naturally ask, under what conditions is $\boldsymbol{\beta}^*$ the same as $\hat{\boldsymbol{\beta}} = (\mathbf{X}'\mathbf{X})^{-1}\mathbf{X}'\mathbf{Y}$; that is, when can we ignore the fact that $\mathscr{D}[\boldsymbol{\varepsilon}]$ may be $\sigma^2\mathbf{V}$ and not $\sigma^2\mathbf{I}_n$? The answer is given in the following theorem (adapted from McElroy [1967]).

THEOREM 3.6 A necessary and sufficient condition for $\boldsymbol{\beta}^*$ and $\hat{\boldsymbol{\beta}}$ to be identical is $\mathscr{R}[\mathbf{V}^{-1}\mathbf{X}] = \mathscr{R}[\mathbf{X}]$.

Proof. $\boldsymbol{\beta}^*$ and $\hat{\boldsymbol{\beta}}$ are identical if and only if

$$\mathbf{X}'\mathbf{V}^{-1}\mathbf{Y} = (\mathbf{X}'\mathbf{V}^{-1}\mathbf{X})(\mathbf{X}'\mathbf{X})^{-1}\mathbf{X}'\mathbf{Y} \tag{3.25}$$

for every $\mathbf{Y}$. Let $\mathbf{Y}$ have the (unique) decomposition $\mathbf{Y} = \mathbf{Y}_1 + \mathbf{Y}_2$ (*B1.1*) where $\mathbf{Y}_1 \in \mathscr{R}[\mathbf{X}]$ and $\mathbf{Y}_2 \perp \mathscr{R}[\mathbf{X}]$. Since $\mathbf{Y}_1 = \mathbf{X}\mathbf{a}$ for some $\mathbf{a}$ we see that (3.25) is satisfied by $\mathbf{Y}_1$ for every $\mathbf{V}^{-1}$. Also $\mathbf{X}'\mathbf{Y}_2 = \mathbf{0}$ so that (3.25) is true for all $\mathbf{Y}_2$ if $\mathbf{X}'\mathbf{V}^{-1}\mathbf{Y}_2 = \mathbf{0}$. Hence a necessary and sufficient condition for (3.25) to hold for every $\mathbf{Y}$ is that $\mathbf{X}'\mathbf{x} = \mathbf{0}$ implies that $\mathbf{X}'\mathbf{V}^{-1}\mathbf{x} = \mathbf{0}$; that is $\mathscr{R}[\mathbf{V}^{-1}\mathbf{X}] \subset \mathscr{R}[\mathbf{X}]$. This implies $\mathscr{R}[\mathbf{V}^{-1}\mathbf{X}] = \mathscr{R}[\mathbf{X}]$ since these two spaces have the same dimension (*A2.2*).

COROLLARY 1 $\boldsymbol{\beta}^*$ and $\hat{\boldsymbol{\beta}}$ are identical if and only if

$$\mathscr{R}[\mathbf{V}\mathbf{X}] = \mathscr{R}[\mathbf{X}]. \tag{3.26}$$

Proof. Given $\mathscr{R}[\mathbf{V}^{-1}\mathbf{X}] = \mathscr{R}[\mathbf{X}]$, let $\mathbf{z} \in \mathscr{R}[\mathbf{V}\mathbf{X}]$. Then there exists $\mathbf{c}$ such that $\mathbf{z} = \mathbf{V}\mathbf{X}\mathbf{b} = \mathbf{V}(\mathbf{V}^{-1}\mathbf{X}\mathbf{c}) = \mathbf{X}\mathbf{c} \in \mathscr{R}[\mathbf{X}]$. Thus $\mathscr{R}[\mathbf{V}\mathbf{X}] \subset \mathscr{R}[\mathbf{X}]$ and, by the dimension argument given above, $\mathscr{R}[\mathbf{V}\mathbf{X}] = \mathscr{R}[\mathbf{X}]$. Conversely, given the latter equality, let $\mathbf{z} \in \mathscr{R}[\mathbf{V}^{-1}\mathbf{X}]$. Then $\mathbf{z} = \mathbf{V}^{-1}\mathbf{X}\mathbf{d} = \mathbf{V}^{-1}(\mathbf{V}\mathbf{X}\mathbf{f}) = \mathbf{X}\mathbf{f}$ for suitable $\mathbf{d}$ and $\mathbf{f}$, and $\mathscr{R}[\mathbf{V}^{-1}\mathbf{X}] \subset \mathscr{R}[\mathbf{X}]$; hence $\mathscr{R}[\mathbf{V}^{-1}\mathbf{X}] = \mathscr{R}[\mathbf{X}]$. Thus $\mathscr{R}[\mathbf{V}\mathbf{X}] = \mathscr{R}[\mathbf{X}]$ if and only if $\mathscr{R}[\mathbf{V}^{-1}\mathbf{X}] = \mathscr{R}[\mathbf{X}]$, and the required result follows from the above theorem. (A general proof of this corollary is given by Kruskal [1968].)

COROLLARY 2 If $\mathbf{Y}=\mathbf{x}\beta+\boldsymbol{\varepsilon}$, then $\boldsymbol{\beta}^*$ and $\hat{\boldsymbol{\beta}}$ are identical for every $\mathbf{x}$ if and only if $\mathbf{V}$ takes the form $c\mathbf{I}_n$ (Watson [1967: p. 1685]).

Proof. When $\mathbf{X}=\mathbf{x}$, $\mathcal{R}[\mathbf{X}]=\{\mathbf{x}c\text{: all } c\}$. Using Corollary 1 we see that $\mathcal{R}[\mathbf{x}]=\mathcal{R}[\mathbf{Vx}]$ for every $\mathbf{x}$ if and only if $\mathbf{Vx}=c_x\mathbf{x}$ for every $\mathbf{x}$. Setting $\mathbf{x}=\boldsymbol{\alpha}_i$ $(i=1,2,\ldots,n)$ and $\mathbf{x}=\mathbf{1}_n$, respectively, where $(\boldsymbol{\alpha}_1,\boldsymbol{\alpha}_2,\ldots,\boldsymbol{\alpha}_n)=\mathbf{I}_n$, we find that $\mathbf{V}$ is diagonal and hence $c_x=c$; thus $\mathbf{V}=c\mathbf{I}_n$ (*A9.1*).

If the first column of $\mathbf{X}$ is $\mathbf{1}_n$ and $\operatorname{tr}\mathbf{V}=n\sigma^2$, then it can be shown that $\boldsymbol{\beta}^*$ and $\hat{\boldsymbol{\beta}}$ are identical if and only if $\mathbf{V}=(1-\rho)\mathbf{I}_n+\rho\mathbf{1}_n\mathbf{1}_n'$, $0\leqslant\rho<1$ (McElroy [1967]).

For a recent discussion of the above topics see Bloomfield and Watson [1975] and Haberman [1975].

EXERCISES 3f

1. Let $Y_i=\beta x_i+\varepsilon_i$ $(i=1,2)$, where $\varepsilon_1\sim N(0,\sigma^2)$, $\varepsilon_2\sim N(0,2\sigma^2)$, and ε_1 and ε_2 are statistically independent. If $x_1=+1$ and $x_2=-1$, obtain the weighted least squares estimate of β and find the variance of your estimate.
2. Let Y_i $(i=1,2,\ldots,n)$ be independent random variables with a common mean θ and variances σ^2/w_i $(i=1,2,\ldots,n)$. Find the linear unbiased estimate of θ with minimum variance, and find this minimum variance.
3. Let $Y_1,Y_2,\ldots,Y_n$ be independent random variables, and let $Y_i\sim N(i\theta,i^2\sigma^2)$ for $i=1,2,\ldots,n$. Find the weighted least squares estimate of θ and prove that its variance is σ^2/n.
4. Let $Y_1,Y_2,\ldots,Y_n$ be random variables with common mean θ and with dispersion matrix $\sigma^2\mathbf{V}$, where $v_{ii}=1$ $(i=1,2,\ldots,n)$ and $v_{ij}=\rho$ $(0<\rho<1;$ $i,j=1,2,\ldots,n;$ $i\neq j)$. Find the generalized least squares estimate of θ and show that it is the same as the ordinary least squares estimate. [Hint: $\mathbf{V}^{-1}$ takes the same form as $\mathbf{V}$.] McElroy [1967]
5. Let $\mathbf{Y}\sim N_n(\mathbf{X}\boldsymbol{\beta},\sigma^2\mathbf{V})$, where $\mathbf{X}$ is $n\times p$ of rank p, and $\mathbf{V}$ is a known positive definite $n\times n$ matrix. If $\boldsymbol{\beta}^*$ is the generalized least squares estimate of $\boldsymbol{\beta}$, prove that
   (a) $Q=(\mathbf{Y}-\mathbf{X}\boldsymbol{\beta}^*)'\mathbf{V}^{-1}(\mathbf{Y}-\mathbf{X}\boldsymbol{\beta}^*)/\sigma^2\sim\chi^2_{n-p}$.
   (b) Q is the quadratic nonnegative unbiased estimate of $(n-p)\sigma^2$ with minimum variance.
   (c) If $\mathbf{Y}^*=\mathbf{X}\boldsymbol{\beta}^*=\mathbf{PY}$, then $\mathbf{P}$ is idempotent but not, in general, symmetric.
6. Suppose that $\mathcal{E}[\mathbf{Y}]=\boldsymbol{\theta}$, $\mathbf{A}\boldsymbol{\theta}=\mathbf{0}$, and $\mathcal{D}[\mathbf{Y}]=\sigma^2\mathbf{V}$, where $\mathbf{A}$ is a $q\times n$ matrix of rank q and $\mathbf{V}$ is a known $n\times n$ positive definite matrix. Let $\boldsymbol{\theta}^*$

be the generalized least squares estimate of $\boldsymbol{\theta}$, that is, $\boldsymbol{\theta}^*$ minimizes $(\mathbf{Y}-\boldsymbol{\theta})'\mathbf{V}^{-1}(\mathbf{Y}-\boldsymbol{\theta})$ subject to $\mathbf{A}\boldsymbol{\theta}=\mathbf{0}$. Show that

$$\mathbf{Y}-\boldsymbol{\theta}^*=\mathbf{V}\mathbf{A}'\boldsymbol{\gamma}^*,$$

where $\boldsymbol{\gamma}^*$ is the generalized least squares estimate of $\boldsymbol{\gamma}$ for the model $\mathcal{E}[\mathbf{Y}]=\mathbf{V}\mathbf{A}'\boldsymbol{\gamma}$, $\mathcal{D}[\mathbf{Y}]=\sigma^2\mathbf{V}$. Wedderburn [1974]

3.7 INTRODUCING FURTHER REGRESSORS

3.7.1 General Theory

Suppose after having fitted the regression model

$$\mathcal{E}[\mathbf{Y}]=\mathbf{X}\boldsymbol{\beta}, \qquad \mathcal{D}[\mathbf{Y}]=\sigma^2\mathbf{I}_n$$

we decide to introduce further x_j's into the model so that the model is now enlarged to

$$\begin{aligned} G\colon \mathcal{E}[\mathbf{Y}] &= \mathbf{X}\boldsymbol{\beta}+\mathbf{Z}\boldsymbol{\gamma} \\ &= (\mathbf{X},\mathbf{Z})\begin{pmatrix}\boldsymbol{\beta}\\ \boldsymbol{\gamma}\end{pmatrix} \\ &= \mathbf{W}\boldsymbol{\delta}, \end{aligned} \tag{3.27}$$

say, where $\mathbf{X}$ is $n\times p$ of rank p, $\mathbf{Z}$ is $n\times t$ of rank t, and the columns of $\mathbf{Z}$ are linearly independent of the columns of $\mathbf{X}$, that is, $\mathbf{W}$ is $n\times(t+p)$ of rank $t+p$. Then to find the least squares estimate $\hat{\boldsymbol{\delta}}_G$ of $\boldsymbol{\delta}$ there are two possible approaches. We can either compute $\hat{\boldsymbol{\delta}}_G$ and its dispersion matrix directly from

$$\hat{\boldsymbol{\delta}}_G=(\mathbf{W}'\mathbf{W})^{-1}\mathbf{W}'\mathbf{Y} \qquad \text{and} \qquad \mathcal{D}[\hat{\boldsymbol{\delta}}_G]=\sigma^2(\mathbf{W}'\mathbf{W})^{-1}$$

or, to reduce the amount of computation, we can utilize the calculations already carried out in fitting the original model, as in Theorem 3.7. A geometrical proof of this theorem, which allows $\mathbf{X}$ to have less than full rank, is given in Section 3.8.3.

LEMMA If $\mathbf{R}=\mathbf{I}_n-\mathbf{X}(\mathbf{X}'\mathbf{X})^{-1}\mathbf{X}'$, then $\mathbf{Z}'\mathbf{R}\mathbf{Z}$ is nonsingular.

Proof. Let $\mathbf{Z}'\mathbf{R}\mathbf{Z}\mathbf{a}=\mathbf{0}$, then by Theorem 3.1 (i)

$$\mathbf{a}'\mathbf{Z}'\mathbf{R}'\mathbf{R}\mathbf{Z}\mathbf{a}=\mathbf{a}'\mathbf{Z}'\mathbf{R}\mathbf{Z}\mathbf{a}=0,$$

or $\mathbf{RZa}=\mathbf{0}$. Hence $\mathbf{Za}=\mathbf{X}(\mathbf{X}'\mathbf{X})^{-1}\mathbf{X}'\mathbf{Za}=\mathbf{Xb}$, say, which implies that $\mathbf{a}=\mathbf{0}$ as the columns of $\mathbf{Z}$ are linearly independent of the columns of $\mathbf{X}$. Because $\mathbf{Z}'\mathbf{RZa}=\mathbf{0}$ implies that $\mathbf{a}=\mathbf{0}$, $\mathbf{Z}'\mathbf{RZ}$ has linearly independent columns and is therefore nonsingular. □

THEOREM 3.7 Let $\mathbf{R}_G=\mathbf{I}_n-\mathbf{W}(\mathbf{W}'\mathbf{W})^{-1}\mathbf{W}'$, $\mathbf{L}=(\mathbf{X}'\mathbf{X})^{-1}\mathbf{X}'\mathbf{Z}$, $\mathbf{M}=[\mathbf{Z}'\mathbf{RZ}]^{-1}$, and

$$\hat{\boldsymbol{\delta}}_G=\begin{pmatrix}\hat{\boldsymbol{\beta}}_G\\ \hat{\boldsymbol{\gamma}}_G\end{pmatrix}.$$

Then:

(i) $\hat{\boldsymbol{\beta}}_G=(\mathbf{X}'\mathbf{X})^{-1}\mathbf{X}'(\mathbf{Y}-\mathbf{Z}\hat{\boldsymbol{\gamma}}_G)=\hat{\boldsymbol{\beta}}-\mathbf{L}\hat{\boldsymbol{\gamma}}_G$.
(ii) $\hat{\boldsymbol{\gamma}}_G=(\mathbf{Z}'\mathbf{RZ})^{-1}\mathbf{Z}'\mathbf{RY}$,
(iii) $\mathbf{Y}'\mathbf{R}_G\mathbf{Y}=(\mathbf{Y}-\mathbf{Z}\hat{\boldsymbol{\gamma}}_G)'\mathbf{R}(\mathbf{Y}-\mathbf{Z}\hat{\boldsymbol{\gamma}}_G)$.
(iv) $\mathbf{Y}'\mathbf{R}_G\mathbf{Y}=\mathbf{Y}'\mathbf{RY}-\hat{\boldsymbol{\gamma}}_G'\mathbf{Z}'\mathbf{RY}$, and
(v)

$$\mathcal{D}\left[\hat{\boldsymbol{\delta}}_G\right]=\sigma^2\begin{pmatrix}(\mathbf{X}'\mathbf{X})^{-1}+\mathbf{LML}', & -\mathbf{LM}\\ -\mathbf{ML}', & \mathbf{M}\end{pmatrix}. \tag{3.28}$$

Proof. (i) Given $\mathbf{Y}=\mathbf{X}\boldsymbol{\beta}+\mathbf{Z}\boldsymbol{\gamma}+\boldsymbol{\varepsilon}$, then

$$\begin{aligned}\boldsymbol{\varepsilon}'\boldsymbol{\varepsilon}&=(\mathbf{Y}-\mathbf{X}\boldsymbol{\beta}-\mathbf{Z}\boldsymbol{\gamma})'(\mathbf{Y}-\mathbf{X}\boldsymbol{\beta}-\mathbf{Z}\boldsymbol{\gamma})\\ &=\mathbf{Y}'\mathbf{Y}-2\boldsymbol{\beta}'\mathbf{X}'\mathbf{Y}-2\boldsymbol{\gamma}'\mathbf{Z}'\mathbf{Y}+2\boldsymbol{\beta}'\mathbf{X}'\mathbf{Z}\boldsymbol{\gamma}\\ &\quad+\boldsymbol{\beta}'\mathbf{X}'\mathbf{X}\boldsymbol{\beta}+\boldsymbol{\gamma}'\mathbf{Z}'\mathbf{Z}\boldsymbol{\gamma},\end{aligned}$$

since $\boldsymbol{\beta}'\mathbf{X}'\mathbf{Z}\boldsymbol{\gamma}=(\boldsymbol{\beta}'\mathbf{X}'\mathbf{Z}\boldsymbol{\gamma})'=\boldsymbol{\gamma}'\mathbf{Z}'\mathbf{X}\boldsymbol{\beta}$, etc. To find $\hat{\boldsymbol{\beta}}_G$ and $\hat{\boldsymbol{\gamma}}_G$ we divide the normal equations $\partial\boldsymbol{\varepsilon}'\boldsymbol{\varepsilon}/\partial\boldsymbol{\delta}=\mathbf{0}$ into two parts and solve $\partial\boldsymbol{\varepsilon}'\boldsymbol{\varepsilon}/\partial\boldsymbol{\beta}=\mathbf{0}$ and $\partial\boldsymbol{\varepsilon}'\boldsymbol{\varepsilon}/\partial\boldsymbol{\gamma}=\mathbf{0}$. Thus using *A6* we have the following equations:

$$-2\mathbf{X}'\mathbf{Y}+2\mathbf{X}'\mathbf{Z}\hat{\boldsymbol{\gamma}}_G+2\mathbf{X}'\mathbf{X}\hat{\boldsymbol{\beta}}_G=\mathbf{0}, \tag{3.29}$$

$$-2\mathbf{Z}'\mathbf{Y}+2\mathbf{Z}'\mathbf{X}\hat{\boldsymbol{\beta}}_G+2\mathbf{Z}'\mathbf{Z}\hat{\boldsymbol{\gamma}}_G=\mathbf{0}. \tag{3.30}$$

From (3.29) we have

$$\hat{\boldsymbol{\beta}}_G=(\mathbf{X}'\mathbf{X})^{-1}\mathbf{X}'(\mathbf{Y}-\mathbf{Z}\hat{\boldsymbol{\gamma}}_G). \tag{3.31}$$

(ii) Substituting (3.31) in (3.30) leads to

$$\mathbf{Z}'\mathbf{Z}\hat{\boldsymbol{\gamma}}_G=\mathbf{Z}'\mathbf{Y}-\mathbf{Z}'\mathbf{X}(\mathbf{X}'\mathbf{X})^{-1}\mathbf{X}'(\mathbf{Y}-\mathbf{Z}\hat{\boldsymbol{\gamma}}_G)$$

so that

$$\mathbf{Z}'\left[\mathbf{I}_n-\mathbf{X}(\mathbf{X}'\mathbf{X})^{-1}\mathbf{X}'\right]\mathbf{Z}\hat{\gamma}_G=\mathbf{Z}'\left[\mathbf{I}_n-\mathbf{X}(\mathbf{X}'\mathbf{X})^{-1}\mathbf{X}'\right]\mathbf{Y}$$

or

$$\mathbf{Z}'\mathbf{R}\mathbf{Z}\hat{\gamma}_G=\mathbf{Z}'\mathbf{R}\mathbf{Y}. \tag{3.32}$$

Hence $\hat{\gamma}_G=(\mathbf{Z}'\mathbf{R}\mathbf{Z})^{-1}\mathbf{Z}'\mathbf{R}\mathbf{Y}$.

(iii)

$$\begin{aligned}\mathbf{Y}-\mathbf{X}\hat{\boldsymbol{\beta}}_G-\mathbf{Z}\hat{\gamma}_G&=\mathbf{Y}-\mathbf{X}(\mathbf{X}'\mathbf{X})^{-1}\mathbf{X}'(\mathbf{Y}-\mathbf{Z}\hat{\gamma}_G)-\mathbf{Z}\hat{\gamma}_G\\&=\left(\mathbf{I}_n-\mathbf{X}(\mathbf{X}'\mathbf{X})^{-1}\mathbf{X}'\right)(\mathbf{Y}-\mathbf{Z}\hat{\gamma}_G)\\&=\mathbf{R}(\mathbf{Y}-\mathbf{Z}\hat{\gamma}_G),\end{aligned}$$

so that

$$\begin{aligned}\mathbf{Y}'\mathbf{R}_G\mathbf{Y}&=(\mathbf{Y}-\mathbf{W}\hat{\boldsymbol{\delta}}_G)'(\mathbf{Y}-\mathbf{W}\hat{\boldsymbol{\delta}}_G)\\&=(\mathbf{Y}-\mathbf{X}\hat{\boldsymbol{\beta}}_G-\mathbf{Z}\hat{\gamma}_G)'(\mathbf{Y}-\mathbf{X}\hat{\boldsymbol{\beta}}_G-\mathbf{Z}\hat{\gamma}_G)\\&=(\mathbf{Y}-\mathbf{Z}\hat{\gamma}_G)'\mathbf{R}'\mathbf{R}(\mathbf{Y}-\mathbf{Z}\hat{\gamma}_G)\\&=(\mathbf{Y}-\mathbf{Z}\hat{\gamma}_G)'\mathbf{R}(\mathbf{Y}-\mathbf{Z}\hat{\gamma}_G),\end{aligned} \tag{3.33}$$

since $\mathbf{R}$ is symmetric and idempotent [Theorem 3.1 (i)].

(iv) From (iii) we have

$$\begin{aligned}\mathbf{Y}'\mathbf{R}_G\mathbf{Y}&=(\mathbf{Y}-\mathbf{Z}\hat{\gamma}_G)'\mathbf{R}(\mathbf{Y}-\mathbf{Z}\hat{\gamma}_G)\\&=\mathbf{Y}'\mathbf{R}\mathbf{Y}-2\hat{\gamma}_G'\mathbf{Z}'\mathbf{R}\mathbf{Y}+\hat{\gamma}_G'\mathbf{Z}'\mathbf{R}\mathbf{Z}\hat{\gamma}_G\\&=\mathbf{Y}'\mathbf{R}\mathbf{Y}-\hat{\gamma}_G'\mathbf{Z}'\mathbf{R}\mathbf{Y}-\hat{\gamma}_G'(\mathbf{Z}'\mathbf{R}\mathbf{Y}-\mathbf{Z}'\mathbf{R}\mathbf{Z}\hat{\gamma}_G)\\&=\mathbf{Y}'\mathbf{R}\mathbf{Y}-\hat{\gamma}_G'\mathbf{Z}'\mathbf{R}\mathbf{Y}\qquad\left[\text{by (3.32)}\right].\end{aligned}$$

(v)

$$\begin{aligned}\mathcal{D}\left[\hat{\gamma}_G\right]&=(\mathbf{Z}'\mathbf{R}\mathbf{Z})^{-1}\mathbf{Z}'\mathbf{R}\mathcal{D}\left[\mathbf{Y}\right]\mathbf{R}\mathbf{Z}(\mathbf{Z}'\mathbf{R}\mathbf{Z})^{-1}\\&=\sigma^2(\mathbf{Z}'\mathbf{R}\mathbf{Z})^{-1}(\mathbf{Z}'\mathbf{R}\mathbf{Z})(\mathbf{Z}'\mathbf{R}\mathbf{Z})^{-1}\\&=\sigma^2(\mathbf{Z}'\mathbf{R}\mathbf{Z})^{-1}=\sigma^2\mathbf{M}.\end{aligned}$$

Now, by Theorem 1.4,

$$\begin{aligned}
\mathcal{C}[\hat{\boldsymbol{\beta}},\hat{\boldsymbol{\gamma}}_G] &= \mathcal{C}\left[(\mathbf{X}'\mathbf{X})^{-1}\mathbf{X}'\mathbf{Y},(\mathbf{Z}'\mathbf{R}\mathbf{Z})^{-1}\mathbf{Z}'\mathbf{R}\mathbf{Y}\right] \\
&= \sigma^2(\mathbf{X}'\mathbf{X})^{-1}\mathbf{X}'\mathbf{R}\mathbf{Z}(\mathbf{Z}'\mathbf{R}\mathbf{Z})^{-1} \\
&= \mathbf{O}, \qquad (3.34)
\end{aligned}$$

since $\mathbf{X}'\mathbf{R}=\mathbf{O}$ [Theorem 3.1 (iii)]. Hence using (i) above we have, from Theorems 1.5 and 1.4,

$$\begin{aligned}
\mathcal{C}[\hat{\boldsymbol{\beta}}_G,\hat{\boldsymbol{\gamma}}_G] &= \mathcal{C}[\hat{\boldsymbol{\beta}}-\mathbf{L}\hat{\boldsymbol{\gamma}}_G,\hat{\boldsymbol{\gamma}}_G] \\
&= \mathcal{C}[\hat{\boldsymbol{\beta}},\hat{\boldsymbol{\gamma}}_G]-\mathbf{L}\mathcal{D}[\hat{\boldsymbol{\gamma}}_G] \\
&= -\sigma^2\mathbf{L}\mathbf{M} \qquad [\text{by } (3.34)]
\end{aligned}$$

and

$$\begin{aligned}
\mathcal{D}[\hat{\boldsymbol{\beta}}_G] &= \mathcal{D}[\hat{\boldsymbol{\beta}}-\mathbf{L}\hat{\boldsymbol{\gamma}}_G] \\
&= \mathcal{D}[\hat{\boldsymbol{\beta}}]-2\mathcal{C}[\hat{\boldsymbol{\beta}},\mathbf{L}\hat{\boldsymbol{\gamma}}_G]+\mathcal{D}[\mathbf{L}\hat{\boldsymbol{\gamma}}_G] \\
&= \mathcal{D}[\hat{\boldsymbol{\beta}}]-2\mathcal{C}[\hat{\boldsymbol{\beta}},\hat{\boldsymbol{\gamma}}_G]\mathbf{L}'+\mathbf{L}\mathcal{D}[\hat{\boldsymbol{\gamma}}_G]\mathbf{L}' \\
&= \sigma^2\left[(\mathbf{X}'\mathbf{X})^{-1}+\mathbf{L}\mathbf{M}\mathbf{L}'\right] \qquad [\text{by } (3.34)]. \qquad \square
\end{aligned}$$

From the above theorem we see that once $\mathbf{X}'\mathbf{X}$ has been inverted we can find $\hat{\boldsymbol{\delta}}_G$ and its variance–covariance matrix by simply inverting the $t\times t$ matrix $\mathbf{Z}'\mathbf{R}\mathbf{Z}$; we need not invert the $(t+p)\times(t+p)$ matrix $\mathbf{W}'\mathbf{W}$. The case $t=1$ is considered below.

3.7.2 *One Extra Variable*

Let the columns of $\mathbf{X}$ be denoted by $\mathbf{x}_j$ $(j=0,1,2,\ldots,p-1)$ so that

$$\begin{aligned}
\mathcal{E}[\mathbf{Y}] &= (\mathbf{x}_0,\mathbf{x}_1,\ldots,\mathbf{x}_{p-1})\boldsymbol{\beta} \\
&= \mathbf{x}_0\beta_0+\mathbf{x}_1\beta_1+\cdots+\mathbf{x}_{p-1}\beta_{p-1}.
\end{aligned}$$

Suppose now that we wish to introduce a further regressor, x_p, say, into the model so that in terms of the above notation we have $\mathbf{Z}\boldsymbol{\gamma}=\mathbf{x}_p\beta_p$. Then by Theorem 3.7 the least squares estimates for the enlarged model are readily calculated as $\mathbf{Z}'\mathbf{R}\mathbf{Z}$ $(=\mathbf{x}_p'\mathbf{R}\mathbf{x}_p)$ is only a 1×1 matrix, that is, a scalar. Hence

$$\hat{\beta}_{p,G}=\hat{\gamma}_G=(\mathbf{Z}'\mathbf{R}\mathbf{Z})^{-1}\mathbf{Z}'\mathbf{R}\mathbf{Y}=\frac{(\mathbf{x}_p'\mathbf{R}\mathbf{Y})}{(\mathbf{x}_p'\mathbf{R}\mathbf{x}_p)},$$

$$\hat{\boldsymbol{\beta}}_G=(\hat{\beta}_{0,G},\ldots,\hat{\beta}_{p-1,G})'=\hat{\boldsymbol{\beta}}-(\mathbf{X}'\mathbf{X})^{-1}\mathbf{X}'\mathbf{x}_p\hat{\beta}_{p,G},$$

$$\mathbf{Y}'\mathbf{R}_G\mathbf{Y}=\mathbf{Y}'\mathbf{R}\mathbf{Y}-\hat{\beta}_{p,G}\mathbf{x}_p'\mathbf{R}\mathbf{Y}, \tag{3.35}$$

and the matrix $\mathfrak{D}[\hat{\boldsymbol{\delta}}_G]$ is readily calculated from $(\mathbf{X}'\mathbf{X})^{-1}$. The ease with which "corrections" can be made to allow for a single additional x variable suggests that if more than one variable is to be added into the regression model, then the variables should be brought in one at a time. We return to this stepwise procedure in Chapter 12.

The above technique for introducing one extra variable was first discussed in detail by Cochran [1938] and generalized to the case of several variables by Quenouille [1950].

3.7.3 *Two-Step Least Squares*

The statements in Theorem 3.7 suggest the following sequence of steps for dealing with situations in which one wishes to "enlarge" the design matrix:

(1) Calculate $\hat{\boldsymbol{\beta}}=(\mathbf{X}'\mathbf{X})^{-1}\mathbf{X}'\mathbf{Y}$ and $\mathbf{Y}'\mathbf{R}\mathbf{Y}$ $(=\mathbf{Y}'\mathbf{Y}-\hat{\boldsymbol{\beta}}'\mathbf{X}'\mathbf{Y})$.

(2) To obtain $\hat{\gamma}_G$, replace $\mathbf{Y}$ by $\mathbf{Y}-\mathbf{Z}\boldsymbol{\gamma}$ in $\mathbf{Y}'\mathbf{R}\mathbf{Y}$ and minimize with respect to $\boldsymbol{\gamma}$. Thus

$$\begin{aligned} r&=(\mathbf{Y}-\mathbf{Z}\boldsymbol{\gamma})'\mathbf{R}(\mathbf{Y}-\mathbf{Z}\boldsymbol{\gamma})\\ &=\mathbf{Y}'\mathbf{R}\mathbf{Y}-2\boldsymbol{\gamma}'\mathbf{Z}'\mathbf{R}\mathbf{Y}+\boldsymbol{\gamma}'\mathbf{Z}'\mathbf{R}\mathbf{Z}\boldsymbol{\gamma},\end{aligned}$$

and setting $\partial r/\partial\boldsymbol{\gamma}=\mathbf{0}$, we have

$$-2\mathbf{Z}'\mathbf{R}\mathbf{Y}+2\mathbf{Z}'\mathbf{R}\mathbf{Z}\boldsymbol{\gamma}=\mathbf{0} \tag{3.36}$$

or

$$\hat{\boldsymbol{\gamma}}_G=(\mathbf{Z}'\mathbf{R}\mathbf{Z})^{-1}\mathbf{Z}'\mathbf{R}\mathbf{Y},$$

the required least squares estimate of $\boldsymbol{\gamma}$.

(3) The residual sum of squares $\mathbf{Y}'\mathbf{R}_G\mathbf{Y}$ for the enlarged model is the above minimum value of r, namely, $(\mathbf{Y}-\mathbf{Z}\hat{\boldsymbol{\gamma}}_G)'\mathbf{R}(\mathbf{Y}-\mathbf{Z}\hat{\boldsymbol{\gamma}}_G)$ [by Theorem 3.7 (iii)].

(4) To obtain $\hat{\boldsymbol{\beta}}_G$, replace $\mathbf{Y}$ by $\mathbf{Y}-\mathbf{Z}\hat{\boldsymbol{\gamma}}_G$ in $\hat{\boldsymbol{\beta}}$; thus

$$\hat{\boldsymbol{\beta}}_G=(\mathbf{X}'\mathbf{X})^{-1}\mathbf{X}'(\mathbf{Y}-\mathbf{Z}\hat{\boldsymbol{\gamma}}_G).$$

(5) The coefficient of 2γ in the equation $\partial r/\partial\boldsymbol{\gamma}=\mathbf{0}$ [Equation (3.36)], namely, $\mathbf{Z}'\mathbf{R}\mathbf{Z}$, enables one to readily obtain $\mathscr{D}[\hat{\gamma}_G]=\sigma^2(\mathbf{Z}'\mathbf{R}\mathbf{Z})^{-1}$.

We call the above procedure the method of *two-step least squares*. This method is widely used in Chapter 10 in the study of analysis of covariance models.

It is of interest to note that the two-stage method of fitting the augmented model is equivalent to fitting the "orthogonalized" model $\mathscr{E}[\mathbf{Y}]=[\mathbf{X},\mathbf{RZ}]\boldsymbol{\lambda}$, which has the orthogonal structure described in Section 3.5 (since $\mathbf{X}'\mathbf{R}=\mathbf{O}$). To see this we use Theorem 3.7 (i) and obtain

$$\begin{aligned}\hat{\mathbf{Y}}_G&=\mathbf{W}\hat{\boldsymbol{\delta}}_G\\&=\mathbf{X}\hat{\boldsymbol{\beta}}_G+\mathbf{Z}\hat{\boldsymbol{\gamma}}_G\\&=\mathbf{X}\big(\hat{\boldsymbol{\beta}}-(\mathbf{X}'\mathbf{X})^{-1}\mathbf{X}'\mathbf{Z}\hat{\boldsymbol{\gamma}}_G\big)+\mathbf{Z}\hat{\boldsymbol{\gamma}}_G\\&=\mathbf{X}\hat{\boldsymbol{\beta}}+\mathbf{RZ}\hat{\boldsymbol{\gamma}}_G &(3.37)\\&=(\mathbf{X},\mathbf{RZ})\hat{\boldsymbol{\lambda}}_G, &(3.38)\end{aligned}$$

say. Here $\hat{\boldsymbol{\gamma}}_G$ is the solution of $\mathbf{Z}'\mathbf{RZ}\boldsymbol{\gamma}=\mathbf{Z}'\mathbf{RY}$, that is, of $(\mathbf{RZ})'(\mathbf{RZ})\boldsymbol{\gamma}=(\mathbf{RZ})'\mathbf{Y}$ (since $\mathbf{R}$ is symmetric and idempotent) so that $\hat{\boldsymbol{\lambda}}_G$ is the least squares estimate of $\boldsymbol{\lambda}$. The above idea is raised again in Section 3.8.3, [Equation (3.53)].

3.7.4 Two-Step Residuals

The residuals for the augmented model are given by

$$\begin{aligned}\mathbf{R}_G\mathbf{Y}&=\mathbf{Y}-\mathbf{W}\hat{\boldsymbol{\delta}}_G\\&=\mathbf{Y}-\mathbf{X}\hat{\boldsymbol{\beta}}-\mathbf{RZ}\hat{\boldsymbol{\gamma}}_G \qquad [\text{by } (3.37)]\\&=\mathbf{RY}-\mathbf{RZ}\hat{\boldsymbol{\gamma}}_G\\&=\mathbf{R}(\mathbf{RY}-\mathbf{Z}\hat{\boldsymbol{\gamma}}_G)\\&=\mathbf{R}\big(\mathbf{RY}-\mathbf{Z}(\mathbf{Z}'\mathbf{RZ})^{-1}\mathbf{Z}'\mathbf{RY}\big) &(3.39)\\&=\mathbf{RSRY},\end{aligned}$$

where $\mathbf{S}=\mathbf{I}_t-\mathbf{Z}(\mathbf{Z}'\mathbf{RZ})^{-1}\mathbf{Z}'$.

The above algebra forms the basis of a recursive algorithm, given by

Wilkinson [1970] (see also James and Wilkinson [1971], Rogers and Wilkinson [1974], and Pearce et al. [1974]) for fitting analysis of variance models by regression methods. The basic steps of the algorithm are as follows:

(1) Compute the residuals **RY**.
(2) Use the operator **S**, which Wilkinson calls "sweep" (not to be confused with the sweep method of Section 12.2.2), to produce a vector of "apparent residuals" $\mathbf{RY}-\mathbf{Z}\hat{\gamma}_G$ $(=\mathbf{SRY})$.
(3) Applying the operator **R** once again, reanalyze the apparent residuals to produce the correct residuals **RSRY**.

If the columns of **Z** are perpendicular to the columns of **X** then $\mathbf{RZ}=\mathbf{Z}$, $\mathbf{RSR}=\mathbf{SR}$ [by Equation (3.39)], and step (3) is unnecessary. We see later (Section 3.8.3) that the above procedure can still be used when the design matrix **X** does not have full rank.

By setting **X** equal to the first k columns of **X**, and **Z** equal to the $(k+1)$th column $(k=1,2,\ldots,p-1)$, this algorithm can be used to fit the regression one column of **X** at a time. Such a stepwise procedure is appropriate in experimental design situations because the columns of **X** then correspond to different components of the model such as the grand mean, main effects, block effects, and interactions, and some of the columns are usually orthogonal. Also the elements of the design matrix **X** are 0 or 1 so that in many of the standard designs the sweep operator **S** amounts to a simple operation like subtracting means, or a multiple of the means, from the residuals.

EXERCISES 3g

1. Prove that

$$\mathbf{Y}'\mathbf{RY}-\mathbf{Y}'\mathbf{R}_G\mathbf{Y}=\sigma^2\hat{\gamma}_G'\left(\mathcal{D}\left[\hat{\gamma}_G\right]\right)^{-1}\hat{\gamma}_G.$$

2. If $\hat{\boldsymbol{\beta}}_G=[(\hat{\beta}_{G,i})]$ and $\hat{\boldsymbol{\beta}}=[(\hat{\beta}_i)]$, prove that

$$\operatorname{var}\left[\hat{\beta}_{G,i}\right]\geqslant\operatorname{var}\left[\hat{\beta}_i\right].$$

3. Given that $Y_1,Y_2,\ldots,Y_n$ are independently distributed as $N(\theta,\sigma^2)$ find the least squares estimate of θ. Use the method of two-step least squares to find the least squares estimates and the residual sum of squares for

the augmented model

$$Y_i = \theta + \gamma x_i + \varepsilon_i \qquad (i = 1, 2, \ldots, n),$$

where the ε_i are independently distributed as $N(0, \sigma^2)$.

3.8 DESIGN MATRIX OF LESS THAN FULL RANK

3.8.1 Least Squares Estimation

When the techniques of regression analysis are used for analyzing data from experimental designs we find that the elements of **X** are 0 or 1 (Chapter 10), and the columns of **X** are usually linearly dependent. For example, consider the randomized block design

$$Y_{ij} = \mu + \alpha_i + \tau_j + \varepsilon_{ij} \qquad (i = 1, 2, \ldots, I;\ j = 1, 2, \ldots, J),$$

where Y_{ij} is the response from the ith treatment in the jth block. Then

$$\left[\begin{array}{c} Y_{11}\\ Y_{12}\\ \cdots\\ Y_{1J}\\ \hline Y_{21}\\ Y_{22}\\ \cdots\\ Y_{2J}\\ \hline \cdots\\ \hline Y_{I1}\\ Y_{I2}\\ \cdots\\ Y_{IJ} \end{array}\right] = \left[\begin{array}{cccccccccc} 1 & 1 & 0 & \cdots & 0 & 1 & 0 & \cdots & 0\\ 1 & 1 & 0 & \cdots & 0 & 0 & 1 & \cdots & 0\\ \cdots & \cdots & \cdots & \cdots & \cdots & \cdots & \cdots & \cdots & \cdots\\ 1 & 1 & 0 & \cdots & 0 & 0 & 0 & \cdots & 1\\ \hline 1 & 0 & 1 & \cdots & 0 & 1 & 0 & \cdots & 0\\ 1 & 0 & 1 & \cdots & 0 & 0 & 1 & \cdots & 0\\ \cdots & \cdots & \cdots & \cdots & \cdots & \cdots & \cdots & \cdots & \cdots\\ 1 & 0 & 1 & \cdots & 0 & 0 & 0 & \cdots & 1\\ \hline \cdots & \cdots & \cdots & \cdots & \cdots & \cdots & \cdots & \cdots & \cdots\\ \hline 1 & 0 & 0 & \cdots & 1 & 1 & 0 & \cdots & 0\\ 1 & 0 & 0 & \cdots & 1 & 0 & 1 & \cdots & 0\\ \cdots & \cdots & \cdots & \cdots & \cdots & \cdots & \cdots & \cdots & \cdots\\ 1 & 0 & 0 & \cdots & 1 & 0 & 0 & \cdots & 1 \end{array}\right] \left[\begin{array}{c} \mu\\ \alpha_1\\ \alpha_2\\ \cdots\\ \alpha_I\\ \tau_1\\ \tau_2\\ \cdots\\ \tau_J \end{array}\right] + \left[\begin{array}{c} \varepsilon_{11}\\ \varepsilon_{12}\\ \cdots\\ \varepsilon_{1J}\\ \hline \varepsilon_{21}\\ \varepsilon_{22}\\ \cdots\\ \varepsilon_{2J}\\ \hline \cdots\\ \hline \varepsilon_{I1}\\ \varepsilon_{I2}\\ \cdots\\ \varepsilon_{IJ} \end{array}\right] \qquad (3.40)$$

or $\mathbf{Y} = \mathbf{X}\boldsymbol{\beta} + \boldsymbol{\varepsilon}$, where, for example, the first column of **X** is linearly dependent on the other columns.

In Section 3.1 we developed a least squares theory which applies whether **X** has full rank or not. We now summarize our findings for the less than full rank case in the following theorem.

THEOREM 3.8 Let $\mathbf{Y}=\boldsymbol{\theta}+\boldsymbol{\varepsilon}$, where $\boldsymbol{\theta}=\mathbf{X}\boldsymbol{\beta}$ and $\mathbf{X}$ is an $n\times p$ matrix of rank r $(r<p)$. Then

(i) subject to $\boldsymbol{\theta}\in\mathcal{R}[\mathbf{X}]$, $\boldsymbol{\varepsilon}'\boldsymbol{\varepsilon}$ is minimized when $\boldsymbol{\beta}=\hat{\boldsymbol{\beta}}$, where $\hat{\boldsymbol{\beta}}$ is any solution of the normal equations $\mathbf{X}'\mathbf{X}\boldsymbol{\beta}=\mathbf{X}'\mathbf{Y}$; and
(ii) $\mathbf{Y}'\mathbf{Y}-\hat{\boldsymbol{\beta}}'\mathbf{X}'\mathbf{Y}$ is unique for all nonzero $\mathbf{Y}$.

Proof. (i) The normal equations always have a solution for $\boldsymbol{\beta}$ as $\mathcal{R}[\mathbf{X}']=\mathcal{R}[\mathbf{X}'\mathbf{X}]$ (*A2.5*). However, the solution is no longer unique as $\mathbf{X}'\mathbf{X}$ is of rank r and is therefore singular. If $\hat{\boldsymbol{\beta}}$ is any solution of the normal equations then, setting $\hat{\boldsymbol{\theta}}=\mathbf{X}\hat{\boldsymbol{\beta}}$, we have $\mathbf{X}'(\mathbf{Y}-\hat{\boldsymbol{\theta}})=\mathbf{0}$. Hence from Equation (3.4) in Section 3.1, $\hat{\boldsymbol{\theta}}$ is the *unique* orthogonal projection of $\mathbf{Y}$ onto Ω, and the minimum value of $\boldsymbol{\varepsilon}'\boldsymbol{\varepsilon}$ is $\text{RSS}=\|\mathbf{Y}-\hat{\boldsymbol{\theta}}\|^2$.

(ii) $\hat{\boldsymbol{\theta}}'(\mathbf{Y}-\hat{\boldsymbol{\theta}})=0$ so that

$$\mathbf{Y}'\mathbf{Y}-\hat{\boldsymbol{\beta}}'\mathbf{X}'\mathbf{Y}=\mathbf{Y}'\mathbf{Y}-\hat{\boldsymbol{\theta}}'\mathbf{Y}=(\mathbf{Y}-\hat{\boldsymbol{\theta}})'(\mathbf{Y}-\hat{\boldsymbol{\theta}})=\text{RSS},$$

which is unique. □

From the discussion following Theorem 3.1 we have that $\hat{\boldsymbol{\theta}}=\mathbf{P}\mathbf{Y}$ where $\mathbf{P}$ is symmetric and idempotent of rank r. Hence, repeating the argument of Theorem 3.3 in Section 3.3 we have that $\text{RSS}=\mathbf{Y}'(\mathbf{I}_n-\mathbf{P})\mathbf{Y}$ is an unbiased estimate of $(n-r)\sigma^2$. Also if $\mathbf{Y}$ is normal then, from Theorem 3.5 (iv) (Section 3.4), $\text{RSS}/\sigma^2\sim\chi^2_{n-r}$.

We note that $\hat{\boldsymbol{\beta}}$ should be regarded as a solution of the normal equations rather than as an estimate of $\boldsymbol{\beta}$. Since $\mathbf{X}$ has less than full rank then, intuitively, $\boldsymbol{\beta}$ cannot be estimated as the representation $\boldsymbol{\theta}=\mathbf{X}\boldsymbol{\beta}$ is not unique for given $\boldsymbol{\theta}\in\mathcal{R}[X]$. However, we see in Section 3.8.2 that certain linear combinations $\mathbf{a}'\boldsymbol{\beta}$ can be estimated for appropriate vectors $\mathbf{a}$.

Three methods for finding a solution $\hat{\boldsymbol{\beta}}$, or finding RSS directly, are now considered.

a REDUCE THE MODEL TO ONE OF FULL RANK

An obvious method for finding RSS is to reduce the regression model to one of full rank. Thus if $\mathbf{X}_1$ is the $n\times r$ matrix of linearly independent columns of $\mathbf{X}$, then $\mathbf{P}=\mathbf{X}_1(\mathbf{X}_1'\mathbf{X}_1)^{-1}\mathbf{X}_1'$ and

$$\text{RSS}=\mathbf{Y}'\mathbf{Y}-\hat{\boldsymbol{\theta}}'\mathbf{Y}=\mathbf{Y}'\mathbf{Y}-\hat{\boldsymbol{\alpha}}'\mathbf{X}_1'\mathbf{Y},$$

where $\hat{\boldsymbol{\alpha}}=(\mathbf{X}_1'\mathbf{X}_1)^{-1}\mathbf{X}_1'\mathbf{Y}$.

We may assume, without loss of generality, that $\mathbf{X}_1$ is the *first* r columns of $\mathbf{X}$ so that $\mathbf{X}=(\mathbf{X}_1,\mathbf{X}_2)$, say. Then $\mathbf{X}_2=\mathbf{X}_1\mathbf{F}$, since the columns of $\mathbf{X}_2$ are linearly dependent on the columns of $\mathbf{X}_1$, and $\mathbf{X}=\mathbf{X}_1(\mathbf{I}_r,\mathbf{F})$. This is a special case of a more general factorization

$$\mathbf{X}=\mathbf{K}\mathbf{L}, \tag{3.41}$$

where $\mathbf{K}$ is $n\times r$ of rank r, and $\mathbf{L}$ is $r\times p$ of rank r. We then write $\mathbf{X}\boldsymbol{\beta}=\mathbf{K}\mathbf{L}\boldsymbol{\beta}=\mathbf{K}\boldsymbol{\alpha}$ and work with $\boldsymbol{\alpha}$.

This reparametrization of the original model forms the basis of a computer program given by Bock [1963, 1965] for analysis of variance (ANOVA) computations; Fowlkes [1969] also uses this reparametrization (called CODE in his article). Bock was one of the first to implement the regression method for ANOVA on a large computer. Other advantages of reparametrization are demonstrated by Johnson [1971].

b IMPOSE IDENTIFIABILITY CONSTRAINTS

This method consists of imposing a set of constraints $\mathbf{H}\boldsymbol{\beta}=\mathbf{0}$ to take up the slack in $\boldsymbol{\beta}$. In particular, we require suitable constraints (called identifiability constraints) so that for *every* $\boldsymbol{\theta}\in\mathcal{R}[\mathbf{X}]$ (i.e., $\mathcal{R}[\mathbf{X}]$ is *not* restricted) there exists a unique $\boldsymbol{\beta}$ satisfying $\boldsymbol{\theta}=\mathbf{X}\boldsymbol{\beta}$ and $\mathbf{0}=\mathbf{H}\boldsymbol{\beta}$, that is, satisfying

$$\begin{pmatrix}\boldsymbol{\theta}\\ \mathbf{0}\end{pmatrix}=\begin{pmatrix}\mathbf{X}\\ \mathbf{H}\end{pmatrix}\boldsymbol{\beta}=\mathbf{G}\boldsymbol{\beta}, \tag{3.42}$$

say. The solution is simple. We choose for the rows of $\mathbf{H}$ a set of $p-r$ linearly independent $p\times 1$ vectors which are linearly independent of the rows of $\mathbf{X}$. Then the $(n+p-r)\times p$ matrix $\mathbf{G}$ has rank p so that $\mathbf{G}'\mathbf{G}=\mathbf{X}'\mathbf{X}+\mathbf{H}'\mathbf{H}$ is $p\times p$ of rank p (*A2.4*) and therefore has an inverse. We have thus effectively made up for the deficiency in the rank of $\mathbf{X}'\mathbf{X}$ by introducing $\mathbf{H}'\mathbf{H}$. Hence adding $\mathbf{H}'\mathbf{H}\hat{\boldsymbol{\beta}}=\mathbf{0}$ to the normal equations gives us $\mathbf{G}'\mathbf{G}\hat{\boldsymbol{\beta}}=\mathbf{X}'\mathbf{Y}$ or $\hat{\boldsymbol{\beta}}=(\mathbf{G}'\mathbf{G})^{-1}\mathbf{X}'\mathbf{Y}$. It is noted, from $\hat{\boldsymbol{\theta}}=\mathbf{X}\hat{\boldsymbol{\beta}}=\mathbf{P}\mathbf{Y}$, that $\mathbf{P}=\mathbf{X}(\mathbf{G}'\mathbf{G})^{-1}\mathbf{X}'$ (since $\mathbf{P}$ is unique).

We now give proofs of the above statements. However, the reader may wish to skip these details at a first reading.

THEOREM 3.9 (Scheffé [1959: p. 17]) If $\mathbf{H}$ is an $s\times p$ matrix, then the constraints $\mathbf{H}\boldsymbol{\beta}=\mathbf{0}$ are identifiability constraints if and only if:

(i) $\mathcal{R}[\mathbf{X}']\cap\mathcal{R}[\mathbf{H}']=\mathbf{0}$ (that is, the rows of $\mathbf{X}$ are linearly independent of the rows of $\mathbf{H}$), and
(ii) the columns of G are linearly independent.

Proof. We show that (i) is necessary and sufficient for the existence of a $\boldsymbol{\beta}$ satisfying (3.42) for every $\boldsymbol{\theta}\in\mathcal{R}[\mathbf{X}]$; (ii) is necessary and sufficient for this $\boldsymbol{\beta}$ to be unique.

Considering (i) first, we see that $\boldsymbol{\beta}$ exists if and only if

$$\boldsymbol{\phi}=\begin{pmatrix}\boldsymbol{\theta}\\ \mathbf{0}\end{pmatrix}\in\mathcal{R}[\mathbf{G}]\quad\text{for every } \boldsymbol{\theta}\in\mathcal{R}[\mathbf{X}].$$

This statement is equivalent to saying that every vector perpendicular to $\mathcal{R}[\mathbf{G}]$ is perpendicular to $\boldsymbol{\phi}$ for every $\boldsymbol{\theta} \in \mathcal{R}[\mathbf{X}]$. Let $\mathbf{u}' = (\mathbf{u}_x', \mathbf{u}_y')$ be any $s+n$ dimensional vector; then we have the following equivalent statements:

$$\mathbf{G}'\mathbf{u} = \mathbf{0} \quad \text{implies that} \quad \boldsymbol{\phi}'\mathbf{u} = 0 \quad \text{for every} \quad \boldsymbol{\theta} \in \mathcal{R}[\mathbf{X}];$$

or $\mathbf{X}'\mathbf{u}_X + \mathbf{H}'\mathbf{u}_Y = \mathbf{0}$ implies that $\boldsymbol{\theta}'\mathbf{u}_X = 0$ for every $\boldsymbol{\theta} \in \mathcal{R}[\mathbf{X}]$; or $\mathbf{X}'\mathbf{u}_X + \mathbf{H}'\mathbf{u}_Y = \mathbf{0}$ implies that $\mathbf{X}'\mathbf{u}_X = \mathbf{0}$ (and therefore $\mathbf{H}'\mathbf{u}_Y = \mathbf{0}$ also). Thus $\boldsymbol{\beta}$ exists if and only if no linear combination of the rows of $\mathbf{X}$ is a linear combination of the rows of $\mathbf{H}$ except $\mathbf{0}$, that is, if and only if $\mathcal{R}[\mathbf{X}'] \cap \mathcal{R}[\mathbf{H}'] = \mathbf{0}$.

Now (ii). Given that a $\boldsymbol{\beta}$ exists for every $\boldsymbol{\phi}$ such that $\boldsymbol{\phi} = \mathbf{G}\boldsymbol{\beta}$, then $\boldsymbol{\beta}$ is unique if and only if the columns of $\mathbf{G}$ are linearly independent. [If the columns of $\mathbf{G}$ are linearly dependent then $\mathbf{G}\boldsymbol{\gamma} = \mathbf{0}$ for some $\boldsymbol{\gamma} \neq \mathbf{0}$, $\boldsymbol{\phi} = \mathbf{G}(\boldsymbol{\beta} - \boldsymbol{\gamma})$, and both $\boldsymbol{\beta}$ and $\boldsymbol{\beta} - \boldsymbol{\gamma}$ satisfy the equation $\mathbf{G}\mathbf{z} = \boldsymbol{\phi}$.] □

COROLLARY From the above theorem we see that if $\mathbf{X}$ is $n \times p$ of rank r, and $\mathbf{H}$ is $s \times p$, then conditions (i) and (ii) are jointly equivalent to (1) rank $\mathbf{G} = p$ and (2) rank $\mathbf{H} = p - r$ [since, by (i), the p independent rows of $\mathbf{G}$ must be made up of r rows from $\mathbf{X}$ and $p - r$ rows from $\mathbf{H}$]. If there are no redundant equations in the set $\mathbf{H}\boldsymbol{\beta} = \mathbf{0}$, that is, the rows of $\mathbf{H}$ are linearly independent, then $s = p - r$.

Since $\hat{\boldsymbol{\theta}} \in \mathcal{R}[\mathbf{X}]$ the above theorem implies that, for suitable $\mathbf{H}$, there exists a unique $\hat{\boldsymbol{\beta}}$ satisfying (3.42); that is, $\mathbf{H}\hat{\boldsymbol{\beta}} = \mathbf{0}$ and $\hat{\boldsymbol{\theta}} = \mathbf{X}\hat{\boldsymbol{\beta}}$ [or, from (3.4), $\mathbf{X}'\mathbf{X}\hat{\boldsymbol{\beta}} = \mathbf{X}'\mathbf{Y}$]. Hence $\hat{\boldsymbol{\beta}} = (\mathbf{G}'\mathbf{G})^{-1}\mathbf{X}'\mathbf{Y}$, as stated above.

It is of interest to note that, subject to $\mathbf{H}\boldsymbol{\beta} = \mathbf{0}$,

$$\begin{aligned} \mathcal{E}[\hat{\boldsymbol{\beta}}] &= (\mathbf{G}'\mathbf{G})^{-1}\mathbf{X}'\mathbf{X}\boldsymbol{\beta} \\ &= (\mathbf{G}'\mathbf{G})^{-1}(\mathbf{X}'\mathbf{X} + \mathbf{H}'\mathbf{H})\boldsymbol{\beta} \\ &= \boldsymbol{\beta}, \end{aligned}$$

and $\hat{\boldsymbol{\beta}}$ is an unbiased estimate of $\boldsymbol{\beta}$. Also if we minimize $\|\mathbf{Y} - \mathbf{X}\boldsymbol{\beta}\|^2$ subject to $\mathbf{H}\boldsymbol{\beta} = \mathbf{0}$ then, from Equation (3.57) in Section 3.9.1, we see that the Lagrange multiplier term $\mathbf{H}'\boldsymbol{\lambda}$ is zero. If $s = p - r$, so that the columns of $\mathbf{H}'$ are linearly independent, then $\boldsymbol{\lambda} = \mathbf{0}$.

The above method using identifiability constraints is particularly useful in analysis of variance models (Chapter 9) because $\mathbf{H}$ is usually readily found. Also the unbiasedness of $\hat{\boldsymbol{\beta}}$ and $\boldsymbol{\lambda} = \mathbf{0}$ are very useful theoretical features. An algorithm for carrying out the computations is described in Section 11.5.4.

c COMPUTE A GENERALIZED INVERSE

If $\mathbf{C}$ is any generalized inverse of $(\mathbf{X}'\mathbf{X})$, then $\hat{\boldsymbol{\beta}}=\mathbf{C}\mathbf{X}'\mathbf{Y}$ is a solution of the normal equations, and $\mathbf{P}=\mathbf{X}\mathbf{C}\mathbf{X}'$.

A *generalized inverse* of an $m\times n$ matrix $\mathbf{B}$ is defined to be any $n\times m$ matrix $\mathbf{B}^-$ that satisfies the condition

(a) $\mathbf{B}\mathbf{B}^-\mathbf{B}=\mathbf{B}$.

Such a matrix always exists (Searle [1971: Chapter 1]). The name "generalized inverse" for $\mathbf{B}^-$ defined by (a) is not universally accepted, although it is used fairly widely (e.g., Rao [1973], Rao and Mitra [1971a, b], Pringle and Rayner [1971], Searle [1971], Kruskal [1975]). Other names such as "conditional inverse," "pseudo inverse," "*g*-inverse," and "*p*-inverse" are also found in the literature, sometimes for $\mathbf{B}^-$ defined above and sometimes for matrices defined as variants of $\mathbf{B}^-$. For example, Graybill [1969] calls $\mathbf{A}^-$ a conditional inverse, and calls $\mathbf{A}^+$, defined below, the generalized inverse.

It should be noted that $\mathbf{B}^-$ is called "a" generalized inverse and not "the" generalized inverse, for $\mathbf{B}^-$ is not unique. Also, taking the transpose of (a), we have

$$\mathbf{B}'(\mathbf{B}^-)'\mathbf{B}'=\mathbf{B}' \tag{3.43}$$

so that $\mathbf{B}^{-\prime}$ is a generalized inverse of $\mathbf{B}'$; we can therefore write

$$(\mathbf{B}^-)'=(\mathbf{B}')^- \tag{3.44}$$

for some $(\mathbf{B}')^-$.

If $\mathbf{B}^-$ also satisfies three more conditions, namely,

(b) $\mathbf{B}^-\mathbf{B}\mathbf{B}^-=\mathbf{B}^-$,
(c) $(\mathbf{B}\mathbf{B}^-)'=\mathbf{B}\mathbf{B}^-$,
(d) $(\mathbf{B}^-\mathbf{B})'=\mathbf{B}^-\mathbf{B}$,

then $\mathbf{B}^-$ is unique and it is called the Moore–Penrose inverse (Albert [1972]); some authors call it the pseudo inverse of the *p*-inverse. We denote this inverse by $\mathbf{B}^+$.

Now setting $\mathbf{B}=\mathbf{X}'\mathbf{X}$ and $\mathbf{c}=\mathbf{X}'\mathbf{Y}$, we have

$$\begin{aligned}\mathbf{c}&=\mathbf{B}\boldsymbol{\beta}\\&=\mathbf{B}\mathbf{B}^-\mathbf{B}\boldsymbol{\beta}\\&=\mathbf{B}(\mathbf{B}^-\mathbf{c})\end{aligned} \tag{3.45}$$

and $\mathbf{B}^-\mathbf{c}$ is a solution of $\mathbf{B}\boldsymbol{\beta}=\mathbf{c}$. (In fact, by *A8*, every solution of $\mathbf{B}\boldsymbol{\beta}=\mathbf{c}$ can be expressed in the form $\mathbf{B}^-\mathbf{c}$ for some $\mathbf{B}^-$.) There are several ways of computing a suitable $\mathbf{B}^-$ for the symmetric matrix $\mathbf{B}$. One method is as follows:

(1) Delete $p-r$ rows and the corresponding columns so as to leave an $r\times r$ matrix that is nonsingular; this can always be done as $\text{rank}[\mathbf{X}'\mathbf{X}]=\text{rank}\,\mathbf{X}=r$.
(2) Invert the $r\times r$ matrix.
(3) Obtain $\mathbf{B}^-$ by inserting zeros into the inverse to correspond to the rows and columns originally deleted. For example, if

$$\mathbf{B}=\begin{pmatrix}\mathbf{B}_{11} & \mathbf{B}_{12}\\ \mathbf{B}_{21} & \mathbf{B}_{22}\end{pmatrix},$$

and $\mathbf{B}_{11}$ is an $r\times r$ nonsingular matrix, then

$$\mathbf{B}^-=\begin{pmatrix}\mathbf{B}_{11}^{-1} & \mathbf{O}\\ \mathbf{O} & \mathbf{O}\end{pmatrix}.$$

Another method for finding $\mathbf{B}^-$ is to use identifiability constraints. If $\mathbf{H}\boldsymbol{\beta}=\mathbf{0}$ are sufficient for identifiability and $s=p-r$, then it can be shown (cf. Exercises 7 and 8 of Exercises 3h below) that $\mathbf{C}_{11}$ in

$$\begin{pmatrix}\mathbf{X}'\mathbf{X} & \mathbf{H}'\\ \mathbf{H} & \mathbf{O}\end{pmatrix}^{-1}=\begin{pmatrix}\mathbf{C}_{11}, & \mathbf{C}_{12}'\\ \mathbf{C}_{21}, & \mathbf{C}_{22}=\mathbf{O}\end{pmatrix}, \tag{3.46}$$

and $(\mathbf{G}'\mathbf{G})^{-1}=(\mathbf{X}'\mathbf{X}+\mathbf{H}'\mathbf{H})^{-1}$ are also generalized inverses of $\mathbf{B}$.

We saw above that

$$\begin{aligned}\hat{\boldsymbol{\beta}}&=\mathbf{B}^-\mathbf{c}\\ &=(\mathbf{X}'\mathbf{X})^-\mathbf{X}'\mathbf{Y}\\ &=\mathbf{X}^*\mathbf{Y},\end{aligned}$$

say, is a solution of the normal equations. Because

$$\begin{aligned}(\mathbf{X}^+)'\mathbf{X}'\mathbf{X}&=(\mathbf{X}\mathbf{X}^+)'\mathbf{X}\\ &=\mathbf{X}\mathbf{X}^+\mathbf{X} \qquad [\text{condition (c)}]\\ &=\mathbf{X} \qquad [\text{condition (a)}]\end{aligned}$$

we can multiply $(\mathbf{X}'\mathbf{X})(\mathbf{X}'\mathbf{X})^-(\mathbf{X}'\mathbf{X})=\mathbf{X}'\mathbf{X}$ on the left by $(\mathbf{X}^+)'$ and obtain

$$\mathbf{X}[(\mathbf{X}'\mathbf{X})^-\mathbf{X}']\mathbf{X}=\mathbf{X}. \tag{3.47}$$

Thus $\mathbf{X}^*$, the matrix in the square brackets, is a generalized inverse of $\mathbf{X}$ as it satisfies condition (a); using similar arguments, we find that it also satisfies (b) and (c). In fact, a generalized inverse of $\mathbf{X}$ satisfies (a), (b), and (c) if and only if it can be expressed in the form $(\mathbf{X}'\mathbf{X})^-\mathbf{X}'$ (Pringle and Rayner [1971: p. 26]). However, any $\mathbf{X}^-$ satisfying just (a) and (c) will do the trick:

$$\begin{aligned}\mathbf{X}'\mathbf{X}(\mathbf{X}^-\mathbf{Y}) &= \mathbf{X}'(\mathbf{X}\mathbf{X}^-)\mathbf{Y} \\ &= \mathbf{X}'(\mathbf{X}\mathbf{X}^-)'\mathbf{Y} \qquad [\text{by (c)}] \\ &= \mathbf{X}'(\mathbf{X}^-)'\mathbf{X}'\mathbf{Y} \\ &= \mathbf{X}'\mathbf{Y} \qquad [\text{by (a) transposed}],\end{aligned}$$

and $\mathbf{X}^-\mathbf{Y}$ is a solution of the normal equations. In particular, $\mathbf{X}^+\mathbf{Y}$ is the unique solution which minimizes $\hat{\boldsymbol{\beta}}'\hat{\boldsymbol{\beta}}$ (Peters and Wilkinson [1970]). Numerical methods for finding $(\mathbf{X}'\mathbf{X})^-$, $\mathbf{X}^*$, and $\mathbf{X}^+$ are described in Chapter 11 (Sections 11.5.1, 11.5.3, and 11.5.5, respectively).

Finally we note that $\hat{\boldsymbol{\theta}} = \mathbf{X}\hat{\boldsymbol{\beta}} = \mathbf{X}(\mathbf{X}'\mathbf{X})^-\mathbf{X}'\mathbf{Y}$ so that by *B1.8*, $\mathbf{P} = \mathbf{X}(\mathbf{X}'\mathbf{X})^-\mathbf{X}'$ is the unique matrix projecting E_n onto Ω. The uniqueness, symmetry, and idempotency of $\mathbf{P}$ can also be proved directly as in Exercise 6 below.

EXERCISES 3h

1. Let $\hat{\boldsymbol{\beta}}_i$ $(i=1,2)$ be any two solutions of the normal equations. Show directly that

$$\|\mathbf{Y} - \mathbf{X}\hat{\boldsymbol{\beta}}_1\|^2 = \|\mathbf{Y} - \mathbf{X}\hat{\boldsymbol{\beta}}_2\|^2.$$

2. If the columns of $\mathbf{X}$ are linearly dependent, prove that there is no matrix $\mathbf{C}$ such that $\mathbf{CY}$ is an unbiased estimate of $\boldsymbol{\beta}$.
3. Prove that a necessary and sufficient condition for the factorization (3.41) to hold is that

$$\operatorname{rank}\begin{pmatrix}\mathbf{X}\\ \mathbf{L}\end{pmatrix} = \operatorname{rank}\mathbf{L} = \operatorname{rank}\mathbf{X}.$$

4. If $\mathbf{B}\mathbf{B}^- = \mathbf{P}$, prove the following statements:
   (a) $\mathbf{P}^2 = \mathbf{P}$.
   (b) $\mathbf{PB} = \mathbf{B}$ and $\operatorname{rank}\mathbf{B} = \operatorname{tr}\mathbf{P}$.

Rao [1973: p. 25]

5. Prove the following statements:
   (a) $\mathbf{B}'\mathbf{B}=\mathbf{O}$ implies $\mathbf{B}=\mathbf{O}$.
   (b) $\mathbf{L}\mathbf{B}'\mathbf{B}=\mathbf{M}\mathbf{B}'\mathbf{B}$ implies $\mathbf{L}\mathbf{B}'=\mathbf{M}\mathbf{B}'$.
   [Hint: show that $(\mathbf{L}\mathbf{B}'\mathbf{B}-\mathbf{M}\mathbf{B}'\mathbf{B})(\mathbf{L}-\mathbf{M})'=(\mathbf{L}\mathbf{B}'-\mathbf{M}\mathbf{B}')(\mathbf{L}\mathbf{B}'-\mathbf{M}\mathbf{B}')'$.]
   Searle [1971: p. 16]

6. If $\mathbf{C}$ is a generalized inverse of $\mathbf{X}'\mathbf{X}$, prove the following:
   (a) $\mathbf{C}'$ is also a generalized inverse of $\mathbf{X}'\mathbf{X}$.
   (b) A symmetric generalized inverse of $\mathbf{X}'\mathbf{X}$ exists.
   (c) $\mathbf{C}\mathbf{X}'$ is a generalized inverse of $\mathbf{X}$.
   (d) $\mathbf{X}\mathbf{C}\mathbf{X}'$ is invariant to $\mathbf{C}$.
   (e) $\mathbf{X}\mathbf{C}\mathbf{X}'$ is symmetric and idempotent.
   (f) $\mathcal{R}[\mathbf{X}\mathbf{C}\mathbf{X}']=\mathcal{R}[\mathbf{X}]$.
   [Hint: (c) and (d) follow from exercise 5; (e) follows from (a) and (d). To prove (f) use $(\mathbf{I}-\mathbf{X}\mathbf{C}\mathbf{X}')\mathbf{X}=\mathbf{O}$ to show that $\mathcal{R}[\mathbf{X}]\subset\mathcal{R}[\mathbf{X}\mathbf{C}\mathbf{X}']$.]
   Searle [1971: p. 20]

7. Let $\mathbf{H}$ be a $(p-r)\times p$ matrix of rank $p-r$ satisfying the conditions of Theorem 3.9, that is, $\mathcal{R}[\mathbf{H}']\cap\mathcal{R}[\mathbf{X}']=\mathbf{0}$.
   (a) By considering $\mathbf{G}'\mathbf{G}(\mathbf{G}'\mathbf{G})^{-1}\mathbf{H}'=\mathbf{H}'$, prove that:
      (1) $\mathbf{H}(\mathbf{G}'\mathbf{G})^{-1}\mathbf{X}'=\mathbf{O}$.
      (2) $\mathbf{H}(\mathbf{G}'\mathbf{G})^{-1}\mathbf{H}'=\mathbf{I}_{p-r}$.
   (b) Deduce that $(\mathbf{G}'\mathbf{G})^{-1}$ is a generalized inverse of $\mathbf{X}'\mathbf{X}$.

8. Prove that the inverse on the left side of Equation (3.46) exists, and hence show that $\mathbf{C}_{11}$ is a generalized inverse of $\mathbf{X}'\mathbf{X}$.
   [Hint: If the left side is $\mathbf{A}^{-1}$, use Theorem 3.9 to show that $\mathbf{A}\boldsymbol{\alpha}=\mathbf{0}$ implies $\boldsymbol{\alpha}=\mathbf{0}$. Then consider the equations $\mathbf{A}\mathbf{A}^{-1}=\mathbf{I}$ and show that $\mathbf{C}_{22}=\mathbf{O}$.]

3.8.2 *Estimable Functions*

Having established that the normal equations can still be used for finding RSS, we now consider the problem of estimating linear combinations of the form $\mathbf{a}'\boldsymbol{\beta}$ when $\mathbf{X}$ is not of full rank.

Definition. *The parametric function* $\mathbf{a}'\boldsymbol{\beta}$ *is said to be estimable if it has a linear unbiased estimate,* $\mathbf{b}'\mathbf{Y}$, *say.*

If $\mathbf{a}'\boldsymbol{\beta}$ is estimable, then $\mathbf{a}'\boldsymbol{\beta}=E[\mathbf{b}'\mathbf{Y}]=\mathbf{b}'\mathbf{X}\boldsymbol{\beta}$ identically in $\boldsymbol{\beta}$ so that $\mathbf{a}'=\mathbf{b}'\mathbf{X}$ or $\mathbf{a}=\mathbf{X}'\mathbf{b}$ (*A9.1*). Hence $\mathbf{a}'\boldsymbol{\beta}$ is estimable if and only if $\mathbf{a}\in\mathcal{R}[\mathbf{X}']$ $(=\mathcal{R}[\mathbf{X}'\mathbf{X}]$ by *A2.5*). We now apply Theorem 3.2 in Section 3.2 to the case when $\mathbf{X}$ has less than full rank and $\mathbf{a}'\boldsymbol{\beta}$ is an estimable function.

THEOREM 3.12 If $\mathbf{a}'\boldsymbol{\beta}$ is estimable, and $\hat{\boldsymbol{\beta}}$ is any solution of the normal equations, then:
(i) $\mathbf{a}'\hat{\boldsymbol{\beta}}$ is unique.
(ii) $\mathbf{a}'\hat{\boldsymbol{\beta}}$ is the BLUE of $\mathbf{a}'\boldsymbol{\beta}$.

Proof. If $\mathbf{a}'\boldsymbol{\beta}$ is estimable then $\mathbf{a}'\boldsymbol{\beta}=\mathbf{b}'\mathbf{X}\boldsymbol{\beta}=\mathbf{b}'\boldsymbol{\theta}$, and $\mathbf{a}'\hat{\boldsymbol{\beta}}=\mathbf{b}'\hat{\boldsymbol{\theta}}$, which is unique because $\hat{\boldsymbol{\theta}}$ is the unique projection of $\mathbf{Y}$ onto Ω. Also, by Theorem 3.2, $\mathbf{b}'\hat{\boldsymbol{\theta}}$ is the BLUE of $\mathbf{b}'\boldsymbol{\theta}$. □

A method of testing for estimability is given in Exercise 6 of Exercises 3i below; namely, $\mathbf{a}'\boldsymbol{\beta}$ is estimable if and only if

$$\mathbf{a}'(\mathbf{X}'\mathbf{X})^{-}\mathbf{X}'\mathbf{X}=\mathbf{a}'.$$

With $\mathbf{X}^*=(\mathbf{X}'\mathbf{X})^{-}\mathbf{X}'$, a computational method for checking this equation is described in Exercise 11 of Miscellaneous Exercises 11.

Suppose that $\mathbf{X}'\mathbf{X}$ has positive eigenvalues $\lambda_1,\lambda_2,\ldots,\lambda_r$ (which may not all be distinct) and corresponding orthonormal eigenvectors $\boldsymbol{\alpha}_1,\boldsymbol{\alpha}_2,\ldots,\boldsymbol{\alpha}_r$ (*A2.7*); that is, $\mathbf{X}'\mathbf{X}\boldsymbol{\alpha}_i=\lambda_i\boldsymbol{\alpha}_i$ $(i=1,2,\ldots,r)$ and $\boldsymbol{\alpha}_i'\boldsymbol{\alpha}_j=\delta_{ij}$. If $\mathbf{a}'\boldsymbol{\beta}$ is estimable then $\mathbf{a}\in\mathcal{R}[\mathbf{X}'\mathbf{X}]$ and, since $\mathcal{R}[\mathbf{X}'\mathbf{X}]$ is the space spanned by the eigenvectors (*A2.7*), $\mathbf{a}$ can be expressed in the form

$$\mathbf{a}=\sum_{i=1}^{r} c_i\boldsymbol{\alpha}_i. \tag{3.48}$$

Hence (Silvey [1969])

$$\begin{aligned}\operatorname{var}[\mathbf{a}'\hat{\boldsymbol{\beta}}]&=\operatorname{var}\left[\sum_i c_i\boldsymbol{\alpha}_i'\hat{\boldsymbol{\beta}}\right]\\&=\sum_i c_i^2\operatorname{var}[\boldsymbol{\alpha}_i'\hat{\boldsymbol{\beta}}]\\&=\sigma^2\sum_i c^2\lambda_i^{-1},\end{aligned} \tag{3.49}$$

since

$$\begin{aligned}\operatorname{cov}[\boldsymbol{\alpha}_i'\hat{\boldsymbol{\beta}},\boldsymbol{\alpha}_j'\hat{\boldsymbol{\beta}}]&=\lambda_i^{-1}\lambda_j^{-1}\operatorname{cov}[\boldsymbol{\alpha}_i'\mathbf{X}'\mathbf{X}\hat{\boldsymbol{\beta}},\boldsymbol{\alpha}_j'\mathbf{X}'\mathbf{X}\hat{\boldsymbol{\beta}}]\\&=(\lambda_i\lambda_j)^{-1}\operatorname{cov}[\boldsymbol{\alpha}_i'\mathbf{X}'\mathbf{Y},\boldsymbol{\alpha}_j'\mathbf{X}'\mathbf{Y}]\\&=(\lambda_i\lambda_j)^{-1}\sigma^2\boldsymbol{\alpha}_i'\mathbf{X}'\mathbf{X}\boldsymbol{\alpha}_j\\&=(\lambda_i\lambda_j)^{-1}\sigma^2\lambda_j\boldsymbol{\alpha}_i'\boldsymbol{\alpha}_j\\&=\sigma^2\delta_{ij}\lambda_i^{-1}.\end{aligned} \tag{3.50}$$

Silvey concludes from (3.49) that relatively precise estimation is possible in the directions of the eigenvectors of $\mathbf{X}'\mathbf{X}$ corresponding to large eigenvalues, whereas relatively imprecise estimation is obtained in the directions corresponding to small eigenvalues.

Suppose $\mathbf{X}$ has full rank but the columns are close to being linearly dependent. Then $\mathbf{X}'\mathbf{X}$ is near singularity, one or more eigenvalues are very small, and estimation in certain directions is very imprecise. If we consider some limiting process in which $\mathbf{X}'\mathbf{X}$ approaches singularity, then estimation in these directions become progressively worse and, in the limit, estimation becomes impossible in the directions of the eigenvectors corresponding to zero eigenvalues. Thus estimable functions have the representation (3.48).

The presence of linear relationships or "near" linear relationships among the regressors is described, in econometrics, by the term *multicollinearity*, the most extreme form of multicollinearity occurring when $\mathbf{X}$ has less than full rank. Silvey [1969] shows that the effect of multicollinearity can be minimized (or eradicated) by taking additional Y observations in certain x-directions, namely, in the directions of eigenvectors corresponding to small (or zero) eigenvalues (see Exercise 8 below). The question of choosing an x-direction for a further observation to improve the precision of estimation of a particular $\mathbf{a}'\boldsymbol{\beta}$ is also discussed in detail by Silvey.

In conclusion mention is made of a paper by Webster et al. [1974] in which the regression analysis is based on the eigenvalues.

EXERCISES 3i

1. Prove that all linear functions $\mathbf{a}'\boldsymbol{\beta}$ are estimable if and only if the columns of $\mathbf{X}$ are linearly independent.
2. Prove that $\mathbf{a}'\mathcal{E}[\hat{\boldsymbol{\beta}}]$ is an estimable function of $\boldsymbol{\beta}$.
3. If $\mathbf{a}_1'\boldsymbol{\beta}, \mathbf{a}_2'\boldsymbol{\beta}, \ldots, \mathbf{a}_k'\boldsymbol{\beta}$ are estimable, prove that any linear combination of these is also estimable.
4. Let $\hat{\boldsymbol{\beta}}$ be any solution of the normal equations. Show that $\hat{\boldsymbol{\beta}}$ can be expressed in the form $\hat{\boldsymbol{\beta}} = \mathbf{b} + \mathbf{c}$ where $\mathbf{b}$ is a unique vector in $\mathcal{R}[\mathbf{X}']$, and $\mathbf{c} \perp \mathcal{R}[\mathbf{X}']$. Prove that $\hat{\mathbf{Y}}$ $(= \mathbf{X}\hat{\boldsymbol{\beta}})$ is unique. Deduce part (i) of Theorem 3.12.
5. Prove the converse of Theorem 3.12 (i), namely, that if $\mathbf{a}'\hat{\boldsymbol{\beta}}$ is invariant with respect to $\hat{\boldsymbol{\beta}}$ then $\mathbf{a}'\boldsymbol{\beta}$ is estimable.
6. Prove that $\mathbf{a}'\boldsymbol{\beta}$ is estimable if and only if

$$\mathbf{a}'(\mathbf{X}'\mathbf{X})^{-}\mathbf{X}'\mathbf{X} = \mathbf{a}'.$$

7. If $\mathbf{a}'\boldsymbol{\beta}$ is an estimable function, prove that

$$\mathcal{D}[\mathbf{a}'\hat{\boldsymbol{\beta}}] = \sigma^2 \mathbf{a}'(\mathbf{X}'\mathbf{X})^{-}\mathbf{a}.$$

8. Suppose that a new observation is added to the model $\mathbf{Y}=\mathbf{X}\boldsymbol{\beta}+\boldsymbol{\varepsilon}$ giving

$$\begin{pmatrix}\mathbf{Y}\\ Y_{n+1}\end{pmatrix}=\begin{pmatrix}\mathbf{X}\\ \mathbf{x}'_{n+1}\end{pmatrix}\boldsymbol{\beta}+\begin{pmatrix}\boldsymbol{\varepsilon}\\ \varepsilon_{n+1}\end{pmatrix}$$

$$=\mathbf{X}_*\boldsymbol{\beta}+\boldsymbol{\varepsilon}_*,$$

say, where $\mathbf{x}_{n+1}=c\boldsymbol{\alpha}$, $\boldsymbol{\alpha}$ being a unit eigenvector of $\mathbf{X}'\mathbf{X}$ with corresponding eigenvalue λ. Show that $\boldsymbol{\alpha}$ is an eigenvector of $\mathbf{X}'_*\mathbf{X}_*$ corresponding to the eigenvalue $\lambda+c^2$. Silvey [1969: p. 544]

3.8.3 Introducing Further Regressors

Consider the enlarged model G given in Section 3.7.1, but with one difference: $\mathbf{X}$ is now of rank r $(r<p)$ and, since the columns of $\mathbf{Z}$ are linearly independent and $\mathcal{R}[\mathbf{X}]\cap\mathcal{R}[\mathbf{Z}]=\mathbf{0}$, $\mathbf{W}$ is $n\times(t+p)$ of rank $t+r$. If $\mathbf{P}$ is the unique idempotent matrix projecting E_n onto $\mathcal{R}[\mathbf{X}]$ (*B1.2*) and $\mathbf{R}=\mathbf{I}_n-\mathbf{P}$, then $\mathbf{Z}'\mathbf{R}\mathbf{Z}$ is still nonsingular (see Exercise 1 of Exercises 3j below). Also, Theorem 3.7 (ii), (iii), and (iv) of Section 3.7.1 still hold and it is instructive to outline several proofs of this generalization.

Proof a. We can reduce G to a full rank model, namely, $\mathscr{E}[\mathbf{Y}]=\mathbf{X}_1\boldsymbol{\alpha}+\mathbf{Z}\boldsymbol{\gamma}$, where $\mathbf{X}_1$ is $n\times r$ of rank r. Since $\mathbf{R}$ is unique, $\mathbf{R}=\mathbf{I}_n-\mathbf{X}_1(\mathbf{X}'_1\mathbf{X}_1)^{-1}\mathbf{X}'_1$ and, following through the steps of Theorem 3.7 with $\mathbf{X}_1$ and $\boldsymbol{\alpha}$ instead of $\mathbf{X}$ and $\boldsymbol{\beta}$, we have

$$\hat{\boldsymbol{\gamma}}_G=(\mathbf{Z}'\mathbf{R}\mathbf{Z})^{-1}\mathbf{Z}'\mathbf{R}\mathbf{Y},$$

$$\text{RSS}_G=(\mathbf{Y}-\mathbf{X}_1\hat{\boldsymbol{\alpha}}_G-\mathbf{Z}\hat{\boldsymbol{\gamma}}_G)'(\mathbf{Y}-\mathbf{X}_1\hat{\boldsymbol{\alpha}}_G-\mathbf{Z}\hat{\boldsymbol{\gamma}}_G)$$

$$=(\mathbf{Y}-\mathbf{Z}\hat{\boldsymbol{\gamma}}_G)'\mathbf{R}(\mathbf{Y}-\mathbf{Z}\hat{\boldsymbol{\gamma}}_G)$$

and

$$\text{RSS}_G=\text{RSS}-\hat{\boldsymbol{\gamma}}'_G\mathbf{Z}'\mathbf{R}\mathbf{Y}. \qquad (3.51)$$

If we rewrite the normal equations (3.29) in the form

$$\mathbf{X}'\mathbf{X}\hat{\boldsymbol{\beta}}_G=\mathbf{X}'(\mathbf{Y}-\mathbf{Z}\hat{\boldsymbol{\gamma}}_G) \qquad (3.52)$$

we see that any solution $\hat{\boldsymbol{\beta}}_G$ can be obtained by replacing $\mathbf{Y}$ by $\mathbf{Y}-\mathbf{Z}\hat{\boldsymbol{\gamma}}_G$ in any solution $\hat{\boldsymbol{\beta}}$ of the equations $\mathbf{X}'\mathbf{X}\boldsymbol{\beta}=\mathbf{X}'\mathbf{Y}$. Hence the method of two-step least squares in Section 3.7.3 still applies, even when $\mathbf{X}$ is not of full rank.

Proof b. Let $\hat{\boldsymbol{\theta}}_G = \mathbf{P}_G\mathbf{Y}$, where $\mathbf{P}_G$ represents the orthogonal projection onto $\mathcal{R}[\mathbf{W}]$. Here

$$\mathcal{R}[\mathbf{W}] = \mathcal{R}[\mathbf{X}, \mathbf{Z}] = \mathcal{R}[\mathbf{X}, \mathbf{RZ}] \tag{3.53}$$

since $\mathbf{Z} = \mathbf{PZ} + \mathbf{RZ}$ and $\mathcal{R}[\mathbf{PZ}] \subset \mathcal{R}[\mathbf{P}] = \mathcal{R}[\mathbf{X}]$. Then $(\mathbf{Y} - \hat{\boldsymbol{\theta}}_G) \perp \mathcal{R}[\mathbf{W}]$ so that

$$\mathbf{X}'(\mathbf{Y} - \hat{\boldsymbol{\theta}}_G) = \mathbf{0} \tag{3.54}$$

and

$$\mathbf{Z}'\mathbf{R}(\mathbf{Y} - \hat{\boldsymbol{\theta}}_G) = \mathbf{0}. \tag{3.55}$$

Now from (3.55)

$$\begin{aligned}
\mathbf{Z}'\mathbf{RY} &= \mathbf{Z}'\mathbf{R}\hat{\boldsymbol{\theta}}_G \\
&= \mathbf{Z}'\mathbf{R}(\mathbf{X}\hat{\boldsymbol{\beta}}_G + \mathbf{Z}\hat{\boldsymbol{\gamma}}_G) \\
&= \mathbf{Z}'\mathbf{RZ}\hat{\boldsymbol{\gamma}}_G,
\end{aligned}$$

and $\hat{\boldsymbol{\gamma}}_G = (\mathbf{Z}'\mathbf{RZ})^{-1}\mathbf{Z}'\mathbf{RY}$. Also, from (3.54),

$$\begin{aligned}
\mathbf{X}'\mathbf{X}\hat{\boldsymbol{\beta}}_G &= \mathbf{X}'(\hat{\boldsymbol{\theta}}_G - \mathbf{Z}\hat{\boldsymbol{\gamma}}_G) \\
&= \mathbf{X}'(\mathbf{Y} - \mathbf{Z}\hat{\boldsymbol{\gamma}}_G).
\end{aligned}$$

Finally, using (3.54) once again,

$$\begin{aligned}
\mathbf{Y} - \hat{\boldsymbol{\theta}}_G &= \mathbf{R}(\mathbf{Y} - \hat{\boldsymbol{\theta}}_G) \\
&= \mathbf{R}(\mathbf{Y} - \mathbf{X}\hat{\boldsymbol{\beta}}_G - \mathbf{Z}\hat{\boldsymbol{\gamma}}_G) \\
&= \mathbf{R}(\mathbf{Y} - \mathbf{Z}\hat{\boldsymbol{\gamma}}_G)
\end{aligned}$$

and

$$\begin{aligned}
\text{RSS}_G &= (\mathbf{Y} - \hat{\boldsymbol{\theta}}_G)'(\mathbf{Y} - \hat{\boldsymbol{\theta}}_G) \\
&= (\mathbf{Y} - \mathbf{Z}\hat{\boldsymbol{\gamma}}_G)'\mathbf{R}(\mathbf{Y} - \mathbf{Z}\hat{\boldsymbol{\gamma}}_G).
\end{aligned}$$

Proof c. We find that all the steps in proving Theorem 3.7 (i)–(iv) and the theory of Sections 3.7.3 and 3.7.4 still hold provided we replace inverses by generalized inverses. This method has one minor advantage. If we now allow the columns of $\mathbf{Z}$ to be linearly dependent as well, or the columns of $\mathbf{Z}$ to be linearly dependent on the columns of $\mathbf{X}$, then the equations still hold provided we use the generalized inverse of $\mathbf{Z}'\mathbf{R}\mathbf{Z}$.

EXERCISES 3j

1. If $\mathbf{P}$ projects onto $\mathcal{R}[\mathbf{X}]$, show that $\mathbf{Z}'(\mathbf{I}_n - \mathbf{P})\mathbf{Z}$ is nonsingular.
2. Given the constraints $\mathbf{H}\boldsymbol{\beta} = \mathbf{0}$ are identifiability constraints for the model $\boldsymbol{\theta} = \mathbf{X}\boldsymbol{\beta}$, prove that they are also identifiability constraints for the model $\boldsymbol{\theta} = \mathbf{X}\boldsymbol{\beta} + \mathbf{Z}\boldsymbol{\gamma}$. Hence show that

$$\hat{\boldsymbol{\beta}}_G = (\mathbf{G}'\mathbf{G})^{-1}\mathbf{X}'(\mathbf{Y} - \mathbf{Z}\hat{\boldsymbol{\gamma}}_G).$$

3.9 ESTIMATION WITH LINEAR RESTRICTIONS

3.9.1 Method of Lagrange Multipliers

Let $\mathbf{Y} = \mathbf{X}\boldsymbol{\beta} + \boldsymbol{\varepsilon}$, where $\mathbf{X}$ is $n \times p$ of rank p. Suppose we wish to find the minimum of $\boldsymbol{\varepsilon}'\boldsymbol{\varepsilon}$ subject to the consistent linear restrictions $\mathbf{A}\boldsymbol{\beta} = \mathbf{c}$, where $\mathbf{A}$ is a known $q \times p$ matrix of rank q and $\mathbf{c}$ is a known $q \times 1$ vector. One method of solving this problem is to use Lagrange multipliers, one for each linear constraint $\mathbf{a}_i'\boldsymbol{\beta} = c_i$ $(i = 1, 2, \ldots, q)$, where $\mathbf{a}_i'$ is the ith row of $\mathbf{A}$. We therefore are interested in the expression

$$\sum_{i=1}^{q} \lambda_i(\mathbf{a}_i'\boldsymbol{\beta} - c_i) = \boldsymbol{\lambda}'(\mathbf{A}\boldsymbol{\beta} - \mathbf{c})$$

$$= (\boldsymbol{\beta}'\mathbf{A}' - \mathbf{c}')\boldsymbol{\lambda},$$

(since the transpose of a 1×1 matrix is itself). To apply the method of Lagrange multipliers we consider the expression $r = \boldsymbol{\varepsilon}'\boldsymbol{\varepsilon} + (\boldsymbol{\beta}'\mathbf{A}' - \mathbf{c}')\boldsymbol{\lambda}$ and solve the equations

$$\mathbf{A}\boldsymbol{\beta} = \mathbf{c} \tag{3.56}$$

and $\partial r / \partial \boldsymbol{\beta} = \mathbf{0}$, that is, (from *A6*)

$$-2\mathbf{X}'\mathbf{Y} + 2\mathbf{X}'\mathbf{X}\boldsymbol{\beta} + \mathbf{A}'\boldsymbol{\lambda} = \mathbf{0}. \tag{3.57}$$

For future reference we denote the solution of these two equations by $\hat{\boldsymbol{\beta}}_H$

and $\hat{\boldsymbol{\lambda}}_H$. Then, from (3.57),

$$\begin{aligned}\hat{\boldsymbol{\beta}}_H &= (\mathbf{X}'\mathbf{X})^{-1}\mathbf{X}'\mathbf{Y} - \tfrac{1}{2}(\mathbf{X}'\mathbf{X})^{-1}\mathbf{A}'\boldsymbol{\lambda}_H \\ &= \hat{\boldsymbol{\beta}} - \tfrac{1}{2}(\mathbf{X}'\mathbf{X})^{-1}\mathbf{A}'\hat{\boldsymbol{\lambda}}_H, \end{aligned} \tag{3.58}$$

and, from (3.56),

$$\begin{aligned}\mathbf{c} &= \mathbf{A}\hat{\boldsymbol{\beta}}_H \\ &= \mathbf{A}\hat{\boldsymbol{\beta}} - \tfrac{1}{2}\mathbf{A}(\mathbf{X}'\mathbf{X})^{-1}\mathbf{A}'\hat{\boldsymbol{\lambda}}_H.\end{aligned}$$

Since $(\mathbf{X}'\mathbf{X})^{-1}$ is positive definite, being the inverse of a positive definite matrix, $\mathbf{A}(\mathbf{X}'\mathbf{X})^{-1}\mathbf{A}'$ is also positive definite (*A4*) and therefore nonsingular. Hence

$$-\tfrac{1}{2}\hat{\boldsymbol{\lambda}}_H = \left[\mathbf{A}(\mathbf{X}'\mathbf{X})^{-1}\mathbf{A}'\right]^{-1}(\mathbf{c} - \mathbf{A}\hat{\boldsymbol{\beta}})$$

and, substituting in (3.58),

$$\hat{\boldsymbol{\beta}}_H = \hat{\boldsymbol{\beta}} + (\mathbf{X}'\mathbf{X})^{-1}\mathbf{A}'\left[\mathbf{A}(\mathbf{X}'\mathbf{X})^{-1}\mathbf{A}'\right]^{-1}(\mathbf{c} - \mathbf{A}\hat{\boldsymbol{\beta}}). \tag{3.59}$$

To prove that $\hat{\boldsymbol{\beta}}_H$ actually minimizes $\boldsymbol{\varepsilon}'\boldsymbol{\varepsilon}$ subject to $\mathbf{A}\boldsymbol{\beta} = \mathbf{c}$, we note first of all that

$$\begin{aligned}\|\mathbf{X}(\hat{\boldsymbol{\beta}} - \boldsymbol{\beta})\|^2 &= (\hat{\boldsymbol{\beta}} - \boldsymbol{\beta})'\mathbf{X}'\mathbf{X}(\hat{\boldsymbol{\beta}} - \boldsymbol{\beta}) \\ &= (\hat{\boldsymbol{\beta}} - \hat{\boldsymbol{\beta}}_H + \hat{\boldsymbol{\beta}}_H - \boldsymbol{\beta})'\mathbf{X}'\mathbf{X}(\hat{\boldsymbol{\beta}} - \hat{\boldsymbol{\beta}}_H + \hat{\boldsymbol{\beta}}_H - \boldsymbol{\beta}) \\ &= (\hat{\boldsymbol{\beta}} - \hat{\boldsymbol{\beta}}_H)'\mathbf{X}'\mathbf{X}(\hat{\boldsymbol{\beta}} - \hat{\boldsymbol{\beta}}_H) + (\hat{\boldsymbol{\beta}}_H - \boldsymbol{\beta})'\mathbf{X}'\mathbf{X}(\hat{\boldsymbol{\beta}}_H - \boldsymbol{\beta}) \qquad (3.60) \\ &= \|\mathbf{X}(\hat{\boldsymbol{\beta}} - \hat{\boldsymbol{\beta}}_H)\|^2 + \|\mathbf{X}(\hat{\boldsymbol{\beta}}_H - \boldsymbol{\beta})\|^2 \qquad (3.61)\end{aligned}$$

since, from (3.58),

$$2(\hat{\boldsymbol{\beta}} - \hat{\boldsymbol{\beta}}_H)'\mathbf{X}'\mathbf{X}(\hat{\boldsymbol{\beta}}_H - \boldsymbol{\beta}) = \hat{\boldsymbol{\lambda}}_H'\mathbf{A}(\hat{\boldsymbol{\beta}}_H - \boldsymbol{\beta}) = \hat{\boldsymbol{\lambda}}_H'(\mathbf{c} - \mathbf{c}) = 0. \tag{3.62}$$

Hence from (3.15) in Section 3.4 and (3.61),

$$\begin{aligned}\boldsymbol{\varepsilon}'\boldsymbol{\varepsilon} &= \|\mathbf{Y} - \mathbf{X}\hat{\boldsymbol{\beta}}\|^2 + \|\mathbf{X}(\hat{\boldsymbol{\beta}} - \boldsymbol{\beta})\|^2 \\ &= \|\mathbf{Y} - \mathbf{X}\hat{\boldsymbol{\beta}}\|^2 + \|\mathbf{X}(\hat{\boldsymbol{\beta}} - \hat{\boldsymbol{\beta}}_H)\|^2 + \|\mathbf{X}(\hat{\boldsymbol{\beta}}_H - \boldsymbol{\beta})\|^2 \end{aligned} \tag{3.63}$$

is a minimum when $\|\mathbf{X}(\hat{\boldsymbol{\beta}}_H - \boldsymbol{\beta})\|^2 = 0$, that is, when $\mathbf{X}(\hat{\boldsymbol{\beta}}_H - \boldsymbol{\beta}) = \mathbf{0}$, or $\boldsymbol{\beta} = \hat{\boldsymbol{\beta}}_H$ (since the columns of $\mathbf{X}$ are linearly independent).

Setting $\boldsymbol{\beta} = \hat{\boldsymbol{\beta}}_H$ we obtain the useful identity

$$\|\mathbf{Y} - \mathbf{X}\hat{\boldsymbol{\beta}}_H\|^2 = \|\mathbf{Y} - \mathbf{X}\hat{\boldsymbol{\beta}}\|^2 + \|\mathbf{X}(\hat{\boldsymbol{\beta}} - \hat{\boldsymbol{\beta}}_H)\|^2 \tag{3.64}$$

or, writing $\hat{\mathbf{Y}} = \mathbf{X}\hat{\boldsymbol{\beta}}$ and $\hat{\mathbf{Y}}_H = \mathbf{X}\hat{\boldsymbol{\beta}}_H$,

$$\|\mathbf{Y} - \hat{\mathbf{Y}}_H\|^2 - \|\mathbf{Y} - \hat{\mathbf{Y}}\|^2 = \|\hat{\mathbf{Y}} - \hat{\mathbf{Y}}_H\|^2. \tag{3.65}$$

This identity can also be derived directly (see Exercise 1 of Exercises 3k at the end of Section 3.9.2).

3.9.2 *Method of Orthogonal Projections*

It is constructive to derive (3.59) using the theory of *B3*. In order to do this we first "remove" $\mathbf{c}$.

Suppose that $\boldsymbol{\beta}_0$ is any solution of $\mathbf{A}\boldsymbol{\beta} = \mathbf{c}$; then

$$\mathbf{Y} - \mathbf{X}\boldsymbol{\beta}_0 = \mathbf{X}(\boldsymbol{\beta} - \boldsymbol{\beta}_0) + \boldsymbol{\varepsilon} \tag{3.66}$$

or $\tilde{\mathbf{Y}} = \mathbf{X}\boldsymbol{\gamma} + \boldsymbol{\varepsilon}$, and $\mathbf{A}\boldsymbol{\gamma} = \mathbf{A}\boldsymbol{\beta} - \mathbf{A}\boldsymbol{\beta}_0 = \mathbf{0}$. Thus we have the model $\tilde{\mathbf{Y}} = \boldsymbol{\theta} + \boldsymbol{\varepsilon}$, where $\boldsymbol{\theta} \in \mathcal{R}[\mathbf{X}]$ $(=\Omega)$ and, since $\mathbf{X}$ has full rank, $\mathbf{A}(\mathbf{X}'\mathbf{X})^{-1}\mathbf{X}'\boldsymbol{\theta} = \mathbf{A}\boldsymbol{\gamma} = \mathbf{0}$. Then setting $\mathbf{A}_1 = \mathbf{A}(\mathbf{X}'\mathbf{X})^{-1}\mathbf{X}'$ and $\omega = \mathcal{N}[\mathbf{A}_1] \cap \Omega$, it follows from *B3.3* that $\omega^{\perp} \cap \Omega = \mathcal{R}[\mathbf{P}_\Omega \mathbf{A}_1']$, where

$$\mathbf{P}_\Omega \mathbf{A}_1' = \mathbf{X}(\mathbf{X}'\mathbf{X})^{-1}\mathbf{X}'\mathbf{X}(\mathbf{X}'\mathbf{X})^{-1}\mathbf{A}' = \mathbf{X}(\mathbf{X}'\mathbf{X})^{-1}\mathbf{A}'$$

is $n \times q$ of rank q (by Exercise 5 below). Therefore, by *B3.2* and *B1.9*

$$\begin{aligned}
\mathbf{P}_\Omega - \mathbf{P}_\omega &= \mathbf{P}_{\omega^{\perp} \cap \Omega} \\
&= (\mathbf{P}_\Omega \mathbf{A}_1')\left[\mathbf{A}_1 \mathbf{P}_\Omega^2 \mathbf{A}_1'\right]^{-1}(\mathbf{P}_\Omega \mathbf{A}_1')' \\
&= \mathbf{X}(\mathbf{X}'\mathbf{X})^{-1}\mathbf{A}'\left[\mathbf{A}(\mathbf{X}'\mathbf{X})^{-1}\mathbf{A}'\right]^{-1}\mathbf{A}(\mathbf{X}'\mathbf{X})^{-1}\mathbf{X}'.
\end{aligned}$$

Hence

$$\begin{aligned}
\mathbf{X}\hat{\boldsymbol{\beta}}_H - \mathbf{X}\boldsymbol{\beta}_0 &= \mathbf{X}\hat{\boldsymbol{\gamma}}_H \\
&= \mathbf{P}_\omega \tilde{\mathbf{Y}} \\
&= \mathbf{P}_\Omega \tilde{\mathbf{Y}} - \mathbf{P}_{\omega^{\perp} \cap \Omega} \tilde{\mathbf{Y}} \\
&= \mathbf{P}_\Omega \mathbf{Y} - \mathbf{X}\boldsymbol{\beta}_0 - \mathbf{X}(\mathbf{X}'\mathbf{X})^{-1}\mathbf{A}'\left[\mathbf{A}(\mathbf{X}'\mathbf{X})^{-1}\mathbf{A}'\right]^{-1}(\mathbf{A}\hat{\boldsymbol{\beta}} - \mathbf{c}),
\end{aligned} \tag{3.67}$$

since $\mathbf{P}_\Omega \mathbf{X}\boldsymbol{\beta}_0 = \mathbf{X}\boldsymbol{\beta}_0$ and $\mathbf{A}\boldsymbol{\beta}_0 = \mathbf{c}$. Therefore cancelling $\mathbf{X}\boldsymbol{\beta}_0$ and multiplying both sides by $(\mathbf{X}'\mathbf{X})^{-1}\mathbf{X}'$ leads to $\hat{\boldsymbol{\beta}}_H$ of (3.59). Clearly this gives a minimum as $\|\mathbf{Y} - \mathbf{X}\hat{\boldsymbol{\beta}}_H\|^2 = \|\tilde{\mathbf{Y}} - \mathbf{X}\hat{\boldsymbol{\gamma}}_H\|^2$.

One advantage of the above approach is that it is readily adapted to the case when $\mathbf{X}$ has rank r $(r<p)$. Since we can only estimate estimable functions we assume that $\mathbf{a}_i'\boldsymbol{\beta}$ $(i=1,2,\ldots,q)$, where $\mathbf{a}_i'$ is the ith row of $\mathbf{A}$, is estimable; that is, $\mathbf{a}_i' = \mathbf{m}_i'\mathbf{X}$ (by Section 3.8.2) and $\mathbf{A} = \mathbf{M}\mathbf{X}$, where $\mathbf{M}$ is $q \times n$. Because $\mathbf{A}$ is $q \times p$ of rank q we must have $q \leqslant r$ and, since $\operatorname{rank} \mathbf{A} \leqslant \operatorname{rank} \mathbf{M}$ (*A2.1*), $\mathbf{M}$ has rank q. Arguing as in (3.66) we again reduce the model to $\tilde{\mathbf{Y}} = \boldsymbol{\theta} + \boldsymbol{\varepsilon}$, where $\boldsymbol{\theta} \in \mathcal{R}[\mathbf{X}]$ and $\mathbf{M}\boldsymbol{\theta} = \mathbf{M}\mathbf{X}\boldsymbol{\gamma} = \mathbf{A}\boldsymbol{\gamma} = \mathbf{0}$. Therefore $\omega = \mathcal{N}[\mathbf{M}] \cap \Omega$ and $\omega^\perp \cap \Omega = \mathcal{R}[\mathbf{P}_\Omega \mathbf{M}']$, where $\mathbf{P}_\Omega \mathbf{M}'$ $(= \mathbf{X}(\mathbf{X}'\mathbf{X})^- \mathbf{X}'\mathbf{M}' = \mathbf{X}(\mathbf{X}'\mathbf{X})^- \mathbf{A}')$ is $n \times q$ of rank q (see Exercise 4 below). Using *B1.9* once again we have

$$\begin{aligned} \mathbf{P}_\Omega - \mathbf{P}_\omega &= (\mathbf{P}_\Omega \mathbf{M}')\left[\mathbf{M}\mathbf{P}_\Omega \mathbf{M}'\right]^{-1}(\mathbf{P}_\Omega \mathbf{M}')' \\ &= \mathbf{X}(\mathbf{X}'\mathbf{X})^- \mathbf{A}'\left[\mathbf{A}(\mathbf{X}'\mathbf{X})^- \mathbf{A}'\right]^{-1}\mathbf{A}(\mathbf{X}'\mathbf{X})^- \mathbf{X}'. \end{aligned} \tag{3.68}$$

Finally, by the same argument that led to (3.67),

$$\begin{aligned} \mathbf{X}'\mathbf{X}\hat{\boldsymbol{\beta}}_H &= \mathbf{X}'\mathbf{P}_\Omega \mathbf{Y} - \mathbf{X}'\mathbf{P}_\Omega \mathbf{M}'\left[\mathbf{M}\mathbf{P}_\Omega \mathbf{M}'\right]^{-1}\mathbf{M}\mathbf{P}_\Omega(\mathbf{Y} - \mathbf{X}\boldsymbol{\beta}_0) \\ &= \mathbf{X}'\mathbf{Y} - \mathbf{X}'\mathbf{M}'\left[\mathbf{M}\mathbf{P}_\Omega \mathbf{M}'\right]^{-1}(\mathbf{A}(\mathbf{X}'\mathbf{X})^- \mathbf{X}'\mathbf{Y} - \mathbf{M}\mathbf{X}\boldsymbol{\beta}_0) \\ &= \mathbf{X}'\mathbf{Y} - \mathbf{A}'\left[\mathbf{A}(\mathbf{X}'\mathbf{X})^- \mathbf{A}'\right]^{-1}(\mathbf{A}\hat{\boldsymbol{\beta}} - \mathbf{c}), \end{aligned}$$

and any solution for $\hat{\boldsymbol{\beta}}_H$ takes the form (*A8*)

$$\hat{\boldsymbol{\beta}} - (\mathbf{X}'\mathbf{X})^- \mathbf{A}'\left[\mathbf{A}(\mathbf{X}'\mathbf{X})^- \mathbf{A}'\right]^{-1}(\mathbf{A}\hat{\boldsymbol{\beta}} - \mathbf{c}),$$

where $\hat{\boldsymbol{\beta}} = (\mathbf{X}'\mathbf{X})^- \mathbf{X}'\mathbf{Y}$.

EXERCISES 3k

1. By considering the identity $\mathbf{Y} - \hat{\mathbf{Y}}_H = \mathbf{Y} - \hat{\mathbf{Y}} + \hat{\mathbf{Y}} - \hat{\mathbf{Y}}_H$, prove that

$$\|\mathbf{Y} - \hat{\mathbf{Y}}_H\|^2 = \|\mathbf{Y} - \hat{\mathbf{Y}}\|^2 + \|\hat{\mathbf{Y}} - \hat{\mathbf{Y}}_H\|^2.$$

2. Prove that

$$\mathcal{D}\left[\hat{\boldsymbol{\beta}}_H\right] = \sigma^2\left\{(\mathbf{X}'\mathbf{X})^{-1} - (\mathbf{X}'\mathbf{X})^{-1}\mathbf{A}'\left[\mathbf{A}(\mathbf{X}'\mathbf{X})^{-1}\mathbf{A}'\right]^{-1}\mathbf{A}(\mathbf{X}'\mathbf{X})^{-1}\right\}.$$

Hence deduce that

$$\operatorname{var}[\hat{\beta}_{Hi}] \leqslant \operatorname{var}[\hat{\beta}_i],$$

where $\hat{\beta}_{Hi}$ and $\hat{\beta}_i$ are the ith elements of $\hat{\boldsymbol{\beta}}_H$ and $\hat{\boldsymbol{\beta}}$, respectively.

3. Show that

$$\|\mathbf{Y}-\hat{\mathbf{Y}}_H\|^2 - \|\mathbf{Y}-\hat{\mathbf{Y}}\|^2 = \sigma^2 \hat{\boldsymbol{\lambda}}_H' \left(\mathcal{D}\left[\hat{\boldsymbol{\lambda}}_H\right]\right)^{-1} \hat{\boldsymbol{\lambda}}_H.$$

4. Using the notation of Section 3.9.2, prove that rank$[\mathbf{P}_\Omega \mathbf{M}'] = q$. [*Hint*: use *B3.4*.]
5. If X is $n \times p$ of rank p and B is $p \times q$ of rank q, show that rank$[\mathbf{XB}] = q$.

3.10 OTHER METHODS OF ESTIMATION

3.10.1 *Biased Estimation*

Given the general linear model $\mathbf{Y} = \mathbf{X}\boldsymbol{\beta} + \boldsymbol{\varepsilon}$, where $\mathcal{E}[\boldsymbol{\varepsilon}] = \mathbf{0}$ and $\mathcal{D}[\boldsymbol{\varepsilon}] = \sigma^2 \mathbf{I}_n$, $\mathbf{a}'\hat{\boldsymbol{\beta}}$ is the minimum variance unbiased estimate of $\mathbf{a}'\boldsymbol{\beta}$ when $\boldsymbol{\varepsilon}$ is normal, and the minimum variance linear unbiased estimate without the normality assumption (Theorem 3.2). Although this implies that $\hat{\beta}_j$ is a minimum variance estimate of β_j, it does not guarantee that its variance will in fact be small. In particular, if $\mathbf{X}'\mathbf{X}$ is near singularity so that its smallest eigenvalue, λ_{p-1}, say, is near zero, then by *A1.5* the "total variance"

$$\begin{aligned}
\sum_{j=0}^{p-1} \operatorname{var}[\hat{\beta}_j] &= \sigma^2 \operatorname{tr}[(\mathbf{X}'\mathbf{X})^{-1}] \\
&= \sigma^2 \sum_{j=0}^{p-1} \lambda_j^{-1} \\
&> \sigma^2 \lambda_{p-1}^{-1}
\end{aligned}$$

may be too large for practical purposes. To get round this problem of an "ill-conditioned" $\mathbf{X}$ matrix, Hoerl and Kennard [1970a, b] introduced the class of estimators

$$\tilde{\boldsymbol{\beta}}_{(k)} = (\mathbf{X}'\mathbf{X} + k\mathbf{I}_n)^{-1}\mathbf{X}'\mathbf{Y} \qquad (0 \leqslant k < \infty),$$

which are known as *ridge estimators*. Since

$$\begin{aligned}\tilde{\boldsymbol{\beta}}_{(k)} &= (\mathbf{X}'\mathbf{X}+k\mathbf{I}_n)^{-1}\mathbf{X}'\mathbf{X}\hat{\boldsymbol{\beta}} \\ &= [\mathbf{I}_n + k(\mathbf{X}'\mathbf{X})^{-1}]^{-1}\hat{\boldsymbol{\beta}} \\ &= \mathbf{K}\hat{\boldsymbol{\beta}},\end{aligned} \tag{3.69}$$

say, we see that $\tilde{\boldsymbol{\beta}}_{(k)}$ is a biased estimate of $\boldsymbol{\beta}$ when $k>0$. The main justification for using ridge estimators has been twofold:

(1) By plotting the components of $\tilde{\boldsymbol{\beta}}_{(k)}$ and the corresponding RSS against k we can obtain some idea as to the ill-conditioning of $\mathbf{X}$. A value of k can then be chosen for which (*a*) the system becomes stable, (*b*) the regression coefficients have reasonable values, and (*c*) RSS is not grossly inflated.

(2) There always exists $k>0$ such that the total mean square error for $\tilde{\boldsymbol{\beta}}_{(k)}$ is less than the total for $\hat{\boldsymbol{\beta}}$. Here the total mean square error is defined to be (Theorem 1.7, Corollary 1)

$$\begin{aligned}\sum_j E\left[\left(\tilde{\beta}_{j(k)}-\beta_j\right)^2\right] &= E\left[(\tilde{\boldsymbol{\beta}}_{(k)}-\boldsymbol{\beta})'(\tilde{\boldsymbol{\beta}}_{(k)}-\boldsymbol{\beta})\right] \\ &= \operatorname{tr}\mathcal{D}\left[\tilde{\boldsymbol{\beta}}_{(k)}\right]+(\mathbf{K}\boldsymbol{\beta}-\boldsymbol{\beta})'(\mathbf{K}\boldsymbol{\beta}-\boldsymbol{\beta}) \\ &= \textit{total variance} + \textit{bias},\end{aligned} \tag{3.70}$$

and Hoerl showed that for a given $\mathbf{X}$ it is always possible to find $k>0$ such that the above expression is less than $\operatorname{tr}\mathcal{D}[\hat{\boldsymbol{\beta}}]$. Thus by allowing a little bias, the total variance in (3.70) can be reduced to such an extent that the total mean square error is reduced.

Using a different characterization for the form of a biased estimator, Banerjee and Carr [1971] suggested that $\tilde{\boldsymbol{\beta}}_{(k)}$ should be compared with a different unbiased estimator rather than with $\hat{\boldsymbol{\beta}}$. They also proved that there exists $k>0$ such that $\tilde{\boldsymbol{\beta}}_{(k)}$ has a smaller total mean square error than that of their unbiased estimator. However, the criterion of total mean square error has been criticized by Nelder [1972] for its use of Euclidean distance to measure the difference between $\tilde{\boldsymbol{\beta}}_{(k)}$ and $\boldsymbol{\beta}$. To answer this criticism Theobald [1974] established a similar result to (2) above using a weighted sum of squares of the form

$$E\left[(\tilde{\boldsymbol{\beta}}_{(k)}-\boldsymbol{\beta})'\mathbf{B}(\tilde{\boldsymbol{\beta}}_{(k)}-\boldsymbol{\beta})\right],$$

where $\mathbf{B}$ is positive semidefinite. However, the objection has been overcome by Goldstein and Smith [1974] and Lowerre [1974], who independently showed that for any $\boldsymbol{\beta}$ there exists $k>0$ such that *each element* of $\tilde{\boldsymbol{\beta}}_{(k)}$ has a smaller mean square error than the corresponding element of $\hat{\boldsymbol{\beta}}$.

Another class of biased estimators considered in the literature are the so-called "shrunken estimators" of the form $\lambda\hat{\boldsymbol{\beta}}$ $(0<\lambda\leqslant 1)$. This "shrinkage" or scale-down technique has been used, for example, by Stein [1960], James and Stein [1961], Sclove [1968], Thompson [1968], Mayer and Willke [1973], and Narula [1974]. Mayer and Willke compare one type of shrunken estimator with the ridge estimator and the rather complex estimators given by Sclove [1968]. It transpires that (Goldstein and Smith [1974]) the ridge estimator can be obtained as a special case of a shrunken estimator for a canonical form of the regression model. The authors use this approach to generate a more general class of ridge estimators.

We note that the ridge estimators belong to the general class of biased estimators of the form

$$\boldsymbol{\beta}^* = (\mathbf{X}'\mathbf{X}+\mathbf{C})^{-1}\mathbf{X}'\mathbf{Y} \tag{3.71}$$

where $\mathbf{C}$ is positive definite, and $\mathbf{C}$ and $\mathbf{X}'\mathbf{X}$ commute. Lowerre [1974] showed that if the eigenvalues of $\mathbf{C}$ are small enough then each element of $\boldsymbol{\beta}^*$ has a smaller mean square error than the corresponding element of $\hat{\boldsymbol{\beta}}$. Setting $\mathbf{C}=k\mathbf{I}_n$, we see that this property holds for ridge estimators provided k is small enough; setting $\mathbf{C}=(\lambda^{-1}-1)\mathbf{X}'\mathbf{X}$, we see it applies to shrunken estimators if λ is close enough to unity. Allen [1974] also examines (3.71) from a prediction viewpoint.

It is of interest to note that for a certain value of k ($k=\sigma^2/\sigma_\beta^2$, where σ_β^2 is the variance of the prior distribution of each β_j), the ridge estimator $\tilde{\boldsymbol{\beta}}_{(k)}$ is also a Bayes estimate of $\boldsymbol{\beta}$ (Lindley and Smith [1972: p. 11], Goldstein and Smith [1974: p. 291]).

3.10.2 Nonnegative Estimation

The problem of least squares estimation subject to the restrictions $\beta_j \geqslant 0$ $(j=0,1,\ldots,p-1)$ has been considered by Waterman [1974]. He shows that the problem can be solved by considering 2^p unrestricted problems.

3.10.3 Censored Data

In some regression problems, data on the response variable Y are censored; that is, the values of some observations are known only to be above or else below some value. Such data often arise in accelerated life testing where life is the response variable and temperature or stress is the regressor variable, and some test units have not failed at the time of analysis. In such situations the standard least squares method cannot be used because the values of the censored observations are not known. However, theory and methods for handling censored data are given by Nelson and Hahn [1972, 1973]. The straight-line case in particular is discussed by Chen and Dixon [1972]. Methods of analysing residuals (cf. Section 6.6) from censored data are described by Nelson [1973].

3.10.4 Robust Estimation

We have seen in this chapter that when Y is normally distributed, the least squares estimate of $\boldsymbol{\beta}$ has a number of desirable properties. However, as pointed out by Andrews et al. [1972: Chapter 7], the method of least squares may be far from optimal in many nonnormal situations in which the distribution has longer tails. For example, a study by these authors clearly demonstrates the inefficiency of least squares relative to more robust estimates of location for a wide variety of distributions (see also Moussa-Hamouda and Leone [1974]). Although residual plots (Section 6.6) may indicate any shortcomings in the model, the interpretation of such plots often requires a skill that is beyond the average user from a non-mathematical discipline. On the other hand, a robust fit may leave several residuals much larger, thereby indicating more clearly that something is wrong. An example of robust fitting in polynomial regression is given by Beaton and Tukey [1974].

One alternative method that has received some attention is that of minimizing $\Sigma_i|\varepsilon_i|$ with respect to $\boldsymbol{\beta}$. This L1-norm minimization problem can be reduced to a general linear programming problem and a procedure similar to the simplex method was given by Davies [1967]. Unfortunately the solution is not always unique and some of the common linear programming (LP) algorithms may lead to biased estimates of $\boldsymbol{\beta}$ (Kiountouzis [1973], Sielken and Hartley [1973]). However, an efficient LP method of finding an unbiased solution has since been proposed by Sielken and Hartley. An iterative technique that is based on the least squares method, and consequently uses less storage, has been given by Schlossmacher [1973]; however, in this case questions of convergence and unbiasedness are not resolved. For further references, and an algorithm for the straight-line case, the reader is referred to Sadovski [1974].

It is of interest to note that this L1-norm method is equivalent to the method of maximum likelihood when ε has a double exponential distribution. When ε has a uniform distribution with an unknown range then the maximum likelihood criterion consists of minimizing the maximum $|\varepsilon_i|$; an algorithm for finding an unbiased estimate in this case is given by Sielken and Hartley [1973].

A natural extension of the above method is to consider minimizing $\Sigma_i |\varepsilon_i|^p$ $(1 \leqslant p \leqslant 2)$; a value of $p = 1.5$ may be a reasonable compromise (cf. Hogg [1974: pp. 915 ff.] for references). Other methods of robust estimation for the regression model are described by Andrews [1974] and Bickel [1975].

3.11 OPTIMAL DESIGN

How does an experimenter choose **X**? Various criteria have been suggested, the two most popular being (1) minimize $|\mathfrak{D}[\hat{\boldsymbol{\beta}}]|$, or equivalently, maximize $|\mathbf{X}'\mathbf{X}|$ (called *D*-optimality), and (2) minimize the total or average variance, that is, minimize $\text{tr}[(\mathbf{X}'\mathbf{X})^{-1}]$. *D*-optimality, introduced by Kiefer [1959], has been studied extensively (cf. St. John and Draper [1975] for a survey and bibliography) and many of the results were first developed for the special case of polynomial regression (Section 8.4). Further support for *D*-optimality was engendered by the so-called equivalence theorem, proved by Kiefer and Wolfowitz [1960] (see also Whittle [1973] and Silvey and Titterington [1974]), which showed that maximizing $|\mathbf{X}'\mathbf{X}|$ over some region $\mathfrak{X}$ was equivalent to minimizing the maximum variance of $\hat{\mathbf{Y}} = \mathbf{x}'\hat{\boldsymbol{\beta}}$ for $\mathbf{x} \in \mathfrak{X}$. A design with this latter property is called *G*-optimal or minimax. Also algorithms are now available for constructing *D*-optimal designs; these are reviewed by St. John and Draper [1975].

Although the trace criterion for optimality leads to **X** matrices with orthogonal columns (Section 3.5, lemma), this criterion has several shortcomings (e.g., a dependence on the scaling of the regressors) so that there is a general preference for *D*-optimality (e.g., Box and Draper [1971]). However, sometimes an experimental design satisfies both criteria—for example, the 2^2 design with $x_1, x_2 = \pm 1$ (see Box and Draper [1971: appendix] for a proof).

Sometimes *D*-optimal designs consist of trials at the same number of experimental conditions as there are parameters, for example, polynomial regression (Section 8.4), so that the design may be useless for testing the adequacy of the model or for comparing competitive models. One approach (Atkinson [1972], Atkinson and Cox [1974]) is to embed the model(s) in a more general model and design to estimate any additional parameters is some optimal fashion. In this case we are interested in optimality criteria for just a subset of the more general model only; the reader is referred to Atkinson [1972], St. John and Draper [1975], and Section 8.4 for general comments about this problem. A related problem is that of discriminating between competing models (Atkinson and Cox [1974]).

MISCELLANEOUS EXERCISES 3

1. Let $Y_i = a_i\beta_1 + b_i\beta_2 + \varepsilon_i$ $(i=1,2,\ldots,n)$ where the a_i, b_i are known, and the ε_i are independently and identically distributed as $N(0,\sigma^2)$. Find a necessary and sufficient condition for the least squares estimates of β_1 and β_2 to be independent.
2. Let $\mathbf{Y} = \boldsymbol{\theta} + \boldsymbol{\varepsilon}$, where $\mathcal{E}[\boldsymbol{\varepsilon}] = \mathbf{0}$. Prove that the value of $\boldsymbol{\theta}$ which minimizes $\|\mathbf{Y} - \boldsymbol{\theta}\|^2$ subject to $\mathbf{A}\boldsymbol{\theta} = \mathbf{0}$, where $\mathbf{A}$ is a known $q \times n$ matrix of rank q, is

$$\hat{\boldsymbol{\theta}} = (\mathbf{I}_n - \mathbf{A}'(\mathbf{A}\mathbf{A}')^{-1}\mathbf{A})\mathbf{Y}.$$

3. Let $\mathbf{Y} = \mathbf{X}\boldsymbol{\beta} + \boldsymbol{\varepsilon}$, where $\mathcal{E}[\boldsymbol{\varepsilon}] = \mathbf{0}$, $\mathcal{D}[\boldsymbol{\varepsilon}] = \sigma^2\mathbf{I}_n$ and $\mathbf{X}$ is $n \times p$ of rank p. If $\mathbf{X}$ and $\boldsymbol{\beta}$ are partitioned in the form

$$\mathbf{X}\boldsymbol{\beta} = (\mathbf{X}_1, \mathbf{X}_2)\begin{pmatrix} \boldsymbol{\beta}_1 \\ \boldsymbol{\beta}_2 \end{pmatrix},$$

prove that the least squares estimate $\hat{\boldsymbol{\beta}}_2$ of $\boldsymbol{\beta}_2$ is given by

$$\hat{\boldsymbol{\beta}}_2 = [\mathbf{X}_2'\mathbf{X}_2 - \mathbf{X}_2'\mathbf{X}_1(\mathbf{X}_1'\mathbf{X}_1)^{-1}\mathbf{X}_1'\mathbf{X}_2]^{-1}[\mathbf{X}_2'\mathbf{Y} - \mathbf{X}_2'\mathbf{X}_1(\mathbf{X}_1'\mathbf{X}_1)^{-1}\mathbf{X}_1'\mathbf{Y}].$$

Find $\mathcal{D}[\hat{\boldsymbol{\beta}}_2]$.
4. Suppose that $\mathcal{E}[\mathbf{Y}] = \mathbf{X}\boldsymbol{\beta}$ and $\mathcal{D}[\mathbf{Y}] = \sigma^2\mathbf{I}_n$. Prove that $\mathbf{a}'\mathbf{Y}$ is the linear unbiased estimate of $E[\mathbf{a}'\mathbf{Y}]$ with minimum variance if and only if $\text{cov}[\mathbf{a}'\mathbf{Y}, \mathbf{b}'\mathbf{Y}] = 0$ for all $\mathbf{b}$ such that $E[\mathbf{b}'\mathbf{Y}] = 0$ (that is, $\mathbf{b}'\mathbf{X} = \mathbf{0}'$).
Rao [1973]
5. If $\mathbf{X}$ has full rank and $\hat{\mathbf{Y}} = \mathbf{X}\hat{\boldsymbol{\beta}}$, prove that

$$\sum_{i=1}^{n} \text{var}[\hat{Y}_i] = \sigma^2 p.$$

6. Estimate the weights β_i $(i=1,2,3,4)$ of four objects from the following weighing data (see Exercise 6 of Exercises 3e at the end of Section 3.5 for notation):

x_1	x_2	x_3	x_4	Weight (Y)
1	1	1	1	20.2
1	−1	1	−1	8.0
1	1	−1	−1	9.7
1	−1	−1	1	1.9

7. Three parcels are weighed at a post office singly, in pairs, and all together, giving weights Y_{ijk} $(i,j,k=0,1)$, the suffix 1 denoting the presence of a particular parcel and the suffix 0 denoting its absence. Find the least squares estimates of the weights. Rahman [1967]
8. An experimenter wishes to estimate the density d of a liquid by weighing known volumes of the liquid. Let Y_i be the weight for volume x_i $(i=1,2,\ldots,n)$ and let $E[Y_i]=dx_i$ and $\text{var}[Y_i]=\sigma^2 f(x_i)$. Find the least squares estimate of d for the following cases:

$$\text{(a) } f(x_i)\equiv 1; \qquad \text{(b) } f(x_i)=x_i; \qquad \text{(c) } f(x_i)=x_i^2.$$

9. Given the factorization $\mathbf{X}=\mathbf{KL}$ in Equation (3.41), prove that the elements of $\mathbf{L}\boldsymbol{\beta}$ are estimable.
10. Let $Y_i=\beta_0+\beta_1 x_i+\varepsilon_i$ $(i=1,2,3)$, where $\mathscr{E}[\boldsymbol{\varepsilon}]=\mathbf{0}$, $\mathscr{D}[\boldsymbol{\varepsilon}]=\sigma^2\mathbf{V}$ with

$$V=\begin{bmatrix} 1 & \rho a & \rho \\ \rho a & a^2 & \rho a \\ \rho & \rho a & 1 \end{bmatrix}, \qquad \begin{matrix} (a,\rho \text{ unknown}) \\ 0<\rho<1 \end{matrix}$$

and $x_1=-1$, $x_2=0$, and $x_3=1$. Show that the generalized least squares estimates of β_0 and β_1 are

$$\begin{pmatrix} \beta_0^* \\ \beta_1^* \end{pmatrix}=\begin{bmatrix} \frac{1}{r}\{(a^2-a\rho)Y_1+(1-2a\rho+\rho)Y_2+(a^2-a\rho)Y_3\} \\ -\frac{1}{2}Y_1+\frac{1}{2}Y_3 \end{bmatrix}$$

where $r=1+\rho+2a^2-4a\rho$. Also prove the following:

(a) If $a=1$ then the fitted regression $Y_i^*=\beta_0^*+\beta_1^* x_i$ cannot lie wholly above or below the values of Y_i (that is, the $Y_i-Y_i^*$ cannot all have the same sign).

(b) If $0<a<\rho<1$ then the fitted regression line can lie wholly above or below the observations. Canner [1969]

11. Given linearly independent identifiability constraints $\mathbf{H}\boldsymbol{\beta}=\mathbf{0}$, and matrices $\mathbf{C}_1$ and $\mathbf{C}_2$ such that

$$(\mathbf{C}_1', \ \mathbf{C}_2')\begin{pmatrix} \mathbf{X}'\mathbf{X} \\ \mathbf{H} \end{pmatrix}=\mathbf{I},$$

prove that $\mathbf{C}_1$ is a generalized inverse of $\mathbf{X}'\mathbf{X}$. Rao [1973]

12. If $\mathbf{X}$ is not of full rank show that any solution $\boldsymbol{\beta}$ of $\mathbf{X}'\mathbf{V}^{-1}\mathbf{X}\boldsymbol{\beta}=\mathbf{X}'\mathbf{V}^{-1}\mathbf{Y}$ minimizes $(\mathbf{Y}-\mathbf{X}\boldsymbol{\beta})'\mathbf{V}^{-1}(\mathbf{Y}-\mathbf{X}\boldsymbol{\beta})$.

13. Let

$$Y_1 = \theta_1 + \theta_2 + \varepsilon_1,$$

$$Y_2 = \theta_1 - 2\theta_2 + \varepsilon_2,$$

and

$$Y_3 = 2\theta_1 - \theta_2 + \varepsilon_3,$$

where $E[\varepsilon_i] = 0$ $(i = 1, 2, 3)$. Find the least squares estimates of θ_1 and θ_2. Using the method of two-step least squares, find the least squares estimate of θ_3 when the above equations are augmented to

$$Y_1 = \theta_1 + \theta_2 + \theta_3 + \varepsilon_1,$$

$$Y_2 = \theta_1 - 2\theta_2 + \theta_3 + \varepsilon_2,$$

$$Y_3 = 2\theta_1 - \theta_2 + \theta_3 + \varepsilon_3.$$

14. Given the usual full rank regression model prove that $\overline{Y}$ and $\Sigma_i (Y_i - \hat{Y}_i)^2$ are statistically independent.
15. Let $Y_i = \beta x_i + u_i$, $x_i > 0$ $(i = 1, 2, \ldots, n)$, where $u_i = \rho u_{i-1} + \varepsilon_i$ and the ε_i are independently distributed as $N(0, \sigma^2)$. If $\hat{\beta}$ is the ordinary least squares estimate of β prove that var$[\hat{\beta}]$ is inflated when $\rho > 0$.
16. Suppose that $E[Y_t] = \beta_0 + \beta_1 \cos(2\pi k_1 t/n) + \beta_2 \sin(2\pi k_2 t/n)$, where $t = 1, 2, \ldots, n$, and k_1 and k_2 are positive integers. Find the least squares estimates of β_0, β_1, and β_2.
17. Suppose that $E[Y_i] = \alpha_0 + \beta_1(x_{i1} - \bar{x}_{.1}) + \beta_2(x_{i2} - \bar{x}_{.2})$, $i = 1, 2, \ldots, n$. Show that the least squares estimates of α_0, β_1, and β_2 can be obtained by the following two-stage procedure:
    (a) Fit the model $E[Y_i] = \alpha_0 + \beta_1(x_{i1} - \bar{x}_{.1})$.
    (b) Regress the residuals from (a) on $(x_{i2} - \bar{x}_{.2})$.

CHAPTER 4

Linear Regression: Hypothesis Testing

4.1 THE F-TEST

4.1.1 Derivation

Consider the linear model $\mathbf{Y}=\mathbf{X}\boldsymbol{\beta}+\boldsymbol{\varepsilon}$, where $\mathbf{X}$ is $n\times p$ of rank p and $\boldsymbol{\varepsilon}\sim N_n(\mathbf{0},\sigma^2\mathbf{I}_n)$. Suppose we wish to test the hypothesis $H:\mathbf{A}\boldsymbol{\beta}=\mathbf{c}$, where $\mathbf{A}$ is a known $q\times p$ matrix of rank q and $\mathbf{c}$ is a known $q\times 1$ vector (cf. Section 1.2 for some motivation of this choice of H). Let

$$\text{RSS}=(\mathbf{Y}-\mathbf{X}\hat{\boldsymbol{\beta}})'(\mathbf{Y}-\mathbf{X}\hat{\boldsymbol{\beta}}) \qquad \left[=(n-p)S^2\right]$$

and

$$\text{RSS}_H=(\mathbf{Y}-\mathbf{X}\hat{\boldsymbol{\beta}}_H)'(\mathbf{Y}-\mathbf{X}\hat{\boldsymbol{\beta}}_H)$$

where, from (3.59),

$$\hat{\boldsymbol{\beta}}_H=\hat{\boldsymbol{\beta}}+(\mathbf{X}'\mathbf{X})^{-1}\mathbf{A}'\left[\mathbf{A}(\mathbf{X}'\mathbf{X})^{-1}\mathbf{A}'\right]^{-1}(\mathbf{c}-\mathbf{A}\hat{\boldsymbol{\beta}}) \tag{4.1}$$

and RSS_H is the minimum value of $\boldsymbol{\varepsilon}'\boldsymbol{\varepsilon}$ subject to $\mathbf{A}\boldsymbol{\beta}=\mathbf{c}$. An F-statistic for testing H is now described in the following theorem.

THEOREM 4.1

(i) $\text{RSS}_H-\text{RSS}=(\mathbf{A}\hat{\boldsymbol{\beta}}-\mathbf{c})'[\mathbf{A}(\mathbf{X}'\mathbf{X})^{-1}\mathbf{A}']^{-1}(\mathbf{A}\hat{\boldsymbol{\beta}}-\mathbf{c})$.

(ii) $E[\text{RSS}_H-\text{RSS}]=\sigma^2q+(\mathbf{A}\boldsymbol{\beta}-\mathbf{c})'[\mathbf{A}(\mathbf{X}'\mathbf{X})^{-1}\mathbf{A}']^{-1}(\mathbf{A}\boldsymbol{\beta}-\mathbf{c})$.

(iii) When H is true

$$F=\frac{\text{RSS}_H-\text{RSS}/q}{\text{RSS}/(n-p)}=\frac{(\mathbf{A}\hat{\boldsymbol{\beta}}-\mathbf{c})'\left[\mathbf{A}(\mathbf{X}'\mathbf{X})^{-1}\mathbf{A}'\right]^{-1}(\mathbf{A}\hat{\boldsymbol{\beta}}-\mathbf{c})}{qS^2}$$

is distributed as $F_{q,n-p}$ (the F-distribution with q and $n-p$ degrees of freedom, respectively).

(iv) When $\mathbf{c}=\mathbf{0}$, F can be expressed in the form

$$F=\frac{n-p}{q}\frac{\mathbf{Y}'(\mathbf{P}-\mathbf{P}_H)\mathbf{Y}}{\mathbf{Y}'(\mathbf{I}_n-\mathbf{P})\mathbf{Y}},$$

where $\mathbf{P}_H$ is symmetric and idempotent, and $\mathbf{P}_H\mathbf{P}=\mathbf{P}\mathbf{P}_H=\mathbf{P}_H$.

Proof. (i) From (3.64) in Section 3.9.1 we have $\text{RSS}_H-\text{RSS}=(\hat{\boldsymbol{\beta}}-\hat{\boldsymbol{\beta}}_H)'\mathbf{X}'\mathbf{X}(\hat{\boldsymbol{\beta}}-\hat{\boldsymbol{\beta}}_H)$; substituting for $\hat{\boldsymbol{\beta}}-\hat{\boldsymbol{\beta}}_H$ using Equation (4.1) leads to the required result.

(ii) Since the rows of $\mathbf{A}$ are linearly independent and $\hat{\boldsymbol{\beta}}\sim N_p(\boldsymbol{\beta},\sigma^2(\mathbf{X}'\mathbf{X})^{-1})$, we have from Theorem 2.2 in Section 2.2, that $\mathbf{A}\hat{\boldsymbol{\beta}}\sim N_q(\mathbf{A}\boldsymbol{\beta},\sigma^2\mathbf{A}(\mathbf{X}'\mathbf{X})^{-1}\mathbf{A}')$. Let $\mathbf{Z}=\mathbf{A}\hat{\boldsymbol{\beta}}-\mathbf{c}$ and $\mathbf{B}=\mathbf{A}(\mathbf{X}'\mathbf{X})^{-1}\mathbf{A}'$; then $\mathcal{E}[\mathbf{Z}]=\mathbf{A}\boldsymbol{\beta}-\mathbf{c}$ and

$$\mathcal{D}[\mathbf{Z}]=\mathcal{D}[\mathbf{A}\hat{\boldsymbol{\beta}}]=\sigma^2\mathbf{B}.$$

Hence, using Theorem 1.7 (Corollary 1) in Section 1.4,

$$\begin{aligned}E[\text{RSS}_H-\text{RSS}]&=E[\mathbf{Z}'\mathbf{B}^{-1}\mathbf{Z}] \qquad [\text{by (i)}]\\&=\text{tr}[\sigma^2\mathbf{B}^{-1}\mathbf{B}]+(\mathbf{A}\boldsymbol{\beta}-\mathbf{c})'\mathbf{B}^{-1}(\mathbf{A}\boldsymbol{\beta}-\mathbf{c})\\&=\text{tr}[\sigma^2\mathbf{I}_q]+(\mathbf{A}\boldsymbol{\beta}-\mathbf{c})'\mathbf{B}^{-1}(\mathbf{A}\boldsymbol{\beta}-\mathbf{c})\\&=\sigma^2q+(\mathbf{A}\boldsymbol{\beta}-\mathbf{c})'\mathbf{B}^{-1}(\mathbf{A}\boldsymbol{\beta}-\mathbf{c}).\end{aligned} \tag{4.2}$$

(iii) From (i), RSS_H-RSS is a continuous function of $\hat{\boldsymbol{\beta}}$ and is therefore independent of RSS (by Theorem 3.5 (iii) in Section 3.4 and Theorem 1.9 in Section 1.5). Also when H is true, $\mathbf{A}\hat{\boldsymbol{\beta}}\sim N_q(\mathbf{c},\sigma^2\mathbf{A}(\mathbf{X}'\mathbf{X})^{-1}\mathbf{A}')$ so that by Theorem 2.1 (iii),

$$\frac{\text{RSS}_H-\text{RSS}}{\sigma^2}=(\mathbf{A}\hat{\boldsymbol{\beta}}-\mathbf{c})'(\mathcal{D}[\mathbf{A}\hat{\boldsymbol{\beta}}])^{-1}(\mathbf{A}\hat{\boldsymbol{\beta}}-\mathbf{c})$$

is χ^2_q. Finally, since $\text{RSS}/\sigma^2\sim\chi^2_{n-p}$ [Theorem 3.5 (iv)], we have that

$$F=\frac{(\text{RSS}_H-\text{RSS})/\sigma^2q}{\text{RSS}/\sigma^2(n-p)}$$

is of the form $[\chi^2_q/q]/[\chi^2_{n-p}/(n-p)]$ when H is true. Hence $F\sim F_{q,n-p}$ when H is true.

(iv) Using Equation (4.1) with $\mathbf{c}=\mathbf{0}$, we have

$$\begin{aligned}\hat{\mathbf{Y}}_H &= \mathbf{X}\hat{\boldsymbol{\beta}}_H \\ &= \left\{\mathbf{X}(\mathbf{X}'\mathbf{X})^{-1}\mathbf{X}' - \mathbf{X}(\mathbf{X}'\mathbf{X})^{-1}\mathbf{A}'\left[\mathbf{A}(\mathbf{X}'\mathbf{X})^{-1}\mathbf{A}'\right]^{-1}\mathbf{A}(\mathbf{X}'\mathbf{X})^{-1}\mathbf{X}'\right\}\mathbf{Y} \\ &= (\mathbf{P}-\mathbf{P}_1)\mathbf{Y} \qquad (4.3) \\ &= \mathbf{P}_H\mathbf{Y}, \qquad (4.4)\end{aligned}$$

say, where $\mathbf{P}_H$ is symmetric. Multiplying the matrices together and cancelling matrices with their inverses where possible, we find that $\mathbf{P}_1$ is symmetric and idempotent and $\mathbf{P}_1\mathbf{P}=\mathbf{P}\mathbf{P}_1=\mathbf{P}_1$. Hence

$$\begin{aligned}\mathbf{P}_H^2 &= \mathbf{P}^2 - \mathbf{P}_1\mathbf{P} - \mathbf{P}\mathbf{P}_1 + \mathbf{P}_1^2 \\ &= \mathbf{P} - 2\mathbf{P}_1 + \mathbf{P}_1 \\ &= \mathbf{P} - \mathbf{P}_1 \\ &= \mathbf{P}_H, \qquad (4.5)\end{aligned}$$

$$\mathbf{P}_H\mathbf{P} = (\mathbf{P}-\mathbf{P}_1)\mathbf{P} = \mathbf{P}-\mathbf{P}_1 = \mathbf{P}_H \qquad (4.6)$$

and, taking transposes, $\mathbf{P}\mathbf{P}_H=\mathbf{P}_H$. To complete the proof we recall that $\text{RSS}=\mathbf{Y}'(\mathbf{I}_n-\mathbf{P})\mathbf{Y}$ and, in a similar fashion, obtain

$$\begin{aligned}\text{RSS}_H &= \|\mathbf{Y}-\mathbf{X}\hat{\boldsymbol{\beta}}_H\|^2 \\ &= \mathbf{Y}'(\mathbf{I}_n-\mathbf{P}_H)^2\mathbf{Y} \\ &= \mathbf{Y}'(\mathbf{I}_n-\mathbf{P}_H)\mathbf{Y}. \qquad (4.7)\end{aligned}$$

Thus $\text{RSS}_H - \text{RSS} = \mathbf{Y}'(\mathbf{P}-\mathbf{P}_H)\mathbf{Y}$. □

We note that if H is true, $\mathbf{A}\hat{\boldsymbol{\beta}}$ (the best linear unbiased estimate of $\mathbf{A}\boldsymbol{\beta}$) is close to $\mathbf{c}$, and $\text{RSS}_H - \text{RSS}$ is "small." However, if $\mathbf{A}\boldsymbol{\beta}$ is very different from $\mathbf{c}$, $\text{RSS}_H - \text{RSS}$ tends to be large. Thus our F-test is a one-tailed test; we reject H if F is significantly large.

When $q>2$ it is usually more convenient to obtain RSS and RSS_H by finding the unrestricted and restricted minimum values of $\boldsymbol{\varepsilon}'\boldsymbol{\varepsilon}$ directly.

However, if $q \leqslant 2$, F can usually be found most readily by applying the general matrix theory above; the matrix $[\mathbf{A}(\mathbf{X}'\mathbf{X})^{-1}\mathbf{A}']$ to be inverted is only of order one or two. Examples are given in Section 4.1.3.

It should be noted that, since RSS_H is unique, it does not matter what method we use for obtaining RSS_H. We could, for example, use the constraints $\mathbf{A}\boldsymbol{\beta}=\mathbf{c}$ to eliminate some of the β_j and then minimize $\boldsymbol{\varepsilon}'\boldsymbol{\varepsilon}$ with respect to the remaining β_j's.

Part (iv) of the above theorem highlights the geometry underlying the F-test. This geometry is described later in Section 4.5.1.

EXERCISES 4a

1. If $H:\mathbf{A}\boldsymbol{\beta}=\mathbf{c}$ is true, show that F can be expressed in the form

$$\frac{n-p}{q}\cdot\frac{\boldsymbol{\varepsilon}'(\mathbf{P}-\mathbf{P}_H)\boldsymbol{\varepsilon}}{\boldsymbol{\varepsilon}'(\mathbf{I}_n-\mathbf{P})\boldsymbol{\varepsilon}}.$$

2. If $\hat{\boldsymbol{\lambda}}_H$ is the "least squares" estimate of the Lagrange multiplier associated with the constraints $\mathbf{A}\boldsymbol{\beta}=\mathbf{c}$ (cf. Section 3.9), show that

$$\text{RSS}_H-\text{RSS}=\sigma^2\hat{\boldsymbol{\lambda}}_H'\left(\mathscr{D}\left[\hat{\boldsymbol{\lambda}}_H\right]\right)^{-1}\hat{\boldsymbol{\lambda}}_H.$$

4.1.2 *Motivation of the F-Test*

The question arises as to how we came to choose F in the first place; we now give two methods of motivating our test.

Suppose we define $S_H^2=(\text{RSS}_H-\text{RSS})/q$; then, by Theorem 4.1 (ii),

$$E\left[S_H^2\right]=\sigma^2+\frac{(\mathbf{A}\boldsymbol{\beta}-\mathbf{c})'\left[\mathbf{A}(\mathbf{X}'\mathbf{X})^{-1}\mathbf{A}'\right]^{-1}(\mathbf{A}\boldsymbol{\beta}-\mathbf{c})}{q}$$

$$=\sigma^2+\delta,\quad \text{say}$$

where $\delta \geqslant 0$ (since $[\mathbf{A}(\mathbf{X}'\mathbf{X})^{-1}\mathbf{A}']^{-1}=\mathscr{D}[\mathbf{A}\hat{\boldsymbol{\beta}}]/\sigma^2$ is positive definite). Also (Theorem 3.3, Section 3.3)

$$E\left[S^2\right]=\sigma^2.$$

When H is true, $\delta=0$ and S_H^2 and S^2 are both unbiased estimates of σ^2, that is, $F=S_H^2/S^2 \approx 1$. When H is false, $\delta > 0$ and $E[S_H^2] > E[S^2]$ so that

$$E[F]=E[S_H^2]E\left[\frac{1}{S^2}\right] > E[S_H^2]/E[S^2] > 1$$

(by the independence of S_H^2 and S^2, and *A11.2*). Thus F gives some indication as to the "true state of affairs"; H is rejected if F is significantly large.

The F-statistic can also be motivated by considering the likelihood ratio test of H. The likelihood function, $L(\boldsymbol{\beta},\sigma^2)$, say, for the underlying model is the probability density function of $\mathbf{Y}$, namely,

$$L(\boldsymbol{\beta},\sigma^2)=(2\pi\sigma^2)^{-(1/2)n}\exp\left\{-\frac{1}{2\sigma^2}(\mathbf{Y}-\mathbf{X}\boldsymbol{\beta})'(\mathbf{Y}-\mathbf{X}\boldsymbol{\beta})\right\}.$$

Solving $\partial \log L/\partial \boldsymbol{\beta}=\mathbf{0}$ and $\partial \log L/\partial \sigma^2 = 0$, we obtain the maximum likelihood estimates

$$\hat{\boldsymbol{\beta}}=(\mathbf{X}'\mathbf{X})^{-1}\mathbf{X}'\mathbf{Y} \qquad \text{and} \qquad \hat{\sigma}^2=\frac{\text{RSS}}{n}, \tag{4.8}$$

and the maximum, $L(\hat{\boldsymbol{\beta}},\hat{\sigma}^2)$, of L is $(2\pi\hat{\sigma}^2)^{-(1/2)n}e^{-(1/2)n}$. (It is not surprising that the maximum likelihood estimate of $\boldsymbol{\beta}$ is the same as the least squares estimate for, in spite of the nuisance parameter σ^2, maximizing L is equivalent to minimizing the quadratic form in the exponent.)

If we use a method almost identical to that used in Section 3.9.1 the maximum likelihood estimates, subject to the constraints $\mathbf{A}\boldsymbol{\beta}=\mathbf{c}$, are $\hat{\boldsymbol{\beta}}_H$ and $\hat{\sigma}_H^2=\text{RSS}_H/n$. The maximum value of L is now

$$L(\hat{\boldsymbol{\beta}}_H,\hat{\sigma}_H^2)=(2\pi\hat{\sigma}_H^2)^{-(1/2)n}e^{-(1/2)n}.$$

The likelihood ratio statistic is

$$l=\frac{L(\hat{\boldsymbol{\beta}}_H,\hat{\sigma}_H^2)}{L(\hat{\boldsymbol{\beta}},\hat{\sigma}^2)}=\left[\frac{\hat{\sigma}^2}{\hat{\sigma}_H^2}\right]^{n/2}$$

and according to the likelihood ratio principle we reject H if l is too small. Since

$$F=\frac{n-p}{q}(l^{-(1/2)n}-1) \tag{4.9}$$

is a monotonic decreasing function of l, we reject H if F is too large.

4.1.3 *Some Examples*

EXAMPLE 4.1 Let

$$Y_1=\alpha_1+\varepsilon_1,$$

$$Y_2=2\alpha_1-\alpha_2+\varepsilon_2,$$

$$Y_3=\alpha_1+2\alpha_2+\varepsilon_3,$$

where $\boldsymbol{\varepsilon}\sim N_3(\mathbf{0},\sigma^2\mathbf{I}_3)$. Derive the F-statistic for testing H: $\alpha_1=\alpha_2$.

Solution.

$$\begin{bmatrix} Y_1 \\ Y_2 \\ Y_3 \end{bmatrix}=\begin{bmatrix} 1 & 0 \\ 2 & -1 \\ 1 & 2 \end{bmatrix}\begin{pmatrix} \alpha_1 \\ \alpha_2 \end{pmatrix}+\begin{bmatrix} \varepsilon_1 \\ \varepsilon_2 \\ \varepsilon_3 \end{bmatrix},$$

or $\mathbf{Y}=\mathbf{X}\boldsymbol{\beta}+\boldsymbol{\varepsilon}$, where $\mathbf{X}$ is 3×2 of rank 2. Also H is equivalent to

$$(1,-1)\begin{pmatrix} \alpha_1 \\ \alpha_2 \end{pmatrix}=0,$$

or $\mathbf{A}\boldsymbol{\beta}=0$, where $\mathbf{A}$ is 1×2 of rank 1. Hence the above theory applies with $n=3$, $p=2$, and $q=1$.

The first step is to find

$$\mathbf{X}'\mathbf{X}=\begin{pmatrix} 1 & 2 & 1 \\ 0 & -1 & 2 \end{pmatrix}\begin{bmatrix} 1 & 0 \\ 2 & -1 \\ 1 & 2 \end{bmatrix}=\begin{pmatrix} 6 & 0 \\ 0 & 5 \end{pmatrix}.$$

Then

$$\hat{\boldsymbol{\beta}}=(\mathbf{X}'\mathbf{X})^{-1}\mathbf{X}'\mathbf{Y}=\begin{pmatrix} \frac{1}{6} & 0 \\ 0 & \frac{1}{5} \end{pmatrix}\begin{pmatrix} Y_1+2Y_2+Y_3 \\ -Y_2+Y_3 \end{pmatrix},$$

$$\begin{pmatrix} \hat{\alpha}_1 \\ \hat{\alpha}_2 \end{pmatrix}=\begin{bmatrix} \frac{1}{6}(Y_1+2Y_2+Y_3) \\ \frac{1}{5}(-Y_2+Y_3) \end{bmatrix}$$

and, from Equation (3.9),

$$\begin{aligned} \mathrm{RSS}&=\mathbf{Y}'\mathbf{Y}-\hat{\boldsymbol{\beta}}'\mathbf{X}'\mathbf{X}\hat{\boldsymbol{\beta}} \\ &=Y_1^2+Y_2^2+Y_3^2-6\hat{\alpha}_1^2-5\hat{\alpha}_2^2. \end{aligned}$$

We have at least two methods of finding the F-statistic.

Method 1

$$\mathbf{A}\hat{\boldsymbol{\beta}}=\hat{\alpha}_1-\hat{\alpha}_2,$$

$$\mathbf{A}(\mathbf{X}'\mathbf{X})^{-1}\mathbf{A}'=(1,-1)\begin{pmatrix}\frac{1}{6} & 0\\ 0 & \frac{1}{5}\end{pmatrix}\begin{pmatrix}1\\ -1\end{pmatrix}=\tfrac{1}{6}+\tfrac{1}{5}=\tfrac{11}{30},$$

and

$$F=\frac{(\mathbf{A}\hat{\boldsymbol{\beta}})'\left[\mathbf{A}(\mathbf{X}'\mathbf{X})^{-1}\mathbf{A}'\right]^{-1}\mathbf{A}\hat{\boldsymbol{\beta}}}{qS^2}$$

$$=\frac{(\hat{\alpha}_1-\hat{\alpha}_2)^2}{\frac{11}{30}S^2}$$

where $S^2=\text{RSS}/(n-p)=\text{RSS}$. When H is true $F\sim F_{q,n-p}=F_{1,1}$.

Method 2

Let $\alpha_1=\alpha_2=\alpha$. When H is true we have

$$\boldsymbol{\varepsilon}'\boldsymbol{\varepsilon}=(Y_1-\alpha)^2+(Y_2-\alpha)^2+(Y_3-3\alpha)^2$$

and $\partial\boldsymbol{\varepsilon}'\boldsymbol{\varepsilon}/\partial\alpha=0$ implies that $\hat{\alpha}_H=\frac{1}{11}(Y_1+Y_2+3Y_3)$. Hence

$$\text{RSS}_H=(Y_1-\hat{\alpha}_H)^2+(Y_2-\hat{\alpha}_H)^2+(Y_3-3\hat{\alpha}_H)^2 \tag{4.10}$$

and

$$F=\frac{\text{RSS}_H-\text{RSS}}{\text{RSS}}.$$

EXAMPLE 4.2 Let $U_1,U_2,\ldots,U_{n_1}$ be independent observations from $N(\mu_1,\sigma^2)$, and let $V_1,V_2,\ldots,V_{n_2}$ be independent observations from $N(\mu_2,\sigma^2)$. Derive a test statistic for H: $\mu_1=\mu_2$.

Solution. We can write

$$U_i=\mu_1+\varepsilon_i \qquad (i=1,2,\ldots,n_1)$$

and

$$V_j=\mu_2+\varepsilon_{n_1+j} \qquad (j=1,2,\ldots,n_2),$$

or, using matrix notation

$$\begin{bmatrix} U_1 \\ U_2 \\ \cdots \\ U_{n_1} \\ \hline V_1 \\ V_2 \\ \cdots \\ V_{n_2} \end{bmatrix} = \begin{bmatrix} 1 & 0 \\ 1 & 0 \\ \cdots & \cdots \\ 1 & 0 \\ \hline 0 & 1 \\ 0 & 1 \\ \cdots & \cdots \\ 0 & 1 \end{bmatrix} \begin{pmatrix} \mu_1 \\ \mu_2 \end{pmatrix} + \begin{bmatrix} \varepsilon_1 \\ \varepsilon_2 \\ \cdots \\ \varepsilon_{n_1} \\ \hline \varepsilon_{n_1+1} \\ \cdots \\ \cdots \\ \varepsilon_n \end{bmatrix} \tag{4.11}$$

where $n = n_1 + n_2$. Thus our model is of the form $\mathbf{Y} = \mathbf{X}\boldsymbol{\beta} + \boldsymbol{\varepsilon}$, where $\mathbf{X}$ is $n \times 2$ of rank 2 and $\boldsymbol{\varepsilon} \sim N_n(\mathbf{0}, \sigma^2\mathbf{I}_n)$. Also, as in Example 4.1, H takes the form $\mathbf{A}\boldsymbol{\beta} = \mathbf{0}$ so that our general regression theory applies with $p = 2$ and $q = 1$.

Now

$$\mathbf{X}'\mathbf{X} = \begin{pmatrix} n_1 & 0 \\ 0 & n_2 \end{pmatrix}$$

so that

$$\hat{\boldsymbol{\beta}} = \begin{pmatrix} \hat{\mu}_1 \\ \hat{\mu}_2 \end{pmatrix} = (\mathbf{X}'\mathbf{X})^{-1}\mathbf{X}'\mathbf{Y} = \begin{bmatrix} \frac{1}{n_1} & 0 \\ 0 & \frac{1}{n_2} \end{bmatrix} \begin{bmatrix} \sum U_i \\ \sum V_j \end{bmatrix} = \begin{pmatrix} \overline{U} \\ \overline{V} \end{pmatrix},$$

$$\mathbf{A}\hat{\boldsymbol{\beta}} = \hat{\mu}_1 - \hat{\mu}_2 = \overline{U} - \overline{V}$$

and

$$\begin{aligned} \text{RSS} &= \mathbf{Y}'\mathbf{Y} - \hat{\boldsymbol{\beta}}'\mathbf{X}'\mathbf{X}\hat{\boldsymbol{\beta}} \\ &= \sum_i U_i^2 + \sum_j V_j^2 - n_1\overline{U}^2 - n_2\overline{V}^2 \\ &= \sum_i \left(U_i - \overline{U}\right)^2 + \sum_j \left(V_j - \overline{V}\right)^2. \end{aligned}$$

Also

$$\mathbf{A}(\mathbf{X}'\mathbf{X})^{-1}\mathbf{A}' = \frac{1}{n_1} + \frac{1}{n_2}$$

so that the F-statistic for H is

$$F=\frac{(\mathbf{A}\hat{\boldsymbol{\beta}})'\left[\mathbf{A}(\mathbf{X}'\mathbf{X})^{-1}\mathbf{A}'\right]^{-1}\mathbf{A}\hat{\boldsymbol{\beta}}}{qS^2}$$

$$=\frac{(\bar{U}-\bar{V})^2}{S^2\left(\dfrac{1}{n_1}+\dfrac{1}{n_2}\right)}, \tag{4.12}$$

where $S^2=\text{RSS}/(n-p)=\text{RSS}/(n_1+n_2-2)$. When H is true $F\sim F_{1,n_1+n_2-2}$.

Since, distribution-wise, we have the identity $F_{1,k}\equiv t_k^2$, the above F-statistic is the square of the usual t-statistic for testing the difference of two normal means (assuming equal variances).

EXAMPLE 4.3 Given the general linear model

$$G: Y_i=\beta_0+\beta_1 x_{i1}+\cdots+\beta_{p-1}x_{i,p-1}+\varepsilon_i \qquad (i=1,2,\ldots,n),$$

obtain a test statistic for H: $\beta_j=c$.

Solution. Suppose we make the following partition:

$$(\mathbf{X}'\mathbf{X})^{-1}=\begin{pmatrix} l & \mathbf{m}' \\ \mathbf{m} & \mathbf{D}\end{pmatrix}$$

where l is 1×1. Now H is of the form $\mathbf{a}'\boldsymbol{\beta}=\mathbf{c}$ where $\mathbf{a}'$ is the row vector with unity in the $(j+1)$th position and zeros elsewhere. Therefore using the general matrix theory, $\mathbf{a}'(\mathbf{X}'\mathbf{X})^{-1}\mathbf{a}=d_{jj}$ (the jth diagonal element of $\mathbf{D}$), $\mathbf{a}'\hat{\boldsymbol{\beta}}-\mathbf{c}=\hat{\beta}_j-c$, and the F-statistic is

$$F=\frac{(\hat{\beta}_j-c)^2}{S^2 d_{jj}}, \tag{4.13}$$

which has the $F_{1,n-p}$ distribution when H is true. As in Example 4.2, F is again the square of the usual t-statistic.

The matrix $\mathbf{D}$ can be identified using the method of *A7* for inverting a partitioned symmetric matrix. Let $\mathbf{1}_n$ be an $n\times 1$ column vector of ones and let $\bar{\mathbf{x}}'=(\bar{x}_{\cdot 1},\bar{x}_{\cdot 2},\ldots,\bar{x}_{\cdot p-1})$. Then $\mathbf{X}=(\mathbf{1}_n,\mathbf{X}_1)$,

$$\mathbf{X}'\mathbf{X}=\begin{pmatrix} n & n\bar{\mathbf{x}}' \\ n\bar{\mathbf{x}} & \mathbf{X}_1'\mathbf{X}_1\end{pmatrix}$$

and, by *A7*,

$$(\mathbf{X}'\mathbf{X})^{-1} = \begin{pmatrix} \frac{1}{n} + \bar{\mathbf{x}}'\mathbf{V}^{-1}\bar{\mathbf{x}}, & -\bar{\mathbf{x}}'\mathbf{V}^{-1} \\ -\mathbf{V}^{-1}\bar{\mathbf{x}}, & \mathbf{V}^{-1} \end{pmatrix}, \tag{4.14}$$

where $\mathbf{V} = [(v_{jk})] = \mathbf{X}_1'\mathbf{X}_1 - n\bar{\mathbf{x}}\bar{\mathbf{x}}'$ and

$$v_{jk} = \sum_i x_{ij}x_{ik} - n\bar{x}_{\cdot j}\bar{x}_{\cdot k}$$

$$= \sum_i (x_{ij} - \bar{x}_{\cdot j})(x_{ik} - \bar{x}_{\cdot k}).$$

Thus $\mathbf{D}$ is the inverse of $\mathbf{V}$, the matrix of corrected sums of squares and products of the x's.

This example is considered in greater detail in Section 11.7.1.

4.1.4 *The Straight Line*

Let $Y_i = \beta_0 + \beta_1 x_i + \varepsilon_i$ $(i = 1, 2, \ldots, n)$, and suppose we wish to test H: $\beta_1 = 0$. Then $\mathbf{X} = (\mathbf{1}_n, \mathbf{x})$,

$$\mathbf{X}'\mathbf{X} = \begin{pmatrix} n, & n\bar{x} \\ n\bar{x}, & \sum x_i^2 \end{pmatrix}, \qquad (\mathbf{X}'\mathbf{X})^{-1} = \frac{1}{\sum (x_i - \bar{x})^2} \begin{bmatrix} \frac{1}{n}\sum x_i^2, & -\bar{x} \\ -\bar{x}, & 1 \end{bmatrix},$$

and

$$\mathbf{X}'\mathbf{Y} = \begin{bmatrix} \sum Y_i \\ \sum x_i Y_i \end{bmatrix}.$$

Also from $\hat{\boldsymbol{\beta}} = (\mathbf{X}'\mathbf{X})^{-1}\mathbf{X}'\mathbf{Y}$ we have, after some simplification,

$$\hat{\beta}_0 = \bar{Y} - \hat{\beta}_1\bar{x},$$

$$\hat{\beta}_1 = \frac{\sum Y_i(x_i - \bar{x})}{\sum (x_i - \bar{x})^2} = \frac{\sum (Y_i - \bar{Y})(x_i - \bar{x})}{\sum (x_i - \bar{x})^2}$$

and

$$\hat{Y}_i = \hat{\beta}_0 + \hat{\beta}_1 x_i$$

$$= \bar{Y} + \hat{\beta}_1 (x_i - \bar{x}).$$

(Actually $\hat{\beta}_0$ and $\hat{\beta}_1$ can be obtained more readily by differentiating $\boldsymbol{\varepsilon}'\boldsymbol{\varepsilon}$ with respect to β_0 and β_1.) Finally, from Example 4.3 above with $p=2$, the F-statistic for testing H is given by

$$F=\frac{\hat{\beta}_1^2}{S^2 d_{11}}=\frac{\hat{\beta}_1^2}{S^2/\sum(x_i-\bar{x})^2}, \tag{4.15}$$

where

$$\begin{aligned}(n-2)S^2&=\sum(Y_i-\hat{Y}_i)^2\\ &=\sum\left[Y_i-\bar{Y}-\hat{\beta}_1(x_i-\bar{x})\right]^2\\ &=\sum(Y_i-\bar{Y})^2-\hat{\beta}_1^2\sum(x_i-\bar{x})^2 &(4.16)\\ &=\sum(Y_i-\bar{Y})^2-\sum(\hat{Y}_i-\bar{Y})^2. &(4.17)\end{aligned}$$

We note from (4.17) that

$$\begin{aligned}\sum(Y_i-\bar{Y})^2&=\sum(Y_i-\hat{Y}_i)^2+\sum(\hat{Y}_i-\bar{Y})^2 &(4.18)\\ &=\sum(Y_i-\hat{Y}_i)^2+r^2\sum(Y_i-\bar{Y})^2, &(4.19)\end{aligned}$$

where

$$\begin{aligned}r^2&=\frac{\sum(\hat{Y}_i-\bar{Y})^2}{\sum(Y_i-\bar{Y})^2}\\ &=\frac{\hat{\beta}_1^2\sum(x_i-\bar{x})^2}{\sum(Y_i-\bar{Y})^2}\\ &=\frac{\left[\sum(Y_i-\bar{Y})(x_i-\bar{x})\right]^2}{\sum(Y_i-\bar{Y})^2\sum(x_i-\bar{x})^2} &(4.20)\end{aligned}$$

is the square of the sample correlation between Y and x. Also r is a measure of the degree of linearity between Y and x since, from (4.19),

$$\begin{aligned}\text{RSS} &= \sum (Y_i - \hat{Y}_i)^2 \\ &= (1-r^2)\sum (Y_i - \bar{Y})^2 \qquad (4.21)\end{aligned}$$

so that the larger the value of r^2, the smaller RSS and the better the fit of the estimated regression line to the observations.

Although $1-r^2$ is a useful measure of fit, the correlation r itself is of doubtful use in making inferences. Tukey [1954] makes the provocative but not unreasonable statement that "correlation coefficients are justified in two and only two circumstances, when they are regression coefficients, or when the measurement of one or both variables on a determinate scale is hopeless." The first part of his statement refers to the situation where X and Y have a bivariate normal distribution; we have (Exercises 2c, no. 2)

$$\begin{aligned}E[Y|X=x] &= \mu_Y + \rho\frac{\sigma_Y}{\sigma_X}(x-\mu_X) \\ &= \beta_0 + \beta_1 x,\end{aligned}$$

and when $\sigma_X^2 = \sigma_Y^2$, $\beta_1 = \rho$. One area where correlation coefficients are widely used, and determinate scales seem hopeless, is in the social sciences. Here measuring scales are often completely arbitrary so that observations are essentially only ranks. A helpful discussion on the question of correlation versus regression is given by Warren [1971].

Finally we note that the F-statistic (4.15) can also be expressed in terms of r^2. From Equation (4.21) we have

$$(n-2)S^2 = (1-r^2)\sum (Y_i - \bar{Y})^2$$

so that

$$\begin{aligned}F &= \frac{\hat{\beta}_1^2 \sum (x_i - \bar{x})^2 (n-2)}{(1-r^2)\sum (Y_i - \bar{Y})^2} \\ &= \frac{r^2(n-2)}{1-r^2}.\end{aligned}$$

The usual t-statistic for testing $\beta_1 = 0$ can also be expressed in the same form; namely,

$$T = \frac{r}{\sqrt{(1-r^2)/(n-2)}} . \tag{4.22}$$

4.1.5 *Significant F-Test*

If the F-test for H: $\mathbf{A}\boldsymbol{\beta} = \mathbf{c}$ is significant, the next step is to decide why H is rejected. One course of action might be to test each of the individual constraints $\mathbf{a}_i'\boldsymbol{\beta} = c_i$ $(i = 1, 2, \ldots, q)$ separately using a t-test to see which constraints are responsible. The appropriate t-test is described below.

By Theorem 2.2, $\mathbf{a}_i'\hat{\boldsymbol{\beta}} \sim N(\mathbf{a}_i'\boldsymbol{\beta}, \sigma^2\mathbf{a}_i'(\mathbf{X}'\mathbf{X})^{-1}\mathbf{a}_i)$ so that

$$U_i = \frac{\mathbf{a}_i'\hat{\boldsymbol{\beta}} - \mathbf{a}_i'\boldsymbol{\beta}}{\sigma\left\{\mathbf{a}_i'(\mathbf{X}'\mathbf{X})^{-1}\mathbf{a}_i\right\}^{1/2}} \sim N(0,1).$$

Also, by Theorem 3.5 (Section 3.4), $V = (n-p)S^2/\sigma^2 \sim \chi^2_{n-p}$ and, since S^2 is statistically independent of $\hat{\boldsymbol{\beta}}$, V is independent of U_i. Hence

$$\begin{aligned} T_i &= \frac{U_i}{\sqrt{V/(n-p)}} \\ &= \frac{\mathbf{a}_i'\hat{\boldsymbol{\beta}} - \mathbf{a}_i'\boldsymbol{\beta}}{S\left\{\mathbf{a}_i'\mathbf{X}'\mathbf{X}\mathbf{a}_i\right\}^{1/2}} \end{aligned} \tag{4.23}$$

has the t_{n-p} distribution. To test H_i: $\mathbf{a}_i'\boldsymbol{\beta} = c_i$ we set $\mathbf{a}_i'\boldsymbol{\beta}$ equal to c_i in T_i and reject H_i at the α level of significance if $|T_i| \geqslant t_{n-p}^{(1/2)\alpha}$; here $t_{n-p}^{(1/2)\alpha}$ is the upper $\frac{1}{2}\alpha$ point of the t_{n-p} distribution, that is, $\mathrm{pr}(T_i > t_{n-p}^{(1/2)\alpha}) = \frac{1}{2}\alpha$.

Alternatively we can construct a $100(1-\alpha)\%$ confidence interval for $\mathbf{a}_i'\boldsymbol{\beta}$, namely,

$$\mathbf{a}_i'\hat{\boldsymbol{\beta}} \pm t_{n-p}^{(1/2)\alpha} S\left\{\mathbf{a}_i'(\mathbf{X}'\mathbf{X})^{-1}\mathbf{a}_i\right\}^{1/2} \tag{4.24}$$

or, since $S^2\{\mathbf{a}_i'(\mathbf{X}'\mathbf{X})^{-1}\mathbf{a}_i\}$ is an unbiased estimate of $\sigma^2\mathbf{a}_i'(\mathbf{X}'\mathbf{X})^{-1}\mathbf{a}_i$ (the variance of $\mathbf{a}'\hat{\boldsymbol{\beta}}$),

$$\mathbf{a}_i'\hat{\boldsymbol{\beta}} \pm t_{n-p}^{(1/2)\alpha}\hat{\sigma}_{\mathbf{a}_i'\hat{\boldsymbol{\beta}}}, \quad \text{say}, \tag{4.25}$$

and see if the above interval contains c_i. Either of the above procedures can be carried out for $i = 1, 2, \ldots, q$. For example, if H: $\beta_1 = \beta_2 = \cdots = \beta_q = 0$ is rejected by the F-test then, noting Equation (4.13), we can test H_i:

$\beta_i=0$ using the statistic

$$T_i=\frac{\hat{\beta}_i}{S\sqrt{d_{ii}}}.$$

The above two-stage test procedure for investigating H, namely, carrying out an overall F-test followed by a series of t-tests when F is significant, is commonly called the least significant difference (LSD) test. The title *least significant difference* stems from the t-critical value, $t_{n-p}^{(1/2)\alpha}$, which is the minimum value that an individual T_i must exceed in order to be judged significant; the term "difference" refers to the fact that the LSD test is generally used for comparing parameters such as population means in a pairwise fashion. As pointed out by Miller [1966: p. 92], the most endearing features of the LSD test are its convenience, simplicity, and versatility. However, the method has its weaknesses. For example, it is quite possible that H is rejected but none of the individual H_i's is rejected. Other difficulties associated with the LSD procedure, and in fact with simultaneous inference in general, are discussed in Chapter 5.

EXERCISES 4b

1. Let $Y_i=\beta_0+\beta_1 x_{i1}+\cdots+\beta_{p-1}x_{i,p-1}+\varepsilon_i$, $i=1,2,\ldots,n$, where the ε_i are independent $N(0,\sigma^2)$. Prove that the F-statistic for testing the hypothesis H: $\beta_q=\beta_{q+1}=\cdots=\beta_{p-1}=0$ $(0<q\leqslant p-1)$ is unchanged if a constant, c, say, is subtracted from each Y_i.
2. Let $Y_i=\beta_0+\beta_1 x_i+\varepsilon_i$, $(i=1,2,\ldots,n)$, where the ε_i are independent $N(0,\sigma^2)$. Derive an F-statistic for testing H: $\beta_0=0$.
3. Given that $\bar{x}=0$, derive an F-statistic for testing the hypothesis H: $\beta_0=\beta_1$ in Exercise 2 above.
4. Let

$$Y_1=\theta_1+\theta_2+\varepsilon_1,$$

$$Y_2=2\theta_2+\varepsilon_2,$$

and

$$Y_3=-\theta_1+\theta_2+\varepsilon_3,$$

where the ε_i $(i=1,2,3)$ are independent $N(0,\sigma^2)$. Derive an F-statistic for testing the hypothesis H: $\theta_1=2\theta_2$.

5. Given $\mathbf{Y}=\boldsymbol{\theta}+\boldsymbol{\varepsilon}$ where $\boldsymbol{\varepsilon}\sim N_4(\mathbf{0},\sigma^2\mathbf{I}_4)$ and $\theta_1+\theta_2+\theta_3+\theta_4=0$, show that

the F-statistic for testing H: $\theta_1 = \theta_3$ is

$$\frac{2(Y_1 - Y_3)^2}{(Y_1 + Y_2 + Y_3 + Y_4)^2}.$$

4.2 MULTIPLE CORRELATION COEFFICIENT

Given the linear model $Y_i = \beta_0 + \beta_1 x_{i1} + \cdots + \beta_{p-1} x_{i,p-1} + \varepsilon_i$ $(i = 1, 2, \ldots, n)$, suppose we wish to test whether the regression on the regressor variables is significant or not; that is, test H: $\beta_1 = \beta_2 = \cdots = \beta_{p-1} = 0$. Then H takes the form $\mathbf{A}\boldsymbol{\beta} = \mathbf{0}$ where $\mathbf{A} = [\mathbf{0}, \mathbf{I}_{p-1}]$ is a $(p-1) \times p$ matrix of rank $p-1$, so that the general regression theory applies with $q = p-1$, $\text{RSS} = \mathbf{Y}'\mathbf{Y} - \hat{\boldsymbol{\beta}}'\mathbf{X}'\mathbf{Y}$, and

$$\begin{aligned} RSS_H &= \underset{\beta_0}{\text{minimum}} \sum_i (Y_i - \beta_0)^2 \\ &= \sum (Y_i - \bar{Y})^2 \\ &= \mathbf{Y}'\mathbf{Y} - n\bar{Y}^2. \end{aligned}$$

Hence

$$\begin{aligned} F &= \frac{(\text{RSS}_H - \text{RSS})/(p-1)}{\text{RSS}/(n-p)} \\ &= \frac{(\hat{\boldsymbol{\beta}}'\mathbf{X}'\mathbf{Y} - n\bar{Y}^2)}{(\mathbf{Y}'\mathbf{Y} - \hat{\boldsymbol{\beta}}'\mathbf{X}'\mathbf{Y})} \cdot \frac{(n-p)}{(p-1)}, \end{aligned} \tag{4.26}$$

and $F \sim F_{p-1,n-p}$ when H is true.

The statistic F provides a test for "overall" regression, and we reject H if $F > F^{\alpha}_{p-1,n-p}$, $F^{\alpha}_{p-1,n-p}$ being the upper α point for the $F_{p-1,n-p}$ distribution. If we reject H we say that there is a significant regression and the x_{ij} values cannot be totally ignored. However, the rejection of H does not mean that the fitted equation $\hat{\mathbf{Y}} = \mathbf{X}\hat{\boldsymbol{\beta}}$ is necessarily adequate, particularly for predictive purposes. In this respect Draper and Smith [1966; p. 64] suggest that, as a working rule, the fitted surface is probably satisfactory for predicting if $F > 4F^{\alpha}_{p-1,n-p}$.

A useful measure of how well an estimated regression fits the observed Y_i is the *sample multiple correlation coefficient* R. This is defined to be the correlation between Y_i and $\hat{Y}_i$, namely,

$$R=\frac{\sum\left(Y_i-\bar{Y}\right)\left(\hat{Y}_i-\bar{\hat{Y}}\right)}{\left\{\sum\left(Y_i-\bar{Y}^2\right)\sum\left(\hat{Y}_i-\bar{\hat{Y}}\right)^2\right\}^{1/2}} \tag{4.27}$$

The quantity R^2 is commonly called the *coefficient of determination*. We now prove a useful theorem that generalizes Equations (4.18) and (4.20).

THEOREM 4.2

(i)

$$\sum_i\left(Y_i-\bar{Y}\right)^2=\sum_i\left(Y_i-\hat{Y}_i\right)^2+\sum_i\left(\hat{Y}_i-\bar{Y}\right)^2.$$

(ii)

$$R^2=\frac{\sum\left(\hat{Y}_i-\bar{Y}\right)^2}{\sum\left(Y_i-\bar{Y}\right)^2}.$$

Proof. (i) $\hat{\mathbf{Y}}=\mathbf{PY}$ so that

$$\hat{\mathbf{Y}}'\hat{\mathbf{Y}}=\mathbf{Y}'\mathbf{P}^2\mathbf{Y}=\mathbf{Y}'\mathbf{PY}=\mathbf{Y}'\hat{\mathbf{Y}}. \tag{4.28}$$

Also, by differentiating $\sum_i(Y_i-\beta_0-\beta_1x_{i1}-\cdots-\beta_{p-1}x_{i,p-1})^2$ with respect to β_0, we have one of the normal equations for $\hat{\boldsymbol{\beta}}$, namely,

$$\sum\left(Y_i-\hat{\beta}_0-\hat{\beta}_1x_{i1}-\cdots-\hat{\beta}_{p-1}x_{i,p-1}\right)=0$$

or

$$\sum_i\left(Y_i-\hat{Y}_i\right)=0. \tag{4.29}$$

Hence

$$\begin{aligned}\sum\left(Y_i-\bar{Y}\right)^2&=\sum\left(Y_i-\hat{Y}_i+\hat{Y}_i-\bar{Y}\right)^2\\&=\sum\left(Y_i-\hat{Y}_i\right)^2+\sum\left(\hat{Y}_i-\bar{Y}\right)^2\end{aligned}$$

since

$$\sum(Y_i-\hat{Y}_i)(\hat{Y}_i-\bar{Y})=\sum(Y_i-\hat{Y}_i)\hat{Y}_i \quad [\text{by Equation (4.29)}]$$

$$=(\mathbf{Y}-\hat{\mathbf{Y}})'\hat{\mathbf{Y}}$$

$$=0 \quad [\text{by Equation (4.28)}].$$

(ii) From Equation (4.29) $\bar{\hat{Y}}=\bar{Y}$ so that

$$\sum(Y_i-\bar{Y})(\hat{Y}_i-\bar{\hat{Y}}_i)=\sum(Y_i-\bar{Y})(\hat{Y}_i-\bar{Y})$$

$$=\sum(Y_i-\hat{Y}_i+\hat{Y}_i-\bar{Y})(\hat{Y}_i-\bar{Y})$$

$$=\sum(\hat{Y}_i-\bar{Y})^2,$$

and the required expression for R^2 immediately follows from (4.27). □

We note that R^2 is simply a generalization of r^2 for the straight line. In particular, Equation (4.21) now becomes

$$\text{RSS}=(1-R^2)\sum(Y_i-\bar{Y})^2 \tag{4.30}$$

and the greater the value of R^2, the closer the fit of the estimated surface to the observed data; if $Y_i=\hat{Y}_i$ we have a perfect fit and $R^2=1$. When there is just a single x-regressor then $R^2=r^2$.

By writing $\mathbf{P}=\mathbf{X}(\mathbf{X}'\mathbf{X})^{-}\mathbf{X}'$, where $(\mathbf{X}'\mathbf{X})^{-}$ is a generalized inverse of $\mathbf{X}'\mathbf{X}$, we find that the above theorem still holds even when $\mathbf{X}$ is not of full rank. Alternatively we can write $\mathbf{P}=\mathbf{X}_1(\mathbf{X}_1'\mathbf{X}_1)^{-1}\mathbf{X}_1'$ where $\mathbf{X}_1$ is the matrix of linearly independent columns of $\mathbf{X}$.

We now show, in the following theorem, that a test of any hypothesis of the form $\mathbf{A}\boldsymbol{\beta}=\mathbf{0}$ which does not involve β_0 (and most tests fall into this category) can be regarded as a test for a significant reduction in R^2.

THEOREM 4.3 Let $\hat{\mathbf{Y}}_H=[(\hat{Y}_{iH})]=\mathbf{X}\hat{\boldsymbol{\beta}}_H$, where $\hat{\boldsymbol{\beta}}_H$ is given by Equation (4.1), and let

$$R_H^2=\frac{\sum(\hat{Y}_{iH}-\bar{Y})^2}{\sum(Y_i-\bar{Y})^2}.$$

Suppose the constraints H: $\mathbf{A}\boldsymbol{\beta}=\mathbf{0}$ do not involve β_0, that is, $\mathbf{A}=[\mathbf{0},\mathbf{A}_1]$, say; then the F-statistic for testing H is

$$F=\frac{(R^2-R_H^2)}{(1-R^2)}\cdot\frac{(n-p)}{q}.$$

Proof. From Equation (4.4) we have $\hat{\mathbf{Y}}_H=\mathbf{P}_H\mathbf{Y}$, where $\mathbf{P}_H$ is symmetric and idempotent. Hence

$$\hat{\mathbf{Y}}_H'\hat{\mathbf{Y}}_H=\mathbf{Y}'\mathbf{P}_H^2\mathbf{Y}=\mathbf{Y}'\mathbf{P}_H\mathbf{Y}=\mathbf{Y}'\hat{\mathbf{Y}}_H.$$

Now the constraints $\mathbf{A}\boldsymbol{\beta}=\mathbf{0}$ do not involve β_0 so that if $r=\boldsymbol{\varepsilon}'\boldsymbol{\varepsilon}+\boldsymbol{\lambda}'\mathbf{A}\boldsymbol{\beta}$, then $\partial r/\partial\beta_0=0$ implies that $\partial\boldsymbol{\varepsilon}'\boldsymbol{\varepsilon}/\partial\beta_0=0$; that is, $\sum(Y_i-\hat{Y}_{iH})=0$. Thus Equations (4.28) and (4.29) still hold under H so that Theorem 4.2, which depends on the validity of those two equations, still holds when H is true. Hence applying both parts of Theorem 4.2 twice,

$$\begin{aligned}\frac{\mathrm{RSS}_H-\mathrm{RSS}}{\mathrm{RSS}}&=\frac{\sum(Y_i-\hat{Y}_{iH})^2-\sum(Y_i-\hat{Y}_i)^2}{\sum(Y_i-\hat{Y}_i)^2}\\&=\frac{\sum(\hat{Y}_i-\bar{Y})^2-\sum(\hat{Y}_{iH}-\bar{Y})^2}{\sum(Y_i-\bar{Y})^2-\sum(\hat{Y}_i-\bar{Y})^2}\\&=\frac{R^2-R_H^2}{1-R^2}.\end{aligned}\tag{4.31}$$

□

Perhaps the most important application of the above theorem is to hypotheses of the form H: $\beta_j=0$ $(j\neq 0)$. In this case, since $\mathrm{RSS}_H-\mathrm{RSS}\geqslant 0$ (and therefore $R^2-R_H^2\geqslant 0$), we see that the residual sum of squares can never be increased, and the coefficient of determination can never be reduced, by adding an extra regressor, x_j, say.

EXAMPLE 4.4 (Goldberger [1964: p. 186]) For the general linear full rank regression model, prove that R^2 and the F-statistic for testing H: $\beta_j=0$ $(j\neq 0)$ are independent of the units in which the Y_i and the x_{ij} are measured.

Solution. For $i=1,2,\ldots,n$; $j=1,2,\ldots,p-1$, let $Z_i=kY_i$ and $w_{ij}=k_jx_{ij}$. Let $\hat{\gamma}$ be the least squares estimate of $\boldsymbol{\beta}$ for the new units. Then if $\mathbf{K}=$

$\text{diag}(1, k_1, \ldots, k_{p-1})$, we have

$$\mathbf{W} = [(w_{ij})] = \mathbf{XK}, \qquad (\mathbf{W}'\mathbf{W})^{-1} = \mathbf{K}^{-1}(\mathbf{X}'\mathbf{X})^{-1}\mathbf{K}^{-1},$$

$$\begin{aligned}\hat{\gamma} &= (\mathbf{W}'\mathbf{W})^{-1}\mathbf{W}'\mathbf{Z}\\ &= \mathbf{K}^{-1}(\mathbf{X}'\mathbf{X})^{-1}\mathbf{K}^{-1}\mathbf{KX}'\mathbf{Y}k\\ &= k\mathbf{K}^{-1}\hat{\boldsymbol{\beta}},\end{aligned}$$

$$\begin{aligned}\hat{\mathbf{Z}} &= \mathbf{W}\hat{\gamma}\\ &= \mathbf{kXKK}^{-1}\hat{\boldsymbol{\beta}}\\ &= k\mathbf{X}\hat{\boldsymbol{\beta}}\\ &= k\hat{\mathbf{Y}},\end{aligned}$$

$$R_Z^2 = \frac{\sum(\hat{Z}_i - \bar{Z})^2}{\sum(Z_i - \bar{Z})^2} = \frac{k^2\sum(\hat{Y}_i - \bar{Y})^2}{k^2\sum(Y_i - \bar{Y})^2} = R^2,$$

and

$$\begin{aligned}\text{RSS}_Z &= \mathbf{Z}'\mathbf{Z} - \hat{\gamma}'\mathbf{W}'\mathbf{Z}\\ &= k^2\mathbf{Y}'\mathbf{Y} - k^2\hat{\boldsymbol{\beta}}'\mathbf{K}^{-1}\mathbf{KX}'\mathbf{Y}\\ &= k^2\text{RSS}.\end{aligned}$$

From Example 4.3, the F-statistic for testing H is

$$F = \frac{\hat{\beta}_j^2(n-p)}{d_{jj}\text{RSS}}$$

where d_{jj} is the $(j+1)$th diagonal element of $(\mathbf{X}'\mathbf{X})^{-1}$. If d_{jj}^* is the corresponding element in $(\mathbf{W}'\mathbf{W})^{-1}$, then $d_{jj}^* = k_j^{-2}d_{jj}$ and

$$F_Z = \frac{\hat{\gamma}_j^2(n-p)}{d_{jj}^*\text{RSS}_Z} = \frac{(kk_j^{-1}\hat{\beta}_j)^2(n-p)}{k_j^{-2}d_{jj}k^2\text{RSS}} = F.$$

EXERCISES 4c

1. Prove that (4.26) can be written in the form

$$F=\frac{R^2(n-p)}{1-R^2}.$$

2. Suppose that $\beta_1=\beta_2=\cdots=\beta_{p-1}=0$. Find the distribution of R^2 and hence prove that

$$E[R^2]=\frac{p}{n-1}.$$

4.3 A CANONICAL FORM FOR H

Suppose we wish to test H: $\mathbf{A}\boldsymbol{\beta}=\mathbf{0}$, where A is $q\times p$ of rank q, for the full rank model $\mathbf{Y}=\mathbf{X}\boldsymbol{\beta}+\boldsymbol{\varepsilon}$. Since $\mathbf{A}$ has q linearly independent columns we can assume, without loss of generality (by relabeling the β_j if necessary), that these are the last q columns; thus $\mathbf{A}=[\mathbf{A}_1,\mathbf{A}_2]$ where $\mathbf{A}_2$ is a $q\times q$ nonsingular matrix. Partitioning $\boldsymbol{\beta}$ in the same way we have

$$\mathbf{0}=\mathbf{A}\boldsymbol{\beta}=\mathbf{A}_1\boldsymbol{\beta}_1+\mathbf{A}_2\boldsymbol{\beta}_2,$$

and multiplying through by $\mathbf{A}_2^{-1}$ leads to

$$\boldsymbol{\beta}_2=-\mathbf{A}_2^{-1}\mathbf{A}_1\boldsymbol{\beta}_1. \tag{4.32}$$

This means that under the hypothesis H, the regression model takes the "canonical" form

$$\begin{aligned}\mathbf{X}\boldsymbol{\beta}&=(\mathbf{X}_1,\mathbf{X}_2)\,\boldsymbol{\beta}\\&=\mathbf{X}_1\boldsymbol{\beta}_1+\mathbf{X}_2\boldsymbol{\beta}_2\\&=(\mathbf{X}_1-\mathbf{X}_2\mathbf{A}_2^{-1}\mathbf{A}_1)\,\boldsymbol{\beta}_1\\&=\mathbf{X}_A\boldsymbol{\gamma},\end{aligned} \tag{4.33}$$

say, where $\mathbf{X}_A$ is $n\times(p-q)$ of rank $p-q$ and $\boldsymbol{\gamma}=\boldsymbol{\beta}_1$. The matrix $\mathbf{X}_A$ has linearly independent columns since

$$\mathbf{X}_A\boldsymbol{\beta}_1=\mathbf{0}\Leftrightarrow\mathbf{X}\boldsymbol{\beta}=\mathbf{0}\Leftrightarrow\boldsymbol{\beta}=\mathbf{0}\Leftrightarrow\boldsymbol{\beta}_1=\mathbf{0}.$$

By expressing the hypothesized model H: $\mathcal{E}[\mathbf{Y}]=\mathbf{X}_A\boldsymbol{\gamma}$ in the same form as the original model $\mathcal{E}[\mathbf{Y}]=\mathbf{X}\boldsymbol{\beta}$, we see that the same "package" computer program can be used for calculating both RSS and RSS_H, provided, of course, that $\mathbf{X}_A$ can be found easily and accurately. If $\mathbf{X}_A$ is not readily found then the numerator of the F-statistic for testing H can be computed directly using the method of Section 11.10.

One very simple application of the above theory is H: $\boldsymbol{\beta}_2=\mathbf{0}$; $\mathbf{X}_A$ is simply the first $p-q$ columns of $\mathbf{X}$ (see also Section 3.7.1 which tackles this problem in the reverse order: $\mathbf{X}_A\boldsymbol{\gamma}$ is fitted first and then $\mathbf{X}_A$ is augmented to $\mathbf{X}$). Further applications are given in Chapter 9 and in the following example.

EXAMPLE 4.5 (Graybill [1961: p. 136]) Suppose we have n_1 observations on $w_1, w_2, \ldots, w_{p-1}$ and U giving the model

$$U_i=\gamma_0^{(1)}+\gamma_1^{(1)}w_{i1}+\cdots+\gamma_{p-1}^{(1)}w_{i,p-1}+\eta_i \qquad (i=1,2,\ldots,n_1),$$

or $\mathbf{U}_1=\mathbf{W}_1\boldsymbol{\gamma}_1+\boldsymbol{\eta}_1$, where $\boldsymbol{\eta}_1 \sim N_{n_1}(\mathbf{0},\sigma^2\mathbf{I}_{n_1})$. We are now given n_2 $(>p)$ additional observations which can be expressed in the same way, namely,

$$U_i=\gamma_0^{(2)}+\gamma_1^{(2)}w_{i1}+\cdots+\gamma_{p-1}^{(2)}w_{i,p-1}+\eta_i,$$

$$(i=n_1+1, n_2+2,\ldots,n_1+n_2),$$

or $\mathbf{U}_2=\mathbf{W}_2\boldsymbol{\gamma}_2+\boldsymbol{\eta}_2$, where $\boldsymbol{\eta}_2 \sim N_{n_2}(\mathbf{0},\sigma^2\mathbf{I}_{n_2})$. Derive an F-statistic for testing the hypothesis H that the additional observations come from the same model.

Solution. Assuming the columns of $\mathbf{W}_i$ $(i=1,2)$ to be linearly independent, we have

$$\begin{pmatrix}\mathbf{U}_1\\ \mathbf{U}_2\end{pmatrix}=\begin{pmatrix}\mathbf{W}_1 & \mathbf{O}\\ \mathbf{O} & \mathbf{W}_2\end{pmatrix}\begin{pmatrix}\boldsymbol{\gamma}_1\\ \boldsymbol{\gamma}_2\end{pmatrix}+\begin{pmatrix}\boldsymbol{\eta}_1\\ \boldsymbol{\eta}_2\end{pmatrix} \qquad (4.36)$$

or $\mathbf{Y}=\mathbf{X}\boldsymbol{\beta}+\boldsymbol{\varepsilon}$, where $\mathbf{X}$ is $n\times 2p$ of rank $2p$, $n=n_1+n_2$, and $\boldsymbol{\varepsilon}\sim N_n(\mathbf{0},\sigma^2\mathbf{I}_n)$. Since H implies $\boldsymbol{\gamma}_1=\boldsymbol{\gamma}_2$ $(=\boldsymbol{\gamma}$, say), or equivalently, $\mathbf{A}\boldsymbol{\beta}=(\mathbf{I}_p, -\mathbf{I}_p)\boldsymbol{\beta}=\mathbf{0}$, we see that the general regression theory can be applied here. In this case the canonical form for H is

$$\mathbf{Y}=\begin{pmatrix}\mathbf{W}_1\\ \mathbf{W}_2\end{pmatrix}\boldsymbol{\gamma}+\boldsymbol{\varepsilon}$$

$$=\mathbf{X}_A\boldsymbol{\gamma}+\boldsymbol{\varepsilon},$$

say, where $\mathbf{X}_A$ is $n\times p$ of rank p (since it contains the p linearly independent rows of $\mathbf{W}_1$). The F-statistic is therefore

$$F=\frac{(\mathrm{RSS}_H-\mathrm{RSS})/p}{\mathrm{RSS}/(n-2p)}$$

where

$$\mathrm{RSS}=\mathbf{Y}'\mathbf{Y}-\sum_{i=1}^{2}\hat{\gamma}_i'\mathbf{W}_i'\mathbf{U}_i$$

$$\mathrm{RSS}_H=\mathbf{Y}'\mathbf{Y}-\hat{\gamma}_H'(\mathbf{W}_1'\mathbf{U}_1+\mathbf{W}_2'\mathbf{U}_2)$$

$$\hat{\gamma}_i=(\mathbf{W}_i'\mathbf{W}_i)^{-1}\mathbf{W}_i'\mathbf{U}_i$$

and

$$\hat{\gamma}_H=(\mathbf{X}_A'\mathbf{X}_A)^{-1}\mathbf{X}_A'\mathbf{Y}.$$

EXERCISES 4d

1. Express the hypotheses in Examples 4.1 and 4.2 in Section 4.1.3 in the form (4.33).
2. Show that the usual full rank regression model and hypothesis H: $\mathbf{A}\boldsymbol{\beta}=\mathbf{0}$ can be transformed to the model $\mathbf{Z}=\boldsymbol{\mu}+\boldsymbol{\eta}$, where $\mu_{p+1}=\mu_{p+2}=\cdots=\mu_n=0$ and $\boldsymbol{\eta}\sim N_n(\mathbf{0},\sigma^2\mathbf{I}_n)$, and the hypothesis H: $\mu_1=\mu_2=\cdots=\mu_q=0$.
[Hint: choose an orthonormal basis of $p-q$ vectors $\{\boldsymbol{\alpha}_{q+1},\boldsymbol{\alpha}_{q+2},\ldots,\boldsymbol{\alpha}_p\}$ for $\mathcal{R}[\mathbf{X}_A]$; extend this to an orthonormal basis $\{\boldsymbol{\alpha}_1,\boldsymbol{\alpha}_2,\ldots,\boldsymbol{\alpha}_p\}$ for $\mathcal{R}[\mathbf{X}]$; and then extend once more to an orthonormal basis $\{\boldsymbol{\alpha}_1,\boldsymbol{\alpha}_2,\ldots,\boldsymbol{\alpha}_n\}$ for E_n. Consider the transformation $\mathbf{Z}=\mathbf{T}'\mathbf{Y}$ where $\mathbf{T}=(\boldsymbol{\alpha}_1,\boldsymbol{\alpha}_2,\ldots,\boldsymbol{\alpha}_n)$ is orthogonal.]
3. A series of $n+1$ observations Y_i $(i=1,2,\ldots,n+1)$ are taken from a normal distribution with unknown variance σ^2. After the first n observations it is suspected that there is a sudden change in the mean of the distribution. Derive a test statistic for testing the hypothesis that the $(n+1)$th observation has the same population mean as the previous observations.

4.4 GOODNESS OF FIT TEST

Suppose that for each set of values taken by the regressors in the model

$$Y=\beta_0+\beta_1x_1+\beta_2x_2+\cdots+\beta_{p-1}x_{p-1}+\varepsilon, \tag{4.34}$$

we have repeated observations on Y, namely,

$$Y_{ir}=\beta_0+\beta_1 x_{i1}+\cdots+\beta_{p-1}x_{i,p-1}+\varepsilon_{ir} \tag{4.35}$$

where $E[\varepsilon_{ir}]=0$, $\text{var}[\varepsilon_{ir}]=\sigma^2$, $r=1,2,\ldots,R_i$, and $i=1,2,\ldots,n$. We assume that the R_i repetitions Y_{ir} for a particular set $(x_{i1},\ldots,x_{i,p-1})$ are genuine replications and not just repetitions of the same reading for Y_i in a given experiment. For example, if $p=2$, Y is yield, and x_1 is temperature, then the replicated observations Y_{ir} $(r=1,2,\ldots,R_i)$ are obtained by having R_i experiments with $x_1=x_{i1}$ in each experiment, and not by having a single experiment with $x_1=x_{i1}$ and measuring the yield R_i times. Draper and Smith [1966] point out that the latter method would supply only information on the variance of the device for measuring yield which is just part of the variance σ^2; our definition of σ^2 also includes the variation in yield between experiments at the same temperature. However, given genuine replications, it is possible to test whether the model (4.34) is appropriate using the F-statistic derived below.

Let $Y_{ir}=\phi_i+\varepsilon_{ir}$, say. Then writing

$$\mathbf{Y}'=(Y_{11},Y_{12},\ldots,Y_{1R_1},\ldots,Y_{n1},Y_{n2},\ldots,Y_{nR_n}), \text{ etc.,}$$

we have $\mathbf{Y}=\mathbf{W}\boldsymbol{\phi}+\boldsymbol{\varepsilon}$ where

$$\mathbf{W}\boldsymbol{\phi}=\begin{bmatrix}\mathbf{1}_{R_1} & \mathbf{0} & \cdots & \mathbf{0}\\ \mathbf{0} & \mathbf{1}_{R_2} & \cdots & \mathbf{0}\\ \cdots & \cdots & \cdots & \cdots\\ \mathbf{0} & \mathbf{0} & \cdots & \mathbf{1}_{R_n}\end{bmatrix}\begin{bmatrix}\phi_1\\ \phi_2\\ \cdots\\ \phi_n\end{bmatrix}. \tag{4.37}$$

Defining $N=\sum_i R_i$, then $\mathbf{W}$ is an $N\times n$ matrix of rank n; we also assume that $\boldsymbol{\varepsilon}\sim N_N(\mathbf{0},\sigma^2\mathbf{I}_N)$. Now testing the adequacy of (4.34) is equivalent to testing the hypothesis

$$H:\phi_i=\beta_0+\beta_1x_{i1}+\cdots+\beta_{p-1}x_{i,p-1} \qquad (i=1,2,\ldots,n)$$

or $H:\boldsymbol{\phi}=\mathbf{X}\boldsymbol{\beta}$, where $\mathbf{X}$ is $n\times p$ of rank p. This hypothesis can be converted into the more familiar "constraint equation" form using the following theorem.

THEOREM 4.4 $\boldsymbol{\phi}\in\mathcal{R}[\mathbf{X}]$ if and only if $\mathbf{A}\boldsymbol{\phi}=\mathbf{0}$ for some $(n-p)\times n$ matrix $\mathbf{A}$ of rank $n-p$.

Proof. Let $\mathbf{P}=\mathbf{X}(\mathbf{X}'\mathbf{X})^{-1}\mathbf{X}'$. If $\boldsymbol{\phi}\in\mathcal{R}[\mathbf{X}]$, that is, $\boldsymbol{\phi}=\mathbf{X}\boldsymbol{\beta}$ for some $\boldsymbol{\beta}$, then $(\mathbf{I}_n-\mathbf{P})\boldsymbol{\phi}=(\mathbf{I}_n-\mathbf{P})\mathbf{X}\boldsymbol{\beta}=\mathbf{0}$ [by Theorem 3.1 (iii)]. Conversely, if $(\mathbf{I}_n-\mathbf{P})\boldsymbol{\phi}=\mathbf{0}$ then $\boldsymbol{\phi}=\mathbf{P}\boldsymbol{\phi}=\mathbf{P}\boldsymbol{\phi}=\mathbf{X}(\mathbf{X}'\mathbf{X})^{-1}\mathbf{X}'\boldsymbol{\phi}=\mathbf{X}\boldsymbol{\gamma}\in\mathcal{R}[\mathbf{X}]$. Hence $\boldsymbol{\phi}\in\mathcal{R}[\mathbf{X}]$ if and only if $(\mathbf{I}_n - \mathbf{P})\boldsymbol{\phi} = \mathbf{0}$. By Theorem 3.1 (ii) the $n \times n$ matrix $\mathbf{I}_n - \mathbf{P}$ has rank $n-p$ and therefore has $n-p$ linearly independent rows which we can take as our required matrix $\mathbf{A}$. □

Using the above theorem we see that the general regression theory applies to H, but with n, p, and q replaced by N, n, and $n-p$, respectively; hence

$$F = \frac{(\mathrm{RSS}_H - \mathrm{RSS})/(n-p)}{\mathrm{RSS}/(N-n)}.$$

Here RSS is found directly by minimizing $\Sigma_i \Sigma_r (Y_{ir} - \phi_i)^2$; thus differentiating partially with respect to ϕ_i we have

$$\hat{\phi}_i = \frac{\sum_r Y_{ir}}{R_i} = \overline{Y}_{i.} \qquad \text{and} \qquad \mathrm{RSS} = \sum\sum \left(Y_{ir} - \overline{Y}_{i.}\right)^2.$$

To find RSS_H we minimize $\Sigma_i \Sigma_r (Y_{ir} - \beta_0 - \beta_1 x_{i1} - \cdots - \beta_{p-1} x_{i,p-1})^2$ $(= d$, say); therefore, setting $\partial d/\partial \beta_0 = 0$ and $\partial d/\partial \beta_j = 0$ $(j \neq 0)$ we have

$$\sum_i \left(\overline{Y}_{i.} - \beta_0 - \beta_1 x_{i1} - \cdots - \beta_{p-1} x_{i,p-1}\right) = 0, \tag{4.38}$$

and

$$\sum_i \sum_r x_{ij} \left(Y_{ir} - \beta_0 - \beta_1 x_{i1} - \cdots - \beta_{p-1} x_{i,p-1}\right) = 0 \qquad (j = 1, 2, \ldots, p-1),$$

that is,

$$\sum_i x_{ij} \left(\overline{Y}_{i.} - \beta_0 - \beta_1 x_{i1} - \cdots - \beta_{p-1} x_{i,p-1}\right) = 0. \tag{4.39}$$

Since Equations (4.38) and (4.39) are identical to the usual normal equations, except that Y_i is replaced by $Z_i = \overline{Y}_{i.}$, we have

$$\hat{\boldsymbol{\beta}}_H = (\mathbf{X}'\mathbf{X})^{-1}\mathbf{X}'\mathbf{Z}$$

and

$$\mathrm{RSS}_H = \sum_i \sum_r \left(Y_{ir} - \hat{\beta}_{0H} - \hat{\beta}_{1H} x_{i1} - \cdots - \hat{\beta}_{p-1,H} x_{i,p-1}\right)^2.$$

4.5 DESIGN MATRIX OF LESS THAN FULL RANK

4.5.1 The F-Test and Projection Matrices

Before considering the case when the $n\times p$ matrix $\mathbf{X}$ has rank r $(r<p)$ we shall find it helpful to give a more general theory of the F-test than that given at the beginning of this chapter.

Suppose we have the model $\mathbf{Y}=\boldsymbol{\theta}+\boldsymbol{\varepsilon}$ where $\boldsymbol{\theta}\in\Omega$ (an r-dimensional subspace of E_n), and we wish to test $H:\boldsymbol{\theta}\in\omega$, where ω is an $r-q$ dimensional subspace of Ω. Then we have the following theorem.

THEOREM 4.5 When H is true and $\boldsymbol{\varepsilon}\sim N_n(\mathbf{0},\sigma^2\mathbf{I}_n)$,

$$F=\frac{(\text{RSS}_H-\text{RSS})/q}{\text{RSS}/(n-r)}=\frac{\boldsymbol{\varepsilon}'(\mathbf{P}_\Omega-\mathbf{P}_\omega)\boldsymbol{\varepsilon}/q}{\boldsymbol{\varepsilon}'(\mathbf{I}_n-\mathbf{P}_\Omega)\boldsymbol{\varepsilon}/(n-r)}\sim F_{q,n-r},$$

where $\mathbf{P}_\Omega$ and $\mathbf{P}_\omega$ are the symmetric idempotent matrices projecting E_n onto Ω and ω, respectively (Appendix B).

Proof. $\hat{\boldsymbol{\theta}}=\mathbf{P}_\Omega\mathbf{Y}$ and $\hat{\boldsymbol{\theta}}_H=\mathbf{P}_\omega\mathbf{Y}$ are the respective least squares estimates of $\boldsymbol{\theta}$. Hence

$$\text{RSS}=\|\mathbf{Y}-\hat{\boldsymbol{\theta}}\|^2=\mathbf{Y}'(\mathbf{I}_n-\mathbf{P}_\Omega)\mathbf{Y}$$

and

$$\text{RSS}_H=\mathbf{Y}'(\mathbf{I}_n-\mathbf{P}_\omega)\mathbf{Y}.$$

Now $(\mathbf{I}_n-\mathbf{P}_\Omega)\boldsymbol{\theta}=\mathbf{0}$ (since $\boldsymbol{\theta}\in\Omega$) so that

$$\text{RSS}=(\mathbf{Y}-\boldsymbol{\theta})'(\mathbf{I}_n-\mathbf{P}_\Omega)(\mathbf{Y}-\boldsymbol{\theta})=\boldsymbol{\varepsilon}'(\mathbf{I}_n-\mathbf{P}_\Omega)\boldsymbol{\varepsilon}.$$

Similarly, when H is true, $\boldsymbol{\theta}\in\omega$ and

$$\text{RSS}_H=\boldsymbol{\varepsilon}'(\mathbf{I}_n-\mathbf{P}_\omega)\boldsymbol{\varepsilon}.$$

Now $(\mathbf{I}_n-\mathbf{P}_\Omega)$ and $(\mathbf{P}_\Omega-\mathbf{P}_\omega)$ project onto $\Omega^\perp$ and $\omega^\perp\cap\Omega$ (by *B1.6* and *B3.2*), so that these matrices are symmetric and idempotent (*B1.4*) and have ranks $n-r$ and $r-(r-q)=q$ (by *B1.5*). Hence by Theorem 2.8 and Example 2.6 (or Theorem 2.9) in Section 2.4, $\boldsymbol{\varepsilon}'(\mathbf{P}_\Omega-\mathbf{P}_\omega)\boldsymbol{\varepsilon}/\sigma^2$ and $\boldsymbol{\varepsilon}(\mathbf{I}_n-\mathbf{P}_\Omega)\boldsymbol{\varepsilon}/\sigma^2$ are independently distributed as χ^2_q and χ^2_{n-r}, respectively. Thus $F\sim F_{q,n-r}$. □

It is readily seen that Theorem 4.1 (iv) is a special case of the above; there $\Omega=\mathcal{R}[\mathbf{X}]$ and, when $\mathbf{c}=\mathbf{0}$, $\omega=\mathcal{N}[\mathbf{A}(\mathbf{X}'\mathbf{X})^{-1}\mathbf{X}']\cap\Omega$. In the next section we generalize the rest of Theorem 4.1.

4.5.2 Testable Hypotheses

Let $\mathbf{Y}=\boldsymbol{\theta}+\boldsymbol{\varepsilon}$, where $\boldsymbol{\theta}=\mathbf{X}\boldsymbol{\beta}$ and $\mathbf{X}$ is $n\times p$ of rank r $(r<p)$. Suppose we wish to test $H:\mathbf{A}\boldsymbol{\beta}=\mathbf{0}$, where $\mathbf{A}$ is a known $q\times p$ matrix of rank q. Since $\mathbf{X}$ is not of full rank, a new problem arises for there is now the possibility that H is not "testable". For example, if the rows of $\mathbf{A}$ are linearly independent of the rows of $\mathbf{X}$ (and consequently $q\leqslant p-r$) then, by Theorem 3.9 in Section 3.8.1, it follows that for every $\boldsymbol{\theta}\in\mathcal{R}[\mathbf{X}]$ there exists a $\boldsymbol{\beta}$ satisfying $\boldsymbol{\theta}=\mathbf{X}\boldsymbol{\beta}$ and $\mathbf{A}\boldsymbol{\beta}=\mathbf{0}$; this $\boldsymbol{\beta}$ would also be unique if $q=p-r$. In this case the equations $\mathbf{A}\boldsymbol{\beta}=\mathbf{0}$ are simply identifiability constraints for $\boldsymbol{\beta}$ so that $\boldsymbol{\theta}$ is not restricted to a proper subset of $\mathcal{R}[\mathbf{X}]$. Hence if $\mathbf{a}_i'$ is the ith row of $\mathbf{A}$ we can ignore any equation $\mathbf{a}_i'\boldsymbol{\beta}=0$ for which $\mathbf{a}_i'$ is linearly independent of the rows of $\mathbf{X}$, and our testable hypothesis consists of the remaining equations. We now have the following definition.

Definition. *The hypothesis* $H:\mathbf{A}\boldsymbol{\beta}=\mathbf{0}$ *is said to be testable if the rows of* $\mathbf{A}$ *are linearly dependent on the rows of* $\mathbf{X}$, *that is, if there exists a* $q\times n$ *matrix* $\mathbf{M}$ *such that*

$$\mathbf{A}=\mathbf{M}\mathbf{X}. \tag{4.40}$$

This definition applies when the rows of $\mathbf{A}$ are not linearly independent. However, if A is $q\times p$ of rank q then $\mathbf{M}$ must have rank q (since rank $\mathbf{A}\leqslant$ rank $\mathbf{M}$ by *A2.1*).

The definition also applies to the more general case $\mathbf{A}\boldsymbol{\beta}=\mathbf{c}$ where $\mathbf{c}\neq\mathbf{0}$ and $\mathbf{c}\in\mathcal{R}[\mathbf{A}]$. To show this we shift the origin (as in Section 3.9.2) and reduce the model and hypothesis to

$$\tilde{\mathbf{Y}}=\mathbf{X}\boldsymbol{\gamma}+\boldsymbol{\varepsilon} \qquad \text{and} \qquad \mathbf{A}\boldsymbol{\gamma}=\mathbf{0}. \tag{4.41}$$

Here $\boldsymbol{\gamma}=\boldsymbol{\beta}-\boldsymbol{\beta}_0$, where $\boldsymbol{\beta}_0$ is a solution of $\mathbf{A}\boldsymbol{\beta}=\mathbf{c}$, and $\tilde{\mathbf{Y}}=\mathbf{Y}-\mathbf{X}\boldsymbol{\beta}_0$. Clearly the original hypothesis is testable if and only if the transformed hypothesis is testable.

We finally note, from Section 3.8.2, that H is testable if and only if each $\mathbf{a}_i'\boldsymbol{\beta}$ is estimable.

THEOREM 4.6 Let $H:\mathbf{A}\boldsymbol{\beta}=\mathbf{c}$ be a testable hypothesis where $\mathbf{A}$ is $k\times n$ of rank q $(q\leqslant r)$, and let RSS and RSS_H be the unrestricted and restricted minimum values of $\boldsymbol{\varepsilon}'\boldsymbol{\varepsilon}=\|\mathbf{Y}-\boldsymbol{\theta}\|^2$, respectively. Then

(i) $F=\dfrac{(\text{RSS}_H-\text{RSS})/q}{\text{RSS}/(n-r)}\sim F_{q,n-r}$
when H is true.

(ii) $\text{RSS}_H-\text{RSS}=(\mathbf{A}\hat{\boldsymbol{\beta}}-\mathbf{c})'[\mathbf{A}(\mathbf{X}'\mathbf{X})^{-}\mathbf{A}']^{-}(\mathbf{A}\hat{\boldsymbol{\beta}}-\mathbf{c})$
where $\hat{\boldsymbol{\beta}}$ is any solution of $\mathbf{X}'\mathbf{X}\boldsymbol{\beta}=\mathbf{X}'\mathbf{Y}$.

Proof. (i) We first transform the model and hypothesis to (4.41). Our model is now $\tilde{\mathbf{Y}}=\boldsymbol{\theta}+\boldsymbol{\varepsilon}$, where $\boldsymbol{\theta}\in\mathcal{R}[\mathbf{X}]\ (=\Omega)$ and, since $\mathbf{M}\boldsymbol{\theta}=\mathbf{MX}\boldsymbol{\gamma}=\mathbf{A}\boldsymbol{\gamma}=\mathbf{0}$, H becomes $\boldsymbol{\theta}\in\omega=\mathcal{N}[\mathbf{M}]\cap\Omega$. From Section 3.9.2 we note that $\omega^{\perp}\cap\Omega=\mathcal{R}[\mathbf{P}_\Omega\mathbf{M}']$ has dimension q. Also, since $(\mathbf{I}_n-\mathbf{P}_\Omega)\mathbf{X}=\mathbf{O}$,

$$(\mathbf{I}_n-\mathbf{P}_\Omega)\tilde{\mathbf{Y}}=(\mathbf{I}_n-\mathbf{P}_\Omega)(\mathbf{Y}-\mathbf{X}\boldsymbol{\beta}+\mathbf{X}(\boldsymbol{\beta}-\boldsymbol{\beta}_0))=(\mathbf{I}_n-\mathbf{P}_\Omega)\boldsymbol{\varepsilon}.$$

Now, when H is true,

$$(\mathbf{P}_\Omega-\mathbf{P}_\omega)\tilde{\mathbf{Y}}=\mathbf{P}_{\omega^{\perp}\cap\Omega}(\mathbf{Y}-\mathbf{X}\boldsymbol{\beta}+\mathbf{X}\boldsymbol{\gamma})=\mathbf{P}_{\omega^{\perp}\cap\Omega}\boldsymbol{\varepsilon}$$

since $(\mathbf{P}_\Omega\mathbf{M}')'\mathbf{X}\boldsymbol{\gamma}=\mathbf{MX}\boldsymbol{\gamma}=\mathbf{A}\boldsymbol{\gamma}=\mathbf{0}$. The required result now follows by evoking Theorem 4.5.

(ii) From Equation (3.68) in Section 3.9.2,

$$\begin{aligned}\text{RSS}_H-\text{RSS}&=\tilde{\mathbf{Y}}'\mathbf{P}_{\omega^{\perp}\cap\Omega}\tilde{\mathbf{Y}}\\&=\tilde{\mathbf{Y}}'\mathbf{X}(\mathbf{X}'\mathbf{X})^{-}\mathbf{A}\left[\mathbf{A}(\mathbf{X}'\mathbf{X})^{-}\mathbf{A}'\right]^{-}\mathbf{A}(\mathbf{X}'\mathbf{X})^{-}\mathbf{X}'\tilde{\mathbf{Y}}\\&=(\mathbf{A}\hat{\boldsymbol{\beta}}-\mathbf{c})'\left[\mathbf{A}(\mathbf{X}'\mathbf{X})^{-}\mathbf{A}'\right]^{-}(\mathbf{A}\hat{\boldsymbol{\beta}}-\mathbf{c})\qquad\text{(by } B1.8)\end{aligned}$$

since

$$\begin{aligned}\mathbf{A}(\mathbf{X}'\mathbf{X})^{-}\mathbf{X}'\tilde{\mathbf{Y}}&=\mathbf{MP}_\Omega(\mathbf{Y}-\mathbf{X}\boldsymbol{\beta}_0)\\&=\mathbf{MP}_\Omega\mathbf{Y}-\mathbf{MX}\boldsymbol{\beta}_0\\&=\mathbf{MX}\hat{\boldsymbol{\beta}}-\mathbf{A}\boldsymbol{\beta}_0\\&=\mathbf{A}\hat{\boldsymbol{\beta}}-\mathbf{c}.\end{aligned}$$

□

The above results can be proved in a long-winded fashion using just the properties of generalized inverses (cf. John and Smith [1974]). In practice part (i) of the theorem is more important as RSS and RSS_H can often be obtained directly by differentiation or even, as in many analysis of variance situations (Chapter 9), by inspection.

EXERCISES 4e

1. Let $H:\mathbf{A}\boldsymbol{\beta}=\mathbf{c}$ be a testable hypothesis. Prove that

$$\mathcal{D}\left[\mathbf{A}\hat{\boldsymbol{\beta}}\right]=\sigma^2\mathbf{A}(\mathbf{X}'\mathbf{X})^{-}\mathbf{A}',$$

and show that the above matrix is nonsingular if the rows of $\mathbf{A}$ are linearly independent.

2. Using the notation of Theorem 4.6 prove that

$$E\left[\text{RSS}_H-\text{RSS}\right]=\sigma^2q+(\mathbf{A}\hat{\boldsymbol{\beta}}-\mathbf{c})'\left[\mathbf{A}(\mathbf{X}'\mathbf{X})^{-}\mathbf{A}'\right]^{-}(\mathbf{A}\hat{\boldsymbol{\beta}}-\mathbf{c}).$$

4.6 HYPOTHESIS TESTING WITH INITIAL CONSTRAINTS

In the first instance we shall consider the full rank model $\mathbf{Y}=\mathbf{X}\boldsymbol{\beta}+\boldsymbol{\varepsilon}$, where $\mathbf{X}$ is $n\times p$ of rank p, but with the additional constraints $\mathbf{C}\boldsymbol{\beta}=\mathbf{0}$, where $\mathbf{C}$ is $k\times p$ of rank k. We wish to test the hypothesis $H:\mathbf{A}\boldsymbol{\beta}=\mathbf{0}$ where $\mathbf{A}$ is $q\times p$ of rank q and the rows of $\mathbf{A}$ are linearly independent of the rows of $\mathbf{C}$ (thus $q+k\leqslant p$). Then, using the notation of Section 4.5.1, we have $\Omega=\mathcal{R}[\mathbf{X}]\cap\mathcal{N}[\mathbf{C}(\mathbf{X}'\mathbf{X})^{-1}\mathbf{X}']$ [since $\mathbf{C}\boldsymbol{\beta}=\mathbf{C}(\mathbf{X}'\mathbf{X})^{-1}\mathbf{X}'\boldsymbol{\theta}$] and $\omega=\mathcal{N}[\mathbf{A}(\mathbf{X}'\mathbf{X})^{-1}\mathbf{X}']\cap\Omega$. Interpreting Ω and $(\mathbf{A}',\mathbf{C}')'$ as the ω and $\mathbf{A}$, respectively, of Section 3.9.2, we see that Ω and ω have dimensions $p-k$ and q, respectively. Hence, from Theorem 4.5 in Section 4.5.1,

$$F=\frac{(\mathrm{RSS}_H-\mathrm{RSS})/q}{\mathrm{RSS}/[n-(p-k)]}\sim F_{q,n-p+k} \tag{4.42}$$

when H is true. This result will still hold for the case $H:\mathbf{A}\boldsymbol{\beta}=\mathbf{c}$ ($\mathbf{c}\neq\mathbf{0}$). We simply find $\boldsymbol{\beta}_0$ such that

$$\begin{pmatrix}\mathbf{A}\\ \mathbf{C}\end{pmatrix}\boldsymbol{\beta}_0=\begin{pmatrix}\mathbf{c}\\ \mathbf{0}\end{pmatrix}$$

and write $\tilde{\mathbf{Y}}=\mathbf{Y}-\mathbf{X}\boldsymbol{\beta}_0$, as in the preceding section.

As with all the variations on the "hypothesis theme" considered thus far, it does not matter how we actually find RSS and RSS_H, for they are unique. However, some care is usually needed in determining the degrees of freedom for the numerator and the denominator of the F-statistic.

What happens when $\mathbf{X}$ is allowed to have less than full rank and the full set of constraints $\mathbf{C}\boldsymbol{\beta}=\mathbf{0}$ and $\mathbf{A}\boldsymbol{\beta}=\mathbf{c}$ are testable? We simply replace p by the rank of $\mathbf{X}$ in (4.42). This can be proved along the lines of Theorem 4.6 in the preceding section.

MISCELLANEOUS EXERCISES 4

1. Aerial observations Y_1, Y_2, Y_3, and Y_4 are made of angles θ_1, θ_2, θ_3, and θ_4, respectively, of a quadrilateral on the ground. If the observations are subject to independent normal errors with zero means and common variance σ^2, derive a test statistic for the hypothesis that the quadrilateral is a parallelogram with $\theta_1=\theta_3$ and $\theta_2=\theta_4$.

   Adapted from Silvey [1970]

2. Let $\mathbf{Y}\sim N_n(\mathbf{X}\boldsymbol{\beta},\sigma^2\mathbf{I}_n)$, where $\mathbf{X}$ is $n\times p$ of rank p, and let $\mathbf{X}\boldsymbol{\beta}=(\mathbf{X}_1,\mathbf{X}_2)(\boldsymbol{\beta}_1',\boldsymbol{\beta}_2')'$, where $\mathbf{X}_1$ is the submatrix consisting of the first p_1 columns of $\mathbf{X}$ ($p_1<p$) and $\boldsymbol{\beta}_1$ consists of the first p_1 elements of $\boldsymbol{\beta}$. If

$\mathbf{P}_2=\mathbf{X}_2(\mathbf{X}_2'\mathbf{X}_2)^{-1}\mathbf{X}_2'$ and H is the hypothesis that $\boldsymbol{\beta}_1=\mathbf{0}$, prove that

$$E[\text{RSS}_H-\text{RSS}]=\sigma^2 p_1+\boldsymbol{\beta}_1'\mathbf{X}_1'(\mathbf{I}_n-\mathbf{P}_2)\mathbf{X}_1\boldsymbol{\beta}_1.$$

3. Let $\beta_1,\beta_2,\ldots,\beta_q$ be a subset of elements of $\boldsymbol{\beta}$ for the model $\mathcal{E}[\mathbf{Y}]=\mathbf{X}\boldsymbol{\beta}$. Prove that the hypothesis $H:\beta_1=\beta_2=\cdots=\beta_q$ is testable if and only if $\sum_{i=1}^q c_i\beta_i$ is estimable for all c_i such that $\sum_{i=1}^q c_i=0$.
4. Let $\hat{\boldsymbol{\beta}}$ be any solution of the normal equations and let $H:\mathbf{AB}=\mathbf{c}$ be a testable hypothesis. If $\mathbf{A}$ is $q\times p$ of rank q, prove the following:
   (a) $\mathbf{A}\hat{\boldsymbol{\beta}}$ has a multivariate normal distribution.
   (b) $(\mathbf{A}\hat{\boldsymbol{\beta}}-\mathbf{c})'[\mathbf{A}(\mathbf{X}'\mathbf{X})^-\mathbf{A}']^-(\mathbf{A}\hat{\boldsymbol{\beta}}-\mathbf{c})/\sigma^2\sim\chi_q^2$, when H is true.
   (c) $\mathbf{A}\hat{\boldsymbol{\beta}}$ is statistically independent of $\|\mathbf{Y}-\mathbf{X}\hat{\boldsymbol{\beta}}\|^2$.
5. Given the two regression lines

$$Y_{ki}=\beta_k x_i+\varepsilon_{ki}\qquad (k=1,2;\ i=1,2,\ldots,n),$$

show that the F-statistic for testing $H:\beta_1=\beta_2$ can be put in the form

$$F=\frac{(\hat{\beta}_1-\hat{\beta}_2)^2}{2S^2\left(\sum_i x_i^2\right)^{-1}}.$$

Obtain RSS and RSS_H and verify that

$$\text{RSS}_H-\text{RSS}=\frac{\sum_i x_i^2(\hat{\beta}_1-\hat{\beta}_2)^2}{2}$$

CHAPTER 5

Confidence Intervals and Regions

5.1 SIMULTANEOUS INTERVAL ESTIMATION

5.1.1 The Problem of Simultaneous Inferences

A common statistical problem is that of finding two-sided confidence intervals for k linear combinations $\mathbf{a}_i'\boldsymbol{\beta}$ $(i=1,2,\ldots,k)$. One solution would be to simply write down k t-intervals of the form given in (4.25) of Section 4.1.5, namely,

$$\mathbf{a}_i'\hat{\boldsymbol{\beta}} \pm t_{n-p}^{(1/2)\alpha}\hat{\sigma}_{\mathbf{a}_i'\hat{\boldsymbol{\beta}}}. \tag{5.1}$$

However, even though we can attach a probability of $1-\alpha$ to each separate interval, the overall probability that the confidence statements are *simultaneously* true is not $1-\alpha$. To see this, suppose that E_i $(i=1,2,\ldots,k)$ is the event that the ith statement is correct, and let $\text{pr}[E_i]=1-\alpha_i$. If $\overline{E}_i$ denotes the complementary event of E_i, then

$$\begin{aligned} 1-\delta=\text{pr}\left[\bigcap_{i=1}^{k} E_i\right] &= 1-\text{pr}\left[\overline{\bigcap_i E_i}\right] = 1-\text{pr}\left[\bigcup_i \overline{E}_i\right] \\ &\geqslant 1-\sum_{i=1}^{k}\text{pr}\left[\overline{E}_i\right] = 1-\sum_{i=1}^{k}\alpha_i. \end{aligned} \tag{5.2}$$

For the case $\alpha_i=\alpha$ $(i=1,2,\ldots,k)$

$$\text{pr}\left[\bigcap_{i=1}^{k} E_i\right] \geqslant 1-k\alpha \tag{5.3}$$

so that the probability of all the statements being correct is not $1-\alpha$ but something greater than $1-k\alpha$. For example, if $\alpha=0.05$ and $k=10$, then $1-k\alpha=0.5$. Furthermore, as pointed out by Miller [1966: p. 8], the inequality (5.3) is not as crude as one might expect, provided k is not too large (say, $k \leqslant 5$) and α is small, say, 0.01.

It is also worth noting that

$$\begin{aligned}\operatorname{pr}\left[\bigcap_i E_i\right] &= \operatorname{pr}[E_1]\operatorname{pr}[E_2|E_1]\cdots\operatorname{pr}[E_k|E_1,\ldots,E_{k-1}]\\ &\cong \operatorname{pr}[E_1]\operatorname{pr}[E_2]\cdots\operatorname{pr}[E_k]\\ &=(1-\alpha_1)(1-\alpha_2)\cdots(1-\alpha_k) \qquad (5.4)\end{aligned}$$

if the dependence between the events E_i is small. This latter situation is quite common in analysis of variance problems where the confidence intervals are frequently based on statistics in which the numerators (here $\mathbf{a}_1'\hat{\boldsymbol{\beta}},\ldots,\mathbf{a}_k'\hat{\boldsymbol{\beta}}$) are mutually independent, or nearly so, and the denominators contain a common random variable such as S^2. In many situations (5.4) provides a lower bound for $\operatorname{pr}[\cap_i E_i]$ (cf. Sidak [1968: p. 1428] and Dykstra et al. [1973]).

There is one other problem associated with the E_i. If $\alpha_i=0.05$ ($i=1,2,\ldots,k$), there is one chance in 20 of making an incorrect statement about $\mathbf{a}_i'\boldsymbol{\beta}$ so that for every 20 statements made we can expect one to be incorrect. In other words, 5% of our k confidence intervals can be expected to be unreliable; there is an expected "error rate" of one in 20.

For the general case when the α_i are not necessarily equal, Miller [1966: p. 8] shows that the expected error rate is $\Sigma_i\alpha_i/k$ ($=\gamma/k$, say). Spjøtvoll [1972a] suggests that simultaneous inference should be based on a prechosen γ rather than on $\delta=1-\operatorname{pr}[\cap_i E_i]=\operatorname{pr}[\cup_i \bar{E}_i]$, the probability of making at least one incorrect statement (commonly called the probability of a nonzero family error rate—Miller [1966]). In any case we see from Equation (5.2) that $\delta \leqslant \gamma$.

We now consider several ways of avoiding some of the problems mentioned above.

a BONFERRONI t-INTERVALS

If we use an individual significance level of α/k instead of α for each of the k confidence intervals, then, from (5.3),

$$\operatorname{pr}\left[\bigcap_{i=1}^{k} E_i\right] \geqslant 1-k\left(\frac{\alpha}{k}\right)=1-\alpha \qquad (5.5)$$

so that the overall probability is at least $1-\alpha$. However, a word of caution. When k is large this method could lead to confidence intervals which are so wide as to be of little practical use. This means that a reasonable compromise may be to increase α, for example, $\alpha=0.10$.

To use the above method we frequently require significance levels for the t-distribution which are not listed in the common t-tables. The following approximation is therefore useful (Scott and Smith [1970]:

$$t_\nu^\alpha \approx z_\alpha\left(1-\frac{z_\alpha^2+1}{4\nu}\right)^{-1}$$

where z_α denotes the upper α point of the $N(0,1)$ distribution. Values of z_α can be found by interpolating in the usual normal tables or using, for example, the extensive Table 1 of the Kelley Statistical Tables (Kelley [1948: p. 37]). Some modern desk calculators have t_ν^α on call. A table of $t_\nu^{\alpha/(2k)}$ (reproduced from Dunn [1961]) for $\alpha=0.05,\ 0.01$; $k=2(1)10(5)50, 100, 250$; $\nu=5,7,10,12,15,20,24,30,40,60,120,\infty$ is given in Appendix D. Related tables which allow different α_i are given by Dayton and Schafer [1973].

Intervals based on the above method of replacing α by α/k are called Bonferroni t-intervals, as (5.2) is a Bonferroni inequality (Feller [1968: p. 110]).

b MAXIMUM MODULUS t-INTERVALS

Let $u_{k,\nu,\rho}^\alpha$ be the upper tail α significance point of the distribution of the maximum absolute value of k Student t-variables, each based on ν degrees of freedom and having a common pairwise correlation ρ; when $\rho=0$ we simply denote this point by $u_{k,\nu}^\alpha$. Now, if the $\mathbf{a}_i'\hat{\boldsymbol\beta}$ $(i=1,2,\ldots,k)$ are mutually independent, the pairwise correlations between the t-variables

$$T_i=\frac{\mathbf{a}_i'\hat{\boldsymbol\beta}-\mathbf{a}_i'\boldsymbol\beta}{\hat\sigma_{\mathbf{a}_i'\hat{\boldsymbol\beta}}},$$

conditional on S^2, are zero. Since S^2 is independent of each $\mathbf{a}_i'\hat{\boldsymbol\beta}$, the unconditional correlations are also zero and we have

$$\begin{aligned}1-\alpha&=\operatorname{pr}\left[\max_{1\leqslant i\leqslant k}|T_i|\leqslant u_{k,n-p}^\alpha\right]\\&=\operatorname{pr}\left[|T_i|\leqslant u_{k,n-p}^\alpha,\ \text{all } i\right].\end{aligned}$$

Hence the set of k intervals

$$\mathbf{a}_i'\hat{\boldsymbol\beta}\pm u_{k,n-p}^\alpha\hat\sigma_{\mathbf{a}_i'\hat{\boldsymbol\beta}} \tag{5.6}$$

will have an overall confidence probability of *exactly* $1-\alpha$; thus $\delta=\alpha$. However, if the $\mathbf{a}_i'\hat{\boldsymbol{\beta}}$ are not independent, which is the more usual situation, then the above intervals (5.6) can still be used, but they will be conservative; the overall probability will be at least $1-\alpha$. (This result follows from a theorem by Sidak [1968]; see Hahn and Hendrickson [1971] and Hahn [1972].)

Hahn [1972] shows that when $k=2$, the intervals

$$\mathbf{a}_i'\hat{\boldsymbol{\beta}} \pm u_{k,n-p,\rho}^{\alpha}\hat{\sigma}_{\mathbf{a}_i'\hat{\boldsymbol{\beta}}} \qquad (i=1,2)$$

where ρ, the correlation coefficient of $\mathbf{a}_1'\hat{\boldsymbol{\beta}}$ and $\mathbf{a}_2'\hat{\boldsymbol{\beta}}$, is given by

$$\rho=\frac{\mathbf{a}_1'(\mathbf{X}'\mathbf{X})^{-1}\mathbf{a}_2}{\left\{\mathbf{a}_1'(\mathbf{X}'\mathbf{X})^{-1}\mathbf{a}_1\mathbf{a}_2'(\mathbf{X}'\mathbf{X})^{-1}\mathbf{a}_2\right\}^{1/2}}, \tag{5.7}$$

have an exact overall probability of $1-\alpha$. This result is useful in straight line regression (see Chapter 7).

A table of $u_{k,\nu,\rho}^{\alpha}$, reproduced from Hahn and Hendrickson [1971], for $\alpha=0,1,0.05,0.01$; $k=1(1)\ 6,8,10,12,15,20$; $\nu=3(1)\ 12,15,20,25,30,40,60$; and $\rho=0.0,0.2,0.4$, and 0.5 is given in Appendix E. A paper by Tong [1970] provides a procedure which can be used to obtain conservative estimates of $u_{k,\nu,\rho}^{\alpha}$ for $k>20$ using the tabulated values for $k=20$.

c SCHEFFÉ'S *S*-METHOD

We may assume, without loss of generality, that the first d vectors of the set $\{\mathbf{a}_1,\mathbf{a}_2,\ldots,\mathbf{a}_k\}$ are linearly independent, and the remaining vectors (if any) are linearly dependent on the first d vectors; thus $d\leqslant\min(k,p)$. Consider the $d\times p$ matrix $\mathbf{A}$ where $\mathbf{A}'=[\mathbf{a}_1,\mathbf{a}_2,\ldots,\mathbf{a}_d]$, and let $\boldsymbol{\phi}=\mathbf{A}\boldsymbol{\beta}$. Now $\mathbf{A}$ is a $d\times p$ matrix of rank d so that using the same argument as that given in proving Theorem 4.1(iii), and setting $\hat{\boldsymbol{\phi}}=\mathbf{A}\hat{\boldsymbol{\beta}}$, we have

$$\frac{(\hat{\boldsymbol{\phi}}-\boldsymbol{\phi})'\left[\mathbf{A}(\mathbf{X}'\mathbf{X})^{-1}\mathbf{A}'\right](\hat{\boldsymbol{\phi}}-\boldsymbol{\phi})}{dS^2}\sim F_{d,n-p}. \tag{5.8}$$

Hence

$$\begin{aligned}1-\alpha&=\operatorname{pr}\left[F_{d,n-p}\leqslant F_{d,n-p}^{\alpha}\right]\\&=\operatorname{pr}\left[(\hat{\boldsymbol{\phi}}-\boldsymbol{\phi})'\left[\mathbf{A}(\mathbf{X}'\mathbf{X})^{-1}\mathbf{A}'\right]^{-1}(\hat{\boldsymbol{\phi}}-\boldsymbol{\phi})\leqslant dS^2F_{d,n-p}^{\alpha}\right]\\&=\operatorname{pr}\left[(\hat{\boldsymbol{\phi}}-\boldsymbol{\phi})'\mathbf{L}^{-1}(\hat{\boldsymbol{\phi}}-\boldsymbol{\phi})\leqslant m\right],\qquad\text{say,}\end{aligned} \tag{5.9}$$

$$=\text{pr}[\mathbf{b}'\mathbf{L}^{-1}\mathbf{b}\leqslant m]$$

$$=\text{pr}\left[\sup_{\mathbf{h}\neq\mathbf{0}}\left\{\frac{(\mathbf{h}'\mathbf{b})^2}{\mathbf{h}'\mathbf{L}\mathbf{h}}\right\}\leqslant m\right], \text{ (by } (A4.11)\text{)}$$

$$=\text{pr}\left[\frac{(\mathbf{h}'\mathbf{b})^2}{\mathbf{h}'\mathbf{L}\mathbf{h}}\leqslant m, \text{ all } \mathbf{h}\ (\neq\mathbf{0})\right]$$

$$=\text{pr}\left[\frac{|\mathbf{h}'\hat{\boldsymbol{\phi}}-\mathbf{h}'\boldsymbol{\phi}|}{S(\mathbf{h}'\mathbf{L}\mathbf{h})^{1/2}}\leqslant(dF^{\alpha}_{d,n-p})^{1/2}, \text{ all } \mathbf{h}\right]. \tag{5.9}$$

We can therefore construct a confidence interval for *any* linear function $\mathbf{h}'\boldsymbol{\phi}$, namely,

$$\mathbf{h}'\hat{\boldsymbol{\phi}}\pm(dF^{\alpha}_{d,n-p})^{1/2}S(\mathbf{h}'\mathbf{L}\mathbf{h})^{1/2}, \tag{5.10}$$

and the overall probability for the whole class of such intervals is exactly $1-\alpha$. We note that the term $S^2\mathbf{h}'\mathbf{L}\mathbf{h}$ involved in the calculation of (5.10) is simply an unbiased estimate of $\text{var}[\mathbf{h}'\hat{\boldsymbol{\phi}}]$; frequently the latter expression can be found directly without the need for any matrix inversion (for example, see Section 9.1.7). The interval (5.10) can therefore be written in the more compact form

$$\mathbf{h}'\hat{\boldsymbol{\phi}}\pm(dF^{\alpha}_{d,n-p})^{1/2}\hat{\sigma}_{\mathbf{h}'\hat{\boldsymbol{\phi}}}. \tag{5.11}$$

Since $\mathbf{h}'\boldsymbol{\phi}=\phi_i$ for certain $\mathbf{h}$, we see that a confidence interval for every $\mathbf{a}_i'\boldsymbol{\beta}=\phi_i$ $(i=1,2,\ldots,d)$ is included in the set of intervals (5.11). In addition, an interval for every ϕ_j $(j=d+1,d+2,\ldots,k)$ is also included in this set owing to the linear dependence of the $\mathbf{a}_j$ $(j=d+1,\ldots,k)$ on the other $\mathbf{a}_i$'s. For example, if $\mathbf{a}_{d+1}=h_1\mathbf{a}_1+\cdots+h_d\mathbf{a}_d$, then $\phi_{d+1}=\mathbf{a}_{d+1}'\boldsymbol{\beta}=\Sigma_{i=1}^d h_i\phi_i=\mathbf{h}'\boldsymbol{\phi}$. Therefore if E_i is the event that $\mathbf{a}_i'\boldsymbol{\beta}$ lies in the interval

$$\mathbf{a}_i'\hat{\boldsymbol{\beta}}\pm(dF^{\alpha}_{d,n-p})^{1/2}\hat{\sigma}_{\mathbf{a}_i'\hat{\boldsymbol{\beta}}} \tag{5.12}$$

then, since the complete set of intervals (5.11) is more than what we asked for,

$$\text{pr}\left[\bigcap_{i=1}^{k}E_i\right]\geqslant 1-\alpha.$$

We note that the class of parametric functions $\mathbf{h}'\boldsymbol{\phi}$ form a linear space, $\mathfrak{L}$, say, with basis $\phi_1,\phi_2,\ldots,\phi_d$. In fact $\mathfrak{L}$ is the smallest linear space containing the k functions ϕ_i $(i=1,2,\ldots,k)$.

The above method is due to Scheffé [1953] and it is called the S-method of multiple comparisons in his book (Scheffé [1959: p. 68]). Other methods for constructing simultaneous confidence intervals for special subsets of $\mathcal{L}$ are discussed in Section 9.1.7. For general references on the subject of multiple comparisons the reader is referred to Miller [1966], O'Neill and Wetherill [1971], and Hahn [1972].

The class of linear functions $\mathcal{L}$ of the form $\mathbf{h}'\boldsymbol{\phi}$ $(=\mathbf{h}'\mathbf{A}\boldsymbol{\beta})$ is only a subclass of all possible linear functions $\mathbf{a}'\boldsymbol{\beta}$ where $\mathbf{a}$ is now any $p\times 1$ vector. However, setting $d=k=p$ and $\mathbf{A}=\mathbf{I}_p$, we have $\boldsymbol{\phi}=\boldsymbol{\beta}$ and the corresponding confidence intervals for the class of all functions $\mathbf{h}'\boldsymbol{\beta}$ take the form [cf. (5.11)]

$$\mathbf{h}'\hat{\boldsymbol{\beta}}\pm\left(pF^{\alpha}_{p,n-p}\right)^{1/2}\hat{\sigma}_{\mathbf{h}'\hat{\boldsymbol{\beta}}} \tag{5.13}$$

An interesting relationship exists between the set of confidence intervals (5.10) and the F-statistic for testing the hypothesis $H:\boldsymbol{\phi}=\mathbf{c}$. From (5.8) and (5.9) we see that the F-statistic is not significant at the α level of significance if and only if

$$F=\frac{(\hat{\boldsymbol{\phi}}-\mathbf{c})'\mathbf{L}^{-1}(\hat{\boldsymbol{\phi}}-\mathbf{c})}{dS^2}\leqslant F^{\alpha}_{d,n-p},$$

which is true if and only if $\boldsymbol{\phi}=\mathbf{c}$ is contained in the region $(\boldsymbol{\phi}-\hat{\boldsymbol{\phi}})\mathbf{L}^{-1}(\boldsymbol{\phi}-\hat{\boldsymbol{\phi}})\leqslant m$, that is, if and only if $\mathbf{h}'\mathbf{c}$ is contained in (5.10) for *every* $\mathbf{h}$. Therefore F is significant if one or more of the intervals (5.10) does not contain $\mathbf{h}'\mathbf{c}$, and the situation can arise where each interval for ϕ_i contains c_i $(i=1,2,\ldots,k)$ but H is rejected. For example, when $k=2$ the separate intervals for ϕ_1 and ϕ_2 form the rectangle given in Fig. 5.1, and the ellipse is the region $(\boldsymbol{\phi}-\hat{\boldsymbol{\phi}})'\mathbf{L}^{-1}(\boldsymbol{\phi}-\hat{\boldsymbol{\phi}})\leqslant m$; a point $\mathbf{c}$ which lies within the rectangle does not necessarily lie within the ellipse.

5.1.2 Comparison of Methods

For k confidence intervals the Bonferroni t-intervals, the maximum modulus t-intervals (5.6), and Scheffé's F-intervals (5.12) all give a lower bound of $1-\alpha$ for $\text{pr}[\cap_i E_i]$. By comparing Tables 5.1 and 5.2 (reproduced from Dunn [1959]) we see that for $\alpha=0.05$, $d\leqslant k$, and k not much greater than d,

$$t_{\nu}^{\alpha/(2k)}<\left(dF^{\alpha}_{d,\nu}\right)^{1/2}. \tag{5.14}$$

When k is much greater than d, the reverse inequality holds. Also it can be shown theoretically (compare, for example, Table 5.1 with Appendix E) that

$$u^{\alpha}_{k,\nu}<t_{\nu}^{\alpha/(2k)}, \tag{5.15}$$

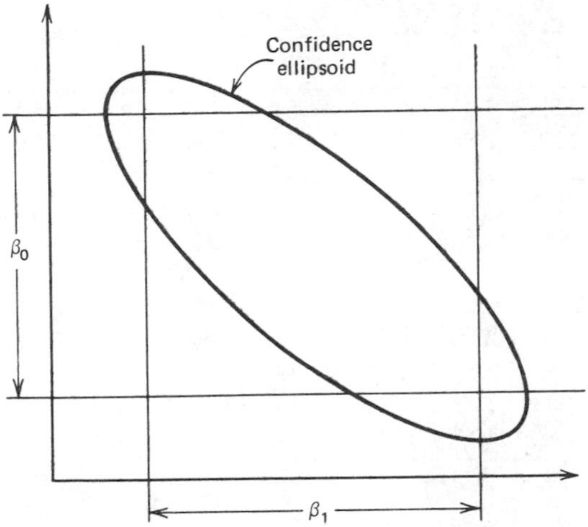

Fig. 5.1 Separate confidence intervals for β_0 and β_1 compared with a joint confidence region.

so that for the common situation of $d=k$, the maximum modulus intervals are the shortest and the F-intervals are the widest. For example, when $\alpha=0.05$, $d=k=5$, $p=6$, and $n=26$, we have

$$\nu=20, \qquad (kF_{k,\nu}^{\alpha})^{1/2}=3.68, \qquad t_{\nu}^{\alpha/(2k)}=2.85, \qquad \text{and } u_{k,\nu}^{\alpha}=2.82.$$

If we were interested in just a single t-interval we would use $t_{\nu}^{(1/2)\alpha}=2.09$, which is much smaller than the previous three numbers.

Table 5.1 Values of $t_{\nu}^{\alpha/(2k)}$ for $\alpha=0.05$

ν \ k	1	2	3	4	5	6	7	8	9	10	15	20	50
5	2.57	3.16	3.54	3.81	4.04	4.22	4.38	4.53	4.66	4.78	5.25	5.60	6.87
10	2.23	2.64	2.87	3.04	3.17	3.28	3.37	3.45	3.52	3.58	3.83	4.01	4.59
15	2.13	2.49	2.70	2.84	2.95	3.04	3.11	3.18	3.24	3.29	3.48	3.62	4.08
20	2.09	2.42	2.61	2.75	2.85	2.93	3.00	3.06	3.11	3.16	3.33	3.46	3.85
24	2.07	2.39	2.58	2.70	2.80	2.88	2.94	3.00	3.05	3.09	3.26	3.38	3.75
30	2.04	2.36	2.54	2.66	2.75	2.83	2.89	2.94	2.99	3.03	3.19	3.30	3.65
40	2.02	2.33	2.50	2.62	2.70	2.78	2.84	2.89	2.93	2.97	3.12	3.23	3.55
60	2.00	2.30	2.47	2.58	2.66	2.73	2.79	2.84	2.88	2.92	3.06	3.16	3.46
120	1.98	2.27	2.43	2.54	2.62	2.68	2.74	2.79	2.83	2.86	3.00	3.09	3.38
∞	1.96	2.24	2.40	2.50	2.58	2.64	2.69	2.74	2.78	2.81	2.94	3.03	3.29

SOURCE. Dunn [1959].

Table 5.2 Values of $(dF_{d,\nu}^{\alpha})^{1/2}$ for $\alpha = 0.05$

ν \ k	1	2	3	4	5	6	7	8
5	2.57	3.40	4.03	4.56	5.02	5.45	5.84	6.21
10	2.23	2.86	3.34	3.73	4.08	4.40	4.69	4.96
15	2.13	2.71	3.14	3.50	3.81	4.09	4.36	4.60
20	2.09	2.64	3.05	3.39	3.68	3.95	4.19	4.43
24	2.06	2.61	3.00	3.34	3.62	3.88	4.12	4.34
30	2.04	2.58	2.96	3.28	3.56	3.81	4.04	4.26
40	2.02	2.54	2.92	3.23	3.50	3.75	3.97	4.18
60	2.00	2.51	2.88	3.18	3.44	3.67	3.90	4.10
120	1.98	2.48	2.84	3.13	3.38	3.62	3.83	4.02
∞	1.96	2.45	2.79	3.08	3.32	3.55	3.75	3.94

SOURCE. Dunn [1959].

A method for comparing the simple-minded t-interval approach with Scheffé's method is given by Anderson [1972]. In some circumstances Bonferroni t-intervals can be used for hypothesis testing (Christensen [1973]).

5.1.3 Hypothesis Testing and Confidence Intervals

Usually interval estimation is preceded by an F-test of some hypothesis $H:\mathbf{A}\boldsymbol{\beta}=\mathbf{c}$; the confidence intervals being calculated and used for inference only if the F-ratio is significant (cf. Section 4.1.5). Two examples of this are given below.

EXAMPLE 5.1 Suppose we test $H:\beta_1=\beta_2=\cdots=\beta_d=0$. If H is rejected, we could then examine each β_j $(j=1,2,\ldots,d)$ separately, using the confidence intervals $\hat{\beta}_j \pm c\hat{\sigma}_{\hat{\beta}_j}$ provided by any one of the three methods given above, though the maximum modulus intervals would normally be preferred if they are the shortest. We hope that those intervals that do not contain zero will indicate which of the β_j are significantly different from zero, and by how much. We can also obtain intervals for all linear combinations $\Sigma_{i=1}^{d} a_i\beta_i$ using Scheffé's method.

EXAMPLE 5.2 Suppose we are given $H:\beta_1=\beta_2=\cdots=\beta_{d+1}$. If H is rejected we will be interested in all k $(=d(d+1)/2)$ pairs $\beta_i-\beta_j$. For example, if $d=5$, $n-p=\nu=20$, and $\alpha=0.05$, then $k=10$, $(dF_{d,\nu}^{\alpha})^{1/2}=3.68$, $t_{\nu}^{\alpha/(2k)}=3.16$, and $u_{k,\nu}^{\alpha}=3.114$, so that the maximum modulus intervals are still the shortest. Now H can also be written in the form $\phi_i=\beta_i-\beta_{d+1}=0$ $(i=1,2,\ldots,d)$ so that Scheffé's method will provide confidence intervals for all linear combinations

$$\sum_{i=1}^{d} h_i\phi_i = \sum_{i=1}^{d} h_i\beta_i - \left(\sum_{i=1}^{d} h_i\right)\beta_{d+1} = \sum_{i=1}^{d+1} c_i\beta_i, \tag{5.16}$$

where $\sum_{i=1}^{d+1}c_i=0$; thus every linear combination of the ϕ_i is a contrast in the β_i. By reversing the above argument we see that every contrast in the β_i is a linear combination of the ϕ_i. Hence Scheffé's method provides a set of multiple confidence intervals for all contrasts in the β_i $(i=1,2,\ldots,d+1)$.

It should be noted that when a preliminary F-test is carried out, then the appropriate probability to be considered is now the *conditional* probability $\text{pr}[\cap_i E_i | F \text{ significant}]$, which may be greater or less than the unconditional probability $\text{pr}[\cap_i E_i]$ (Olshen [1973]).

5.1.4 Confidence Regions

Suppose that $d=k$. Then from (5.9) we have

$$1-\alpha=\text{pr}\left[(\boldsymbol{\phi}-\hat{\boldsymbol{\phi}})'\mathbf{L}^{-1}(\boldsymbol{\phi}-\hat{\boldsymbol{\phi}})\leqslant m\right]$$

where the region $(\boldsymbol{\phi}-\hat{\boldsymbol{\phi}})'\mathbf{L}^{-1}(\boldsymbol{\phi}-\hat{\boldsymbol{\phi}})\leqslant m$ is a solid ellipsoid (since $\mathbf{L}$, and therefore $\mathbf{L}^{-1}$, is positive definite) with center $\hat{\boldsymbol{\phi}}$. This ellipsoid gives us a $100(1-\alpha)\%$ confidence *region* for $\boldsymbol{\phi}$. However, unless k is small, say, 2 or 3, such a region will not be readily computed, nor easily interpreted. In this respect suitable contour lines or surfaces may be sufficient to give a reasonable description of the region. For example, if $k=3$ the region may be pictured in two dimensions by means of a contour map as in Fig. 5.2;

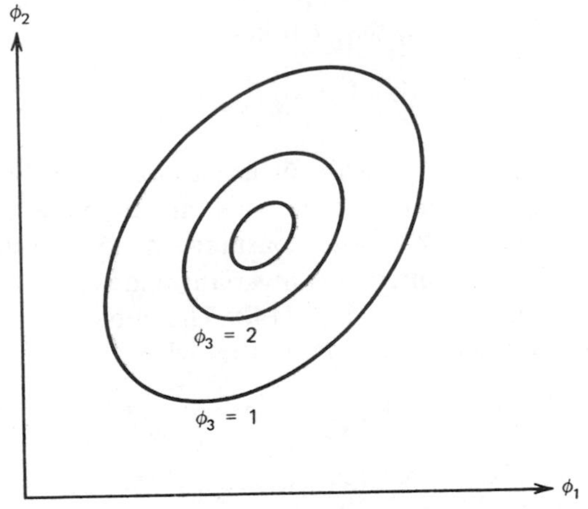

Fig. 5.2 Contour map of the confidence region for $\boldsymbol{\phi}'=(\phi_1,\phi_2,\phi_3)$.

here we have a plot of ϕ_1 versus ϕ_2 for three values of ϕ_3.

For $k>3$ it is still possible to convey the general shape of the confidence region by using a set of contour maps. However, generally speaking, the contour region approach is of limited value.

5.2 CONFIDENCE BANDS FOR THE REGRESSION SURFACE

Once we have estimated $\boldsymbol{\beta}$ from n observations $\mathbf{Y}$ we can use the predictor

$$\hat{Y}=\hat{\beta}_0+\hat{\beta}_1x_1+\cdots+\hat{\beta}_{p-1}x_{p-1} \qquad (=\mathbf{x}'\hat{\boldsymbol{\beta}}, \text{ say})$$

for studying the shape of the regression surface

$$f(x_1,x_2,\ldots,x_{p-1})=\beta_0+\beta_1x_1+\cdots+\beta_{p-1}x_{p-1}=\mathbf{x}'\boldsymbol{\beta}$$

over a range of values of the regressors x_j. In particular we can construct a two-sided $100(1-\alpha)\%$ confidence interval for the value of f at a particular value of $\mathbf{x}'$, say, $\mathbf{x}_*'=(1,x_{*1},x_{*2},\ldots,x_{*,p-1})$, using $\hat{Y}_*=\mathbf{x}_*'\hat{\boldsymbol{\beta}}$. Thus from (4.24) we have the interval

$$\hat{Y}_*\pm t_{n-p}^{(1/2)\alpha}S\sqrt{v_*} \tag{5.17}$$

where $v_*=\mathbf{x}_*'(\mathbf{X}'\mathbf{X})^{-1}\mathbf{x}_*$.

If we are interested in k particular values of $\mathbf{x}$, say, $\mathbf{x}=\mathbf{a}_i$ $(i=1,2,\ldots,k)$, then we can use any of the three methods discussed in Section 5.1 to obtain k two-sided confidence intervals with a joint confidence probability of at least $1-\alpha$. (The application of the Bonferroni and the Scheffé intervals to this problem seems to be due to Liebermann [1961]).

However, if we are interested in all values of $\mathbf{x}$ then, using Scheffé's method, we have from (5.13) that $\mathbf{x}'\boldsymbol{\beta}$ lies in

$$\mathbf{x}'\hat{\boldsymbol{\beta}}\pm(pF_{p,n-p}^{\alpha})^{1/2}S\left\{\mathbf{x}'(\mathbf{X}'\mathbf{X})^{-1}\mathbf{x}\right\}^{1/2}, \tag{5.18}$$

for all $\mathbf{x}'=(1,x_1,x_2,\ldots,x_{p-1})$, with an exact overall probability of $1-\alpha$. (Although the first element of $\mathbf{x}$ is constrained to be unity, this does not mean that the appropriate constant in (5.18) should now be $[(p-1)F_{p-1,n-p}^{\alpha})]^{1/2}$; the interval is invariant under a scale change of any element of $\mathbf{x}$—Miller [1966: pp. 110–114]). The above expression gives two surfaces defined by the functions f^* and f_*, where

$$\text{pr}\big[f^*(x_1,x_2,\ldots,x_{p-1})\geqslant f(x_1,x_2,\ldots,x_{p-1}) \geqslant f_*(x_1,x_2,\ldots,x_{p-1}),\ \text{all } x_1,x_2,\ldots,x_{p-1}\big]=1-\alpha.$$

The region between f^* and f_* is commonly called a *confidence band.* As pointed out by Miller [1966], the band over that part of the regression surface that is not of interest, or is physically meaningless, is ignored. This means that the probability associated with the regression band over a limited region exceeds $1-\alpha$, and the intervals given by (5.18) will be somewhat conservative. The question of constructing a confidence band over a limited region, with an *exact* probability $1-\alpha$, is discussed by Wynn and Bloomfield [1971]. A solution is given by Halperin and Gurian [1968] for the case of an ellipsoidal region centered on the vector of means $(\bar{x}_{.1},\bar{x}_{.2},\ldots,\bar{x}_{.p-1})$. Various solutions for the straight line case are considered in detail in Section 7.2.3.

Scheffé's method described by (5.18) is a special case of a more general procedure developed by Bowden [1970]. Let

$$\begin{aligned}\|\mathbf{a}\|_m &= \left(\sum_{i=1}^{p}|a_i|^m\right)^{1/m} \qquad 1\leqslant m<\infty\\ &= \max_i |a_i| \qquad m=\infty;\end{aligned}$$

then, as Bowden proves,

$$\text{pr}\left[|\mathbf{x}'\hat{\boldsymbol{\beta}}-\mathbf{x}'\boldsymbol{\beta}|\leqslant S\|\mathbf{x}\|_m z_m^\alpha,\ \text{all } \mathbf{x}\right]=1-\alpha, \tag{5.19}$$

where z_m^α is the upper α significant point of the distribution of $\|(\hat{\boldsymbol{\beta}}-\boldsymbol{\beta})/S\|_m$. By taking $m=1,2$, or ∞ and varying the value of $\mathbf{x}$, different types of regression bands can be obtained; when $p=2$ (the straight-line case), the band has uniform or trapezoidal width ($m=1$), or is hyperbolic ($m=2$), or is bounded by straight-line segments ($m=\infty$). However, it transpires that for $p>2$, Scheffé's method ($m=2$) and its corresponding one-sided analogue have certain optimal properties (Bohrer [1973]).

When k is large it is natural to ask whether the maximum modulus t-intervals of (5.6) are still shorter than the intervals given by the confidence band of (5.18), particularly when k is much greater than p. Hahn [1972] has calculated

$$r=\frac{u_{k,n-p}^\alpha}{\left(pF_{p,n-p}^\alpha\right)^{1/2}}, \tag{5.20}$$

the ratio of the interval widths, for $\alpha=0.1$, 0.05, 0.01, and for different values of k, p, and $n-p$. Table 5.3, reproduced from Hahn [1972: Table 3], gives the maximum value of k (for $\alpha=0.05$, $p=2,3,5$ and $n-p=5,10,20,40,60$) for which $r<1$. Hahn also found that, for the above values of α, r increased slightly as α decreased.

Table 5.3 Maximum Value of k for which $r<1$ [r is Defined by Equation (5.20)].

$\alpha=0.05$

$n-p$ \ p	2	3	5
5	3	6	20+
10	3	8	20+
20	3	8	20+
40	3	9	20+
60	3	9	20+

SOURCE. Hahn [1972].

5.3 PREDICTION INTERVALS FOR THE RESPONSE

In the preceding section we discussed the problem of predicting the value of a regression surface $\mathbf{x}'\boldsymbol{\beta}$ at a given value of $\mathbf{x}=\mathbf{x}_*$, say. However, in practice we are generally more interested in predicting the value, Y_*, say, of the random variable Y, where

$$Y_*=\mathbf{x}_*'\boldsymbol{\beta}+\varepsilon_*.$$

If we assume that $\varepsilon_*\sim N(0,\sigma^2)$ and that ε_* is independent of $\boldsymbol{\varepsilon}'=(\varepsilon_1,\varepsilon_2,\ldots,\varepsilon_n)$, then

$$E\left[\hat{Y}_*-Y_*\right]=\mathbf{x}_*'\boldsymbol{\beta}-\mathbf{x}_*'\boldsymbol{\beta}=0,$$

$$\begin{aligned}\operatorname{var}\left[\hat{Y}_*-Y_*\right]&=\operatorname{var}\left[\hat{Y}_*\right]+\operatorname{var}\left[Y_*\right]\\&=\sigma^2\mathbf{x}_*'(\mathbf{X}'\mathbf{X})^{-1}\mathbf{x}_*+\sigma^2\\&=\sigma^2\left(v_*+1\right),\end{aligned}\tag{5.21}$$

and $(\hat{Y}_*-Y_*)\sim N(0,\sigma^2(v_*+1))$. We can therefore construct a t-statistic and obtain the following $100(1-\alpha)\%$ confidence interval for Y_*

$$\hat{Y}_*\pm t_{n-p}^{(1/2)\alpha}S\left(v_*+1\right)^{1/2},\tag{5.22}$$

which may be compared with the interval (5.17).

If we are interested in predicting Y for k different values of $\mathbf{x}$, say, $\mathbf{x}=\mathbf{a}_i$ $(i=1,2,\ldots,k)$, then we can use any of the three methods discussed in Section 5.1.1 to obtain k confidence intervals with an overall confidence probability of at least $1-\alpha$. For $k=2$ Hahn [1972] shows that the intervals

$$\hat{Y}_*^{(i)} \pm u_{2,n-p,\rho}^{\alpha} S\left(v_*^{(i)}+1\right)^{1/2} \qquad (i=1,2),$$

where $\hat{Y}_*^{(i)}=\mathbf{a}_i'\hat{\boldsymbol{\beta}}$, $v_*^{(i)}=\mathbf{a}_i'(\mathbf{X}'\mathbf{X})^{-1}\mathbf{a}_i$, and

$$\rho=\frac{\mathbf{a}_1'(\mathbf{X}'\mathbf{X})^{-1}\mathbf{a}_2}{\left\{\left(v_*^{(1)}+1\right)\left(v_*^{(2)}+1\right)\right\}^{1/2}}$$

have an *exact* overall probability of $1-\alpha$.

Fitted regressions are used for two types of prediction (Box [1966]): (1) to predict Y in the future from passive observation of the x_j's. We assume that the system is not interfered with so that the proposed regression model is still appropriate in the future. (2) To discover how deliberate changes in the x_j's will affect Y, with the intention of actually modifying the system to get a better value of Y. The need to distinguish between these two situations is borne out by the following example adapted from Box [1966].

In a chemical process it is found that undesirable frothing can be reduced by increasing the pressure (x_1); it is also known that the yield (Y) is unaffected directly by a change in pressure. The standard operating procedure then consists of increasing the pressure whenever frothing occurs. Suppose, however, that the frothing is actually caused by the presence of an unsuspected impurity (x_2), and that, unknown to the experimenter, an increase in concentration of impurity causes an increase in frothing and a decrease in Y. If x_1 and x_2 are positively correlated so that an increase in pressure causes an increase in impurity then, although Y is unaffected directly by changes in x_1, there is a spurious negative correlation between Y and x_1 as Y and x_1 are both affected by x_2, but in opposite directions. This means that there will be a significant regression of Y on x_1, and the fitted regression can be used for adequately predicting Y, provided the system continues to run in the same fashion as when the data were recorded. However, this regression does not indicate the true causal situation. We are mistaken if we think that we can increase Y by decreasing x_1.

5.4 ENLARGING THE REGRESSION MATRIX

Suppose that our original regression model is enlarged by the addition of an extra regressor x_p, say, so that our model is now

$$G: Y_i=\beta_0+\beta_1 x_{i1}+\cdots+\beta_p x_{ip}+\varepsilon_i \qquad (i=1,2,\ldots,n).$$

What effect will this have on the width of the confidence intervals given in Sections 5.2 and 5.3? Surprisingly the answer is that the intervals will be at least as wide and, in fact, almost invariably wider! To see this, we use the general theory of Section 3.7 to show that $\sigma^2 v$, the variance of the predictor $\hat{Y}$, cannot decrease when another regressor is added to the model. Setting

$$\beta_p = \gamma, \qquad [(x_{ip})] = \mathbf{x}_p = \mathbf{z}, \qquad \text{and } \mathbf{W} = [\mathbf{X}, \mathbf{z}],$$

we can write the model G in the form

$$\begin{aligned}\mathbf{Y} &= \mathbf{X}\boldsymbol{\beta} + \mathbf{z}\gamma + \boldsymbol{\varepsilon} \\ &= \mathbf{W}\boldsymbol{\delta} + \boldsymbol{\varepsilon},\end{aligned}$$

and the least squares estimate of $\boldsymbol{\delta}$ is

$$\hat{\boldsymbol{\delta}}_G = (\mathbf{W}'\mathbf{W})^{-1}\mathbf{W}'\mathbf{Y}.$$

For G, the new predictor at $(\mathbf{x}'_*, x_{*p})$ is

$$\hat{Y}_{*G} = (\mathbf{x}'_*, x_{*p})\hat{\boldsymbol{\delta}}_G,$$

and, from Theorem 3.7 (v) in Section 3.7.1,

$$\begin{aligned}\text{var}[\hat{Y}_{*G}] &= \mathcal{D}[\hat{Y}_{*G}] \\ &= (\mathbf{x}'_*, x_{*p})\mathcal{D}[\hat{\boldsymbol{\delta}}_G](\mathbf{x}'_*, x_{*p})' \\ &= \sigma^2(\mathbf{x}'_*, x_{*p})(\mathbf{W}'\mathbf{W})^{-1}(\mathbf{x}'_*, x_{*p})' \\ &= \sigma^2(\mathbf{x}'_*, x_{*p})\begin{pmatrix}(\mathbf{X}'\mathbf{X})^{-1} + m\mathbf{k}\mathbf{k}', & -m\mathbf{k} \\ -m\mathbf{k}', & m\end{pmatrix}\begin{bmatrix}\mathbf{x}_* \\ x_{*p}\end{bmatrix},\end{aligned}$$

where $m = (\mathbf{z}'\mathbf{R}\mathbf{z})^{-1}$ and $\mathbf{k} = (\mathbf{X}'\mathbf{X})^{-1}\mathbf{X}'\mathbf{z}$. Multiplying the above matrix expression out, and completing the square on $\mathbf{k}'\mathbf{x}_*$, we have

$$\begin{aligned}\text{var}[\hat{Y}_{*G}] &= \sigma^2\mathbf{x}'_*(\mathbf{X}'\mathbf{X})^{-1}\mathbf{x}_* + m\sigma^2(\mathbf{k}'\mathbf{x}_* - x_{*p})^2 \\ &\geqslant \sigma^2\mathbf{x}'_*(\mathbf{X}'\mathbf{X})^{-1}\mathbf{x}_* \; (= \sigma^2 v_*) \\ &= \text{var}[\hat{Y}_*], \qquad (5.23)\end{aligned}$$

with equality if and only if $x_{*p} = \mathbf{k}'\mathbf{x}_* = \mathbf{z}'\mathbf{X}(\mathbf{X}'\mathbf{X})^{-1}\mathbf{X}'\mathbf{x}_*$. Since variances and covariances are independent of a change of origin we see that the above result still holds even if $\mathcal{E}[\mathbf{Y}]$ is not equal to either $\mathbf{X}\boldsymbol{\beta}$ or $\mathbf{W}\boldsymbol{\delta}$; in this case both predictors $\hat{Y}_{*G}$ and $\hat{Y}_*$ are biased estimates of $\mathcal{E}[\mathbf{Y}]$. We conclude, therefore, that although we may sometimes reduce the bias and improve the fit by enlarging the regression model, the variance of the predictor is not reduced. Walls and Weeks [1969] give an example in which the variance of prediction at a particular point is increased tenfold when the model is enlarged from a straight line to a quadratic.

If we use the mean square error (MSE) of prediction as our criterion, then the MSE may increase or decrease when extra regressors are added to the model. Mallows' C_p statistic (see Section 12.2.3c), which is used for comparing different regression models, is based on an "average" MSE criterion.

By setting $x_{*p} = 0$ and setting $\mathbf{x}_*$ equal to the column vector with unity in the $(j+1)$th position and zeros elsewhere in the above theory, we have $\text{var}[\hat{\beta}_{jG}] \geqslant \text{var}[\hat{\beta}_j]$ with equality if and only if $\mathbf{z}'\mathbf{X}(\mathbf{X}'\mathbf{X})^{-1}\mathbf{X}'\mathbf{x}_* = 0$. Equality holds if $\mathbf{z}$ is orthogonal to the columns of $\mathbf{X}$. However, in general the variance of the least squares estimate of β_j increases when the model is enlarged.

The lesson to be learned from the above discussion is that we should avoid "overfitting" regression models.

MISCELLANEOUS EXERCISE 5

1. Given the predictor $\hat{Y} = \hat{\beta}_0 + \hat{\beta}_1 x_1 + \cdots + \hat{\beta}_{p-1} x_{p-1}$, show that $\hat{Y}$ has a minimum variance of σ^2/n at the x point $x_j = \bar{x}_{.j}$ $(j = 1, 2, \ldots, p-1)$. [Hint: consider the model $Y_i = \alpha_0 + \beta_1(x_{i1} - \bar{x}_{.1}) + \cdots + \beta_{p-1}(x_{i,p-1} - \bar{x}_{.p-1}) + \varepsilon_i$.] Kupper [1972: p. 52]
2. Generalize the argument given in Section 5.4; namely, show that the addition of several regressors to a regression model cannot decrease the variance of the prediction $\hat{Y}$. [Such a proof is, of course, not necessary as we can add in the regressors just one at a time and evoke (5.23).]
3. Let $Y_i = \beta_0 + \beta_1 x_i + \varepsilon_i$ $(i = 1, 2, \ldots, n)$, where the ε_i are independently distributed as $N(0, \sigma^2)$. Obtain a set of multiple confidence intervals for all linear combinations $a_0\beta_0 + a_1\beta_1$ (a_0, a_1 not both zero) such that the overall confidence for the set is $100(1-\alpha)\%$.

CHAPTER 6

Departures from Underlying Assumptions

The basic multiple regression model that we have been studying thus far is $\mathbf{Y}=\mathbf{X}\boldsymbol{\beta}+\boldsymbol{\varepsilon}$, where $\mathbf{X}$ is $n\times p$ of rank p. We assume that the elements of $\boldsymbol{\varepsilon}$

(1) are unbiased;
(2) have constant variance;
(3) are uncorrelated, and
(4) are normally distributed.

Assumption (1) implies $\mathcal{E}[\boldsymbol{\varepsilon}]=\mathbf{0}$, (2) and (3) imply $\mathcal{D}[\boldsymbol{\varepsilon}]=\sigma^2\mathbf{I}_n$, and (3) and (4) imply the independence of the ε_i. It is also assumed, implicitly, that the regressor variables x_j are not random variables but are predetermined constants. If the regressors are random and are measured without error, then the regression can be regarded as being conditional on the observed values of the regressors (this problem is discussed in Sections 6.1.3 and 6.5). In this chapter we examine each of the above five assumptions in detail.

It should be noted that there is a tendency for errors that occur in many real situations to be normally distributed owing to the central limit theorem. If ε is a sum of n errors from different sources then, as n increases, ε tends to normality irrespective of the probability distributions of the individual errors. This argument applies to small errors δ_i, say, in a nonlinear system since

$$\varepsilon=f(\delta_1+a_1,\ldots,\delta_n+a_n)-f(a_1,\ldots,a_n)\approx\delta_1\frac{\partial f}{\partial a_1}+\cdots+\delta_n\frac{\partial f}{\partial a_n},$$

and ε is once again a (weighted) sum of errors.

6.1 BIAS

6.1.1 Bias due to Underfitting

If $\mathscr{E}[\boldsymbol{\varepsilon}]=\mathbf{0}$, then $\mathscr{E}[\mathbf{Y}]=\mathbf{X}\boldsymbol{\beta}$ and the least squares estimate $\hat{\boldsymbol{\beta}}=(\mathbf{X}'\mathbf{X})^{-1}\mathbf{X}'\mathbf{Y}$ is an unbiased estimate of $\boldsymbol{\beta}$. However, if the model is underfitted so that the true model is actually

$$\mathscr{E}[\mathbf{Y}]=\mathbf{X}\boldsymbol{\beta}+\mathbf{Z}\boldsymbol{\gamma}, \tag{6.1}$$

where the columns of $\mathbf{Z}$ are linearly independent of the columns of $\mathbf{X}$, then $\boldsymbol{\varepsilon}$ is biased and

$$\begin{aligned}\mathscr{E}[\hat{\boldsymbol{\beta}}]&=(\mathbf{X}'\mathbf{X})^{-1}\mathbf{X}'(\mathbf{X}\boldsymbol{\beta}+\mathbf{Z}\boldsymbol{\gamma})\\ &=\boldsymbol{\beta}+(\mathbf{X}'\mathbf{X})^{-1}\mathbf{X}'\mathbf{Z}\boldsymbol{\gamma}\\ &=\boldsymbol{\beta}+\mathbf{C}\boldsymbol{\gamma}, \end{aligned}\tag{6.2}$$

say. Thus $\hat{\boldsymbol{\beta}}$ is now a biased estimate of $\boldsymbol{\beta}$ with bias $\mathbf{C}\boldsymbol{\gamma}$. This bias term depends on both the postulated and the true models, and $\mathbf{C}$ can be interpreted as the matrix of regression coefficients of the omitted variables regressed on the x variables actually included in the model. A good choice of design may keep the bias to a minimum even if the wrong model has been postulated and fitted. For example, if the columns of $\mathbf{Z}$ are orthogonal to the columns of $\mathbf{X}$ then $\mathbf{X}'\mathbf{Z}=\mathbf{O}$, $\mathbf{C}=\mathbf{O}$, and $\hat{\boldsymbol{\beta}}$ is unbiased. Under some circumstances the orthogonality of the columns of $\mathbf{X}$ and $\mathbf{Z}$ may be described as zero correlation between a pair of x and z regressors (Malinvaud [1970]). In this case inadvertently omitting an uncorrelated regressor may not be serious.

If Equation (6.1) represents the true model then, provided $\mathscr{D}[\boldsymbol{\varepsilon}]=\sigma^2\mathbf{I}_n$, we still have $\mathscr{D}[\hat{\boldsymbol{\beta}}]=\sigma^2(\mathbf{X}'\mathbf{X})^{-1}$. However, given $\mathbf{P}=\mathbf{X}(\mathbf{X}'\mathbf{X})^{-1}\mathbf{X}'$ and $S^2=\mathbf{Y}'(\mathbf{I}_n-\mathbf{P})\mathbf{Y}/(n-p)$, then

$$E[S^2]=\sigma^2+\frac{\boldsymbol{\gamma}'\mathbf{Z}'(\mathbf{I}_n-\mathbf{P})\mathbf{Z}\boldsymbol{\gamma}}{n-p}>\sigma^2$$

since $(\mathbf{I}_n-\mathbf{P})$ is idempotent and therefore positive semidefinite, and $(\mathbf{I}_n-\mathbf{P})\mathbf{Z}\boldsymbol{\gamma}\neq\mathbf{0}$ (as $\mathbf{Z}\boldsymbol{\gamma}\notin\mathscr{R}[\mathbf{X}]$). Hence S^2 is a biased estimate of σ^2.

To examine the effect of underfitting on prediction we note that

$$\begin{aligned}\hat{\mathbf{Y}}&=\mathbf{X}\hat{\boldsymbol{\beta}}\\ &=\mathbf{X}(\mathbf{X}'\mathbf{X})^{-1}\mathbf{X}'(\mathbf{X}\boldsymbol{\beta}+\mathbf{Z}\boldsymbol{\gamma}+\boldsymbol{\varepsilon})\\ &=\mathbf{X}\boldsymbol{\beta}+\hat{\mathbf{Z}}\boldsymbol{\gamma}+\boldsymbol{\eta}, \end{aligned}\tag{6.3}$$

where $\mathcal{E}[\boldsymbol{\eta}]=\mathcal{E}[(\mathbf{X}'\mathbf{X})^{-1}\mathbf{X}'\boldsymbol{\varepsilon}]=\mathbf{0}$ and $\hat{\mathbf{Z}}=\mathbf{XC}$. Therefore, the effect of ignoring $\mathbf{Z}\boldsymbol{\gamma}$ in the regression amounts to using the "estimate" $\hat{\mathbf{Z}}$ instead of $\mathbf{Z}$.

As far as the residuals are concerned we see that

$$\begin{aligned}\mathcal{E}[\mathbf{e}]&=\mathcal{E}[\mathbf{Y}]-\mathcal{E}[\mathbf{X}\hat{\boldsymbol{\beta}}]\\&=\mathbf{X}\boldsymbol{\beta}+\mathbf{Z}\boldsymbol{\gamma}-(\mathbf{X}\boldsymbol{\beta}+\mathbf{XC}\boldsymbol{\gamma})\\&=(\mathbf{I}_n-\mathbf{P})\mathbf{Z}\boldsymbol{\gamma}\qquad[\text{by (6.2)}]\end{aligned}\tag{6.4}$$

and

$$\mathcal{D}[\mathbf{e}]=\mathcal{D}[(\mathbf{I}_n-\mathbf{P})\mathbf{Y}]=\sigma^2(\mathbf{I}_n-\mathbf{P})^2=\sigma^2(\mathbf{I}_n-\mathbf{P}),$$

so that the effect of misspecification is to bias $\mathbf{e}$; however, $\mathcal{D}[\mathbf{e}]$ is unchanged. Ramsey [1969] makes use of this fact to provide several tests for this kind of misspecification.

Another approach to this question of omitted or "hidden" regressors is given in Section 6.1.3.

EXAMPLE 6.1 (Draper and Smith [1966: p. 82]) Suppose we postulate the model $E[Y]=\beta_0+\beta_1 x$ when the true model is $E[Y]=\beta_0+\beta_1 x+\beta_2 x^2$. If we use observations of Y at $x_1=-1$, $x_2=0$, and $x_3=1$ to estimate β_0 and β_1 in the postulated model, what biases will be introduced?

Solution. The true model is

$$\mathcal{E}\begin{pmatrix}Y_1\\Y_2\\Y_3\end{pmatrix}=\begin{pmatrix}1&x_1&x_1^2\\1&x_2&x_2^2\\1&x_3&x_3^2\end{pmatrix}\begin{pmatrix}\beta_0\\\beta_1\\\beta_2\end{pmatrix}=\begin{pmatrix}1&-1&1\\1&0&0\\1&1&1\end{pmatrix}\begin{pmatrix}\beta_0\\\beta_1\\\beta_2\end{pmatrix}$$

$$=\begin{pmatrix}1&-1\\1&0\\1&1\end{pmatrix}\begin{pmatrix}\beta_0\\\beta_1\end{pmatrix}+\begin{pmatrix}1\\0\\1\end{pmatrix}\beta_2=\mathbf{X}\boldsymbol{\beta}+\mathbf{Z}\beta_2.$$

Now

$$(\mathbf{X}'\mathbf{X})^{-1}=\begin{pmatrix}\frac{1}{3}&0\\0&\frac{1}{2}\end{pmatrix}$$

$$\mathbf{X}'\mathbf{Z}=\begin{pmatrix}1&1&1\\-1&0&1\end{pmatrix}\begin{pmatrix}1\\0\\1\end{pmatrix}=\begin{pmatrix}2\\0\end{pmatrix}$$

and, from Equation (6.2), the bias of $\hat{\boldsymbol{\beta}}$ is

$$\mathbf{C}\boldsymbol{\gamma}=\begin{pmatrix}\frac{1}{3} & 0\\ 0 & \frac{1}{2}\end{pmatrix}\begin{pmatrix}2\\0\end{pmatrix}\beta_2=\begin{pmatrix}\frac{2}{3}\\0\end{pmatrix}\beta_2.$$

Thus $\hat{\beta}_0$ has bias $\frac{2}{3}\beta_2$ and $\hat{\beta}_1$ is unbiased.

6.1.2 *Bias due to Overfitting*

Suppose that the true model is $\mathcal{E}[\mathbf{Y}]=\mathbf{X}_1\boldsymbol{\beta}_1$, where $\mathbf{X}_1$ consists of the first k columns of $\mathbf{X}$; thus $\mathbf{X}=(\mathbf{X}_1,\mathbf{X}_2)$, say. Then

$$\begin{aligned}\mathcal{E}[\hat{\boldsymbol{\beta}}]&=(\mathbf{X}'\mathbf{X})^{-1}\mathbf{X}'\mathbf{X}_1\boldsymbol{\beta}_1\\&=(\mathbf{X}'\mathbf{X})^{-1}\mathbf{X}'\mathbf{X}\begin{pmatrix}\boldsymbol{\beta}_1\\ \mathbf{0}\end{pmatrix}\\&=\begin{pmatrix}\boldsymbol{\beta}_1\\ \mathbf{0}\end{pmatrix},\end{aligned} \tag{6.5}$$

and $\hat{\boldsymbol{\beta}}_1$, consisting of the first k elements of $\hat{\boldsymbol{\beta}}$, is an unbiased estimate of $\boldsymbol{\beta}_1$. Also

$$\mathcal{E}[\hat{\mathbf{Y}}]=\mathcal{E}[\mathbf{X}\hat{\boldsymbol{\beta}}]=\mathbf{X}\begin{pmatrix}\boldsymbol{\beta}_1\\ \mathbf{0}\end{pmatrix}=\mathbf{X}_1\boldsymbol{\beta}_1 \tag{6.6}$$

so that the fitted model is an unbiased estimate of the true model. However, the familiar formula $\sigma^2(\mathbf{X}'\mathbf{X})^{-1}$ leads to inflated expressions for the variances of the elements of $\hat{\boldsymbol{\beta}}_1$. From Equation (3.28) in Section 3.7, with $\mathbf{X}$ and $\mathbf{Z}$ set equal to $\mathbf{X}_1$ and $\mathbf{X}_2$, respectively, we have

$$(\mathbf{X}'\mathbf{X})^{-1}=\begin{pmatrix}(\mathbf{X}_1'\mathbf{X}_1)^{-1}+\mathbf{LML}', & -\mathbf{LM}\\ -\mathbf{ML}', & \mathbf{M}\end{pmatrix}$$

where $\mathbf{LML}'$ is positive definite. Hence, by *A4.8*,

$$\begin{aligned}\text{“apparent” var}[\hat{\beta}_i]&=\text{“true” var}[\hat{\beta}_i]+(\mathbf{LML}')_{ii}\\&>\text{“true” var}[\hat{\beta}_i].\end{aligned}$$

Since $(\mathbf{I}_n-\mathbf{P})(\mathbf{X}_1,\mathbf{X}_2)=\mathbf{O}$ [Theorem 3.1 (iii)], we note that

$$E[\mathbf{Y}'(\mathbf{I}_n-\mathbf{P})\mathbf{Y}]=(n-p)\sigma^2+\boldsymbol{\beta}_1'\mathbf{X}_1'(\mathbf{I}_n-\mathbf{P})\mathbf{X}_1\boldsymbol{\beta}_1=(n-p)\sigma^2$$

and S^2 is still an unbiased estimate of σ^2.

6.1.3 Bias and Random Regressor Variables

Consider the model

$$\begin{aligned} Y &= \beta_0 + \beta_1 X_1 + \cdots + \beta_s X_s \\ &= \beta_0 + \beta_1 X_1 + \cdots + \beta_r X_r + \delta \qquad (r<s) \end{aligned}$$

where the X_j $(j=1,2,\ldots,s)$ are random variables with $E[X_j]=\theta_j$, and the "error" δ is assumed to be due to further "hidden" variables $X_{r+1},\ldots,X_s$. Suppose that for the ith (random) repetition of the underlying experiment X_j is denoted by X_{ij} with observed value x_{ij}. Then

$$\begin{aligned} Y_i &= \left(\beta_0 + \beta_1 \overline{X}_{.1} + \cdots + \beta_r \overline{X}_{.r} + \beta_{r+1}\theta_{r+1} + \cdots + \beta_s\theta_s\right) \\ &\quad + \beta_1\left(X_{i1} - \overline{X}_{.1}\right) + \cdots + \beta_r\left(X_{ir} - \overline{X}_{.r}\right) \\ &\quad + \left[\beta_{r+1}(X_{i,r+1} - \theta_{r+1}) + \cdots + \beta_s(X_{is} - \theta_s)\right] \\ &= \alpha_0 + \beta_1\left(X_{i1} - \overline{X}_{.1}\right) + \cdots + \beta_r\left(X_{ir} - \overline{X}_{.r}\right) + \varepsilon \end{aligned}$$

where α_0 is a random variable, and $E[\varepsilon]=0$ (since $E[X_{ij}]=\theta_j$). If we now consider the regression as being conditional on $X_{ij}=x_{ij}$ for just $j=1,2,\ldots,r$, then we have the model

$$E\left[Y_i|\{x_{ij}\}, j=1,2,\ldots,r\right] = \alpha_0 + \beta_1(x_{i1} - \bar{x}_{.1}) + \cdots + \beta_r(x_{ir} - \bar{x}_{.r}) \qquad (6.7)$$

where α_0 is now (conditionally) a constant. Since r is arbitrary we will always have $E[\varepsilon]=0$ *irrespective* of the number of x variables we include in the model. This means that when we have no control over the X_j variables there is no question of finding the "true" model. What we are looking for is an adequate model, that is, one which reduces ε to a reasonable level. For this type of model it can be argued, therefore, that the question of bias due to overfitting or underfitting does not arise.

6.2 INCORRECT DISPERSION MATRIX

6.2.1 General Case

If we assume that $\mathcal{D}[\boldsymbol{\varepsilon}]=\sigma^2\mathbf{I}_n$ when in fact $\mathcal{D}[\boldsymbol{\varepsilon}]=\sigma^2\mathbf{V}$, then $\hat{\boldsymbol{\beta}}$ is still an unbiased estimate of $\boldsymbol{\beta}$. However,

$$\mathcal{D}[\hat{\boldsymbol{\beta}}] = \mathcal{D}\left[(\mathbf{X}'\mathbf{X})^{-1}\mathbf{X}'\mathbf{Y}\right] = \sigma^2(\mathbf{X}'\mathbf{X})^{-1}\mathbf{X}'\mathbf{V}\mathbf{X}(\mathbf{X}'\mathbf{X})^{-1}$$

is, in general, not equal to $\sigma^2(\mathbf{X}'\mathbf{X})^{-1}$ and, since (cf. Theorem 3.3 in Section 3.3)

$$E[S^2]=\frac{\sigma^2}{n-p}E[\mathbf{Y}'(\mathbf{I}_n-\mathbf{P})\mathbf{Y}]$$

$$=\frac{\sigma^2}{n-p}\operatorname{tr}[\mathbf{V}(\mathbf{I}_n-\mathbf{P})], \tag{6.8}$$

S^2 is generally a biased estimate of σ^2. It follows that

$$\hat{v}=S^2\mathbf{a}'(\mathbf{X}'\mathbf{X})^{-1}\mathbf{a} \tag{6.9}$$

will normally be a biased estimate of

$$\operatorname{var}[\mathbf{a}'\hat{\boldsymbol{\beta}}]=\sigma^2\mathbf{a}'(\mathbf{X}'\mathbf{X})^{-1}\mathbf{X}'\mathbf{V}\mathbf{X}(\mathbf{X}'\mathbf{X})^{-1}\mathbf{a}. \tag{6.10}$$

In fact, Swindel [1968] has shown that if

$$E[\hat{v}]=\operatorname{var}[\mathbf{a}'\hat{\boldsymbol{\beta}}]+b,$$

then

$$\frac{\{\text{mean of } (n-p) \text{ least eigenvalues of } \mathbf{V}\}-\{\text{greatest eigenvalue of } \mathbf{V}\}}{\{\text{greatest eigenvalue of } \mathbf{V}\}}$$

$$\leqslant\frac{b}{\operatorname{var}[\mathbf{a}'\hat{\boldsymbol{\beta}}]}$$

$$\leqslant\frac{\{\text{mean of } (n-p) \text{ greatest eigenvalues of } \mathbf{V}\}-\{\text{least eigenvalue of } \mathbf{V}\}}{\{\text{least eigenvalue of } \mathbf{V}\}}$$

and the bounds are attainable. The case $\mathbf{X}'\mathbf{X}=\mathbf{I}_p$ was considered by Watson [1955].

Another question of interest is, under what conditions is $\hat{v}$ unbiased for all $\mathbf{a}$? This question is partly answered by the following example.

EXAMPLE 6.2 Show that a sufficient condition for $\hat{v}$ to be unbiased for all $\mathbf{a}$ is that $\boldsymbol{\beta}^*=\hat{\boldsymbol{\beta}}$ for all $\mathbf{Y}$, where $\boldsymbol{\beta}^*=(\mathbf{X}'\mathbf{V}^{-1}\mathbf{X})^{-1}\mathbf{X}'\mathbf{V}^{-1}\mathbf{Y}$ is the generalized least squares estimate of $\boldsymbol{\beta}$.

Solution. Given $\hat{\boldsymbol{\beta}}=\boldsymbol{\beta}^*$ for all $\mathbf{Y}$, then

$$(\mathbf{X}'\mathbf{X})^{-1}\mathbf{X}'=(\mathbf{X}'\mathbf{V}^{-1}\mathbf{X})^{-1}\mathbf{X}'\mathbf{V}^{-1}$$

or

$$\mathbf{X}'\mathbf{V}\mathbf{X}=(\mathbf{X}'\mathbf{X})(\mathbf{X}'\mathbf{V}^{-1}\mathbf{X})^{-1}\mathbf{X}'\mathbf{X}.$$

Also, by Equation (3.22) of Section 3.6,

$$(\mathbf{X}'\mathbf{X})^{-1}=\mathscr{D}[\hat{\boldsymbol{\beta}}]=\mathscr{D}[\boldsymbol{\beta}^*]=\sigma^2(\mathbf{X}'\mathbf{V}^{-1}\mathbf{X})^{-1}$$

and combining the above two equations we have

$$\mathbf{X}'\mathbf{V}\mathbf{X}=(\mathbf{X}'\mathbf{X})(\mathbf{X}'\mathbf{X})^{-1}\mathbf{X}'\mathbf{X}=\mathbf{X}'\mathbf{X}. \tag{6.11}$$

Hence

$$(\mathbf{X}'\mathbf{X})^{-1}\mathbf{X}'\mathbf{V}\mathbf{X}(\mathbf{X}'\mathbf{X})^{-1}=(\mathbf{X}'\mathbf{X})^{-1} \tag{6.12}$$

and comparing Equations (6.9) and (6.10) we see that $\hat{v}$ is an unbiased estimate of var$[\mathbf{a}'\hat{\boldsymbol{\beta}}]$ for all $\mathbf{a}$ if $E[S^2]=\sigma^2$.

Now, as far as $\sigma^2\mathbf{V}$ is concerned, σ^2 is only an arbitrary scale factor so that we can assume without loss of generality that $\operatorname{tr}\mathbf{V}=n$. Hence

$$\begin{aligned}\operatorname{tr}\mathbf{VP}&=\operatorname{tr}\left[\mathbf{VX}(\mathbf{X}'\mathbf{X})^{-1}\mathbf{X}'\right]\\&=\operatorname{tr}\left[\mathbf{X}'\mathbf{VX}(\mathbf{X}'\mathbf{X})^{-1}\right] \qquad (\text{by } A1.2)\\&=\operatorname{tr}\left[\mathbf{X}'\mathbf{X}(\mathbf{X}'\mathbf{X})^{-1}\right] \qquad [\text{by Equation (6.11)}]\\&=p\end{aligned}$$

and, by Equation (6.8),

$$\begin{aligned}E[S^2]&=\frac{\sigma^2}{n-p}\{\operatorname{tr}\mathbf{V}-\operatorname{tr}\mathbf{VP}\}\\&=\sigma^2. \end{aligned}\tag{6.13}$$

□

Necessary and sufficient conditions for the BLUE $\boldsymbol{\beta}^*$ to equal $\hat{\boldsymbol{\beta}}$ are given in Theorem 3.6 (Section 3.6). However, these do not help if $\mathbf{V}$ is completely unknown, though in some cases the structure of $\mathbf{V}$ may be known.

6.2.2 Diagonal Dispersion Matrix

Suppose that the true dispersion matrix of $\boldsymbol{\varepsilon}$ is diagonal, namely,

$$\mathscr{D}[\boldsymbol{\varepsilon}]=\boldsymbol{\Sigma}=\operatorname{diag}(\sigma_{11},\sigma_{22},\ldots,\sigma_{nn}).$$

Then maximum likelihood estimates of $\boldsymbol{\beta}$ and the σ_{ii} can be obtained using the methods of Hartley and Jayatillake [1973]; the maximization is carried out subject to the σ_{ii} being strictly positive. Several procedures for testing whether the ε_i are all independent normal with common variance, σ^2, say, are mentioned in Section 6.6.5. However, for the general linear regression model the graphic methods described in Section 6.6 below are usually more informative.

A common special case of the above theory is when the variances of the ε_i are known to be equal within certain groups. Suppose, for example, that the variances in the kth group are all equal to σ_k^2, say $(k=1,2,\ldots,K)$, and we wish to test the hypothesis $H:\sigma_1^2=\sigma_2^2=\cdots=\sigma_K^2$ $(=\sigma^2$, say). Then if we have K mutually independent statistics S_k^2, where $f_k S_k^2/\sigma_k^2 \sim \chi_{f_k}^2$, we can test H using the following methods.

a BARTLETT'S TEST

This test, due to Bartlett [1937a], requires the computation of

$$T_1 = \frac{\left(\sum_k f_k\right)\log S^2 - \sum_k \left(f_k \log S_k^2\right)}{C} \tag{6.14}$$

where

$$S^2 = \frac{\sum_k f_k S_k^2}{\sum_k f_k}$$

and

$$C = 1 + \frac{\sum_k f_k^{-1} - \left(\sum_k f_k\right)^{-1}}{3(K-1)}.$$

When H is true T_1 is distributed approximately as χ_{K-1}^2 and the approximation is satisfactory for quite small samples ($f_k \geqslant 3$). Unfortunately this test is quite sensitive to any nonnormality of the variables which make up each S_k^2. A significant T_1 statistic may be an indication of nonnormality rather than lack of homogeneity of variance.

b COCHRAN'S TEST

If the f_k are all equal, Cochran [1941] has proposed the test statistic

$$T_2 = \frac{\max(S_1^2, S_2^2, \ldots, S_K^2)}{S_1^2 + S_2^2 + \cdots + S_K^2}. \tag{6.15}$$

This test is particularly sensitive to the case where all the variances σ_k^2 are expected to be equal, except for one variance which could be larger (Gartside [1972]). Percentage points for the distribution of T_2 are given in Dixon and Massey [1969: p. 536] or Pearson and Hartley [1970: p. 203]; critical values not tabled can be obtained by quadrature from the approximation to the distribution function given by Cochran [1941].

c HARTLEY'S TEST

When the f_k are all equal (to ν, say) H can be tested using Hartley's [1950]

$$F_{\text{MAX}} = \frac{\max(S_1^2, S_2^2, \ldots, S_K^2)}{\min(S_1^2, S_2^2, \ldots, S_K^2)} = \frac{S_{\text{MAX}}^2}{S_{\text{MIN}}^2}, \tag{6.16}$$

say. Significance points for this test are given in David [1952] or Pearson and Hartley [1970: p. 202]. David [1956] has also used this ratio along with the following procedure, analogous to Duncan's multiple range test, for ordering a set of variances into distinct groups. By relabeling let $s_K^2 \geqslant s_{K-1}^2 \geqslant \cdots \geqslant s_1^2$, where s_k^2 is the observed value of S_k^2. The sequence of testing is to compare the largest s_k^2 with each of the others, beginning with the smallest; then the next largest is compared with each of the others, again beginning with the smallest; etc. More formally (Tietjen and Beckman [1972]), define $R_{ij} = s_i^2/s_j^2$, and let $C_\alpha(K,\nu)$ be the upper $100[1-(1-\alpha)^{K-1}]$ percentage points of F_{MAX}. Begin testing with R_{K1}. If $R_{K1} > C_\alpha(K,\nu)$ declare $\sigma_K^2 > \sigma_1^2$. Then test $R_{K2}, R_{K3}, \ldots, R_{Kj_1}$ until, for some j_1, $R_{Kj_1} \leqslant C_\alpha(K - j_1 + 1, \nu)$ and declare $\sigma_K^2 > \sigma_i^2$ for all $i = 1, 2, \ldots, j_1 - 1$. Repeat the process using $R_{K-1,1}, R_{K-1,2}, \ldots$ until $R_{K-1,j_2} \leqslant C_\alpha(K - j_2, \nu)$ for some j_2 and declare $\sigma_{K-1}^2 > \sigma_i^2$ for all $i = 1, 2, \ldots, j_2 - 1$. Continue in this fashion until $R_{K-m,1} \leqslant C_\alpha(K-m,\nu)$ for some m. Values of $C_\alpha(K,\nu)$ are given by Tietjen and Beckman [1972] for $\alpha = 0.10, 0.05, 0.01$; $K = 1(1)15, 20, 30, 40(20)100$; and $\nu = 2(1)15, 20, 30, 40, 50, 60, 100$.

Of the three tests given above, Bartlett's test is the most powerful when the alternative hypothesis is unknown and normality can be relied on. However, all three tests are sensitive to nonnormality.

A useful application of the above methods is the case of replicated data (cf. Section 4.4). Let Y_{ir} be the rth replicate $(r = 1, 2, \ldots, R_i)$ at the ith x-data point; thus

$$\begin{aligned} Y_{ir} &= \mu_i + \varepsilon_{ir} \qquad (i = 1, 2, \ldots, n; \quad r = 1, 2, \ldots, R_i) \\ &= \beta_0 + \beta_1 x_{i1} + \cdots + \beta_{p-1} x_{i,p-1} + \varepsilon_{ir}, \end{aligned} \tag{6.17}$$

where the ε_{ir} $(r = 1, 2, \ldots, R_i)$ are independently and identically distributed as $N(0, \sigma_i^2)$. Setting $S_i^2 = \Sigma_r (Y_{ir} - \overline{Y}_{i\cdot})^2/(R_i - 1)$ and $f_i = R_i - 1$ we have $f_i S_i^2/\sigma_i^2 \sim \chi_{f_i}^2$ and the above tests can be used to test $H: \sigma_1^2 = \sigma_2^2 = \cdots = \sigma_n^2$.

In particular, if $R_i \geqslant 4$ we can use Bartlett's test with $K=n$ and $S^2=\Sigma\Sigma(Y_{ir}-\bar{Y}_{i.})^2/(\Sigma_i R_i - n)$. However, if the ε_{ir} are not normally distributed but, under H, have a common kurtosis γ $(=E[(Y-\mu)^4/\sigma^4]-3)$ then, for large R_i, Box [1953] showed that T_1 tends in distribution to $(1+\frac{1}{2}\gamma)\chi^2_{n-1}$. If $\gamma>0$ (for example, $\gamma=3$ for the double exponential) then assuming that T_1 is chi-square gives too many significant results, whereas if $\gamma<0$ (for example, $\gamma=-1.2$ for the uniform distribution) too few significant results are obtained (Layard [1973: Table 1]). From the chi-square tables we see that the distortion of significance levels becomes worse as n increases.

Although the above example has been couched in a regression framework, we see from Equation (6.17) that it is basically a problem of comparing the variances of n populations with R_i observations on the ith population—a problem that arises in one-way analysis of variance (Section 9.1). When the R_i are large (say, greater than 10), which is more the case in one-way classification than in regression, a number of robust procedures are available, namely, Scheffé's [1959: pp. 83–87] approximate F-test (due to Box [1953] who assumed equal R_i and the same γ for the n populations), an approximate F-test based on absolute deviations (Levene [1960], Draper and Hunter [1969]), Layard's chi-square test (Layard [1973]), and a jackknife test (Layard [1973]). Slight modifications of these procedures based on more robust estimates of central location such as the median are given by Brown and Forsythe [1974].

6.3 ROBUSTNESS OF THE *F*-TEST TO NONNORMALITY

6.3.1 *Effect of the Regressor Variables*

Box and Watson [1962] showed that the sensitivity of the F test to normality depends very much on the numerical values of the regression variables. In terms of the experimental design situation in which the elements of the design matrix $\mathbf{X}$ are 0 or 1, this means that some designs will have more robust tests associated with them. For example, Box and Watson show, by an appropriate choice of $\mathbf{X}$, that almost the same regression model can be made to reproduce on the one hand a test to compare means which is little affected by nonnormality and on the other a comparison of variances test which is notoriously sensitive to nonnormality.

Let $Y_i=\beta_0+\beta_1 x_{i1}+\cdots+\beta_k x_{ik}+\varepsilon_i$, and consider $H:\beta_1=\beta_2=\cdots=\beta_k=0$. When H is true and the regression assumptions are valid, then

$$F=\frac{n-k-1}{k}\cdot\frac{\text{RSS}_H-\text{RSS}}{\text{RSS}}\sim F_{k,n-k-1}.$$

However, if we now relax the distributional assumptions and assume that the ε_i are independently distributed with some common—not necessarily normal—distribution, then Box and Watson [1962: p. 101] show that, when H is true, F is approximately distributed as F_{ν_1,ν_2}, with $\nu_1=\delta k, \nu_2=\delta(n-k-1)$, and

$$\delta^{-1}=1+\frac{(n+1)\alpha_2}{n-1-2\alpha_2},$$

where

$$\alpha_2=\frac{n-3}{2n(n-1)}\cdot C_X\Gamma_Y;$$

or (to order n^{-1})

$$\delta^{-1}=1+\frac{C_X\Gamma_Y}{(2n)}. \qquad (6.18)$$

Here $\Gamma_Y=E[k_4/k_2^2]$, where k_2 and k_4 are the sample cumulants for the n values of Y, and C_X is a multivariate analogue of k_4/k_2^2 for the x variables. When ε, and therefore Y, has a normal distribution, then $\Gamma_Y=0$, $\delta=1$, and $F_{\nu_1,\nu_2}=F_{k,n-k-1}$.

We see that the effect of any nonnormality in Y depends on C_X in the term δ. Box and Watson show that

$$-2\leqslant\frac{n-3}{n-1}C_X\leqslant n-1; \qquad (6.19)$$

the lower bound is obtainable whereas the upper bound is approached but cannot be attained in finite samples. When the regressors can be regarded as being approximately "normal," then $C_X\approx 0$ and the F-test is insensitive to nonnormality. Thus we may sum up by saying that it is the "extent of nonnormality" in the regressors which determines the sensitivity of F to nonnormality in the Y observations.

Let $\tilde{x}_{ij}=x_{ij}-\bar{x}_{\cdot j}$ $(i=1,2,\ldots,n;\ j=1,2,\ldots,k)$ and let $\tilde{X}=[(\tilde{x}_{ij})]$. If $\mathbf{M}=[(m_{rs})]=\tilde{\mathbf{X}}(\tilde{\mathbf{X}}'\tilde{\mathbf{X}})^{-1}\tilde{\mathbf{X}}'$, and $m=\sum_{r=1}^{n}m_{rr}^2$, Box and Watson show that

$$C_X=\frac{n(n^2-1)}{k(n-k-1)(n-3)}\left\{m-\frac{k^2}{n}-\frac{2k(n-k-1)}{n(n+1)}\right\}. \qquad (6.20)$$

Now applying Theorem 3.1 (ii) to the $n\times n$ matrix $\mathbf{M}$ we have $\operatorname{tr}\mathbf{M}=k$. If the diagonal elements of $\mathbf{M}$ are all equal, we have $m_{rr}=k/n$ $(r=1,2,\ldots,n)$,

$m=k^2/n$, and

$$C_X=\frac{n(n^2-1)}{k(n-k-1)(n-3)}\left\{-\frac{2k(n-k-1)}{n(n+1)}\right\}$$

$$=-\frac{2(n-1)}{n-3}.$$

Hence in this case the lower bound of (6.19) is attained, $\delta^{-1}=1-(\Gamma_Y/n)\approx 1$, and for large n the F-test is insensitive to nonnormality. From symmetry conditions it is not hard to show that any cross-classification with equal cell frequencies in every cell (for example, the models of Chapter 9), and any hierarchical classification with equal cell frequencies at each stage of the hierarchy (Section 9.5) are designs with equal elements m_{rr}.

The above theory refers only to the case $H:\beta_1=\cdots=\beta_k=0$. However, an alternative approach, which allows a more general hypothesis, has been given by Atiqullah [1962] and we now consider his method in detail.

6.3.2 *Quadratically Balanced F-tests*

Let $Y_1,Y_2,\ldots,Y_n$ be independent random variables with means $\theta_1,\theta_2,\ldots,\theta_n$, respectively, common variance σ^2, and common third and fourth moments about their means; let $\gamma_2=(\mu_4-3\sigma^4)/\sigma^4$ be their common kurtosis. Then from Atiqullah [1962] we have the following theorems.

THEOREM 6.1 Let $\mathbf{P}_i$ $(i=1,2)$ be a symmetric idempotent matrix of rank f_i such that $E[\mathbf{Y}'\mathbf{P}_i\mathbf{Y}]=\sigma^2 f_i$, and let $\mathbf{P}_1\mathbf{P}_2=\mathbf{O}$. If $\mathbf{p}_i$ is the column vector of the diagonal elements of $\mathbf{P}_i$ then:

(*i*) $\text{var}[\mathbf{Y}'\mathbf{P}_i\mathbf{Y}]=2\sigma^4(f_i+\tfrac{1}{2}\gamma_2\mathbf{p}_i'\mathbf{p}_i)$.

(*ii*) $\text{cov}[\mathbf{Y}'\mathbf{P}_1\mathbf{Y},\mathbf{Y}'\mathbf{P}_2\mathbf{Y}]=\sigma^4\gamma_2\mathbf{p}_1'\mathbf{p}_2$.

Proof. (i) Since $\mathbf{P}_i$ is symmetric and idempotent, $\text{tr}\,\mathbf{P}_i=\text{rank}\,\mathbf{P}_i=f_i$ (*A5.2*). Also $E[\mathbf{Y}'\mathbf{P}_i\mathbf{Y}]=\sigma^2\,\text{tr}\,\mathbf{P}_i+\boldsymbol{\theta}'\mathbf{P}_i\boldsymbol{\theta}=\sigma^2 f_i$ (Theorem 1.7, Section 1.4), so that $\boldsymbol{\theta}'\mathbf{P}_i^2\boldsymbol{\theta}=\boldsymbol{\theta}'\mathbf{P}_i\boldsymbol{\theta}=0$ for all $\boldsymbol{\theta}$; that is, $\mathbf{P}_i\boldsymbol{\theta}=\mathbf{0}$ for all $\boldsymbol{\theta}$. Therefore, substituting $\mathbf{A}=\mathbf{P}_i$ in Theorem 1.8, we have

$$\begin{aligned}\text{var}[\mathbf{Y}'\mathbf{P}_i\mathbf{Y}]&=2\sigma^4\text{tr}\,\mathbf{P}_i^2+(\mu_4-3\sigma^4)\mathbf{p}_i'\mathbf{p}_i\\&=2\sigma^4(\text{tr}\,\mathbf{P}_i+\tfrac{1}{2}\gamma_2\mathbf{p}_i'\mathbf{p}_i)\\&=2\sigma^4(f_i+\tfrac{1}{2}\gamma_2\mathbf{p}_i'\mathbf{p}_i).\end{aligned}$$

(ii) Given $\mathbf{P}_1\mathbf{P}_2 = \mathbf{O}$ we have

$$\begin{aligned}(\mathbf{P}_1+\mathbf{P}_2)^2 &= \mathbf{P}_1^2+\mathbf{P}_1\mathbf{P}_2+\mathbf{P}_2\mathbf{P}_1+\mathbf{P}_2^2\\ &= \mathbf{P}_1+\mathbf{P}_1\mathbf{P}_2+(\mathbf{P}_1\mathbf{P}_2)'+\mathbf{P}_2\\ &= \mathbf{P}_1+\mathbf{P}_2.\end{aligned}$$

Therefore $\mathbf{P}_1+\mathbf{P}_2$ is idempotent and, by (i),

$$\begin{aligned}\text{var}\left[\mathbf{Y}'\mathbf{P}_1\mathbf{Y}+\mathbf{Y}'\mathbf{P}_2\mathbf{Y}\right] &= \text{var}\left[\mathbf{Y}'(\mathbf{P}_1+\mathbf{P}_2)\mathbf{Y}\right]\\ &= 2\sigma^4\left[\text{tr}(\mathbf{P}_1+\mathbf{P}_2)+\tfrac{1}{2}\gamma_2(\mathbf{p}_1+\mathbf{p}_2)'(\mathbf{p}_1+\mathbf{p}_2)\right]\\ &= 2\sigma^4\left[\text{tr}\,\mathbf{P}_1+\text{tr}\,\mathbf{P}_2+\tfrac{1}{2}\gamma_2(\mathbf{p}_1'\mathbf{p}_1+2\mathbf{p}_1'\mathbf{p}_2+\mathbf{p}_2'\mathbf{p}_2)\right]\\ &= \text{var}\left[\mathbf{Y}'\mathbf{P}_1\mathbf{Y}\right]+\text{var}\left[\mathbf{Y}'\mathbf{P}_2\mathbf{Y}\right]+2\sigma^4\gamma_2\mathbf{p}_1'\mathbf{p}_2.\end{aligned}$$

Hence $\text{cov}[\mathbf{Y}'\mathbf{P}_1\mathbf{Y}, \mathbf{Y}'\mathbf{P}_2\mathbf{Y}] = \sigma^4\gamma_2\mathbf{p}_1'\mathbf{p}_2$. □

THEOREM 6.2 Suppose $\mathbf{P}_1$ and $\mathbf{P}_2$ satisfy the conditions of Theorem 6.1 and let $Z = \frac{1}{2}\log F$, where

$$F = \frac{\mathbf{Y}'\mathbf{P}_1\mathbf{Y}/f_1}{\mathbf{Y}'\mathbf{P}_2\mathbf{Y}/f_2}\left(=\frac{S_1^2}{S_2^2}, \text{ say}\right).$$

Then for large f_1 and f_2 we have, asymptotically,

$$E[Z] \sim \tfrac{1}{2}(f_2^{-1}-f_1^{-1})\left[1+\tfrac{1}{2}\gamma_2(f_1\mathbf{p}_2-f_2\mathbf{p}_1)'(f_1\mathbf{p}_2+f_2\mathbf{p}_1)\{f_1f_2(f_1-f_2)\}^{-1}\right] \tag{6.21}$$

and

$$\begin{aligned}\text{var}[Z] \sim \tfrac{1}{2}(f_1^{-1}+f_2^{-1})&\left[1+\tfrac{1}{2}\gamma_2(f_1\mathbf{p}_2-f_2\mathbf{p}_1)'(f_1\mathbf{p}_2-f_2\mathbf{p}_1)\right.\\ &\left.\times\{f_1f_2(f_1+f_2)\}^{-1}\right].\end{aligned} \tag{6.22}$$

Proof. Using a Taylor expansion of $\log S_i^2$ about $\log\sigma^2$ we have

$$\log S_i^2 \sim \log\sigma^2 + \frac{(S_i^2-\sigma^2)}{\sigma^2} - \frac{(S_i^2-\sigma^2)^2}{2\sigma^4}. \tag{6.23}$$

Taking expected values, and using $E[S_i^2]=\sigma^2$, we have

$$E[\log S_i^2] \sim \log\sigma^2 - \frac{1}{2\sigma^4}\operatorname{var}[S_i^2],$$

where, from Theorem 6.1,

$$\operatorname{var}[S_i^2] = \frac{\operatorname{var}[\mathbf{Y}'\mathbf{P}_i\mathbf{Y}]}{f_i^2} = 2\sigma^4\left(f_i^{-1} + \tfrac{1}{2}\gamma_2 f_i^{-2}\mathbf{p}_i'\mathbf{p}_i\right).$$

Substituting in

$$E[Z] = \tfrac{1}{2}\left\{E[\log S_1^2] - E[\log S_2^2]\right\}$$

leads to Equation (6.21).

To find an asymptotic expression for $\operatorname{var} Z$ we note first of all that

$$\operatorname{var}[Z] = \tfrac{1}{4}\left\{\operatorname{var}[\log S_1^2] + \operatorname{var}[\log S_2^2] - 2\operatorname{cov}[\log S_1^2, \log S_2^2]\right\}. \quad (6.24)$$

Then, ignoring the third term in (6.23), we have $E[\log S_i^2] \sim \log\sigma^2$ and

$$\begin{aligned}\operatorname{var}[\log S_i^2] &\sim E\left[(\log S_i^2 - \log\sigma^2)^2\right]\\ &\sim \frac{E\left[(S_i^2-\sigma^2)^2\right]}{\sigma^4}\\ &= \frac{\operatorname{var}[S_i^2]}{\sigma^4}.\end{aligned}$$

Similarly,

$$\begin{aligned}\operatorname{cov}[\log S_1^2, \log S_2^2] &\sim E\left[(\log S_1^2 - \log\sigma^2)(\log S_2^2 - \log\sigma^2)\right]\\ &\sim \frac{E\left[(S_1^2-\sigma^2)(S_2^2-\sigma^2)\right]}{\sigma^4}\\ &= \frac{\operatorname{cov}[S_1^2, S_2^2]}{\sigma^4}.\end{aligned}$$

Finally, substituting in

$$\operatorname{var}[Z] \sim \frac{1}{4\sigma^4}\left\{\operatorname{var}[S_1^2] + \operatorname{var}[S_2^2] - 2\operatorname{cov}[S_1^2, S_2^2]\right\}$$

and using Theorem 6.1 leads to Equation (6.22). □

We can now apply the above theory to the usual F statistic for testing $H: \mathbf{A}\boldsymbol{\beta}=\mathbf{0}$. From Theorem 4.1 (iv) we have

$$F=\frac{\mathbf{Y}'(\mathbf{P}-\mathbf{P}_H)\mathbf{Y}/q}{\mathbf{Y}'(\mathbf{I}_n-\mathbf{P})\mathbf{Y}/(n-p)}$$

$$=\frac{\mathbf{Y}'\mathbf{P}_1\mathbf{Y}/q}{\mathbf{Y}'\mathbf{P}_2\mathbf{Y}/(n-p)}$$

$$=\frac{S_1^2}{S_2^2}, \tag{6.25}$$

say, where $\mathbf{P}_1\mathbf{P}_2=(\mathbf{P}-\mathbf{P}_H)(\mathbf{I}_n-\mathbf{P})=\mathbf{P}_H-\mathbf{P}_H\mathbf{P}=\mathbf{O}$. Suppose we now relax the distributional assumptions underlying F and assume only that the ε_i are independently and identically distributed; in particular, $\mathcal{E}[\boldsymbol{\varepsilon}]=\mathbf{0}$ and $\mathfrak{D}[\boldsymbol{\varepsilon}]=\sigma^2\mathbf{I}_n$. Then $E[S_2^2]=E[S^2]=\sigma^2$ (Theorem 3.3 in Section 3.3) and, *when H is true*, $E[S_1^2]=\sigma^2$ [by Theorem 4.1 (ii) with $\mathbf{A}\boldsymbol{\beta}=\mathbf{c}=\mathbf{0}$; the assumption of normality is not used in the proof]. Also the Y_i satisfy the conditions stated at the beginning of this section (with $[(\theta_i)]=\boldsymbol{\theta}=\mathbf{X}\boldsymbol{\beta}$) so that, when H is true, Theorem 6.2 can be applied directly to the F-statistic (6.25) with $f_1=q$ and $f_2=n-p$.

When the ε_i, and therefore the Y_i, are normally distributed it is known that, for large f_1 and f_2, $Z=\frac{1}{2}\log F$ is approximately normally distributed with mean and variance given by setting $\gamma_2=0$ in Equations (6.21) and (6.22), when H is true. As this approximation is evidently quite good even when f_1 and f_2 are as small as four, it is not unreasonable to accept Atiqullah's proposition that, for a moderate amount of nonnormality, Z is still approximately normal with mean and variance given by (6.21) and (6.22). On this assumption Z, and therefore F, will be approximately independent of γ_2 if the coefficient of γ_2 in (6.21) and (6.22) is zero; that is, if

$$f_1\mathbf{p}_2=f_2\mathbf{p}_1. \tag{6.26}$$

Now, using Atiqullah's terminology, we say that F is quadratically balanced if the diagonal elements of $\mathbf{P}_i$ $(i=1,2)$ are equal; most of the usual F-tests for balanced experimental designs belong to this category. In this case, since $\operatorname{tr}\mathbf{P}_i=f_i$, we have

$$\mathbf{p}_i=\left(\frac{f_i}{n}\right)\mathbf{I}_n \qquad \text{and} \qquad f_1\mathbf{p}_2=\left(\frac{f_1f_2}{n}\right)\mathbf{I}_n=f_2\mathbf{p}_1.$$

Thus a sufficient condition for (6.26) to hold is that F is quadratically balanced.

Atiqullah [1962: p. 88] also states that even if γ_2 varies among the Y_i, quadratic balance is still sufficient for $E[Z]$ and var$[Z]$ to be independent of kurtosis effects, to the order involved in Theorem 6.2.

Finally we note that if γ_2 can be estimated, Equations (6.21) and (6.22) can be used to modify the degrees of freedom and improve the correspondence between the distribution of the F-ratio and an F-distribution (Prentice [1974]).

6.4 REGRESSOR VARIABLES MEASURED WITH ERROR

6.4.1 Random Errors

The true regression model is assumed to be

$$\begin{aligned} Y_i &= \beta_0 + \beta_1 u_{i1} + \cdots + \beta_{p-1} u_{i,p-1} + \varepsilon_i \\ &= \mathbf{u}_i' \boldsymbol{\beta} + \varepsilon_i, \end{aligned}$$

say, or

$$\mathbf{Y} = \mathbf{U}\boldsymbol{\beta} + \boldsymbol{\varepsilon}, \tag{6.27}$$

where $\mathbf{U}' = [\mathbf{u}_1, \mathbf{u}_2, \ldots, \mathbf{u}_n]$. Suppose, however, that the data "point" $\mathbf{u}_i$ is measured with an unbiased error of $\boldsymbol{\delta}_i$ so that what is observed is actually $\mathbf{X}_i = \mathbf{u}_i + \boldsymbol{\delta}_i$ or $\mathbf{X} = \mathbf{U} + \boldsymbol{\Delta}$, where $\mathbf{X}' = [\mathbf{X}_1, \mathbf{X}_2, \ldots, \mathbf{X}_n]$, $\boldsymbol{\Delta}' = [\boldsymbol{\delta}_1, \boldsymbol{\delta}_2, \ldots, \boldsymbol{\delta}_n]$ and $\mathcal{E}[\boldsymbol{\Delta}] = \mathbf{O}$. It is assumed that the $\boldsymbol{\delta}_i$ are uncorrelated and have the same dispersion matrix; thus

$$\begin{aligned} \mathcal{E}[\boldsymbol{\delta}_i \boldsymbol{\delta}_j'] &= \mathbf{D}, \qquad i = j \\ &= \mathbf{O}, \qquad i \neq j. \end{aligned}$$

Since the first element of each $\mathbf{u}_i$ and $\mathbf{X}_i$ is unity, the first element of $\boldsymbol{\delta}_i$ is zero, and the first row and column of $\mathbf{D}$ consist of zeros. We also assume that $\boldsymbol{\Delta}$ is independent of $\boldsymbol{\varepsilon}$.

The usual least squares estimate of $\boldsymbol{\beta}$ is now

$$\hat{\boldsymbol{\beta}}_\Delta = (\mathbf{X}'\mathbf{X})^{-1}\mathbf{X}'\mathbf{Y}$$

instead of $\hat{\boldsymbol{\beta}} = (\mathbf{U}'\mathbf{U})^{-1}\mathbf{U}'\mathbf{Y}$ so that $\hat{\boldsymbol{\beta}}_\Delta$ is no longer unbiased. The properties of $\hat{\boldsymbol{\beta}}_\Delta$ were discussed in detail by Hodges and Moore [1972] for the common special case of $\mathbf{D} = \text{diag}(0, \sigma_1^2, \sigma_2^2, \ldots, \sigma_{p-1}^2)$. However, using a more rigorous approximation theory, Davies and Hutton [1975] have extended this work to the case of a general matrix $\mathbf{D}$ (in their notation $\mathbf{U} \to \mathbf{X}'$, $\mathbf{X} \to \mathbf{W}'$, $\mathbf{D} \to \mathbf{S}$, and $\boldsymbol{\Delta} \to \boldsymbol{\Delta}'$).

a BIAS

Since $\boldsymbol{\Delta}$ is independent of $\boldsymbol{\varepsilon}$ (and $\mathbf{Y}$),

$$\begin{aligned}
\mathscr{E}[\hat{\boldsymbol{\beta}}_\Delta] &= \mathop{\mathscr{E}}_{\Delta} \mathscr{E}[\hat{\boldsymbol{\beta}}_\Delta | \boldsymbol{\Delta}] \\
&= \mathop{\mathscr{E}}_{\Delta}\left[(\mathbf{X}'\mathbf{X})^{-1}\mathbf{X}'\mathbf{U}\boldsymbol{\beta}\right] \qquad [\text{from } (6.27)] \\
&= \mathop{\mathscr{E}}_{\Delta}\left[(\mathbf{X}'\mathbf{X})^{-1}\mathbf{X}'(\mathbf{X}-\boldsymbol{\Delta})\boldsymbol{\beta}\right] \\
&= \boldsymbol{\beta} - \mathop{\mathscr{E}}_{\Delta}\left[(\mathbf{X}'\mathbf{X})^{-1}\mathbf{X}'\boldsymbol{\Delta}\boldsymbol{\beta}\right] \\
&= \boldsymbol{\beta} - \mathbf{b}, \qquad (6.28)
\end{aligned}$$

say. When n is large, Davies and Hutton [1975: Theorem 4.1] show that

$$\begin{aligned}
\mathbf{b} &\approx \left(\frac{1}{n}\mathbf{U}'\mathbf{U}+\mathbf{D}\right)^{-1}\mathbf{D}\boldsymbol{\beta} \\
&= n(\mathbf{U}'\mathbf{U}+n\mathbf{D})^{-1}\mathbf{D}\boldsymbol{\beta}. \qquad (6.29)
\end{aligned}$$

(In fact, if $\operatorname{plim}_{n\to\infty}\{(1/n)\mathbf{X}'\mathbf{X}\} = \mathbf{A}$, say, then $\hat{\boldsymbol{\beta}}_\Delta$ is a consistent estimate of $(\mathbf{A}+\mathbf{D})^{-1}\mathbf{A}\boldsymbol{\beta} = \boldsymbol{\beta} - (\mathbf{A}+\mathbf{D})^{-1}\mathbf{D}\boldsymbol{\beta}$). Since

$$\begin{aligned}
\mathscr{E}[\mathbf{X}'\mathbf{X}] &= \mathscr{E}[\mathbf{U}'\mathbf{U}+\boldsymbol{\Delta}'\mathbf{U}+\mathbf{U}'\boldsymbol{\Delta}+\boldsymbol{\Delta}'\boldsymbol{\Delta}] \\
&= \mathbf{U}'\mathbf{U} + \mathscr{E}[\boldsymbol{\Delta}'\boldsymbol{\Delta}] \\
&= \mathbf{U}'\mathbf{U} + \mathscr{E}\left[\sum_{i=1}^{n} \boldsymbol{\delta}_i\boldsymbol{\delta}_i'\right] \\
&= \mathbf{U}'\mathbf{U} + n\mathbf{D}, \qquad (6.30)
\end{aligned}$$

an obvious estimate of the bias $\mathbf{b}$ is

$$\hat{\mathbf{b}} = n(\mathbf{X}'\mathbf{X})^{-1}\hat{\mathbf{D}}\hat{\boldsymbol{\beta}},$$

where $\hat{\mathbf{D}}$ is a rough estimate of $\mathbf{D}$ available, we hope, from other experiments. When $\mathbf{D} = \operatorname{diag}(0, \sigma_1^2, \ldots, \sigma_{p-1}^2)$, the approximations used by Hodges and Moore lead to a similar estimate of $\mathbf{b}$ (with $n-p-1$ instead of n).

Davies and Hutton show that the magnitude of $\mathbf{b}$ is related to how close $\mathbf{X}'\mathbf{X}$ is to being singular. If the errors are such that they may bring $\mathbf{X}'\mathbf{X}$

close to being singular then the bias could be large. Using the central limit theorem they also show that $\sqrt{n}\,\hat{\boldsymbol{\beta}}_\Delta$ is asymptotically normal.

b STANDARD ERRORS

Davies and Hutton [1975: Equation 4.2] show that when $\mathbf{D}$ is close to the zero matrix,

$$\mathscr{D}[\hat{\boldsymbol{\beta}}_\Delta] \approx \frac{1}{n}\left\{\left(\frac{1}{n}\mathbf{U}'\mathbf{U}+\mathbf{D}\right)^{-1}(\sigma^2+\boldsymbol{\beta}'\boldsymbol{\Delta}\boldsymbol{\beta})+\mathbf{O}(\mathbf{D}^2)\right\}.$$

The usual estimate of this variance–covariance matrix is $\hat{\mathbf{V}}=S^2(\mathbf{X}'\mathbf{X})^{-1}$, where

$$\begin{aligned}(n-p)S^2 &= (\mathbf{Y}-\mathbf{X}\hat{\boldsymbol{\beta}}_\Delta)'(\mathbf{Y}-\mathbf{X}\hat{\boldsymbol{\beta}}_\Delta)\\ &= \mathbf{Y}'(\mathbf{I}_n-\mathbf{X}(\mathbf{X}'\mathbf{X})^{-1}\mathbf{X}')\mathbf{Y}\\ &= \mathbf{Y}'(\mathbf{I}_n-\mathbf{P}_X)\mathbf{Y},\end{aligned}$$

say. The question now is, does $\hat{\mathbf{V}}$ still provide an unbiased estimate of $\mathscr{D}[\hat{\boldsymbol{\beta}}_\Delta]$?

Since $\boldsymbol{\Delta}$ is independent of $\boldsymbol{\varepsilon}$, $\mathbf{X}'(\mathbf{I}_n-\mathbf{P}_X)=\mathbf{O}$, and $\operatorname{tr}[\mathbf{I}_n-\mathbf{P}_X]=n-p$, we have

$$\begin{aligned}E\left[(n-p)S^2|\boldsymbol{\Delta}\right] &= E\left[(n-p)\sigma^2+\boldsymbol{\beta}'\mathbf{U}'(\mathbf{I}_n-\mathbf{P}_X)\mathbf{U}\boldsymbol{\beta}|\boldsymbol{\Delta}\right]\\ &= E\left[(n-p)\sigma^2+\boldsymbol{\beta}'(\mathbf{X}'-\boldsymbol{\Delta}')(\mathbf{I}_n-\mathbf{P}_X)(\mathbf{X}-\boldsymbol{\Delta})\boldsymbol{\beta}|\boldsymbol{\Delta}\right]\\ &= E\left[(n-p)\sigma^2+\boldsymbol{\beta}'\boldsymbol{\Delta}'(\mathbf{I}_n-\mathbf{P}_X)\boldsymbol{\Delta}\boldsymbol{\beta}|\boldsymbol{\Delta}\right]. \qquad (6.31)\end{aligned}$$

Now for any matrix $\mathbf{C}$

$$\begin{aligned}\underset{\Delta}{\mathscr{E}}[\boldsymbol{\Delta}'\mathbf{C}\boldsymbol{\Delta}] &= \sum_i\sum_j c_{ij}\mathscr{E}[\boldsymbol{\delta}_i\boldsymbol{\delta}_j']\\ &= \sum_i c_{ii}\mathbf{D}\\ &= \mathbf{D}\operatorname{tr}\mathbf{C},\end{aligned}$$

so that from (6.31)

$$\begin{aligned}
\mathcal{E}[\hat{\mathbf{V}}] &= \mathop{\mathcal{E}}_{\Delta} \mathcal{E}\left[S^2(\mathbf{X}'\mathbf{X})^{-1}|\boldsymbol{\Delta}\right] \\
&= \mathop{\mathcal{E}}_{\Delta}\left[\left\{\sigma^2 + \frac{1}{n-p}\boldsymbol{\beta}'\boldsymbol{\Delta}'(\mathbf{I}_n - \mathbf{P}_X)\boldsymbol{\Delta}\boldsymbol{\beta}\right\}\{(\mathbf{X}'\mathbf{X})^{-1}\}\right] \\
&\approx \left\{\sigma^2 + \frac{1}{n-p}\boldsymbol{\beta}'\mathcal{E}[\boldsymbol{\Delta}'(\mathbf{I}_n - \mathbf{P}_X)\boldsymbol{\Delta}]\boldsymbol{\beta}\right\}\{\mathcal{E}[\mathbf{X}'\mathbf{X}]\}^{-1} \\
&\approx (\sigma^2 + \boldsymbol{\beta}'\mathbf{D}\boldsymbol{\beta})(\mathbf{U}'\mathbf{U} + n\mathbf{D})^{-1} \\
&\approx \mathcal{D}[\boldsymbol{\beta}_\Delta]. \qquad (6.32)
\end{aligned}$$

Hence, for large n and small $\mathbf{D}$, $\hat{\mathbf{V}}$ is still approximately unbiased.

6.4.2 *Roundoff Errors*

Using the above notation we suppose, once again, that $\mathbf{U}$ is the correct data matrix (that is, $\mathcal{E}[\mathbf{Y}] = \mathbf{U}\boldsymbol{\beta}$), and we observe $\mathbf{X} = \mathbf{U} + \boldsymbol{\Delta}$. However, following Swindel and Bower [1972], we now assume that the measurements are *accurate* but they are rounded off according to some consistent rule to give $x_{ij} = u_{ij} + \Delta_{ij}$. In this case the rounding error Δ_{ij} can be regarded as an (unknown) constant, and *not* a random variable; Δ_{ij} is determined solely by the u_{ij} and the rounding rule. The matrix $\mathbf{X}$ is now a matrix of constants rather than a random matrix as in the preceding section.

The bias of $\hat{\boldsymbol{\beta}}_\Delta = (\mathbf{X}'\mathbf{X})^{-1}\mathbf{X}'\mathbf{Y}$ is

$$\begin{aligned}
\mathcal{E}[\hat{\boldsymbol{\beta}}_\Delta - \boldsymbol{\beta}] &= (\mathbf{X}'\mathbf{X})^{-1}\mathbf{X}'(\mathbf{U} - \mathbf{X})\boldsymbol{\beta} \\
&= -(\mathbf{X}'\mathbf{X})^{-1}\mathbf{X}'\boldsymbol{\Delta}\boldsymbol{\beta}. \qquad (6.33)
\end{aligned}$$

By writing $\boldsymbol{\Delta}\boldsymbol{\beta} = \sum_{j=0}^{p-1}\boldsymbol{\Delta}_j\beta_j$, where the $\boldsymbol{\Delta}_j$ are columns of $\boldsymbol{\Delta}$, we see that the bias does not depend on β_j if $\boldsymbol{\Delta}_j = \mathbf{0}$; the bias depends only on the regressors containing rounding errors.

We see from

$$\begin{aligned}
E[S^2] &= \sigma^2 + \frac{\boldsymbol{\beta}'\mathbf{U}'(\mathbf{I}_n - \mathbf{P}_X)\mathbf{U}\boldsymbol{\beta}}{n-p} \\
&= \sigma^2 + \frac{\boldsymbol{\beta}'(\mathbf{X} - \boldsymbol{\Delta})'(\mathbf{I}_n - \mathbf{P}_X)(\mathbf{X} - \boldsymbol{\Delta})\boldsymbol{\beta}}{n-p} \\
&= \sigma^2 + \frac{\boldsymbol{\beta}'\boldsymbol{\Delta}'(\mathbf{I}_n - \mathbf{P}_X)\boldsymbol{\Delta}\boldsymbol{\beta}}{n-p} \\
&\geqslant \sigma^2, \qquad (6.34)
\end{aligned}$$

(since $\mathbf{I}_n - \mathbf{P}_X$ is positive semidefinite), that S^2 will tend to overestimate σ^2. However, $\sigma^2(\mathbf{X}'\mathbf{X})^{-1}$ is the correct dispersion matrix of $\hat{\boldsymbol{\beta}}_\Delta$.

Using eigenvalues Swindel and Bower [1972] prove that for any $\mathbf{a}$ the estimate $\mathbf{a}'\hat{\boldsymbol{\beta}}_\Delta$ of $\mathbf{a}'\boldsymbol{\beta}$ has the property that

$$0 \leqslant \mathrm{RB}(\mathbf{a}'\hat{\boldsymbol{\beta}}_\Delta) \leqslant \frac{1}{\sigma}(\boldsymbol{\beta}'\boldsymbol{\Delta}'\boldsymbol{\Delta}\boldsymbol{\beta})^{1/2},$$

where RB is the relative bias, that is, $|bias|/(standard\ deviation)$.

6.4.3 Some Working Rules

Davies and Hutton [1975] consider both random and roundoff error in their analysis and give the following working rules.

For the roundoff type situation define r_j to be the square root of the jth diagonal element of $\boldsymbol{\Delta}'\boldsymbol{\Delta}/n$, and suppose that m_1 of the r_j's are nonzero. We first compute

$$\rho_1 = \left\{ \sum_j r_j^2 \left[(\mathbf{X}'\mathbf{X})^{-1} \right]_{jj} \right\}^{-1/2}. \tag{6.35}$$

If ρ_1 is not somewhat larger than $(m_1 n)^{1/2}$, or possibly $n^{1/2}$ if n is large, then at least some of the elements of $\hat{\boldsymbol{\beta}}_\Delta$ are likely to have little meaning. (In practice r_j will not be known exactly and will be replaced by, say, an upper bound.) If this test is passed then

$$\frac{n \sum_j r_j |\hat{\beta}_{j,\Delta}|}{S}$$

should be evaluated. If this quantity is markedly less than 1 then the errors in $\mathbf{U}$ can be ignored. However, if this test is failed and the situation of Section 6.4.1 prevails (with random Δ_{ij}), then the next step is to compute

$$\frac{n \left(\sum_j r_j^2 \hat{\beta}_{j,\Delta}^2 \right)^{1/2}}{(\rho_1 S)}$$

where, in the above formula and the definition of ρ_1 in (6.35), r_j is now the square root of the jth diagonal element of $\mathbf{D}$. If the above quantity is markedly less than 1 then the effects of the errors are probably negligible, particularly if n is large. On the other hand, if this term is larger than 1 then the bias is likely to constitute a major part in the error of at least some of the estimates.

The authors also suggest that the diagonal elements of $\mathbf{P}_X$ be calculated in order to check whether any single regressor observation has an undue effect on the estimates. In particular if any diagonal element is greater than about 0.2, it is possible for a moderate error in the corresponding regressor to affect the estimates significantly and yet go undetected when the residuals are checked (Section 6.6).

6.5 MODELS WITH RANDOM REGRESSORS

In Section 6.4 [Equation (6.27)] we used a model of the form

$$\begin{aligned} Y &= \beta_0 + \beta_1 u_1 + \cdots + \beta_{p-1} u_{p-1} + \varepsilon \\ &= \beta_0 + \beta_1 E[X_1] + \cdots + \beta_{p-1} E[X_{p-1}] + \varepsilon, \end{aligned} \tag{6.36}$$

where ε is assumed to be independent of the X_j, and $E[\varepsilon]=0$. Equation (6.36) can also be expressed in terms of expected values, namely,

$$v = E[Y] = \beta_0 + \beta_1 u_1 + \cdots + \beta_{p-1} u_{p-1}. \tag{6.37}$$

This lawlike relationship between the expected values is often called a *functional relationship*.

Sometimes the relationship comes from some physical law (suitably transformed to achieve linearity) with the randomness in the model arising from experimental errors in measuring the mathematical variables v and $\{u_j\}$. For this reason the model is sometimes called the "errors-in-variables" model or, in the straight-line case, the model for "regression with both variables subject to error." It should be distinguished from the model

$$V = \beta_0 + \beta_1 U_1 + \cdots + \beta_{p-1} U_{p-1}$$

in which the lawlike relationship exists between random variables V and $\{U_j\}$ rather than between mathematical variables; Sprent [1969] and others call this relationship a "structural relationship." In practice the U_j are *observed* but V is unknown (for example, due to experimental error) so that Y $(= V + \varepsilon)$ is actually observed. Thus the appropriate model is now

$$Y = \beta_0 + \beta_1 U_1 + \cdots + \beta_{p-1} U_{p-1} + \varepsilon \tag{6.38}$$

or

$$E[Y | \{U_j\}] = \beta_0 + \beta_1 U_1 + \cdots + \beta_{p-1} U_{p-1}.$$

If the U_j are also measured with (unbiased) error so that X_j is observed instead of U_j, then

$$E[Y|\{U_j\}]=\beta_0+\beta_1E[X_1|U_1]+\cdots+\beta_{p-1}E[X_{p-1}|U_{p-1}]. \quad (6.39)$$

Since this model is analogous to (6.36) it can be analyzed in the same way, that is, by treating the $\{U_j\}$ as though they were (conditionally) constants.

We also note that (6.38) can be written in the form

$$\begin{aligned} Y&=\beta_0+\beta_1E[U_1]+\cdots+\beta_{p-1}E[U_{p-1}]+\varepsilon+\sum_{j=1}^{p-1}\beta_j\{U_j-E[U_j]\} \\ &=\beta_0+\beta_1E[U_1]+\cdots+\beta_{p-1}E[U_{p-1}]+\varepsilon', \end{aligned} \quad (6.40)$$

where $E[\varepsilon']=0$. This looks like (6.36) but there is a difference: ε' is not independent of the U_j.

There exists an extensive literature on methods of analyzing the models (6.37), (6.38), and (6.39), and the reader is referred, for example, to Sprent [1969], Hodges and Moore [1972], and Narula [1974] for references; further references are given in Section 7.7 where the straight-line case is considered in detail. However, these methods all tend to be more complicated than the ordinary least squares technique, and most of them require additional information in the form of extra data (for example, the technique of "instrumental variables"), or estimates concerning error variances (or their ratios) as in the case of estimating **b** in Section 6.4. For these reasons the methods have tended to be neglected in favor of the popular technique of regressing Y *conditionally* on the observed values of the regressors, a technique that is appropriate only in the case of (6.38). However, in both (6.37) and (6.39), one can correct for the bias in the least squares estimate using the method outlined in Section 6.4.1. It seems, therefore, that provided care is exercised in determining the most appropriate model, the ordinary least squares approach can be used in all three cases.

There are two other models worth mentioning: the so-called "components of variance" model, and the "controlled variables" model of Berkson [1950]. In the components of variance model (or so-called regression model of the second kind) $\boldsymbol{\beta}$ is regarded as a random variable (see Sprent [1969: pp. 54, 82] and Searle [1971] for further references). In Berkson's model the regressors are random but their observed values are controlled, a common situation when investigating lawlike relationships in the physical sciences. For example, suppose we wish to study Ohm's law $v=\beta u$, where $v=$ voltage in volts, $u=$current in amperes, and $\beta=$resistance in ohms. Then, for a given resistance, a natural experimental procedure would be to adjust the current through the circuit so that the ammeter reads a certain prescribed or "target" value x_i, for example, $x_i = 1$ A, and then measure the voltage Y_i with a voltmeter. The ammeter will have a random error so that the current actually flowing through the circuit is an unknown random variable U_i, say. Similarly the true voltage will also be an unknown random variable V_i so that our model for this experiment is now

$$Y_i = V_i + \varepsilon_i = \beta U_i + \varepsilon_i,$$

which is of the form (6.38). However, the above model reduces to a "standard" least squares model

$$\begin{aligned} Y_i &= \beta x_i + \varepsilon_i + \beta(U_i - x_i) \\ &= \beta x_i + \varepsilon_i' \end{aligned}$$

where the error or fluctuation term is now ε_i' instead of ε_i. What the above discussion implies is that, in the controlled regressors situation, the model may be analyzed as though the regressors are nonrandom and error free.

6.6 ANALYSIS OF RESIDUALS

6.6.1 *Definition and Properties*

The electronic computer has provided us with an excellent tool for computing the departure of each observed Y_i from the fitted regression $\hat{Y}_i$. These differences are called residuals and are denoted by

$$e_i = Y_i - \hat{Y}_i \qquad (i = 1, 2, \ldots, n)$$

or

$$\mathbf{e} = \mathbf{Y} - \hat{\mathbf{Y}} = \mathbf{Y} - \mathbf{X}\hat{\boldsymbol{\beta}} = (\mathbf{I}_n - \mathbf{P})\mathbf{Y},$$

where $\mathbf{P} = \mathbf{X}(\mathbf{X}'\mathbf{X})^{-1}\mathbf{X}' = [(p_{ij})]$. (If $\mathbf{X}$ has less than full rank we can use a generalized inverse, $(\mathbf{X}'\mathbf{X})^-$, to define $\mathbf{P}$.) These residuals are mathematically dependent since, from Equation (4.29) in Section 4.2,

$$\sum_i e_i = 0. \tag{6.41}$$

When $\mathscr{E}[\boldsymbol{\varepsilon}] = \mathbf{0}$ and $\mathscr{D}[\boldsymbol{\varepsilon}] = \sigma^2 \mathbf{I}_n$ $(= \mathscr{D}[\mathbf{Y}])$, then $\mathscr{E}[\mathbf{e}] = \mathbf{0}$ and $\mathscr{D}[\mathbf{e}] = \sigma^2(\mathbf{I}_n - \mathbf{P})^2 = \sigma^2(\mathbf{I}_n - \mathbf{P})$. We recall that S^2 $[= \sum_i e_i^2/(n-p)]$ is an unbiased estimate of σ^2 and, since $(\mathbf{I}_n - \mathbf{P})\mathbf{X} = \mathbf{O}$, $\mathbf{e} = (\mathbf{I}_n - \mathbf{P})\boldsymbol{\varepsilon}$.

If $\boldsymbol{\varepsilon} \sim N_n(\mathbf{0}, \sigma^2 \mathbf{I}_n)$, then $\mathbf{e}$ has a "singular" multivariate normal distribution (singular because $\mathbf{I}_n - \mathbf{P}$ is an $n \times n$ positive semidefinite matrix), and the marginal distribution of e_i is $N(0, \sigma^2(1-p_{ii}))$. Since $\mathbf{P}'\mathbf{P} = \mathbf{P}$, $p_{ii} > 0$ and hence $0 \leqslant 1 - p_{ii} < 1$; equality occurs, that is, $\text{var}[e_i] = 0$, in certain specially constructed designs only (Behnken and Draper [1972: p. 102].

Finally we note that (Theorem 3.5 (iii), Section 3.4)

$$\mathcal{C}[\hat{\mathbf{Y}}, \mathbf{e}] = \mathbf{X}\mathcal{C}[\hat{\boldsymbol{\beta}}, \mathbf{Y} - \mathbf{X}\hat{\boldsymbol{\beta}}] = \mathbf{O}, \tag{6.42}$$

and for $j = 1, 2, \ldots, p-1$,

$$\begin{aligned}
\sum_i (e_i - \bar{e}_{\cdot})(x_{ij} - \bar{x}_{\cdot j}) &= \sum_i e_i (x_{ij} - \bar{x}_{\cdot j}) \\
&= \sum_i e_i x_{ij} \qquad [\text{by } (6.41)] \\
&= \sum_i (Y_i - \hat{\beta}_0 - \hat{\beta}_1 x_{i1} - \cdots - \hat{\beta}_{p-1} x_{i,p-1}) x_{ij} \\
&= 0.
\end{aligned} \tag{6.43}$$

This last step follows from the normal equations; to see this differentiate $(Y_i - \beta_0 - \beta_1 x_{i1} - \cdots - \beta_{p-1} x_{i,p-1})^2$ with respect to β_j. This step also follows from $(\mathbf{I}_n - \mathbf{P})\mathbf{X} = \mathbf{O}$.

We now consider a number of graphic methods, based on the residuals, for investigating departures from the underlying model and distributional assumptions. The rationale behind the following plots is that any departures from the distributional assumptions of $\boldsymbol{\varepsilon}$ are reflected in $\mathbf{e}$; different plots reflect different departures.

6.6.2 *Residual Plots*

The first step is to scale the residuals so that they have approximately unit variances. One method of doing this uses the notion of average variance:

$$\begin{aligned}
\frac{1}{n} \sum_{i=1}^{n} \text{var}[e_i] &= \frac{1}{n} \text{tr}\, \mathcal{D}[\mathbf{e}] \\
&= \frac{1}{n} \sigma^2 \text{tr}[\mathbf{I}_n - \mathbf{P}] \\
&= \frac{\sigma^2 (n-p)}{n}.
\end{aligned} \tag{6.44}$$

Estimating σ^2 by S^2 leads to the scaled residual (Daniel and Wood [1971: p. 28], Behnken and Draper [1972: p. 102])

$$c_i = \frac{e_i}{\{S^2(n-p)/n\}^{1/2}}$$

$$= \frac{e_i}{\{(1/n)\mathbf{e}'\mathbf{e}\}^{1/2}}. \qquad (6.45)$$

However, since $\text{var}[e_i] = \sigma^2(1-p_{ii})$, the natural choice is the "Studentized" residual

$$d_i = \frac{e_i}{S(1-p_{ii})^{1/2}}. \qquad (6.46)$$

Although the residuals or their scaled variants c_i and d_i are correlated, it seems that the correlation has little effect on the graphic plots described below. We can therefore treat the d_i and, to some extent, the c_i as being approximately independently and identically distributed as $N(0,1)$.

General practice indicates that, in many regression situations, the p_{ii} can be ignored so that we can use e_i/S or, when p/n is not small, c_i (Behnken and Draper [1972: 1.3 and 2.5]).

a NORMAL PROBABILITY PLOT

Let $d_{(1)} < d_{(2)} < \cdots < d_{(n)}$ represent the n ranked Studentized residuals. For n moderately large (see Appendix C) a plot of $d_{(i)}$ versus $(i-\frac{1}{2})/n$ on probability paper will show up any marked departures from normality; these points should lie roughly on the line $y = x$. Such a plot is also useful for detecting outliers or "dubious" observations. These show up as points that are well away from the general linear trend indicated by the other points. However, such points should be treated with caution. It has been suggested that an observation should be rejected as an outlier only if there is strong nonstatistical evidence that it is abnormal. For example, the measuring device may have been faulty when the particular observation was made or a number may have been copied or punched on a card incorrectly. Sometimes a "peculiar" point is more important than the rest of the plot because it may indicate serious shortcomings in the model. For example, it may correspond to an extreme value of one of the x_j's, thus suggesting a change in the model as one moves away from the usual experimental range. It should be noted that, when n is small, it is almost impossible to tell whether or not a single observation is a true outlier.

b PLOT RESIDUAL VERSUS FITTED VALUE

Three common defects may be revealed by plotting d_i against $\hat{Y}_i$.

(1) Outliers: a few of the residuals may be much larger in absolute magnitude than all the others. One test procedure would be to reject the most extreme residual, say, d_{max}, if $|d_{max}|>C$, where C is to be determined. For $n>20$, a rough and ready rule would be to reject any residual for which $|d_i|>3$. The joint distribution of the $\{d_i\}$ is known (Ellenberg [1973]) so that, theoretically, the distribution of $|d_{max}|$ could be found and C determined; an empirical solution of this problem for the straight-line case is given by Tietjen et al. [1973]. The marginal distribution of d_i is also known (Beckman and Trussell [1974]). However, an approximate test based on the maximum of the normed residuals $|e_i|/(\mathbf{e}'\mathbf{e})^{1/2}$ is available (see Stefansky [1971, 1972], Goldsmith and Boddy [1973], and, in particular, Williams [1973]).

(2) Progressive change in the variance: if the variance of ε_i is constant then we would expect the variability of the residuals to be fairly uniform. This situation is indicated by a plot like Fig. 6.1*a* which roughly represents a "band" of uniform width. However, if the plot is wedge-shaped like Fig. 6.1*b*, then there is a strong indication that

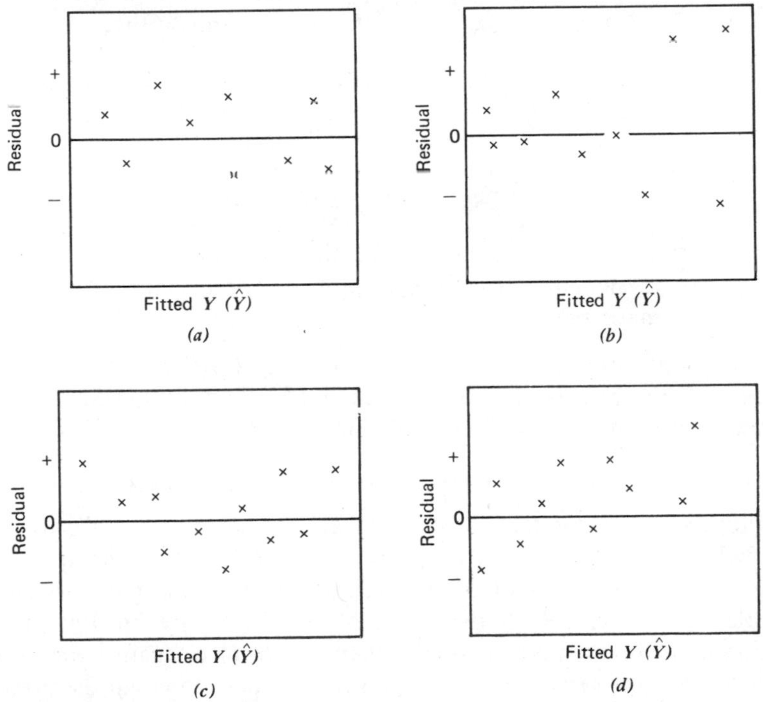

Fig. 6.1 Possible defects detected by residual plots: (*a*) Satisfactory plot; (*b*) variance increases with Y; (*c*) presence of curvature indicating an inadequate model; (*d*) linear trend indicating incorrect calculation.

the variance of ε_i is increasing with i. For example, this would be the case if the observations were taken in chronological order and the variance increased with time. Further, if the error is multiplicative rather than additive, that is,

$$Y = \theta(1+\varepsilon') = \theta + \varepsilon, \tag{6.47}$$

say, where $E[\varepsilon'] = 0$ and $\text{var}[\varepsilon'] = \sigma^2$, then the variance of ε $(=\sigma^2\theta^2)$ will vary with θ, the mean of Y.

Sometimes the variance can be stabilized by a suitable transformation (see Sections 6.7 and 7.1). Alternatively we can carry out a weighted least squares analysis. An iterative method for estimating weights when $\text{var}[\varepsilon]$ is a function of $E[Y]$ is given by Hill and Box [1974]; a robust method of weighting based on the residuals is described by Beaton and Tukey [1974].

(3) Inadequacy of the model: a curved plot like that of Fig. 6.1*c* indicates that the model is inadequate. For example, suppose that the model fitted is $Y_i = \beta_0 + \beta_1 x_{i1} + \varepsilon_i$ when the true model is actually $E[Y_i] = \beta_0 + \beta_1 x_{i1} + \beta_2 x_{i2}$. Then it is readily shown that

$$\begin{aligned} E[e_i] &= E[Y_i - \hat{Y}_i] \\ &= E[Y_i - \hat{\beta}_0 - \hat{\beta}_1 x_{i1}] \\ &= \beta_2(x_{i2} + g x_{i1} + h), \end{aligned} \tag{6.48}$$

where g and h are functions of the x_{ij}. Since $\hat{Y}_i$ and $E[e_i]$ both change systematically with x_{i1}, a plot of e_i or d_i against $\hat{Y}_i$ also shows a systematic pattern.

Finally we note that a linear trend like Fig. 6.1*d* indicates something wrong with the calculations; as pointed out by Draper and Smith [1966], we must have $\text{cov}[e_i, \hat{Y}_i] = 0$ [see Equation (6.42)].

C PLOT RESIDUAL AGAINST AN OMITTED FACTOR (E.G., TIME)

In practice any factor that is likely to affect the response Y should be included as an x variable in the regression model. However, if a likely factor is overlooked, a plot of the residual against this factor (which is possible, of course, only if the levels of the factor are known) may be illuminating. For example, a time sequence plot of d_i against time order, which is often the same as a plot of d_i versus i, may show up the presence of any correlation between time-consecutive ε_i (in Fig. 6.2 we have positive and negative correlations, respectively), or may indicate that the variance

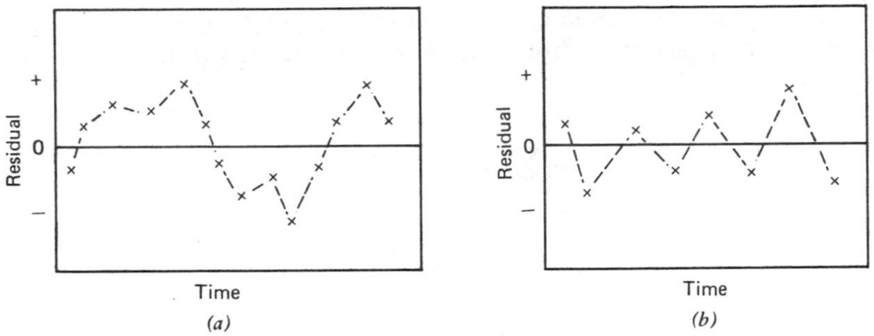

Fig. 6.2 Residual plots showing correlation between time-consecutive ε_i. (*a*) Positive correlation; (*b*) negative correlation.

changes with time (e.g., Fig. 6.1*b* but with the *x* axis representing time). If the plot looks like a band of uniform width but with a linear or curvilinear trend as in Fig. 6.3, then we would need to include in the model linear or nonlinear terms in time. In this situation it would also be helpful to plot Y_i and $\hat{Y}_i$ against time to see how close the fitted regression follows the observed response. One other useful plot consists of dividing up time-ordered residuals into consecutive pairs and plotting one member of the pair against the other.

There are a number of tests available for testing for correlation in a time series. For example, the simplest test is the so-called runs test based on the sequence of signs of the time-ordered residuals (see Brunk [1965: p. 354] for an excellent discussion), though this test is only approximate as the residuals are slightly correlated. However, perhaps the most popular test for serial correlation is the *d*-test proposed by Durbin and Watson [1950, 1951, 1971]; this test is described below.

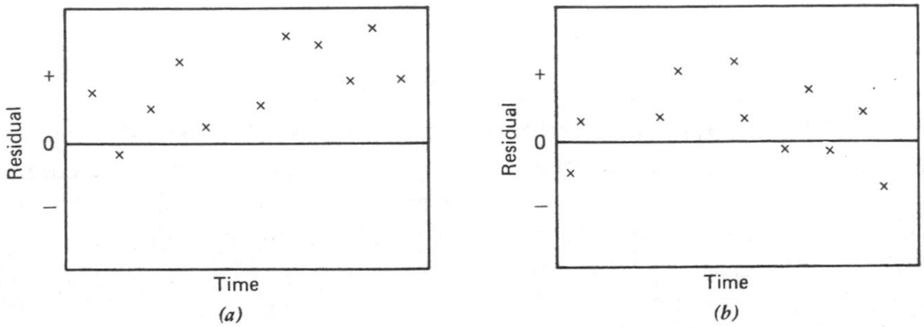

Fig. 6.3 Residual plots showing the presence of a time trend in the model. (*a*) Linear trend; (*b*) curvilinear trend.

Suppose that the ε follow a first-order autoregressive model; that is, $\varepsilon_i = \rho\varepsilon_{i-1} + \delta_i$, where the δ_i are independently and identically distributed as $N(0,\sigma^2)$. Let

$$D = \frac{\sum_{i=2}^{n} (e_i - e_{i-1})^2}{\sum_{i=1}^{n} e_i^2} = \frac{\mathbf{e}'\mathbf{A}\mathbf{e}}{\mathbf{e}'\mathbf{e}}, \tag{6.49}$$

say. Then Durbin and Watson [1971] show that the critical region $D < d_\alpha$ for testing the null hypothesis $H_0: \rho = 0$ against the one-sided alternative $H_1: \rho > 0$ has certain optimal properties; for example, it is the locally most powerful invariant critical region. Unfortunately, when H_0 is true, the null distribution of D depends on the data matrix $\mathbf{X}$ so that d_α has to be specially computed for each $\mathbf{X}$. However, Durbin and Watson give several approximate procedures which seem to work very well in practice. In the first instance they proved, in their 1950 paper, that $D_L \leqslant D \leqslant D_U$, where the distributions of D_U and D_L do not depend on $\mathbf{X}$; the significance points of these distributions are tabulated in their 1951 paper for different n and k' $(=p-1)$, and in Koerts and Abrahamse [1969: pp. 176–178]. They also showed that, when H_0 is true, $R = \frac{1}{4}D$ can be satisfactorily approximated by a beta random variable with the same mean and variance; that is, the null density function of R is approximately

$$f(r) = \frac{r^{p_0-1}(1-r)^{q_0-1}}{B(p_0, q_0)}, \qquad 0 \leqslant r \leqslant 1, \tag{6.50}$$

where

$$p_0 + q_0 = \frac{E[D]\{4 - E[D]\}}{\operatorname{var}[D]} - 1,$$

$$p_0 = \tfrac{1}{4}(p_0 + q_0)E[D],$$

and $E[D]$ and $\operatorname{var}[D]$ are given by equations 3.1 to 3.4 in their 1971 paper. On the basis of these approximations they suggest the following procedure: let d be the observed value of D, let α be the size of the test, and let $d_{L\alpha}$ and $d_{U\alpha}$ be the lower tail α significance points for the distributions of D_L and D_U, respectively. If $d < d_{L\alpha}$, reject H_0; if $d > d_{U\alpha}$, "accept" H_0; and if $d_{L\alpha} \leqslant d \leqslant d_{U\alpha}$, evaluate

$$\int_0^d f(r)\,dr$$

numerically and accept or reject H_0 according as this integral is greater or less than α. (Package computer programs are generally available for computing this beta distribution function.)

To test for negative correlation, that is, $H_1: \rho < 0$, we simply use the statistic $4 - D$; the quantity $4 - d$ may now be treated as though it is the observed value of a D statistic to be tested for positive correlation. Two-sided tests for $H_1: \rho \neq 0$ are obtained by combining the two one-sided tests and using a significance level of $\frac{1}{2}\alpha$ in each case.

Durbin and Watson [1971: p. 18] give one other approximate test procedure based on the critical value $d_\alpha = a + bd_{U\alpha}$, where a and b are calculated so that D and $a + bD_U$ have the same mean and variance. In the same paper they also describe briefly a number of other test statistics that have null distributions that are functionally independent of $\mathbf{X}$. The so-called BLUS test (Koerts and Abrahamse [1969]) belongs to this category, and this is an exact test based on another type of residual, the BLUS residual; unfortunately BLUS residuals require considerable computation.

Another exact test, based on recursive-type residuals, has been proposed by Phillips and Harvey [1974]. This test appears to compare favorably with the BLUS test, though both tests are less powerful than the full Durbin and Watson procedure described above.

A useful graphic procedure for detecting serial correlation, which uses the cumulated periodogram based on the residuals e_i, has been proposed by Durbin [1969]. The reader is referred to his paper for details and further test procedures based on the periodogram.

d PLOT RESIDUAL AGAINST EACH REGRESSOR

These plots are useful for detecting a curvilinear relationship with the variable x_j; it may be appropriate to include a term x_{ij}^2 in the original model (or transform x_{ij} to $\log x_{ij}$, for example). To illustrate this suppose, for example, that $x_{i1} = x_i$ and $x_{i2} = x_i^2$ in (6.48); then $E[e_i] = \beta_2(x_i^2 + gx_i + h)$. Hence for the common situation $x_1 < x_2 < \cdots < x_n$ we see that the residuals e_i from the straight-line model will be close to the parabola $y = \beta_2(x^2 + gx + h)$; for $\beta_2 > 0$ the plot of e_i (or d_i) versus x_i will be similar in shape to Fig. 6.1*c*. We note that there should be no linear trend as the sample covariance of the pairs (d_i, x_{ij}) is zero [Equation (6.43)].

Our linear model postulates that there is no interaction between the regressors so that the effect of changing one regressor is independent of the values taken by any other. To check this assumption we can plot d_i against the product $x_{ij}x_{ik}$; if such a term is needed in the model, the residual tends to be correlated with the product and there is a trend in the plot.

The above plots also show up any marked changes in variance.

Finally, a word of caution. Plots of Y versus each x_j are seldom helpful and can be completely misleading (e.g., Daniel and Wood [1971: p. 53]).

e PLOT x_j AGAINST x_k

If two regressors x_j and x_k are highly correlated, then generally it is not necessary to include both variables in the original regression. The inclusion of one of the variables means that the other is effectively taken care of at the same time.

Other types of plots between regressors, along with some useful references, are given by Anscombe [1973].

6.6.3 *Statistical Tests Based on Residuals*

All the above plots have associated statistical tests, and these are described in Anscombe and Tukey [1963], Anscombe [1961, 1967], Shapiro and Wilk [1965], and Andrews [1971a]. Other tests, based on transformed residuals, are described briefly in Section 6.6.5.

Cox [1968] points out that although formal tests of significance for the adequacy of a model are valuable, they do need correct interpretation. For example, a very significant lack of fit means there is decisive evidence of systematic departures from the model, yet the model may account for enough of the variation to be very valuable (Nelder [1968]). A nonsignificant test result means that the aspect of the model tested is reasonably consistent with the data; nevertheless there may be other reasons for regarding the model as inadequate. However, there seems to be a general consensus that the plots are more informative so that further testing may be somewhat unnecessary. On the other hand, some skill is needed in interpreting probability plots; plots based on small samples can be misleading (see Appendix C). The article by Feder [1974] is an interesting example of the interplay between plot and test.

6.6.4 *Partial Residual Plots*

Suppose we wish to examine more closely the relationship between the residuals and the regressor x_j. One way of doing this is to plot the "partial residual"

$$\begin{aligned} e_i' &= Y_i - \hat{\beta}_0 - \hat{\beta}_1 x_{i1} - \cdots - \hat{\beta}_{j-1} x_{i,j-1} - \hat{\beta}_{j+1} x_{i,j+1} - \cdots - \hat{\beta}_{p-1} x_{i,p-1} \\ &= e_i + \hat{\beta}_j x_{ij} \end{aligned} \tag{6.51}$$

against x_{ij} for $i = 1, 2, \ldots, n$ (Ezekiel and Fox [1959] and, in particular Larsen and McCleary [1972]). Such a plot essentially examines the relationship between Y and x_j after the effect of the other regressors on Y has been removed.

An interesting property of the plot is that if we consider the linear regression, through the origin, of e_i' on x_{ij}, then the least squares estimate of the slope turns out to be $\hat{\beta}_j$, that is, $\Sigma_i(e_i' - \beta_j x_{ij})^2$ is minimized when $\beta_j = \hat{\beta}_j$. A slope of $\hat{\beta}_j$ rather than zero, as in the usual plot of d_i versus x_{ij}, allows the experimenter to assess the extent and the direction of linearity in relationship to outliers, heterogeneity of variance, and, in particular, nonlinearity. Also a partial residual plot generally indicates more precisely how to transform Y to achieve linearity of the plot; a good example of this is pictured in Larsen and McCleary [1972: p. 787].

An obvious question to ask is, should we standardize the e_i''s so that they have approximately unit variances? Since $\Sigma_i(e_i' - \hat{\beta}_j x_{ij})^2 = \Sigma_i e_i^2$, an unbiased estimate of var$[\hat{\beta}_j]$ obtained from fitting a straight line to the partial residual plot is [cf. (4.15) in Section 4.1.4]

$$v_{1j} = \frac{\sum_i e_i^2}{(n-2)\sum_i (x_{ij} - \bar{x}_{.j})^2},$$

while the actual minimum variance unbiased estimate obtained when fitting the complete model is [cf. (11.48) in Section 11.7.1]

$$v_{2j} = \frac{\sum_i e_i^2}{\left\{(n-p)(1-R_j^2)\sum_i (x_{ij} - \bar{x}_{.j})^2\right\}},$$

where R_j is the multiple correlation between x_j and the other regressors. Because $v_{1j} < v_{2j}$, the experimenter tends to (visually) overestimate both the stability of $\hat{\beta}_j$ and the importance of x_j in terms of predicting Y, by examining the partial plot. This overestimation is not serious unless R_j^2 is large. However, if we scale the partial residuals to correct this defect the resulting plot will exaggerate any nonlinearity that is present. Therefore, following Larsen and McCleary [1972], it seems preferable to use unscaled partial residuals.

Since we usually fit β_0 in our regression model we can also plot the modified residual

$$\begin{aligned} e_i'' &= Y_i - \bar{Y} - \sum_{r \neq j} \hat{\beta}_r (x_{ir} - \bar{x}_{.r}) \\ &= e_i + \hat{\beta}_j (x_{ij} - \bar{x}_{.j}) \\ &= e_i + C_{ij} \qquad (6.52) \end{aligned}$$

against x_{ij} for each $i=1,2,\ldots,n$. The term C_{ij} is called the *component effect* of x_j on Y_i (Daniel and Wood [1971: Section 7.4]), and Wood [1973] calls the plot a component-plus-residual plot; examples are given in Wood's paper.

It should be noted that the above partial residual plots (or component-plus-residuals plots) are meant to supplement the usual residual plots and not replace them.

6.6.5 Transformed Residuals

Since $\mathbf{e}=(\mathbf{I}_n-\mathbf{P})\boldsymbol{\varepsilon}$ and $\mathfrak{D}[\mathbf{e}]\ (=\sigma^2(\mathbf{I}_n-\mathbf{P}))$ is of rank $n-p$, it is theoretically possible to transform the n mathematically independent residuals e_i into $n-p$ orthogonal functions of the ε_i; that is, there exists an $(n-p)\times n$ matrix $\mathbf{C}$ of rank $n-p$ such that $\mathbf{f}=\mathbf{C}\boldsymbol{\varepsilon}\sim N_{n-p}(\mathbf{0},\sigma^2\mathbf{I}_{n-p})$. Here $\mathbf{C}$ satisfies the equations (Putter [1967])

$$\mathbf{CC}'=\mathbf{I}_{n-p} \qquad \text{and} \qquad \mathbf{CX}=\mathbf{O}, \tag{6.53}$$

so that $\mathbf{C}\boldsymbol{\varepsilon}=\mathbf{Ce}$. One method of choosing $\mathbf{C}$, commonly called the BLUS procedure (Theil [1965, 1968], Koerts [1967], and, in particular, Abrahamse and Koerts [1971, Chapter 3]), has received some attention in the literature. With this particular choice of $\mathbf{C}$ the f_i, called the Best Linear Unbiased Scaled (BLUS) residuals, have certain optimal properties (Grossman and Styan [1972]). Another method of deriving an $\mathbf{f}$, which is close to being BLUS (Golub and Styan [1974]), is described in Section 11.2.4 [cf. Equations (11.20) and (11.22)]; there $\mathbf{f}=\mathbf{t}$ and $\mathbf{C}=\mathbf{Q}'_{n-p}$.

Having found a set of $(n-p)$ transformed residuals f_i which are independently and identically distributed as $N(0,\sigma^2)$, we can use these new residuals to investigate indirectly any departures from the underlying distributional assumptions of $\boldsymbol{\varepsilon}$. For example, we can carry out various standard tests for normality such as the seven tests listed by Dyer [1974]: Kolmogorov-Smirnov (K_n or D_n), Cramèr-Smirnov (W_n^2), Anderson-Darling (A_n^2), Watson (U_n^2), Kuiper (V_n), modified Kolmogorov (E_n), and Wilk-Shapiro (W). Kuiper's test (Kuiper [1960]) is discussed in some detail by Koerts and Abramhamse [1969], who prefer this test to D_n. However, W, due to Shapiro and Wilk [1965] (see, in particular, Hahn and Shapiro [1967: p. 295]), seems to generally provide the most powerful test for a reasonable class of alternatives (Dyer [1974], Shapiro et al. [1968], and Huang and Bolch [1974]): a large sample version of this test is given by Shapiro and Francia [1972]. Other references of interest relating to this problem are Putter [1967] and Kowalski [1970]. Another approach to the problem is to transform the f_i residuals in such a way as to eliminate σ^2; for example, Csorgo et al. [1973] provide a number of exact tests for normality by transforming the residuals to independent t variables.

BLUS residuals can be used to test for specification errors (Ramsey [1969]) and for serial correlation (Koerts and Abrahamse [1969]). However, in the latter case the Durbin-Watson test of Section 6.6.2c is more powerful (Durbin and Watson [1971]).

Although any transformation **C** satisfying (6.53) provides suitable residuals, most transformations are not meaningful in the sense that detailed inferences about the f_i are not easy to interpret. For example, if one of the f_i's appears to be an outlier, what does this mean in terms of the original model? Which observation Y_i is responsible? One particular transformation that endeavors to identify each f_i with a particular design point is described by Hedayat and Robson [1970] and Brown et al. [1975]. The latter authors give methods for testing the constancy of $\boldsymbol{\beta}$ with respect to the data points.

6.7 TRANSFORMING THE DATA

If the usual normality assumptions do not appear to be satisfied, a nonlinear transformation of the data may improve matters. For example, if theory suggests an approximate relationship $y=\alpha e^{\beta x}$, then $\log y=\log\alpha+\beta x$ and we would expect an approximate linear relationship between $\log y$ and x. However, when carrying out such transformations we must pay careful attention to the "error" term. If, for example, the error is multiplicative so that

$$\begin{aligned}Y&=\alpha e^{\beta x}(1+\varepsilon_0)\\&=\alpha e^{\beta x}+\varepsilon,\end{aligned}$$

where $E[\varepsilon_0]=0$ and $\mathrm{var}[\varepsilon_0]=\sigma_0^2$, then $\mathrm{var}[\varepsilon]$ $(=\sigma_0^2\{E[Y]\}^2)$ varies with $E[Y]$. However, if we take logarithms then

$$\begin{aligned}\log Y&=\log\alpha+\beta x+\log(1+\varepsilon_0)\\&=(\log\alpha+\alpha_0)+\beta x+\varepsilon,\end{aligned}$$

where $E[\log(1+\varepsilon_0)]=\alpha_0$, $E[\varepsilon]=0$, and $\mathrm{var}[\varepsilon]=\sigma^2$, say. If ε_0 is normally distributed then ε is not, and vice versa.

On the other hand, if the error in Y is additive so that

$$\begin{aligned}Y&=\alpha e^{\beta x}+\varepsilon_0\\&=\alpha e^{\beta x}\left\{1+\frac{\varepsilon_0}{E[Y]}\right\}\\&=\alpha e^{\beta x}\{1+\eta_0\},\end{aligned}$$

say, then

$$\log Y = \log\alpha + \beta x + \log(1+\eta_0),$$

and the variance of $\log(1+\eta_0)$ varies with $E[Y]$. Thus in the first case taking logarithms stabilizes the variance of the error, while in the second case the variance of the error becomes dependent on x through $E[Y]$. The effect of any transformation can be examined by plotting the residuals once again.

A useful family of transformations is the following:

$$\begin{aligned} y^{(\lambda)} &= \quad y^{\lambda} \quad \lambda \neq 0 \\ &= \log y \quad \lambda = 0, \quad y > 0. \end{aligned} \tag{6.54}$$

This particular family, which was studied in detail by Tukey [1957] for $|\lambda| \leqslant 1$, contains the well-known log, square-root, and inverse transformations. To avoid a discontinuity at $\lambda = 0$, Box and Cox [1964] considered the modification

$$\begin{aligned} y^{(\lambda)} &= \frac{y^{\lambda}-1}{\lambda} \quad \lambda \neq 0 \\ &= \log y \quad \lambda = 0, \end{aligned} \tag{6.55}$$

which is essentially identical to (6.54) when the regression model contains a constant term β_0 (Schlesselman [1971]). They assume that, for some λ, the transformed observations $\mathbf{Y}^{(\lambda)} = [(Y_i^{(\lambda)})]$ satisfy the normal theory assumptions, namely, $\mathbf{Y}^{(\lambda)} \sim N_n(\mathbf{X}\boldsymbol{\beta}, \sigma^2 \mathbf{I}_n)$. Under this assumption the likelihood function for the original observations $\mathbf{Y}$ is

$$(2\pi\sigma^2)^{-(1/2)n} \exp\left\{ -\frac{1}{2\sigma^2} (\mathbf{y}^{(\lambda)} - \mathbf{X}\boldsymbol{\beta})'(\mathbf{y}^{(\lambda)} - \mathbf{X}\boldsymbol{\beta}) \right\} J \tag{6.56}$$

where

$$J = \prod_{i=1}^{n} \left| \frac{dy_i^{(\lambda)}}{dy_i} \right| = \prod_{i=1}^{n} y_i^{\lambda-1},$$

the absolute value of the Jacobian. For λ fixed, (6.56) is the likelihood corresponding to a standard least squares problem, except for the constant factor J. From Section 4.1.2 the maximum value of this likelihood function is $(2\pi\hat{\sigma}^2)^{-(1/2)n} e^{-(1/2)n} J$, where

$$n\hat{\sigma}^2 = \mathbf{y}^{(\lambda)\prime}\left(\mathbf{I}_n - \mathbf{X}(\mathbf{X}'\mathbf{X})^{-1}\mathbf{X}'\right)\mathbf{y}^{(\lambda)} = \mathrm{RSS}(\lambda; \mathbf{y}),$$

say. Hence, apart from a constant, the maximum log likelihood is

$$L_{\max}(\lambda) = -\tfrac{1}{2}n\log\{\mathrm{RSS}(\lambda;\mathbf{y})\} + \log J$$
$$= -\tfrac{1}{2}n\log\{\mathrm{RSS}(\lambda;\mathbf{z})\},$$

where

$$z_i^{(\lambda)} = \frac{y_i^{(\lambda)}}{J^{1/n}}$$
$$= \frac{y_i^{(\lambda)}}{\dot{y}^{\lambda-1}}, \qquad \lambda \neq 0$$
$$= \dot{y}\log y_i, \qquad \lambda = 0,$$

and $\dot{y}$ is the geometric mean of the y_i; that is, $\dot{y} = (\Pi_i y_i)^{1/n}$. Box and Cox suggest plotting $L_{\max}(\lambda)$ against λ for a trial series of values and reading off the maximizing value $\hat{\lambda}$. A more accurate value of $\hat{\lambda}$ can be obtained by solving the equations $dL_{\max}(\lambda)/d\lambda = 0$ (see Equation (12) in their paper, or Equation (9) in Schlesselman [1971]); some properties of $\hat{\lambda}$ are discussed further in Draper and Cox [1969]. Box and Cox also discuss the estimation of λ from a Bayesian viewpoint.

To test the suitability of a specific transformation $\lambda = \lambda_0$, the authors suggest using the likelihood ratio test statistic $-2\{L_{\max}(\lambda_0) \quad L_{\max}(\hat{\lambda})\}$, which is asymptotically distributed as χ_1^2 when the null hypothesis $H: \lambda = \lambda_0$ is true. Andrews [1971b] proposed an "exact" test for H, though the empirical study of Atkinson [1973] suggests that this test is less powerful than the likelihood ratio test.

An approximate $100(1-\alpha)\%$ confidence region for the true value of λ is the set of all λ satisfying

$$L_{\max}(\hat{\lambda}) - L_{\max}(\lambda) \leqslant \tfrac{1}{2}\chi_{1,\alpha}^2,$$

where $\mathrm{pr}[\chi_1^2 \geqslant \chi_{1,\alpha}^2] = \alpha$.

In practice we may wish to transform the x-variables as well as y, as in the case of the model $y = \alpha e^{\beta x}$ discussed above. The appropriate procedure then may be to apply a transformation suggested by prior knowledge and then use the above theory, which involves only y, to see if any further modifications are needed. The above method can be combined with Box and Tidwell's [1962] procedure for transforming just the regressor variables. If y can be negative then we work with $y + \lambda_2$, where λ_2 is chosen such that $y + \lambda_2 > 0$.

Two interesting applications of the above theory are given by Box and Cox [1964] and their paper should be consulted for further details. A further example, involving the test of $\lambda=0$ ($\log Y$ versus x), is given by Sclove [1972]. The question of deriving transformations for the case of a single x-variable is discussed in Section 7.1.

Sometimes it is helpful to have a simple transformation which achieves normality and homogeneity of variance but without necessarily obtaining linearity: a method for doing this is described by Wood [1974].

MISCELLANEOUS EXERCISES 6

1. Suppose that the postulated regression model is

$$E[Y]=\beta_0+\beta_1 x$$

when, in fact, the true model is

$$E[Y]=\beta_0+\beta_1 x+\beta_2 x^2+\beta_3 x^3.$$

If we use observations of Y at $x=-3,-2,-1,0,1,2,3$ to estimate β_0 and β_1 in the postulated model, what bias will be introduced in these estimates?

Draper and Smith [1966: p. 84]

2. Fill in the details of the proof of Theorem 6.2.
3. Show that the theory of Section 6.3.2 can also be applied to the case $H:\mathbf{A}\boldsymbol{\beta}=\mathbf{c}$ where $\mathbf{c}\neq\mathbf{0}$.
4. Verify Equation (6.48) in Section 6.6.2.
5. Consider the full rank regression model

$$Y_i=\beta_0+\beta_1 x_{i1}+\cdots+\beta_k x_{ik}+\varepsilon_i \qquad (i=1,2,\ldots,n),$$

and suppose we wish to test $H:\beta_1=\beta_2=\cdots=\beta_k=0$. Assuming that H is true, find an approximate expression for $E[Z]$, where $Z=\frac{1}{2}\log F$, in terms of the diagonal elements of $\mathbf{X}(\mathbf{X}'\mathbf{X})^{-1}\mathbf{X}'$.

6. Suppose we wish to test the hypothesis H that the means of two populations are equal, given n_i observations from the ith population $(i=1,2)$. Assuming that the populations have the same variance and kurtosis (γ_2), find approximate expressions for $E[Z]$ and var$[Z]$ on the assumption that H is true. Show that to the order of approximation used, these expressions are independent of γ_2 if $n_1=n_2$.

CHAPTER 7

Straight-Line Regression

7.1 INTRODUCTION

The simplest regression model is that of a straight line, namely,

$$Y_i = \beta_0 + \beta_1 x_i + \varepsilon_i \qquad (i = 1, 2, \ldots, n),$$

where the ε_i are independently and identically distributed as $N(0, \sigma^2)$. If the regressor X is also random then we can argue conditionally on the x-values actually observed, provided the x_i by themselves give no information about $\boldsymbol{\beta}$. Our model is then effectively

$$E[Y_i | X_i = x_i] = \beta_0 + \beta_i x_i.$$

When X and Y are both random we also have the model

$$E[X_i | Y_i = y_i] = \beta_0' + \beta_1' y_i,$$

and we would choose this model if X is to be predicted from Y. The question of a random regressor is discussed further in Section 7.7.

Although a fitted regression line may be useful as a means of summarizing the bivariate data, it is desirable that the relation should be stable and reproducible (Cox [1968]). By stability we mean that the linear "form" is preserved when the experiment is repeated under different conditions, that is, (1) the same regression equation holds even though other aspects of the data change; or (2) parallel regression lines are obtained; or (3) satisfactory regression lines are always obtained but with different positions and slopes. Tests for (1) and (2) are described in Section 7.5.

Frequently the regression of Y on x is nonlinear and we would seek a suitable transformation of Y, and possibly x, such that the "transformed

Y" satisfy the usual assumptions, and the regression of "transformed Y" on "transformed x" is linear (cf. Section 6.7). A preliminary plot of the data (x_i, y_i), called a scatter diagram, generally gives some idea as to the type of model that might be appropriate. The collection of theoretical curves graphed in Daniel and Wood [1971: pp. 20–24] is useful for this purpose. In choosing a model one would naturally take into account any theoretical analysis, including dimensional analysis, of the system and any limiting behavior (for example, $y \to 0$ as $x \to 0$, or $y \to c$ as $x \to \infty$, etc). A number of biological examples of this are given by Seber [1973: 128, 141, 145, 150–151, 254, 260–266, 276ff., 297, 325ff.]; however, the following example from Cox [1968] is illuminating.

Suppose we are studying the relationship between breaking load (Y) and diameter (x) for a certain fiber, where x has relatively little variation. Then since many curves are approximately linear over a narrow range of x values we might not be surprised to find that a linear regression of Y on x gives a reasonable fit (model I). However, a linear regression of $\log Y$ on $\log x$ may also fit the data equally well (model II). Which model do we choose? Obviously $y \to 0$ as $x \to 0$, and a reasonable assumption might be that the breaking load is proportional to the area of cross-section, that is, $y \propto x^2$. One would then choose model II because (1) it permits easier comparison with the theoretical model $y = \beta x^2$ (or $\log y = \log \beta + 2 \log x$), (2) it ensures that $y \to 0$ as $x \to 0$ (in model I we would have to fit a straight line, $y = \beta_1 x$, say, through the origin—a rather strong restriction on the model), and (3) the slope of the regression line is a dimensionless power, that is, a constant that does not depend on the units used for measuring load and diameter. If we wished we could carry out a significance test of model II versus model I using the methods of Cox [1961, 1962]. Alternatively we could consider a comprehensive model which contains both models as special cases. For example, we could assume that there exist λ_1 and λ_2 such that the regression of

$$Y^{(\lambda_2)} = \frac{Y^{\lambda_2} - 1}{\lambda_2} \qquad \text{on} \qquad x^{(\lambda_1)} = \frac{x^{\lambda_1} - 1}{\lambda_1}$$

is linear and the $Y_i^{(\lambda_2)}$ satisfy the usual normal theory assumptions; when $\lambda = 0$ the power transformation becomes a log transformation [cf. Equation (6.55)]. All the parameters, including λ_1 and λ_2, can now be estimated and tested by maximum likelihood (Box and Tidwell [1962], Box and Cox [1964]).

In practice such a pair of transformations may not exist. However, in choosing between linearizing the regression and stabilizing the variance the first usually has preference. For example, in Section 6.7 it was shown that taking logarithms for the "additive error" model $Y = \alpha e^{\beta x} + \varepsilon$ leads to a linear model in which the error variance depends on $E[Y]$. In this case a weighted least squares analysis (cf. Section 7.4) can be carried out, though the changes in variance have to be quite substantial before there is an appreciable gain in precision in the least squares estimates (Cox [1968]).

When we are looking for a suitable transformation, various types of graph paper can be used. For example, if we are comparing log transformations such as $\log y$, $\log x$, or $\log y$ and $\log x$, we can use specially ruled graph paper with one or two log scales. We can first of all plot y versus x, draw a freehand curve, and then select an appropriate transformation by simply plotting a few well-selected points from this curve (Smith [1972]). Occasionally we know the distribution of Y; for example, the log transformation is commonly used for Poisson count data.

The adequacy of a particular model can be examined using the residual plots described in Section 6.6. Summing up, the main types of departure to look for are as follows:

(1) the presence of outliers (an approximate test procedure based on the maximum Studentized residual is described by Tietjen et al. [1973]);
(2) nonlinear regression, detected by plotting the Studentized residual d_i against x_i and obtaining a curved relationship;
(3) nonconstancy of variance, detected by plotting d_i against x_i, or d_i against $\hat{Y}_i$;
(4) correlations between different ε_i's, detected by the Durbin–Watson test for serial correlation, or by plotting consecutive pairs of time-ordered residuals; and
(5) nonnormality of the distribution of the ε_i's, detected by plotting $d_{(i)}$, the ith ordered residual (Appendix C), against the expected order statistic $\Phi^{-1}((i-\frac{1}{2})/n)$ from the standard (unit) normal distribution.

7.2 CONFIDENCE INTERVALS AND BANDS

7.2.1 *Confidence Intervals for Slope and Intercept*

From Section 4.1.4 we have

$$(\mathbf{X}'\mathbf{X})^{-1} = \frac{1}{n\sum(x_i-\bar{x})^2}\begin{pmatrix} \sum x_i^2 & -n\bar{x} \\ -n\bar{x} & n \end{pmatrix}, \tag{7.1}$$

$$\hat{\beta}_0 = \bar{Y} - \hat{\beta}_1\bar{x},$$

$$\hat{\beta}_1 = \frac{\sum(Y_i-\bar{Y})(x_i-\bar{x})}{\sum(x_i-\bar{x})^2} = \frac{\sum Y_i(x_i-\bar{x})}{\sum(x_i-\bar{x})^2},$$

and

$$S^2=\frac{1}{n-2}\left\{\sum\left(Y_i-\bar{Y}\right)^2-\hat{\beta}_1^2\sum\left(x_i-\bar{x}\right)^2\right\}.$$

Using the maximum modulus method of Section 5.1.1. [Equation (5.6)] with $\mathbf{a}_1'=(1,0)$ and $\mathbf{a}_2'=(0,1)$, we have an *exact* overall confidence probability of $1-\alpha$ for the following confidence intervals for β_0 and β_1:

$$\hat{\beta}_0\pm u_{2,n-2,\rho}^{\alpha}S\left\{\frac{\sum x_i^2}{n\sum\left(x_i-\bar{x}\right)^2}\right\}^{1/2}$$

and

$$\hat{\beta}_1\pm u_{2,n-2,\rho}^{\alpha}S\left\{\frac{1}{\sum\left(x_i-\bar{x}\right)^2}\right\}^{1/2},$$

where

$$\rho=\frac{-n\bar{x}}{\left(n\sum x_i^2\right)^{1/2}}.$$

Conservative intervals are obtained by setting $\rho=0$.

The two intervals can also be used for *jointly* testing a hypothesis about β_0 and a hypothesis about β_1. However, if we are interested in just a single hypothesis, say, H: $\beta_1=c$, then we use the usual t-statistic

$$T=\frac{\hat{\beta}_1-c}{S/\left\{\sum\left(x_i-\bar{x}\right)^2\right\}^{1/2}}, \tag{7.2}$$

and reject H at the α level of significance if $|T|>t_{n-2}^{(1/2)\alpha}$. This statistic can be derived directly from the fact that $\hat{\beta}_1\sim N(\beta_1,\sigma^2/\Sigma(x_i-\bar{x})^2)$ and S^2 is independent of $\hat{\beta}_1$.

7.2.2 *Confidence Interval for* $-\beta_0/\beta_1$

We now derive a confidence interval for the ratio $\phi=-\beta_0/\beta_1$ using a technique due to Fieller [1940]. Let

$$\begin{aligned}\delta&=\frac{E[\bar{Y}]}{E[\hat{\beta}_1]}\\&=\frac{\beta_0+\beta_1\bar{x}}{\beta_1}\\&=-\phi+\bar{x};\end{aligned} \tag{7.3}$$

then $E[\bar{Y}-\delta\hat{\beta}_1]=0$. Also

$$\begin{aligned}\operatorname{cov}[\bar{Y},\hat{\beta}_1]&=\operatorname{cov}[\mathbf{a}'\mathbf{Y},\mathbf{b}'\mathbf{Y}]\\&=\mathbf{a}'\mathscr{D}[\mathbf{Y}]\mathbf{b}\\&=\sigma^2\mathbf{a}'\mathbf{b}\\&=\sigma^2\frac{\sum_i(x_i-\bar{x})}{\sum_i(x_i-\bar{x})^2n}\\&=0,\end{aligned}$$

so that

$$\begin{aligned}\operatorname{var}[(\bar{Y}-\delta\hat{\beta}_1)]&=\operatorname{var}[\bar{Y}]+\delta^2\operatorname{var}[\hat{\beta}_1]\\&=\sigma^2\left\{\frac{1}{n}+\frac{\delta^2}{\sum(x_i-\bar{x})^2}\right\}\\&=\sigma^2w,\end{aligned}$$

say. Now $\bar{Y}-\delta\hat{\beta}_1$ is of the form $\mathbf{c}'\mathbf{Y}$ so that it is univariate normal, namely, $N(0,\sigma^2w)$. Also S^2 is independent of $(\hat{\beta}_0,\hat{\beta}_1)$ (Theorem 3.5 (iii), Section 3.4) and therefore of $\bar{Y}-\delta\hat{\beta}_1$ $[=\hat{\beta}_0+\hat{\beta}_1(\bar{x}-\delta)]$. Hence, by the usual argument for constructing t-variables (Section 4.1.5)

$$T=\frac{\bar{Y}-\delta\hat{\beta}_1}{S\sqrt{w}}\sim t_{n-2},$$

and a $100(1-\alpha)\%$ confidence set for δ is given by

$$T^2\leqslant(t_{n-2}^{(1/2)\alpha})^2=F_{1,n-2}^{\alpha}.$$

It transpires that this set reduces to the simple interval $d_1\leqslant\delta\leqslant d_2$, where d_1 and d_2 are the roots of the quadratic

$$d^2\left\{\hat{\beta}_1^2-\frac{S^2F_{1,n-2}^{\alpha}}{\sum(x_i-\bar{x})^2}\right\}-2d\bar{Y}\hat{\beta}_1+\left\{\bar{Y}^2-\frac{1}{n}S^2F_{1,n-2}^{\alpha}\right\}=0, \qquad (7.4)$$

if and only if the coefficient of d^2 in the above equation is positive (that is, the line is not too flat). In this case, from Equation (7.3), the corresponding interval for ϕ is $(\bar{x}-d_2,\bar{x}-d_1)$, and $\hat{\phi}=-\hat{\beta}_0/\hat{\beta}_1$ lies in this interval.

When $E[Y]=0$, $0=\beta_0+\beta_1x$ and we see that the above method gives a confidence interval for the intercept $x=-\beta_0/\beta_1$ of the regression line with the x-axis. This is a special case of inverse prediction discussed in Section 7.2.6 (with $Y_* = 0$).

A model that often arises in animal population studies (c.f. Seber [1973: 11, 128, 145–148, 150, 298–299, 325]) is the following:

$$\begin{aligned} E[Y] &= \gamma(\phi - x) \\ &= \gamma\phi - \gamma x \\ &(= \beta_0 + \beta_1 x, \text{ say}). \end{aligned}$$

In such applications we are interested in finding a confidence interval for $\phi = -\beta_0/\beta_1$ so that the above theory can be used here.

We note that $\hat{\phi}$ is the ratio of two correlated normal random variables; the exact distribution of such a ratio is given by Hinkley [1969a].

7.2.3 *Prediction Intervals and Bands*

The fitted regression line is

$$\begin{aligned} \hat{Y} &= \hat{\beta}_0 + \hat{\beta}_1 x \\ &= \overline{Y} + \hat{\beta}_1 (x - \bar{x}), \end{aligned}$$

which passes through the point $(\bar{x}, \overline{Y})$. From the general theory of Section 5.2 we see that we can use the prediction $\hat{Y}_* = \mathbf{x}_*'\hat{\boldsymbol{\beta}} = (1, x_*)\hat{\boldsymbol{\beta}}$ to obtain a $100(1-\alpha)\%$ confidence interval for $E[Y_*] = (1, x_*)\boldsymbol{\beta}$, the expected value of Y at $x = x_*$. This interval is

$$\hat{Y}_* \pm t_{n-2}^{(1/2)\alpha} S\sqrt{v_*} \tag{7.5}$$

where, from Equation (7.1),

$$\begin{aligned} v_* &= \mathbf{x}_*'(\mathbf{X}'\mathbf{X})^{-1}\mathbf{x}_* \\ &= \frac{\left\{\sum x_i^2 - 2x_* n\bar{x} + nx_*^2\right\}}{n\sum (x_i - \bar{x})^2} \\ &= \frac{\left\{\sum x_i^2 - n\bar{x}^2 + n(x_* - \bar{x})^2\right\}}{n\sum (x_i - \bar{x})^2} \\ &= \frac{1}{n} + \frac{(x_* - \bar{x})^2}{\sum (x_i - \bar{x})^2}. \end{aligned} \tag{7.6}$$

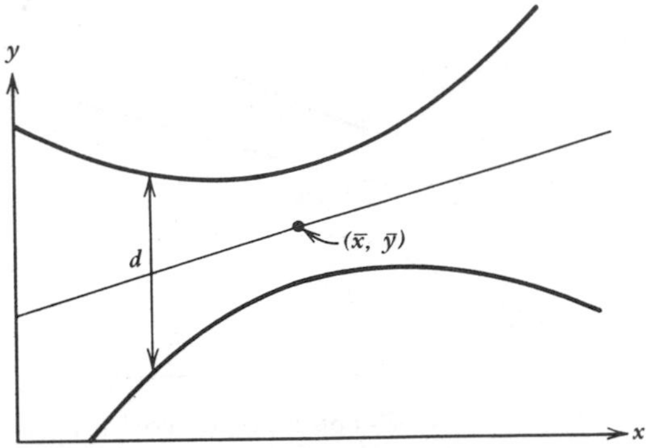

Fig. 7.1 Working–Hotelling confidence band.

(Here v_* can also be obtained directly—see Exercise 1 at the end of the chapter). We note that v_* is a minimum when $x_* = \bar{x}$; the further we are from $\bar{x}$, the wider our confidence interval.

If we require k confidence intervals then our critical constant $t_{n-2}^{(1/2)\alpha}$ in (7.5) is replaced by $t_{n-2}^{\alpha/(2k)}$, $(2F_{2,n-2}^{\alpha})^{1/2}$ and $u_{k,n-2}^{\alpha}$ for the Bonferroni, Scheffé, and maximum modulus methods, respectively. However, if k is unknown or is so large that the intervals are too wide, we can construct a confidence band for the whole regression line and thus obtain an unlimited number of confidence intervals with an overall confidence probability of at least $1-\alpha$. From Equation (5.18) this infinite band is the region between the two curves (Fig. 7.1)

$$y = \bar{Y} + \hat{\beta}_1 (x - \bar{x}) \pm \lambda S \left\{ \frac{1}{n} + \frac{(x-\bar{x})^2}{\sum (x_i - \bar{x})^2} \right\}^{1/2} \tag{7.7}$$

where $\lambda = (2F_{2,n-2}^{\alpha})^{1/2}$. This band, commonly called the Working–Hotelling confidence band (Working and Hotelling [1929]), is of variable vertical width d, d being a minimum at the point $(\bar{x}, \bar{Y})$. The intervals obtained from this band are simply the Scheffé F-intervals.

An alternative confidence band with straight sides (Fig. 7.2) has been proposed by Graybill and Bowden [1967], namely,

$$y = \bar{Y} + \hat{\beta}_1 (x - \bar{x}) \pm u_{2,n-2}^{\alpha} S \frac{1}{\sqrt{n}} \left\{ 1 + \frac{|x - \bar{x}|}{s_x} \right\} \tag{7.8}$$

where $s_x^2 = \Sigma(x_i - \bar{x})^2/n$. This band has two advantages over (7.7): (1) it is easier to graph, and (2) it has a smaller average width, though this is

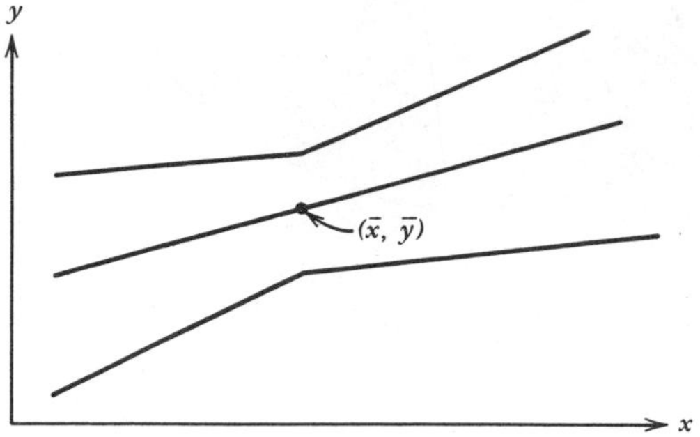

Fig. 7.2 Graybill–Bowden confidence band.

misleading since the average is taken over the whole band including extreme values of x. However, Dunn [1968] and Halperin and Gurian [1968: p. 1027] showed that, for $\alpha=0.05$, (7.7) provides narrower intervals than (7.8) when x satisfies (approximately)

$$0.1 \leqslant \frac{|x-\bar{x}|}{s_x} \leqslant 9.$$

Since, in practice, one would not expect the experimental range of $|x-\bar{x}|$ to exceed $5s_x$, the Working–Hotelling band is preferred. A similar conclusion holds for 90% confidence levels ($\alpha=0.1$). Both bands can be derived as special cases of a general procedure given by Bowden [1970] [cf. Equation (5.19) and the following discussion].

The problem of obtaining an exact confidence band for the regression line when x_* is restricted to the finite interval $[a,b]$ was first solved by Gafarian [1964]. He showed how to construct a band of uniform width 2δ, and provided appropriate tables for the case $\bar{x}=\frac{1}{2}(a+b)$ and even n. Miller [1966: p. 121] gave a useful discussion of this method and pointed out that the two conditions necessary for the use of the tables are not very restrictive: the interval $[a,b]$ is usually sufficiently ill-defined to permit adjustment so that $\bar{x}$ is the middle point, and interpolation in the tables gives approximate results for odd values of n. However, Bowden and Graybill [1966] later provided tables for *any* finite interval $[a,b]$ and even n. Their tables can also be used for computing exact trapezoidal confidence bands, which may be more appropriate than uniform width bands when $\bar{x}$ lies outside $[a,b]$.

Dunn [1968] provided a truncated modification of (7.8) which gave a conservative confidence band. Halperin et al. [1967] and Halperin and Gurian [1968] gave an exact confidence band of the form (7.7) but with a different value of λ and truncated at $x=a$ and $x=b$. However, their tables can only be used for the case $\bar{x}=\frac{1}{2}(a+b)$; λ is tabulated in Halperin et al. [1967] for different values of Q^{-1}, where

$$Q=1+\frac{(b-a)^2}{4s_x^2}.$$

Wynn and Bloomfield [1971], however, tackled the problem from a different viewpoint and provided tables (reproduced in Appendix F) for any interval $[a,b]$. One simply calculates a "standardized" version of the interval width, namely,

$$c=\frac{(b-a)s_x}{\left[\left\{s_x^2+(a-\bar{x})^2\right\}\left\{s_x^2+(b-\bar{x})^2\right\}\right]^{1/2}+s_x^2+(a-\bar{x})(b-\bar{x})} \tag{7.9}$$

and looks up the corresponding value of λ in Appendix F. When $\bar{x}=\frac{1}{2}(a+b)$ we note that $c=(b-a)/2s_x$ and $Q=1+c^2$, thus linking the tables in Halperin et al. [1967] with Appendix F. Letting $a\to-\infty$, $b\to\infty$, we have $c=\infty$ and $\lambda=(2F^{\alpha}_{2,n-2})^{1/2}$, as expected. Calculations given by Halperin and Gurian [1968] suggest that this modification of the Working–Hotelling band generally provides narrower confidence intervals than either the uniform or trapezoidal bands mentioned above. In conclusion, therefore, we recommend the general use of (7.7) but with λ obtained from Appendix F in the case $x\in[a,b]$.

Finally some mention should be made of one-sided confidence intervals. Bohrer and Francis [1972] give an (upper) one-sided analogue of (7.7), namely, (modifying their model slightly so that $x\in[a,b]$ instead of $x-\bar{x}\in[a,b]$)

$$1-\alpha=\operatorname{pr}\left[\beta_0+\beta_1 x\leqslant \bar{y}+\hat{\beta}_1(x-\bar{x})+\lambda S\left\{\frac{1}{n}+\frac{x-\bar{x}}{\sum(x_i-\bar{x})^2}\right\}^{1/2},\right.$$

$$\left.\text{all } x\in[a,b]\right],$$

where λ ($=c\#$ in their notation) is tabulated for different n, ϕ^* ($=\arctan[(b-\bar{x})/s_x]-\arctan[(a-\bar{x})/s_x]$) and α ($=1-\alpha$ in their notation). Lower one-sided intervals are obtained by reversing the inequality and replacing λ by $-\lambda$.

7.2.4 Prediction Intervals for the Response

From the general theory of Section 5.3 we can use the predictor $\hat{Y}_*$ to obtain a $100(1-\alpha)\%$ confidence interval for the random variable Y_*, namely,

$$\hat{Y}_* \pm t_{n-2}^{(1/2)\alpha} S \left(1+v_*\right)^{1/2},$$

where v_* is given by Equation (7.6). If k intervals are required at $x = x_*^{(i)}$ $(i=1,2,\ldots,k)$, then we can use

$$\hat{Y}_*^{(i)} \pm \lambda S \left(1+v_*^{(i)}\right)^{1/2} \qquad (i=1,2,\ldots,k),$$

where λ is $t_{n-2}^{\alpha/(2k)}$, $(kF_{k,n-2}^{\alpha})^{1/2}$, and $u_{k,n-2}^{\alpha}$ for the Bonferroni, Scheffé, and maximum modulus methods, respectively. However, if k is so large that the intervals are hopelessly wide, or k is unknown, then we can use the simultaneous tolerance intervals of Liebermann and Miller [1963]; this method is also described in Miller [1966: p. 123].

7.2.5 Optimal Allocation of Observations

A good prediction interval requires that we choose the x_i's such that

$$v_* = \frac{1}{n} + \frac{\left(x_* - \bar{x}\right)^2}{\sum \left(x_i - \bar{x}\right)^2}$$

is as small as possible. If the range of interest for x is scaled to be $[-1,1]$, and n is even, then it is known that the maximum value of v_* with respect to x_* over $[-1,1]$ is a minimum when half the Y observations are taken at $x=-1$ and the other half at $x=+1$. This result follows from Section 8.4 since such an allocation gives a minimax (D-optimal) design (see also Gaylor and Sweeny [1965], Herzberg and Cox [1972: p. 553]). However, this design (called D_1, say) is optimal only if we are certain that the straight line model is correct and that the variances of the ε_i are equal; it would obviously be the worst possible design if the regression was actually quadratic. Because in practice we may want to check this assumption, a more appropriate procedure might be to choose a design that would enable us to investigate the coefficient β_2 in $E[Y]=\beta_0+\beta_1 x+\beta_2 x^2$ in an "optimal" manner. For example, the design (D_2, say) that minimizes var$[\hat{\beta}_2]$ allocates $\frac{1}{4}n$ observations to each of $x=\pm 1$, and $\frac{1}{2}n$ observations to $x=0$. However, this design may lead to a rather inefficient estimate of β_1 if β_2 is in fact zero, so that a compromise might be to find D which minimizes var[$\hat{\beta}_2$] subject to a specified efficiency for estimating β_1. The efficiency of D with respect to β_1 can be defined as

$$f = \frac{\operatorname{var}\left[\tilde{\beta}_1 | D_1\right]}{\operatorname{var}\left[\tilde{\beta}_1 | D\right]}$$

where $\tilde{\beta}_1$ is the least squares estimate of β_1 assuming $\beta_2 = 0$. For symmetrical designs Atkinson [1972] shows that, for a given f, var[$\hat{\beta}_2$] is minimized when $\frac{1}{2}fn$ observations are made at $x = \pm 1$ and $(1-f)n$ observations at $x = 0$. When $f = \frac{1}{2}$, which is actually the minimum value, $D = D_2$; as f increases, f tends to 1 and D tends to D_1.

Two other solutions to the above problem have been proposed: Stigler [1971] finds a D-optimal design subject to the condition var[$\hat{\beta}_2$] $\leqslant \sigma^2 C/n$ for a prechosen C, and Atwood [1971] uses an appropriate "linear combination" of D_1 and D_2.

7.2.6 *Inverse Prediction (Discrimination)*

Suppose we wish to calibrate an instrument, say, a pressure gauge, and we know that the gauge reading is a linear function of the pressure, namely,

$$\text{"gauge reading"} = \beta_0 + \beta_1 \text{ "pressure"} + \text{"error"}$$

or

$$Y = \beta_0 + \beta_1 x + \varepsilon.$$

In order to calibrate the gauge we subject it to two or more (say, n) controlled pressures x_i $(i = 1, 2, \ldots, n)$ and note the gauge readings Y_i. Using these data we obtain the fitted equation $\hat{Y} = \hat{\beta}_0 + \hat{\beta}_1 x$ which can be used for estimating (predicting) the unknown pressure x_* for a given gauge reading Y_*. This is the inverse problem to the one considered in Section 7.2.4 of predicting Y_* for a given $x = x_*$.

A natural estimate of x_* (which is also the maximum likelihood estimate) is found by solving the fitted equation $Y_* = \hat{\beta}_0 + \hat{\beta}_1 x$, namely,

$$\hat{x}_* = \frac{Y_* - \hat{\beta}_0}{\hat{\beta}_1} = \bar{x} + \left[\frac{Y_* - \bar{Y}}{\hat{\beta}_1}\right], \tag{7.10}$$

though this ratio-type estimate is biased because, in general,

$$E\left[\hat{x}_*\right] \neq \frac{E\left[Y_* - \hat{\beta}_0\right]}{E\left[\hat{\beta}_1\right]} = x_*.$$

However, a confidence interval for x_* can be constructed using the method of Section 7.2.2. From Equation (5.21)

$$Y_* - \hat{Y}_* = Y_* - \hat{\beta}_0 - \hat{\beta}_1 x_* \sim N\left(0, \sigma^2(1+v_*)\right)$$

so that

$$T = \frac{Y_* - \hat{Y}_*}{S\sqrt{1+v_*}} = \frac{Y_* - \bar{Y} - \hat{\beta}_1\left(x_* - \bar{x}\right)}{S\sqrt{1+v_*}} \sim t_{n-2}.$$

Since

$$\begin{aligned} 1-\alpha &= \text{pr}\left[|T| \leqslant t_{n-2}^{(1/2)\alpha}\right] \\ &= \text{pr}\left[T^2 \leqslant \left(t_{n-2}^{(1/2)\alpha}\right)^2\right], \end{aligned}$$

the set of all values of x satisfying the inequality

$$\left\{Y_* - \bar{Y} - \hat{\beta}_1(x-\bar{x})\right\}^2 \leqslant \lambda^2 S^2 \left\{1 + \frac{1}{n} + \frac{(x-\bar{x})^2}{\sum_i (x_i - \bar{x})^2}\right\}, \qquad (7.11)$$

where $\lambda = t_{n-2}^{(1/2)\alpha}$ (and $\lambda^2 = F_{1,n-2}^{\alpha}$), will provide a $100(1-\alpha)\%$ confidence region for the unknown x_*. This set of points, commonly called the discrimination interval, may give a finite interval, two semi-infinite lines, or the entire real line (see Miller [1966: pp. 118–119; Figures 2, 3 and 4] and Hoadley [1970]). One obtains a finite interval if and only if $\hat{\beta}_1^2 > \lambda^2 S^2/\Sigma(x_i - \bar{x})^2$ (that is, the F test for $\beta_1 = 0$ is significant). In this case the interval contains the estimate $\hat{x}_*$ and is given by $[d_1 + \bar{x}, d_2 + \bar{x}]$, where d_1 and d_2 are the (real unequal) roots of

$$d^2\left\{\hat{\beta}_1^2 - \frac{\lambda^2 S^2}{\Sigma(x_i - \bar{x})^2}\right\} - 2d\hat{\beta}_1\left(Y_* - \bar{Y}\right) + \left\{\left(Y_* - \bar{Y}\right)^2 - \lambda^2 S^2\left(1 + \frac{1}{n}\right)\right\} = 0. \qquad (7.12)$$

[This equation follows from (7.11) by setting $d = x - \bar{x}$.] If $\hat{x}_*$ does not lie in $[d_1 + \bar{x}, d_2 + \bar{x}]$ then the confidence region for x_* is the union of two semi-infinite lines. However, if (7.12) has no real roots then the region is the entire real line.

The above theory is readily extended to the problem of finding k simultaneous discrimination intervals corresponding to k different values of Y_*, say, $Y_*^{(i)}$ $(i=1,2,\ldots,k)$. One simply substitutes $Y_*^{(i)}$ in (7.12) and sets λ equal to $t_{n-2}^{\alpha/(2k)}$, $(kF_{k,n-2}^{\alpha})^{1/2}$ and $u_{k,n-2}^{\alpha}$ for the Bonferroni, Scheffé, and maximum modulus intervals, respectively.

Unfortunately this method cannot be used when k is unknown. Such will be the case in calibration problems where the estimated calibration line is used to correct an unlimited number of future readings taken with the instrument; for example, in bioassay a standard curve is constructed for making future assays (discriminations). If k is large, λ may be so large as to render the above discrimination intervals useless. However, when k is large or unknown we can use two methods described in Liebermann et al. [1967]. The first method, suggested apparently by Miller [1966: pp. 125–128], uses the Bonferroni inequality (5.5) (with $k=2$) to combine, for a given Y_*, the confidence interval on $E[Y_*]$ with the Working–Hotelling confidence band on the line $\beta_0+\beta_1 x$. The second method effectively uses a Scheffé-type argument and is called the augmented F method. Both methods give conservative discrimination intervals and Libermann et al. feel that in most problems where these methods may be useful, the Bonferroni approach will give shorter intervals, especially when the future Y_* values are not expected to be far from $\hat{\beta}_0+\hat{\beta}_1\bar{x}$. A modification of the Bonferroni method is given by Odén [1973], who refines one of Miller's inequalities and gives a shorter interval.

a ALTERNATIVE ESTIMATORS

We saw above that the estimate $\hat{x}_*$ given by (7.10) is biased; it also has infinite mean square error $E[(\hat{x}_*-x_*)^2]$. Williams [1969] has shown that no unbiased estimate of x_* has finite variance and recommends that $\hat{x}_*$ be used on the grounds that it is based on a set of sufficient statistics for the unknown parameters. An alternative method based on regressing x on Y (even when x is not a random variable) was resurrected by Krutchoff [1967, 1969], and the usual prediction for x from this "inverse" model, say, $\dot{x}_*$, was compared with $\hat{x}_*$ using simulation. However, Williams [1969] and Halperin [1970] give arguments showing that Krutchoff's criterion for comparing the two estimates was unsatisfactory. Halperin gives a theoretical analysis which supports the use of $\hat{x}_*$ in preference to $\dot{x}_*$. Hoadley [1970] has also looked at this problem, but from a Bayesian viewpoint. He mentions the following unsatisfactory feature of $\hat{x}_*$. The coefficient of d^2 in (7.12) is

$$\frac{S^2}{\sum(x_i-\bar{x})^2}\left[\frac{\hat{\beta}_1^2\sum(x_i-\bar{x})^2}{S^2}-F_{1,n-2}^{\alpha}\right]$$

where the first term in the square bracket is the usual F-ratio for testing $\beta_1=0$. Therefore when F is much larger than $F^{\alpha}_{1,n-2}$ the discrimination interval is narrow and the estimation of x_* is fairly precise; on the other hand, if F is only just significant (the straight line is fairly flat), then the estimation is imprecise. In other words, the data contain information about the precision of $\hat{x}_*$ so that it would seem reasonable to have some way of giving $\hat{x}_*$ less weight when it is known to be unreliable; this is precisely what a Bayes estimator does. Hoadley [1970: p. 365] proves that

$$\dot{x}_* = \left[\frac{F}{F+(n-2)}\right]\hat{x}_*, \tag{7.13}$$

and that $\dot{x}_*$ is a Bayes solution with respect to a *particular* prior distribution on x_*. He also gives a confidence interval based on $\dot{x}_*$ when the particular prior can be justified.

In practice there is not a great deal of difference between $\hat{x}_*$ and $\dot{x}_*$ when the data are close to a straight line; S^2 is then small and F in (7.13) is large (see Exercise 4 at the end of the chapter).

b REPLICATED OBSERVATIONS

Suppose we have m replications Y_{*j} $(j=1,2,\ldots,m;\ m>1)$, with sample mean $\overline{Y}_*$, at the unknown value $x=x_*$. In this situation we have two estimates of σ^2, namely, S^2 and $\Sigma_j(Y_{*j}-\overline{Y}_*)^2/(m-1)$, which can be combined to give a confidence interval for x_* as follows.

Following Graybill [1961: pp. 125–127], let $U=\overline{Y}_*-\overline{Y}-\hat{\beta}_1(x_*-\bar{x})$. Then $E[U]=0$,

$$\operatorname{var}[U]=\sigma^2\left\{\frac{1}{m}+\frac{1}{n}+\frac{(x_*-\bar{x})^2}{\Sigma(x_i-\bar{x})^2}\right\}=\sigma_U^2,$$

say, and $U/\sigma_U \sim N(0,1)$. If

$$V_1=\sum_i\left[Y_i-\overline{Y}-\hat{\beta}_1(x_i-\bar{x})\right]^2=\text{RSS}$$

and

$$V_2=\sum_j\left(Y_{*j}-\overline{Y}_*\right)^2,$$

then U, V_1, and V_2 are mutually independent and

$$\frac{(n+m-3)\hat{\sigma}^2}{\sigma^2}=\frac{V_1+V_2}{\sigma^2}\sim\chi^2_{n-2+m-1}.$$

Therefore

$$T=\frac{U/\sigma_u}{\hat{\sigma}/\sigma}$$

$$=\frac{U}{\hat{\sigma}\left\{\frac{1}{m}+\frac{1}{n}+\frac{(x_*-\bar{x})^2}{\sum(x_i-\bar{x})^2}\right\}^{1/2}}\sim t_{n+m-3}$$

and (7.12) now becomes

$$d^2\left\{\hat{\beta}_1^2-\frac{\mu^2\hat{\sigma}^2}{\sum(x_i-\bar{x})^2}\right\}-2d\hat{\beta}_1\left(\bar{Y}_*-\bar{Y}\right)$$

$$+\left\{\left(\bar{Y}_*-\bar{Y}\right)^2-\mu^2\hat{\sigma}^2\left(\frac{1}{m}+\frac{1}{n}\right)\right\}=0, \qquad (7.14)$$

where $\mu^2=(t_{n+m-3}^{(1/2)\alpha})^2=F_{1,n+m-3}^{\alpha}$. We note that $\hat{\sigma}^2$, based on $n+m-3$ degrees of freedom, has a smaller sampling variance than S^2 with $n-2$ degrees of freedom; also $\mu^2<\lambda^2$. These two facts imply that (Cox [1971]) (1) the intervals given by (7.13) will, on the average, be narrower than those given by (7.12), and (2) the coefficient of d^2 in (7.14) is generally larger than that in (7.12) so that the probability of obtaining a *finite* confidence interval for x_* is greater when there are replications of Y_*.

In conclusion we mention two further methods. Kalotay [1971] proposed a structural solution to the problem, and Perng and Tong [1974] give a sequential method in which m is a random variable. The latter give a stopping rule for constructing a fixed-width confidence interval for x_*.

7.3 STRAIGHT LINE THROUGH THE ORIGIN

In many situations it is known that $E[Y]=0$ when $x=0$ so that the appropriate regression line is $Y_i=\beta_1 x_i+\varepsilon_i$. The least squares estimate of β_1 is now

$$\tilde{\beta}_1=\frac{\sum_i Y_i x_i}{\sum_i x_i^2},$$

and the unbiased estimate of σ^2 is

$$S^2 = \frac{1}{n-1}\left\{ \sum Y_i^2 - \tilde{\beta}_1^2 \sum x_i^2 \right\}. \tag{7.15}$$

Because $\tilde{\beta}_1 \sim N(\beta_1, \sigma^2/\Sigma x_i^2)$, a t-confidence interval for β_1 is

$$\tilde{\beta}_1 \pm t_{n-1}^{(1/2)\alpha} S\left(\sum_i x_i^2\right)^{-1/2}. \tag{7.16}$$

We can use the predictor $\tilde{Y}_* = x_* \tilde{\beta}_1$ to obtain a confidence interval for $E[Y_*] = x_* \beta_1$ at $x = x_*$, namely,

$$\tilde{Y}_* \pm t_{n-1}^{(1/2)\alpha} S\sqrt{v_*}\ , \tag{7.17}$$

where $v_* = x_*^2/\Sigma x_i^2$; this interval gets wider as we move away from the origin. Since β_1 lies in the interval (7.16) if and only if $x_*\beta_1$ lies in (7.17) for every x_*, a $100(1-\alpha)\%$ confidence band for the whole regression line is the region between the two lines

$$y = \tilde{\beta}_1 x \pm t_{n-1}^{(1/2)\alpha} S|x|\left(\sum x_i^2\right)^{-1/2}.$$

Prediction intervals for Y_*, or k values of Y_*, are obtained as in Section 7.2.4; however, v_* is defined as above and the appropriate degrees of freedom are now $n-1$ instead of $n-2$.

Inverse prediction is also straightforward. Following the method of Section 7.2.6 we find that x_* is estimated by $\tilde{x}_* = Y_*/\tilde{\beta}_1$, and the corresponding confidence interval for x_* is given by the roots of

$$x^2\left(\tilde{\beta}_1^2 - \frac{\lambda^2 S^2}{\sum x_i^2}\right) - 2x\tilde{\beta}_1 Y_* + Y_*^2 - \lambda^2 S^2 = 0, \tag{7.18}$$

where $\lambda = t_{n-1}^{(1/2)\alpha}$ and S^2 is given by (7.15). For m replications Y_{*j} at $x = x_*$, the corresponding quadratic is (cf. Cox [1971])

$$x^2\left[\tilde{\beta}_1^2 - \frac{\mu^2\hat{\sigma}^2}{\sum x_i^2}\right] - 2x\tilde{\beta}_1 \overline{Y}_* + \overline{Y}_*^2 - \frac{\mu^2\hat{\sigma}^2}{m} = 0$$

where $\mu = t_{n+m-2}^{(1/2)\alpha}$, and

$$\hat{\sigma}^2 = \frac{1}{n+m-2}\left\{ \sum_i \left(Y_i - \tilde{\beta}_1 x_i\right)^2 + \sum_j \left(Y_{*j} - \overline{Y}_*\right)^2 \right\}.$$

7.4 WEIGHTED LEAST SQUARES

7.4.1 Known Weights

Let $Y_i = \beta_0 + \beta_1 x_i + \varepsilon_i$, $(i = 1, 2, \ldots, n)$, where the ε_i are independently distributed as $N(0, \sigma^2 w_i^{-1})$, and the w_i are known positive numbers. Then, from Section 3.6, the weighted least squares estimates β_0^* and β_1^* of β_0 and β_1, respectively, are obtained by minimizing $\sum w_i (Y_i - \beta_0 - \beta_1 x_i)^2$. Therefore, differentiating this expression partially with respect to β_0 and β_1, we have

$$\beta_0^* \sum w_i + \beta_1^* \sum w_i x_i = \sum w_i Y_i \tag{7.19}$$

and

$$\beta_0^* \sum w_i x_i + \beta_1^* \sum w_i x_i^2 = \sum w_i Y_i x_i. \tag{7.20}$$

Dividing (7.19) by $\sum w_i$ and defining the weighted means $\overline{Y}_w = \sum w_i Y_i / \sum w_i$, etc., we have

$$\beta_0^* = \overline{Y}_w - \beta_1^* \overline{x}_w. \tag{7.21}$$

Substituting (7.21) in (7.20) leads to

$$\begin{aligned} \beta_1^* &= \frac{\sum w_i Y_i x_i - \sum w_i x_i \overline{Y}_w}{\sum w_i x_i^2 - \sum w_i x_i \overline{x}_w} \\ &= \frac{\sum w_i (Y_i - \overline{Y}_w)(x_i - \overline{x}_w)}{\sum w_i (x_i - \overline{x}_w)^2}. \end{aligned}$$

From the alternative expression

$$\beta_1^* = \frac{\sum w_i Y_i (x_i - \overline{x}_w)}{\sum w_i (x_i - \overline{x}_w)^2} \tag{7.22}$$

it readily follows that

$$\text{var}[\beta_1^*] = \frac{\sigma^2}{\sum w_i (x_i - \overline{x}_w)^2}.$$

Using the general theory of Section 3.6 we can show that

$$S_w^2 = \frac{1}{n-2}\left\{\sum w_i\left[Y_i - \bar{Y}_w - \beta_1^*(x_i - \bar{x}_w)\right]^2\right\} \tag{7.23}$$

$$= \frac{1}{n-2}\left\{\sum w_i\left(Y_i - \bar{Y}_w\right)^2 - (\beta_1^*)^2 \sum w_i (x_i - \bar{x}_w)^2\right\} \tag{7.24}$$

is an unbiased estimate of σ^2, and a $100(1-\alpha)\%$ confidence interval for β_1 is given by

$$\beta_1^* \pm t_{n-2}^{(1/2)\alpha}\left\{\frac{S_w^2}{\sum w_i (x_i - \bar{x})^2}\right\}^{1/2}. \tag{7.25}$$

When $\beta_0 = 0$ and $\beta_1 = \beta$ we have, from Example 3.2 in Section 3.6,

$$\beta^* = \frac{\sum w_i Y_i x_i}{\sum w_i x_i^2}, \tag{7.26}$$

and the appropriate confidence interval for β is now

$$\beta^* \pm t_{n-1}^{(1/2)\alpha}\left\{\frac{S_w^2}{\sum w_i x_i^2}\right\}^{1/2},$$

where

$$S_w^2 = \frac{1}{n-1}\left\{\sum w_i Y_i^2 - (\beta^*)^2 \sum w_i x_i^2\right\}. \tag{7.27}$$

[We note these formulas follow from those given by Equations (7.22) to (7.25) by setting $\bar{Y}_w = \bar{x}_w = 0$ and replacing $n-2$ by $n-1$.] Under the normality assumptions β^* is the maximum likelihood estimate of β. However, Turner [1960] has shown that, for certain w_i, β^* can still be the maximum likelihood estimate when Y is not normally distributed (cf. Exercise 5 at the end of this chapter). Inverse prediction (discrimination) for this model is discussed by Cox [1971].

7.4.2 *Unknown Weights*

Let

$$\begin{aligned} Y_i &= \theta_i + \varepsilon_i \\ &= \beta_0 + \beta_1 x_i + \varepsilon_i \qquad (i = 1, 2, \ldots, n) \end{aligned}$$

where the ε_i are independently distributed as $N(0, vg(\theta_i))$; here $v = \sigma^2$, g is a known positive function, and the weights $w_i = 1/g(\theta_i)$ are now unknown. Two methods are available for estimating β_0 and β_1.

a MAXIMUM LIKELIHOOD METHOD

If $g_i = g(\theta_i)$ then L, the logarithm of the likelihood function, is given by

$$L = -\tfrac{1}{2}n\log 2\pi - \tfrac{1}{2}\sum \log(vg_i) - \frac{\tfrac{1}{2}\sum (Y_i - \beta_0 - \beta_1 x_i)^2}{vg_i}.$$

Now

$$\frac{\partial \log g}{\partial \theta} = \frac{1}{g}\cdot\frac{\partial g}{\partial \theta} = h,$$

say, so that

$$\frac{\partial g}{\partial \beta_0} = \frac{\partial g}{\partial \theta}\cdot\frac{\partial \theta}{\partial \beta_0} = gh$$

and

$$\frac{\partial g}{\partial \beta_1} = \frac{\partial g}{\partial \theta}\cdot\frac{\partial \theta}{\partial \beta_1} = ghx.$$

The maximum likelihood estimates $\tilde{\beta}_0$, $\tilde{\beta}_1$, and $\tilde{v}$ are obtained by solving $\partial L/\partial\beta_0 = \partial L/\partial\beta_1 = \partial L/\partial v = 0$, namely,

$$-\tfrac{1}{2}\sum_i \tilde{h}_i + \tfrac{1}{2}\sum_i \left\{\frac{\tilde{h}_i(Y_i - \tilde{\theta}_i)^2}{\tilde{v}\tilde{g}_i}\right\} + \sum_i \left\{\frac{(Y_i - \tilde{\theta}_i)}{\tilde{v}\tilde{g}_i}\right\} = 0,$$

$$-\tfrac{1}{2}\sum_i \tilde{h}_i x_i + \tfrac{1}{2}\sum_i \left\{\frac{\tilde{h}_i x_i(Y_i - \tilde{\theta}_i)^2}{\tilde{v}\tilde{g}_i}\right\} + \sum_i \left\{\frac{x_i(Y_i - \tilde{\theta}_i)}{\tilde{v}\tilde{g}_i}\right\} = 0,$$

and

$$-\tfrac{1}{2}\frac{n}{\tilde{v}} + \tfrac{1}{2}\sum \frac{(Y_i - \tilde{\theta}_i)^2}{\tilde{v}^2\tilde{g}_i} = 0,$$

where $\tilde{h}_i$, $\tilde{g}_i$, and $\tilde{\theta}_i$ are functions of $\tilde{\beta}_0$ and $\tilde{\beta}_1$. Multiplying through by $\tilde{v}$, setting $\tilde{Y}_i = \tilde{\theta}_i = \tilde{\beta}_0 + \tilde{\beta}_1 x_i$, and $\tilde{w}_i = 1/\tilde{g}_i$, we can reduce the above equations to

$$\tilde{\beta}_0 \sum \tilde{w}_i + \tilde{\beta}_1 \sum \tilde{w}_i x_i = \sum \tilde{w}_i Y_i + \tfrac{1}{2} \sum \tilde{h}_i \left\{ \tilde{w}_i \left(Y_i - \tilde{Y}_i \right)^2 - \tilde{v} \right\}, \tag{7.28}$$

$$\tilde{\beta}_0 \sum \tilde{w}_i x_i + \tilde{\beta}_1 \sum \tilde{w}_i x_i^2 = \sum \tilde{w}_i x_i Y_i + \tfrac{1}{2} \sum \tilde{h}_i x_i \left\{ \tilde{w}_i \left(Y_i - \tilde{Y}_i \right)^2 - \tilde{v} \right\} \tag{7.29}$$

and

$$\tilde{v} = \frac{1}{n} \sum \tilde{w}_i \left(Y_i - \tilde{Y}_i \right)^2. \tag{7.30}$$

Equations (7.28) and (7.29) may be compared with (7.19) and (7.20). Therefore, given initial approximations to $\tilde{\beta}_0$ and $\tilde{\beta}_1$ (say, unweighted least squares estimates), we can evaluate the corresponding values of w_i, h_i, and v, solve (7.28) and (7.29), and obtain new approximations for $\tilde{\beta}_0$ and $\tilde{\beta}_1$. This process is then repeated.

When n is large the variance–covariance matrix of the maximum likelihood estimates is approximately

$$\begin{bmatrix} -E\left[\dfrac{\partial^2 L}{\partial \beta_0^2}\right], & -E\left[\dfrac{\partial^2 L}{\partial \beta_0 \partial \beta_1}\right], & -E\left[\dfrac{\partial^2 L}{\partial \beta_0 \partial v}\right] \\ -E\left[\dfrac{\partial^2 L}{\partial \beta_0 \partial \beta_1}\right], & -E\left[\dfrac{\partial^2 L}{\partial^2 \beta_1}\right], & -E\left[\dfrac{\partial^2 L}{\partial \beta_1 \partial v}\right] \\ -E\left[\dfrac{\partial^2 L}{\partial \beta_0 \partial v}\right], & -E\left[\dfrac{\partial^2 L}{\partial \beta_1 \partial v}\right], & -E\left[\dfrac{\partial^2 L}{\partial^2 v}\right] \end{bmatrix}^{-1}$$

$$\approx v \begin{bmatrix} \sum_i a_i & \sum_i x_i a_i & \tfrac{1}{2} \sum h_i \\ \sum_i x_i a_i & \sum_i x_i^2 a_i & \tfrac{1}{2} \sum_i h_i x_i \\ \tfrac{1}{2} \sum_i h_i & \tfrac{1}{2} \sum_i h_i x_i & \tfrac{1}{2}\left(\dfrac{n}{v}\right) \end{bmatrix}^{-1},$$

where

$$\begin{aligned} a_i &= g_i^{-1} + \tfrac{1}{2} h_i^2 v \\ &= \frac{1}{g_i} \left\{ 1 + \frac{v}{2g_i} \left(\frac{\partial g_i}{\partial \theta_i} \right)^2 \right\}. \end{aligned}$$

Frequently the second term of the above equation is small. For example, if $g_i = \theta_i^2$ then the second term is $2v$ which can be neglected if v is much smaller than $\frac{1}{2}$. In this case the variance–covariance matrix for $\tilde{\beta}_0$ and $\tilde{\beta}_1$ is approximately

$$v\begin{bmatrix} \sum_i w_i & \sum_i x_i w_i \\ \sum_i x_i w_i & \sum_i x_i^2 w_i \end{bmatrix}^{-1}, \tag{7.31}$$

which is the variance–covariance matrix of β_0^* and β_1^* in Section 7.4.1.

The above treatment is based on Williams [1959: pp. 67–70], though with the following differences: we have worked with σ^2 instead of σ, and Williams' g_i^2 is our g_i (his g_i may be negative).

b LEAST SQUARES METHOD

This technique consists of estimating the weights w_i $[=1/g(\beta_0+\beta_1 x_i)]$ from trial estimates of β_0 and β_1, say, the unweighted least squares estimates (which are unbiased), and then solving Equations (7.19) and (7.20) for new estimates of β_0 and β_1. These new values may be used for recalculating the w_i and the process can be repeated. Williams [1959] suggests that only two cycles of iteration are generally required as a great accuracy in the weights is not necessary for giving accurate estimates of β_0 and β_1.

Ignoring the fact that the estimated w_i are strictly random variables, the variance–covariance matrix of the least squares estimates is approximately given by (7.31). By the same argument approximate tests and confidence intervals can be obtained using the theory of Section 7.4.1, but with the w_i estimated. For this reason, and for computational simplicity, the least squares method is often preferred to the maximum likelihood approach.

7.5 COMPARING STRAIGHT LINES

7.5.1 *General Model*

Suppose we wish to compare K regression lines

$$Y = \alpha_k + \beta_k x_k + \varepsilon \qquad (k = 1, 2, \ldots, K),$$

where $E[\varepsilon] = 0$, and $\operatorname{var}[\varepsilon]$ $(=\sigma^2$, say) is the same for each line. If we are given n_k pairs of observations (x_{ki}, Y_{ki}) $(i = 1, 2, \ldots, n_k)$ on the kth line then we have the model

$$Y_{ki} = \alpha_k + \beta_k x_{ki} + \varepsilon_{ki} \qquad (i = 1, 2, \ldots, n_k), \tag{7.32}$$

where the ε_{ki} are independently and identically distributed as $N(0, \sigma^2)$. Writing

$$\mathbf{Y}' = (Y_{11}, Y_{12}, \ldots, Y_{1n_1}, \ldots, Y_{K1}, Y_{K2}, \ldots, Y_{Kn_K})$$

etc., we have $\mathbf{Y} = \mathbf{X}\boldsymbol{\gamma} + \boldsymbol{\varepsilon}$, where

$$\mathbf{X}\boldsymbol{\gamma} = \left[\begin{array}{cccccccc} 1 & 0 & \cdots & 0 & x_{11} & 0 & \cdots & 0 \\ 1 & 0 & \cdots & 0 & x_{12} & 0 & \cdots & 0 \\ \cdots & \cdots & \cdots & \cdots & \cdots & \cdots & \cdots & \cdots \\ 1 & 0 & \cdots & 0 & x_{1n_1} & 0 & \cdots & 0 \\ \hline 0 & 1 & \cdots & 0 & 0 & x_{21} & \cdots & 0 \\ 0 & 1 & \cdots & 0 & 0 & x_{22} & \cdots & 0 \\ \cdots & \cdots & \cdots & \cdots & \cdots & \cdots & \cdots & \cdots \\ 0 & 1 & \cdots & 0 & 0 & x_{2n_2} & \cdots & 0 \\ \hline \cdots & \cdots & \cdots & \cdots & \cdots & \cdots & \cdots & \cdots \\ \hline 0 & 0 & \cdots & 1 & 0 & 0 & \cdots & x_{K1} \\ 0 & 0 & \cdots & 1 & 0 & 0 & \cdots & x_{K2} \\ \cdots & \cdots & \cdots & \cdots & \cdots & \cdots & \cdots & \cdots \\ 0 & 0 & \cdots & 1 & 0 & 0 & \cdots & x_{Kn_K} \end{array}\right] \left[\begin{array}{c} \alpha_1 \\ \alpha_2 \\ \cdots \\ \alpha_K \\ \beta_1 \\ \beta_2 \\ \cdots \\ \beta_K \end{array}\right] \tag{7.33}$$

Here $\mathbf{X}$ is an $N \times 2K$ matrix of rank $2K$, where $N = \sum_{k=1}^{K} n_k$, so that we can test any hypothesis of the form H: $\mathbf{A}\boldsymbol{\gamma} = \mathbf{c}$ using the general regression theory of Chapter 4; three such hypotheses are considered below. The following problems can also be handled using the analysis of covariance method (Section 10.1).

7.5.2 *Test for Parallelism*

If we wish to test whether the K lines are parallel, then our hypothesis is H_1: $\beta_1 = \beta_2 = \cdots = \beta_K$ $(=\beta$, say) or $\beta_1 - \beta_K = \beta_2 - \beta_K = \cdots = \beta_{K-1} - \beta_K = 0$; in matrix form this is

$$\left[\begin{array}{c|cccccc} & 1 & 0 & 0 & \cdots & 0 & -1 \\ & 0 & 1 & 0 & \cdots & 0 & -1 \\ \mathbf{O} & \cdots & \cdots & \cdots & \cdots & \cdots & \cdots \\ & 0 & 0 & 0 & 0 & 1 & -1 \end{array}\right] \begin{pmatrix} \boldsymbol{\alpha} \\ \boldsymbol{\beta} \end{pmatrix} = \mathbf{0}$$

or $\mathbf{A}\boldsymbol{\gamma} = \mathbf{0}$, where $\mathbf{A}$ is $(K-1) \times 2K$ of rank $K-1$. Applying the general regression theory with $q = K-1$, $n = N$, and $p = 2K$, the test statistic for H_1

is

$$F = \frac{(\mathrm{RSS}_{H_1} - \mathrm{RSS})/(K-1)}{\mathrm{RSS}/(N-2K)}. \tag{7.34}$$

Minimizing $\boldsymbol{\varepsilon}'\boldsymbol{\varepsilon}$ we find that RSS is simply the sum of the residual sums of squares for each regression, namely,

$$\mathrm{RSS} = \sum_{k=1}^{K} \left\{ \sum_{i=1}^{n_k} \left(Y_{ki} - \overline{Y}_{k\cdot}\right)^2 - \hat{\beta}_k^2 \sum_{i=1}^{n_k} (x_{ki} - \bar{x}_{k\cdot})^2 \right\} \tag{7.35}$$

where

$$\hat{\beta}_k = \frac{\sum_i \left(Y_{ki} - \overline{Y}_{k\cdot}\right)(x_{ki} - \bar{x}_{k\cdot})}{\sum_i (x_{ki} - \bar{x}_{k\cdot})^2}. \tag{7.36}$$

Also

$$\hat{\alpha}_k = \overline{Y}_{k\cdot} - \hat{\beta}_k \bar{x}_{k\cdot}. \tag{7.37}$$

To find RSS_{H_1} we minimize $\sum_k \sum_i (Y_{ki} - \alpha_k - \beta x_{ki})^2$ with respect to the α_k and β. Taking partial derivatives we obtain the equations

$$\sum_i \left(Y_{ki} - \tilde{\alpha}_k - \tilde{\beta} x_{ki}\right) = 0 \qquad (k = 1, 2, \ldots, K), \tag{7.38}$$

and

$$\sum_k \sum_i x_{ki}\left(Y_{ki} - \tilde{\alpha}_k - \tilde{\beta} x_{ki}\right) = 0. \tag{7.39}$$

From (7.38) we have

$$\tilde{\alpha}_k = \overline{Y}_{k\cdot} - \tilde{\beta} \bar{x}_{k\cdot}, \tag{7.40}$$

and substituting in (7.39) leads to

$$\tilde{\beta} = \frac{\sum_k \sum_i x_{ki}\left(Y_{ki} - \overline{Y}_{k\cdot}\right)}{\sum_k \sum_i x_{ki}(x_{ki} - \bar{x}_{k\cdot})}$$

$$= \frac{\sum_k \sum_i \left(Y_{ki} - \overline{Y}_{k\cdot}\right)(x_{ki} - \bar{x}_{k\cdot})}{\sum_k \sum_i (x_{ki} - \bar{x}_{k\cdot})^2},$$

which is a "pooled" estimate of the common slope. Finally

$$\begin{aligned}
\mathrm{RSS}_{H_1} &= \sum_k \sum_i \left(Y_{ki} - \tilde{\alpha}_k - \tilde{\beta} x_{ki}\right)^2 \\
&= \sum_k \sum_i \left\{ Y_{ki} - \bar{Y}_{k\cdot} - \tilde{\beta}\left(x_{ki} - \bar{x}_{k\cdot}\right)\right\}^2 \\
&= \sum_k \sum_i \left(Y_{ki} - \bar{Y}_{k\cdot}\right)^2 - \tilde{\beta}^2 \sum_k \sum_i \left(x_{ki} - \bar{x}_{k\cdot}\right)^2,
\end{aligned}$$

and

$$\mathrm{RSS}_{H_1} - \mathrm{RSS} = \sum_k \hat{\beta}_k^2 \sum_i \left(x_{ki} - \bar{x}_{k\cdot}\right)^2 - \tilde{\beta}^2 \sum_k \sum_i \left(x_{ki} - \bar{x}_{k\cdot}\right)^2. \qquad (7.41)$$

7.5.3 *Test for Coincidence*

To test whether the K lines are coincident we consider H_2: $\alpha_1 = \alpha_2 = \cdots = \alpha_K$ $(=\alpha$, say) and $\beta_1 = \beta_2 = \cdots = \beta_K$ $(=\beta$, say). Arguing as in Section 7.5.2, we see readily that H_2 is of the form $\mathbf{A}\boldsymbol{\gamma} = \mathbf{0}$ where $\mathbf{A}$ is now $(2K-2)\times 2K$ of rank $2K-2$. Minimizing $\boldsymbol{\varepsilon}'\boldsymbol{\varepsilon}$ with respect to α and β, we obtain

$$\begin{aligned}
\mathrm{RSS}_{H_2} &= \sum_k \sum_i \left(Y_{ki} - \alpha^* - \beta^* x_{ki}\right)^2 \\
&= \sum_k \sum_i \left\{ Y_{ki} - \bar{Y}_{\cdot\cdot} - \beta^*\left(x_{ki} - \bar{x}_{\cdot\cdot}\right)\right\}^2 \\
&= \sum_k \sum_i \left(Y_{ki} - \bar{Y}_{\cdot\cdot}\right)^2 - (\beta^*)^2 \sum_k \sum_i \left(x_{ki} - \bar{x}_{\cdot\cdot}\right)^2,
\end{aligned}$$

where

$$\beta^* = \frac{\sum_k \sum_i \left(Y_{ki} - \bar{Y}_{\cdot\cdot}\right)\left(x_{ki} - \bar{x}_{\cdot\cdot}\right)}{\sum_k \sum_i \left(x_{ki} - \bar{x}_{\cdot\cdot}\right)^2}$$

and

$$\bar{Y}_{\cdot\cdot} = \frac{\sum_k \sum_i Y_{ki}}{N},$$

etc. Our F-statistic for testing H_2 is

$$F=\frac{(\mathrm{RSS}_{H_2}-\mathrm{RSS})/(2K-2)}{\mathrm{RSS}/(N-2K)} \tag{7.42}$$

In practice we would probably test for parallelism first and then, if H_1 is not rejected, test for H_2 (given that H_1 is true) using

$$F=\frac{(\mathrm{RSS}_{H_2}-\mathrm{RSS}_{H_1})/(K-1)}{\mathrm{RSS}_{H_1}/(N-K-1)}$$

If this also is not significant then we can check this nested procedure using (7.42) as a final test statistic.

7.5.4 *Test for Concurrence*

a x-COORDINATE KNOWN

Suppose we wish to test the hypothesis H_3 that all the lines meet at a point on the y-axis ($x=0$), that is, H_3: $\alpha_1=\alpha_2=\cdots=\alpha_K(=\alpha$, say). Taking partial derivatives of $\Sigma\Sigma(Y_{ki}-\alpha-\beta_k x_{ki})^2$ with respect to α and the β_k, the least squares estimates, α' and β_k', say, are the solutions of

$$\sum_k\sum_i(Y_{ki}-\alpha'-\beta_k' x_{ki})=0 \tag{7.43}$$

and

$$\sum_i x_{ki}(Y_{ki}-\alpha'-\beta_k' x_{ki})=0 \qquad (k=1,2,\ldots,K), \tag{7.44}$$

This last set of equations can be written in the matrix form:

$$\begin{bmatrix} N & x_{1\cdot} & x_{2\cdot} & \cdots & x_{K\cdot} \\ x_{1\cdot} & \sum_i x_{1i}^2 & 0 & \cdots & 0 \\ x_{2\cdot} & 0 & \sum_i x_{2i}^2 & \cdots & 0 \\ \cdots & \cdots & \cdots & \cdots & \cdots \\ x_{K\cdot} & 0 & 0 & \cdots & \sum_i x_{Ki}^2 \end{bmatrix} \begin{bmatrix} \alpha' \\ \beta_1' \\ \beta_2' \\ \cdots \\ \beta_K' \end{bmatrix} = \begin{bmatrix} Y_{\cdot\cdot} \\ \sum_i Y_{1i}x_{1i} \\ \sum_i Y_{2i}x_{2i} \\ \cdots \\ \sum_i Y_{Ki}x_{Ki} \end{bmatrix} \tag{7.45}$$

where $x_{k\cdot}=\Sigma_i x_{ki}$ and $Y_{\cdot\cdot}=\Sigma_k\Sigma_i y_{ki}$. Therefore, subtracting suitable multiples of all the rows of (7.45) from the first so as to annihilate all the elements in the first row except the (1, 1) element, we have

$$\left[N - \frac{x_{1\cdot}^2}{\sum_i x_{1i}^2} - \cdots - \frac{x_{K\cdot}^2}{\sum_i x_{Ki}^2}\right]\alpha' = \left[Y_{\cdot\cdot} - \frac{x_{1\cdot}\sum_i Y_{1i}x_{1i}}{\sum_i x_{1i}^2} - \cdots - \frac{x_{K\cdot}\sum_i Y_{Ki}x_{Ki}}{\sum_i x_{Ki}^2}\right] \tag{7.46}$$

and

$$\beta_k' = \frac{\sum_i (Y_{ki} - \alpha')x_{ki}}{\sum_i x_{ki}^2} \qquad (k = 1, 2, \ldots, K).$$

Finally,

$$\text{RSS}_{H_3} = \sum_k \sum_i (Y_{ki} - \alpha' - \beta_k' x_{ki})^2 \tag{7.47}$$

and the F-statistic for testing H_3 is

$$F = \frac{(\text{RSS}_{H_3} - \text{RSS})/(K-1)}{\text{RSS}/(N-2K)}.$$

The y coordinate of the point of concurrence on the y axis is estimated by α'.

When the values of x are the same for each line so that $n_k = n$ and $x_{ki} = x_i$ ($k = 1, 2, \ldots, K$), then (7.46) becomes

$$\left[Kn - \frac{Kx_{\cdot}^2}{\sum x_i^2}\right]\alpha' = Y_{\cdot\cdot} - \frac{x_{\cdot}}{\sum x_i^2}\sum Y_{\cdot i}x_i$$

which, after some algebra, leads to

$$\begin{aligned}\alpha' &= \overline{Y}_{\cdot\cdot} - \frac{\bar{x}\sum\sum Y_{ki}(x_i - \bar{x})}{K\sum (x_i - \bar{x})^2}\\ &= \overline{Y}_{\cdot\cdot} - \bar{x}\frac{\sum_k \hat{\beta}_k}{K}.\end{aligned} \tag{7.48}$$

Here $\hat{\beta}_k$ is the least squares estimate of the slope of the kth line so that α' may be regarded as simply the intercept of the "mean regression line" (Williams [1959: p. 139], Sprent [1969: p. 104]). In this case it can be shown that

$$\text{RSS}_{H_3}=\sum_k\sum_i Y_{ki}^2-\frac{\left(\sum\sum Y_{ki}x_i\right)^2}{\sum x_i^2}-(\alpha')^2Kn\frac{\sum(x_i-\bar{x})^2}{\sum x_i^2}. \qquad (7.49)$$

Since $\bar{Y}_{..}$ and $\hat{\beta}_k$ are uncorrelated we have, from (7.48),

$$\begin{aligned}\text{var}[\alpha']&=\text{var}[\bar{Y}_{..}]+K\bar{x}^2\frac{\text{var}[\hat{\beta}_k]}{K^2}\\&=\frac{\sigma^2}{K}\left\{\frac{1}{n}+\frac{\bar{x}^2}{\sum(x_i-\bar{x})^2}\right\}\\&=\frac{\sigma^2\sum x_i^2}{nK\sum(x_i-\bar{x})^2},\end{aligned}$$

and a confidence interval for α can be obtained from the following t-variable

$$T=(\alpha'-\alpha)\left\{\frac{nK\sum(x_i-\bar{x})^2}{S^2\sum x_i^2}\right\}^{1/2}\sim t_{K(n-2)}$$

where $S^2=\text{RSS}/(nK-2K)$.

If we wish to test whether the lines meet on the line $x=c$ we simply replace x_{ki} by $x_{ki}-c$ in the above theory; we shift the origin from $(0,0)$ to $(c,0)$. In this case the y-coordinate of the concurrence point is still estimated by α'.

b x-COORDINATE UNKNOWN

The hypothesis that the lines meet at $x=\phi$, where ϕ is now unknown, takes the form H: $\alpha_k+\beta_k\phi=$ *constant* for $k=1,2,\ldots,K$, or, eliminating ϕ,

$$H: \frac{\alpha_1-\bar{\alpha}}{\beta_1-\bar{\beta}}=\cdots=\frac{\alpha_K-\bar{\alpha}}{\beta_K-\bar{\beta}}.$$

Since H is no longer a linear hypothesis we cannot use the general regression theory to derive a test statistic. However, an approximate test is provided by Saw [1966].

7.5.5 Use of Dummy Categorical Regressors

Suppose we wish to compare just two regression lines

$$Y_{ki}=\alpha_k+\beta_k x_{ki}+\varepsilon_{ki} \quad (k=1,2;\ i=1,2,\ldots,n_k).$$

By introducing the dummy variable d, where

$$\begin{aligned} d&=1 \quad \text{if the observation comes from the second line}\\ &=0 \quad \text{otherwise,}\end{aligned}$$

we can combine these two lines into a single model, namely,

$$\begin{aligned} Y_i&=\alpha_1+\beta_1 x_i+(\alpha_2-\alpha_1)d_i+(\beta_2-\beta_1)(dx)_i+\varepsilon_i\\ &=\gamma_0+\gamma_1 z_{i1}+\gamma_2 z_{i2}+\gamma_3 z_{i3}+\varepsilon_i, \end{aligned} \tag{7.50}$$

where

$$\begin{aligned} (x_i,Y_i)&=(x_{1i},Y_{1i}), \qquad i=1,2,\ldots,n_1\\ &=(x_{2i},Y_{2i}), \qquad i=n_1+1,\ldots,n_1+n_2 \end{aligned}$$

and

$$\begin{aligned} d_i&=0, \qquad i=1,2,\ldots,n_1\\ &=1, \qquad i=n_1+1,\ldots,n_1+n_2. \end{aligned}$$

We note that the model (7.50) is simply a reparametrization of (7.33) (with $K=2$); the parameters $\alpha_1,\alpha_2,\beta_1$, and β_2 are now replaced by $\gamma_0=\alpha_1,\gamma_1=\beta_1,\gamma_2=\alpha_2-\alpha_1$, and $\gamma_3=\beta_2-\beta_1$. For this new model the various tests discussed above reduce to the following: $\gamma_3=0$ (parallelism), $\gamma_2=0$ (common intercept on the y-axis), and $\gamma_2=\gamma_3=0$ (coincidence).

In the case of three straight lines we introduce two dummy variables:

$$\begin{aligned} d_1&=1 \quad \text{if the observation comes from the second line,}\\ &=0 \quad \text{otherwise;}\\ d_2&=1 \quad \text{if the observation comes from the third line,}\\ &=0 \quad \text{otherwise,}\end{aligned}$$

and obtain

$$Y_i=\alpha_1+\beta_1 x_i+(\alpha_2-\alpha_1)d_{i1}+(\alpha_3-\alpha_1)d_{i2}+(\beta_2-\beta_1)(d_1x)_i$$

$$+(\beta_3-\beta_1)(d_2x)_i+\varepsilon_i.$$

Further generalizations are straightforward (see, for example, Gujarati [1970]).

7.5.6 *Test for Equal Variances*

The error variances, say, σ_k^2 $(k=1,2,\ldots,K)$, for the K straight lines can be tested for equality using the methods of Section 6.2.2. For each line we compute a residual sum of squares

$$S_k^2=\frac{\sum_i\left\{Y_{ki}-\overline{Y}_{k\cdot}-\hat{\beta}_k(x_{ki}-\bar{x}_{k\cdot})^2\right\}}{n_k-2}$$

where $\hat{\beta}_k$ is given by (7.36). Assuming normality, $f_kS_k^2/\sigma_k^2\sim\chi_{f_k}^2$ where $f_k=n_k-2$, and we can test H: $\sigma_1^2=\cdots=\sigma_K^2$ using the methods of Section 6.2.2.

7.6 TWO-PHASE LINEAR REGRESSION

Sometimes a regression of Y on x can be reasonably represented by two intersecting straight lines, one being appropriate when $x<\gamma$ and the other when $x>\gamma$; thus

$$E[Y]=\alpha_1+\beta_1x \qquad x\leqslant\gamma$$

$$E[Y]=\alpha_2+\beta_2x \qquad x\geqslant\gamma$$

and

$$\alpha_1+\beta_1\gamma=\alpha_2+\beta_2\gamma \qquad (=\theta). \tag{7.51}$$

For example, x may be an increasing function of time and at time t_c a treatment is applied that may possibly affect the slope of the regression line either immediately or after a time lag. Following Sprent [1961] we call $x=\gamma$ the *changeover point* and θ the *changeover value*.

7.6.1 Least Squares Estimation with Known Changeover Point

Suppose we wish to fit the two-phase model

$$Y_{1i} = \alpha_1 + \beta_1 x_{1i} + \varepsilon_{1i} \qquad (i = 1, 2, \ldots, n_1)$$

$$Y_{2i} = \alpha_2 + \beta_2 x_{2i} + \varepsilon_{2i} \qquad (i = 1, 2, \ldots, n_2)$$

by least squares, where

$$x_{11} < x_{12} < \cdots < x_{1n_1} < \gamma < x_{21} < x_{22} < \cdots < x_{2n_2},$$

and γ is known. Then we must minimize $\boldsymbol{\varepsilon}'\boldsymbol{\varepsilon} = \sum_k \sum_i \varepsilon_{ki}^2$ with respect to the linear constraint (7.51) as follows (Sprent [1961: pp. 637–638]). Consider

$$r = \sum_{k=1}^{2} \sum_{i=1}^{n_k} (Y_{ki} - \alpha_k - \beta_k x_{ki})^2 + 2\lambda[\alpha_2 - \alpha_1 + \gamma(\beta_2 - \beta_1)],$$

where -2λ is the Lagrange multiplier associated with (7.51). Setting the partial derivatives of r with respect to α_k, β_k $(k = 1, 2)$ equal to zero, we have the least squares equations:

$$-2\left(\sum_i Y_{1i} - n_1\tilde{\alpha}_1 - \sum_i x_{1i}\tilde{\beta}_1\right) - 2\lambda = 0, \tag{7.52}$$

$$-2\left(\sum_i Y_{2i} - n_2\tilde{\alpha}_2 - \sum_i x_{2i}\tilde{\beta}_2\right) + 2\lambda = 0, \tag{7.53}$$

$$-2\left(\sum_i x_{1i}(Y_{1i} - \tilde{\alpha}_1 - \tilde{\beta}_1 x_{1i})\right) - 2\lambda\gamma = 0, \tag{7.54}$$

and

$$-2\left(\sum_i x_{2i}(Y_{2i} - \tilde{\alpha}_2 - \tilde{\beta}_2 x_{2i})\right) + 2\lambda\gamma = 0. \tag{7.55}$$

From (7.52) and (7.53) we have

$$\tilde{\alpha}_k = \overline{Y}_{k\cdot} - \tilde{\beta}_k \bar{x}_{k\cdot} + (-1)^{k-1}\lambda n_k^{-1} \qquad (k = 1, 2), \tag{7.56}$$

and substituting in

$$\tilde{\alpha}_1 - \tilde{\alpha}_2 + \gamma(\tilde{\beta}_1 - \tilde{\beta}_2) = 0 \tag{7.57}$$

leads to

$$\lambda = w\left\{\bar{Y}_{2\cdot} - \bar{Y}_{1\cdot} + \tilde{\beta}_1(\bar{x}_{1\cdot} - \gamma) - \tilde{\beta}_2(\bar{x}_{2\cdot} - \gamma)\right\} \tag{7.58}$$

where $w = n_1 n_2/(n_1 + n_2)$. Finally, substituting (7.56) and (7.58) into (7.54) and (7.55), we have

$$c_{11}\tilde{\beta}_1 + c_{12}\tilde{\beta}_2 = c_{13}$$

$$c_{21}\tilde{\beta}_1 + c_{22}\tilde{\beta}_2 = c_{23},$$

where

$$c_{kk} = \sum_i (x_{ki} - \bar{x}_{k\cdot})^2 + w(\bar{x}_{k\cdot} - \gamma)^2 \qquad (k = 1, 2),$$

$$c_{12} = c_{21} = -w(\bar{x}_{1\cdot} - \gamma)(\bar{x}_{2\cdot} - \gamma),$$

and

$$c_{k3} = \sum_i (Y_{ki} - \bar{Y}_{k\cdot})(x_{ki} - \bar{x}_{k\cdot}) + (-1)^k w(\bar{Y}_{2\cdot} - \bar{Y}_{1\cdot})(\bar{x}_{k\cdot} - \gamma) \qquad (k = 1, 2).$$

Solving for $\tilde{\beta}_1$ and $\tilde{\beta}_2$ we can then find λ from (7.58), and $\tilde{\alpha}_k$ from (7.56).

We note that the minimum value of $\boldsymbol{\varepsilon}'\boldsymbol{\varepsilon}$ is

$$\begin{aligned}
\sum_{k=1}^{2}\sum_{i=1}^{n_k}(Y_{ki} - \tilde{\alpha}_k - \tilde{\beta}_k x_{ki})^2
&= \sum\sum\left\{Y_{ki} - \bar{Y}_{k\cdot} - \tilde{\beta}_k(x_{ki} - \bar{x}_{k\cdot}) + (-1)^k \lambda n_k^{-1}\right\}^2 \\
&= \sum\sum (Y_{ki} - \bar{Y}_{k\cdot})^2 - 2\sum\sum \tilde{\beta}_k (Y_{ki} - \bar{Y}_{k\cdot})(x_{ki} - \bar{x}_{k\cdot}) \\
&\quad + \sum\sum \tilde{\beta}_k^2 (x_{ki} - \bar{x}_{k\cdot})^2 + \frac{\lambda^2}{w} \\
&\quad (= \text{RSS}_H, \text{ say}).
\end{aligned} \tag{7.59}$$

7.6.2 *Hypothesis Test that Changeover Point is a Given Value*

If we can assume that a two-phase linear regression model is appropriate, we may wish to test H: $\gamma = c$ where c lies between a pair of x-values, say, $x_{1n_1} < c < x_{21}$. Then H is equivalent to testing whether the two lines concur at $x=c$, and this can be done using the method of Section 7.5.4a. However, because only two lines are involved there are some algebraic simplifications. For example, RSS_{H_3} of (7.47) (with x_{ki} replaced by $x_{ki}-c$) is the same as RSS_H of (7.59) (with $\gamma=c$), so that the test for H is

$$F=\frac{\mathrm{RSS}_H-\mathrm{RSS}}{\mathrm{RSS}/(N-4)},$$

where $N=n_1+n_2$ and RSS is given by (7.35). It can also be shown that (Sprent [1961: equation (7)])

$$\mathrm{RSS}_H-\mathrm{RSS}=\sum_{k=1}^{2}\left(\hat{\beta}_k-\tilde{\beta}_k\right)\sum_i\left(Y_{ki}-\overline{Y}_{k\cdot}\right)(x_{ki}-\bar{x}_{k\cdot})+\lambda\left(\overline{Y}_{2\cdot}-\overline{Y}_{1\cdot}\right).$$

Sprent also discusses a number of related problems and the reader is referred to his article for details.

7.6.3 Unknown Changeover Point

If it is known that $x_{1n_1}<\gamma<x_{21}$, γ can be estimated by [cf. Equation (7.51)]

$$\hat{\gamma}=-\left(\frac{\hat{\alpha}_1-\hat{\alpha}_2}{\hat{\beta}_1-\hat{\beta}_2}\right)$$

where $\hat{\alpha}_k$ and $\hat{\beta}_k$ are the usual least squares estimates for the kth line ($k=1,2$). Since $\hat{\gamma}$ is the ratio of two correlated normal variables we can use Fieller's method for finding a confidence interval for γ as follows.

Consider $U=(\hat{\alpha}_1-\hat{\alpha}_2)+\gamma(\hat{\beta}_1-\hat{\beta}_2)$. Then $E[U]=0$ and, from (7.6) in Section 7.2.3 with $x_*=\gamma$, we have

$$\begin{aligned}\mathrm{var}[U]&=\sigma^2\left\{\frac{1}{n_1}+\frac{(\bar{x}_{1\cdot}-\gamma)^2}{\sum(x_{1i}-\bar{x}_{1\cdot})^2}+\frac{1}{n_2}+\frac{(\bar{x}_{2\cdot}-\gamma)^2}{\sum(x_{2i}-\bar{x}_{2\cdot})^2}\right\}\\&=\sigma^2 w,\end{aligned}$$

say. Arguing as in Section 7.2.2, a $100(1-\alpha)\%$ confidence interval for γ is given by the roots of

$$\left[\hat{\alpha}_1-\hat{\alpha}_2+\gamma\left(\hat{\beta}_1-\hat{\beta}_2\right)\right]^2-F^{\alpha}_{1,N-4}S^2w=0,$$

that is, of

$$\gamma^2\left[(\hat\beta_1-\hat\beta_2)^2-F^{\alpha}_{1,N-4}S^2\left\{\sum_{k=1}^{2}\left[\frac{1}{\sum_i (x_{ki}-\bar{x}_{k\cdot})^2}\right]\right\}\right]$$

$$+2\gamma\left[(\hat\alpha_1-\hat\alpha_2)(\hat\beta_1-\hat\beta_2)+F^{\alpha}_{1,N-4}S^2\left\{\sum_{k=1}^{2}\left[\frac{\bar{x}_{k\cdot}}{\sum_i (x_{ki}-\bar{x}_{k\cdot})^2}\right]\right\}\right]$$

$$-F^{\alpha}_{1,N-4}S^2\left\{\sum_{k=1}^{2}\left[\frac{\bar{x}^2_{k\cdot}}{\sum_i (x_{ki}-\bar{x}_{k\cdot})^2}+\frac{1}{n_k}\right]\right\}+(\hat\alpha_1-\hat\alpha_2)^2,$$

where $S^2=\text{RSS}/(N-4)$ and $N=n_1+n_2$.

If $\hat\gamma$ does not lie in the interval (x_{1n_1},x_{21}), then the experimenter must decide whether to attribute this to sampling errors (and the above confidence interval for γ will shed some light on this), or to an incorrect assumption about the position of γ. When the position of γ is unknown the problem becomes much more difficult as it is now nonlinear. In this case the two-phase model can be written in the form (Hinkley [1971])

$$Y_i=\theta+\beta_1(x_i-\gamma)+\varepsilon_i \qquad (i=1,2,\ldots,n)$$

$$Y_i=\theta+\beta_2(x_i-\gamma)+\varepsilon_i \qquad (i=n+1,\ldots,N)$$

where $x_1<\cdots<x_n\leqslant\gamma<x_{n+1}<\cdots<x_N$, θ is the changeover value, and n is now unknown and has to be estimated. Hinkley summarizes the maximum likelihood estimation procedure for estimating γ, θ, β_1, β_2, and n: this is described in detail in Hudson [1966] and Hinkley [1969b]. He also provides approximate large sample confidence intervals for the parameters and gives large sample tests for the hypotheses $\beta_1=\beta_2$ (no change in slope) and $\beta_2=0$. Another approach to testing $\beta_1=\beta_2$ is given by Farley and Hinich [1970].

Finally we note that Hudson's technique was generalized by Williams [1970] to the case of three-phase linear regression. An interesting variation on this particular model, in which the first and third lines are assumed to be horizontal, is described by Curnow [1973]. The problem of piecewise linear regression in general is discussed by Hudson [1966], Bellman and Roth [1969], and McGee and Carleton [1970] (see also Feder [1975] for a discussion of some of the theoretical problems).

7.7 RANDOM REGRESSOR VARIABLE

The problem of random regressor variables was discussed in general terms in Section 6.5. However, because the straight-line case has received considerable attention in the literature it is appropriate to summarize in more detail some of the difficulties involved; this summary is partly based on Moran [1970]. The reader is referred to Sprent [1969] for further comments and references†.

Suppose that $(U_1, V_1), \ldots, (U_n, V_n)$ are unobserved random variables connected by the structural relationship $V = \beta_0 + \beta_1 U$. These are observed with independent random errors δ_i and ε_i which are normally distributed with zero means and unknown variances σ_δ^2 and σ_ε^2, respectively. The observations are now (X_i, Y_i) $(i = 1, 2, \ldots, n)$ where $X_i = U_i + \delta_i$ and $Y_i = V_i + \varepsilon_i$. Two sorts of assumptions are usually made about U:

(1) The U_i are normally distributed with unknown mean m_U and variance σ_U^2.
(2) The U_i are fixed (unknown) quantities so that the V_i are also fixed: we now have a functional relationship $v = \beta_0 + \beta_1 u$.

Case 1. For this situation there are six unknown parameters in the model: m_U, σ_U^2, β_0, β_1, σ_δ^2, and σ_ε^2. However, because the observed quantities (X_i, Y_i) are jointly normally distributed, and a bivariate normal has only five parameters, we cannot expect to estimate all six parameters of the model. In fact only m_U can be estimated; the remaining parameters are not identifiable and the structural relation $V = \alpha + \beta U$ cannot be estimated. If the distribution of V is not normal, and is *known* to be nonnormal, it is possible to devise methods that will identify all the parameters. However, in practice, we never know the distribution of U, and the nearer the distribution to normality the worse the estimates.

One obvious approach to the problem is to impose some constraint on the parameters, thus effectively reducing the number of unknown parameters by one. Three types of constraint have been studied:

(1) σ_δ^2 or σ_ε^2 is known: all the parameters are now identifiable and can be estimated.
(2) $\sigma_\delta / \sigma_\varepsilon$ $(= k$, say) is known: all the parameters can be estimated consistently.
(3) Both σ_δ and σ_ε are known: this leads to "overidentification" of the model.

†See also Moran [1971], *J. Multivariate Anal.*, **1**, 232–255.

In conclusion we might ask the following question: what happens if we use the usual least squares estimates, for example, $\hat{\beta}_1 = \sum Y_i(X_i - \bar{X})/\sum(X_i - \bar{X})^2$, and ignore the effect of error in the regressors (that is, we use the $\{X_i\}$ instead of the true values $\{U_i\}$)? From Richardson and Wu [1970: p. 732] we have

$$E[\hat{\beta}_1] = \beta_1 \frac{(\sigma_U^2/\sigma_\delta^2)}{1 + (\sigma_U^2/\sigma_\delta^2)}$$

and

$$\operatorname{var}[\hat{\beta}_1] = \frac{1}{n-2}\left[\frac{\sigma_\varepsilon^2}{\sigma_U^2 + \sigma_\delta^2} + \beta_1^2 \frac{\sigma_U^2 \sigma_\delta^2}{(\sigma_U^2 + \sigma_\delta^2)^2}\right].$$

Case 2. We now have the model

$$Y_i = \beta_0 + \beta_1 u_i + \varepsilon_i, \qquad X_i = u_i + \delta_i,$$

which is the one considered in Section 6.4. Case 2 has a theory which closely parallels that of case 1; questions of nonidentifiability still remain, for we now have n unknown "parameters" $u_1, u_2, \ldots, u_n$ instead of m_U and σ_U^2.

For large n we have, from Section 6.4.1 with $\mathbf{D} = \operatorname{diag}(0, \sigma_\delta^2)$:

$$E[\hat{\beta}_{0\Delta}] \approx \beta_0 + \frac{\bar{u}\sigma_\delta^2 \beta_1 n}{\sum (u_i - \bar{u})^2 + n\sigma_\delta^2}$$

and

$$E[\hat{\beta}_{1\Delta}] \approx \beta_1 - \frac{\beta_1 \sigma_\delta^2 n}{\sum (u_i - \bar{u})^2 + n\sigma_\delta^2}.$$

However, the exact values of $E[\hat{\beta}_{1\Delta}]$ and $E[(\hat{\beta}_{1\Delta} - \beta_1)^2]$, along with more accurate large sample approximations, are given by Richardson and Wu [1970]. Their results are generalized by Halperin and Gurian [1971] to the case where δ_i and ε_i are correlated (their model should read $y_i = \alpha + \beta\xi_i + \varepsilon_i$, $x_i = \xi_i + \delta_i$).

A number of "grouping" methods for estimating β_1 have been suggested; these are described by Richardson and Wu [1970].

MISCELLANEOUS EXERCISES 7

1. In fitting the straight line $Y_i = \beta_0 + \beta_1 x_i + \varepsilon_i$ $(i = 1, 2, \ldots, n)$, prove that $\bar{Y}$ and $\hat{\beta}_1$ are uncorrelated. If $\hat{Y}_* = \hat{\beta}_0 + \hat{\beta}_1 x_*$, deduce that

$$\operatorname{var}[\hat{Y}_*] = \sigma^2 \left\{ \frac{1}{n} + \frac{(x_* - \bar{x})^2}{\sum (x_i - \bar{x})^2} \right\}.$$

2. Using the notation of Section 7.2.2, prove that $\hat{\phi} = -\hat{\beta}_0 / \hat{\beta}_1$ is the maximum likelihood estimate of ϕ.
3. Given a general linear regression model, show how to find a confidence interval for the ratio $\mathbf{a}_1' \boldsymbol{\beta} / \mathbf{a}_2' \boldsymbol{\beta}$ of two linear parametric functions.
4. Using the notation of Section 7.2.6, show that when $\bar{x} = 0$

$$\frac{\hat{x}_* - \dot{x}_*}{\hat{x}_*} = 1 - r^2,$$

where r is the correlation coefficient of the pairs (x_i, Y_i).

5. Let $Y_1, Y_2, \ldots, Y_n$ be independent random variables such that for $i = 1, 2, \ldots, n$

$$E[Y_i | X = x_i] = \beta_1 x_i$$

and

$$\operatorname{var}[Y_i | X = x_i] = \sigma^2 w_i^{-1}, \quad (w_i > 0).$$

(a) If the conditional distribution of Y given x is the Type III (scaled gamma) distribution,

$$f(y|x) = \frac{1}{a_x^p \Gamma(p)} y^{p-1} e^{-y/a_x} \qquad 0 \leqslant y < \infty, \quad p > 0,$$

where a_x is a function of x, and $w_i^{-1} = x_i^2$, prove that the maximum likelihood estimate of β_1 is also the weighted least squares estimate.

(b) If the conditional distribution of Y given x is Poisson, and $w_i^{-1} = x_i$, show that the maximum likelihood estimate is again the weighted least squares estimate.

Turner [1960]

6. Given the model $Y_i = \beta_1 x_i + \varepsilon_i$ $(i = 1, 2, \ldots, n)$, where the ε_i are independently distributed as $N(0, \sigma^2 w_i^{-1})$, $w_i > 0$, show how to predict x_* for a given value Y_* of Y. Describe briefly a method for constructing a confidence interval for x_*.

7. Verify Equations (7.48) and (7.49).
8. Derive an F-statistic for testing the hypothesis that two straight lines intersect at the point (a,b).
9. Obtain an estimate and a confidence interval for the horizontal distance between two parallel lines.
10. Show how to transform the following equation into a straight line so that α and β can be estimated by least squares:

$$y=\frac{\alpha\beta}{\alpha\sin^2\theta+\beta\cos^2\theta}$$

Williams [1959: p. 19]

11. Given the regression line

$$Y_i=\beta_0+\beta_1 x_i+\varepsilon_i \qquad (i=1,2,\ldots,n)$$

where the ε_i are independent with $E[\varepsilon_i]=0$ and $\text{var}[\varepsilon_i]=\sigma^2 x_i^2$, show that weighted least squares estimation is equivalent to ordinary least squares estimation for the model

$$\left(\frac{Y_i}{x_i}\right)=\beta_1+\left(\frac{\beta_0}{x_i}\right)+\delta_i.$$

CHAPTER 8

Polynomial Regression

8.1 POLYNOMIALS IN ONE VARIABLE

8.1.1 The Problem of Ill-Conditioning

If we set $x_{ij}=x_i^j$ and $k=p-1$ ($\leqslant n-1$) in the general multiple linear regression model we have the kth-degree (order) polynomial model

$$Y_i=\beta_0+\beta_1 x_i+\beta_2 x_i^2+\cdots+\beta_k x_i^k+\varepsilon_i \qquad (i=1,2,\ldots,n). \tag{8.1}$$

Although it is theoretically possible to fit a polynomial of degree up to $n-1$, a number of practical difficulties arise when k is large. First, for k greater than about six, we find that the regression matrix $\mathbf{X}$ associated with (8.1) becomes ill-conditioned (Section 11.4). For example, assuming x_i is distributed approximately uniformly on [0, 1], then for large n we have (Forsythe [1957])

$$\begin{aligned}(\mathbf{X}'\mathbf{X})_{rs} &= n\sum_{i=1}^{n} x_i^r x_i^s\cdot\frac{1}{n}\\ &\approx n\int_0^1 x^r x^s\,dx\\ &= n\int_0^1 x^{r+s}\,dx\\ &= \frac{n}{r+s+1}.\end{aligned} \tag{8.2}$$

Hence $\mathbf{X}'\mathbf{X}$ is something like n times the matrix $[(1/r+s+1)]$, ($r,s=0,1,\ldots,k$), which is the $(k+1)\times(k+1)$ principal minor of the so-called

Hilbert matrix

$$\mathbf{H}=\begin{bmatrix} 1 & \frac{1}{2} & \frac{1}{3} & \cdots \\ \frac{1}{2} & \frac{1}{3} & \frac{1}{4} & \cdots \\ \frac{1}{3} & \frac{1}{4} & \frac{1}{5} & \cdots \\ \cdots & \cdots & \cdots & \cdots \end{bmatrix}.$$

It is well-known that $\mathbf{H}$ is very ill-conditioned (Todd [1954, 1961]); for example, when $k=9$ the inverse of $\mathbf{H}_{10}$, the 10×10 principal minor of $\mathbf{H}$, has elements of magnitude 3×10^{10} (Savage and Lukacs [1954]). Thus a small error of 10^{-10} in one element of $\mathbf{X}'\mathbf{Y}$ will lead to an error of about 3 in an element of $\hat{\boldsymbol{\beta}}=(\mathbf{X}'\mathbf{X})^{-1}\mathbf{X}'\mathbf{Y}$. The extent of ill-conditioning is measured by the condition number of $\mathbf{X}'\mathbf{X}$ (defined in Section 11.4), which in the case of $\mathbf{H}_m$ is of order $e^{3.5m}$ (Marcus [1964: p. 23]). It is interesting to note that $|\mathbf{H}_m^{-1}|\sim 2^{-2m^2}$ (Todd [1954]).

One way of reducing the effect of ill-conditioning is to work with Chebyshev polynomials and fit a model of the form

$$E[Y]=\tfrac{1}{2}\gamma_0 T_0(x)+\gamma_1 T_1(x)+\cdots+\gamma_k T_k(x), \tag{8.3}$$

where $T_r(x)$ is a Chebyshev polynomial of the first kind of degree r. (It is customary to associate one-half with $T_0(x)$ when using a Chebyshev series expansion, though in this context it is not necessary.) These polynomials may be generated by the recurrence relation

$$T_{r+1}(x)=2xT_r(x)-T_{r-1}(x) \qquad (r=1,2,\ldots) \tag{8.4}$$

starting with $T_0(x)=1$ and $T_1(x)=x$: thus

$$T_2(x)=2x^2-1,$$

$$T_3(x)=4x^3-3x,$$

$$T_4(x)=8x^4-8x^2+1$$

$$T_5(x)=16x^5-20x^3+5x, \text{ etc.}$$

Computationally (8.4) is almost as simple to use as the usual recursion $x^{j+1}=x.x^j$ for monomials. The Chebyshev expansions for monomials are as follows:

$$1 = T_0(x)$$

$$x = T_1(x)$$

$$x^2 = \tfrac{1}{2}\{T_2(x) + T_0(x)\}$$

$$x^3 = \tfrac{1}{4}\{T_3(x) + 3T_1(x)\}$$

$$x^4 = \tfrac{1}{8}\{T_4(x) + 4T_2(x) + 3T_0(x)\}$$

$$x^5 = \tfrac{1}{16}\{T_5(x) + 5T_3(x) + 10T_1(x)\}, \text{ etc.}$$

In practice, for the sake of numerical stability, the x_i are "normalized" so that they run from -1 to $+1$. The normalized x is given by

$$x' = \frac{2x - \max(x_i) - \min(x_i)}{\max(x_i) - \min(x_i)}.$$

In this case we can write $x' = \cos\theta$, for some θ in $[0, \pi]$, and we then have $T_r(x') = \cos(r\theta)$, $r = 0, 1, 2, \ldots$.

The Chebyshev polynomials have a number of interesting properties. When $r \neq s$ we have, for example,

$$\int_{-1}^{1} T_r(x) T_s(x)(1-x^2)^{-1/2}\, dx = 0$$

and

$$\sum_{i=1}^{n} T_r(z_i) T_s(z_i) = 0$$

where the $z_i \{= \cos[(i - \tfrac{1}{2})(\pi/n)]\}$ are the zeros of $T_n(x)$. Both these properties suggest that, for reasonably spaced x-values, the matrix $\mathbf{X}$ for the model (8.3) (expressed in the form $\mathscr{E}[\mathbf{Y}] = \mathbf{X}\boldsymbol{\gamma}$) will have columns which are approximately orthogonal so that $\mathbf{X}'\mathbf{X}$ will have relatively small off-diagonal elements; such matrices are generally well-conditioned. A recommended procedure, therefore, for fitting polynomials is to use Chebyshev polynomials along with one of the accurate orthogonal decomposition methods of Section 11.2.4. In this respect the modified Gram–Schmidt algorithm, which seems to be the most accurate, would be a popular choice.

In addition to the problem of ill-conditioning, there is also the question of interpreting the form of a fitted polynomial when Chebyshev polynomials are used and k is large. Where physical interpretation is important it may therefore be appropriate to reexpress the fitted polynomial in terms of monomials. However, for some accurate double precision programs there may be little to be gained by working with Chebyshev polynomials and then transforming back to monomials, rather than working with monomials directly (e.g., Beaton and Tukey [1974: p. 150]). However, if monomials are used the x_i's should of course be normalized; Hayes [1970a: Section 8, Examples B and C] illustrates some of the difficulties that can arise with data that are not normalized.

8.1.2 Choosing the Degree

One guide for choosing k is to consider RSS_{r+1}, the residual sum of squares for a polynomial of degree r, as r increases ($r+1$ parameters are estimated in fitting RSS_{r+1}). Ideally RSS_{r+1} decreases consistently at first and then levels off to a fairly constant value, at which stage it is usually clear when to stop (for example, Hayes [1970a: Section 8, Example A]). In cases of doubt we can test for significance the coefficient of the last monomial added to the model; this is the so-called forward selection procedure with a predetermined order for the regressors (though it is used only at an appropriate stage of the fitting and not necessarily right from the beginning). However, this test procedure should be used cautiously because it may lead to stopping prematurely. For example, in fitting a polynomial to an almost symmetric function the coefficients of the odd powers will be small so that we may have the situation in which terms of odd order are not significant and significant terms of even order remain to be computed. We also have the possibility that RSS_{r+1} may level off for a while before decreasing again. To safeguard against these contingencies it is preferable to go several steps beyond the first insignificant term, and then look carefully at RSS.

Another test procedure that can be used is the so-called backward elimination procedure. In this case the maximum degree that will be fitted is determined in advance and then the highest degree monomial terms are eliminated one at a time using the F-test; the process stops when there is a significant F-statistic. This procedure is more efficient than forward selection, and it is suggested that the best significance level to use at each step is $\alpha \approx 0.10$ (Kennedy and Bancroft [1971: p. 1281]). However, there still remains the problem of deciding the maximum degree to be fitted. Unfortunately the forward and backward procedures do not necessarily lead to the same answer. Another procedure which begins with the maximum degree has been given by Hoel [1968].

The adequacy of a given model can be examined by making various residual plots such as Y_i versus $\hat{Y}_i$ and, in particular, e_i (or one of its scaled versions: Section 6.6.2) versus x_i. Suppose, for example, that the fitted polynomial is of degree k_1 and the true model is of degree k_2. If $k_1 < k_2$, then, from Equation (6.2)

$$E[e_i] = E\left[\hat{Y}_i - \hat{\beta}_0 - \hat{\beta}_1 x_i - \cdots - \hat{\beta}_{k_1} x_i^{k_1}\right]$$

$$= \sum_{r=0}^{k_2} \beta_r x_i^r - \sum_{r=0}^{k_1} (\beta_r + \delta_r) x_i^r$$

where the δ_r will tend to be small. In this case a plot of e_i versus x_i will exhibit systematic rather than random behavior and will have the characteristics of a (high-degree) polynomial plot. However, if $k_1 > k_2$ then, from Equation (6.6), Section 6.1.2, $E[e_i]=0$ and there will be no pattern or trend in the residual plot. Overfitting, however, can be detected by plotting Y_i versus $\hat{Y}_i$ and looking for "wobbles" between the data points; the fit should be satisfactory *between* adjacent data points as well as *at* the data points.

8.2 ORTHOGONAL POLYNOMIALS

8.2.1 General Statistical Properties

Some of the computational difficulties mentioned in the preceding section can be avoided by using orthogonal polynomials. For example, consider the model

$$Y_i = \gamma_0 \phi_0(x_i) + \gamma_1 \phi_1(x_i) + \cdots + \gamma_k \phi_k(x_i) + \varepsilon_i,$$

where $\phi_r(x_i)$ is an rth-degree polynomial in x_i $(r=0,1,\ldots,k)$, and the polynomials are orthogonal over the x-set; namely,

$$\sum_{i=1}^{n} \phi_r(x_i)\phi_s(x_i) = 0 \qquad (\text{all } r,s,r\neq s). \tag{8.5}$$

Then $\mathbf{Y} = \mathbf{X}\boldsymbol{\gamma} + \boldsymbol{\varepsilon}$, where

$$\mathbf{X} = \begin{bmatrix} \phi_0(x_1) & \phi_1(x_1) & \cdots & \phi_k(x_1) \\ \phi_0(x_2) & \phi_1(x_2) & \cdots & \phi_k(x_2) \\ \cdots & \cdots & \cdots & \cdots \\ \phi_0(x_n) & \phi_1(x_n) & \cdots & \phi_k(x_n) \end{bmatrix},$$

has mutually orthogonal columns, and

$$\mathbf{X}'\mathbf{X}=\begin{bmatrix} \sum_i \phi_0^2(x_i) & 0 & \cdots & 0 \\ 0 & \sum_i \phi_1^2(x_i) & \cdots & 0 \\ \cdots & \cdots & \cdots & \\ 0 & 0 & \cdots & \sum_i \phi_k^2(x_i) \end{bmatrix}.$$

Hence, from $\hat{\gamma}=(\mathbf{X}'\mathbf{X})^{-1}\mathbf{X}'\mathbf{Y}$, we have

$$\hat{\gamma}_r=\frac{\sum_i \phi_r(x_i)Y_i}{\sum_i \phi_r^2(x_i)} \qquad (r=0,1,\ldots,k), \tag{8.6}$$

which holds for *all* k. The orthogonal structure of $\mathbf{X}$ implies that the least squares estimate of $\gamma_r (r \leqslant k)$ is independent of the degree k of the polynomial (cf. Section 3.5)—a very desirable property.

Since $\phi_0(x_i)$ is a polynomial of degree zero we can set $\phi_0(x)\equiv 1$ and obtain

$$\hat{\gamma}_0=\frac{\sum_i 1\cdot Y_i}{\sum_i 1}=\overline{Y}.$$

The residual sum of squares is then

$$\begin{aligned}\text{RSS}_{k+1}&=(\mathbf{Y}-\mathbf{X}\hat{\gamma})'(\mathbf{Y}-\mathbf{X}\hat{\gamma})\\ &=\mathbf{Y}'\mathbf{Y}-\hat{\gamma}'\mathbf{X}'\mathbf{X}\hat{\gamma}\\ &=\sum_i Y_i^2-\sum_{r=0}^{k}\left[\sum_i \phi_r^2(x_i)\right]\hat{\gamma}_r^2\\ &=\sum_i \left(Y_i-\overline{Y}\right)^2-\sum_{r=1}^{k}\left[\sum_i \phi_r^2(x_i)\right]\hat{\gamma}_r^2. \end{aligned} \tag{8.7}$$

If we wish to test $H:\gamma_k=0$ [which is equivalent to testing $\beta_k=0$ in

Equation (8.1)] then the residual sum of squares for the model H is

$$\begin{aligned}\text{RSS}_k &= \sum_{i=1}^{n}\left(Y_i-\overline{Y}\right)^2-\sum_{r=1}^{k-1}\left[\sum_i \phi_r^2(x_i)\right]\hat{\gamma}_r^2\\ &=\text{RSS}_{k+1}+\left[\sum_i \phi_k^2(x_i)\right]\hat{\gamma}_k^2,\end{aligned}$$

and the appropriate F-statistic is

$$\begin{aligned}F&=\frac{\text{RSS}_k-\text{RSS}_{k+1}}{\text{RSS}_{k+1}/(n-k-1)}\\ &=\frac{\sum_i \phi_k^2(x_i)\hat{\gamma}_k^2}{\text{RSS}_{k+1}/(n-k-1)}.\end{aligned}$$

As already mentioned in the preceding section, we can use either the forward selection or backward elimination procedures for determining the appropriate degree. Provided we can readily determine the maximum degree to be entertained, the backward procedure is more efficient and it also avoids the possibility associated with the forward procedure of stopping too soon. The backward procedure has been studied, from a decision theory point of view, by Anderson [1962] (see also Anderson [1971: Section 3.2.2]).

8.2.2 *Generating Orthogonal Polynomials*

Orthogonal polynomials can be obtained in a number of ways. Following Forsythe [1957], a pioneer in this field, Hayes [1974] suggests using the three-term recurrence relationship

$$\phi_{r+1}(x)=2(x-a_{r+1})\phi_r(x)-b_r\phi_{r-1}(x), \tag{8.8}$$

beginning with initial polynomials

$$\phi_0(x)=1 \qquad \text{and} \qquad \phi_1(x)=2(x-a_1).$$

Here x is normalized so that $-1\leqslant x\leqslant +1$, and the a_{r+1} and b_r are chosen to make the orthogonal relations (8.5) hold, namely,

$$a_{r+1}=\frac{\sum_{i=1}^{n} x_i\phi_r^2(x_i)}{\sum_{i=1}^{n} \phi_r^2(x_i)}, \tag{8.9}$$

and

$$b_r=\frac{\sum_{i=1}^{n} \phi_r^2(x_i)}{\sum_{i=1}^{n} \phi_{r-1}^2(x_i)}, \tag{8.10}$$

where $r=0,1,2,\ldots,k-1$, $b_0=0$, and $a_1=\bar{x}$. (Forsythe used the range -2 to $+2$ and the factor unity instead of the factor 2 given in Equation (8.8). These two differences in detail are essentially compensatory, for there is an arbitrary constant factor associated with each orthogonal polynomial; cf. Hayes [1969].) We note that the method of generating the ϕ_r is similar to Gram–Schmidt orthogonalization, with the difference that only the preceding two polynomials are involved at each stage. A computer program based on Forsythe's method is given by Cooper [1968, 1971a, b].

Each $\phi_r(x)$ can be represented in the computer by its values at the (normalized) points x_i or by its a's and b's. However, Clenshaw [1960] has given a useful modification of the above method in which each $\phi_r(x)$ is represented by the coefficients $\{c_j^{(r)}\}$ in its Chebyshev series form, namely,

$$\phi_r(x)=\tfrac{1}{2}c_0^{(r)}T_0(x)+c_1^{(r)}T_1(x)+\cdots+c_r^{(r)}T_r(x). \tag{8.11}$$

The recurrence (8.8) is now carried out in terms of the coefficients $\{c_j^{(r)}\}$, and the fitted polynomial can be expressed in terms of Chebyshev polynomials, namely,

$$\begin{aligned}\hat{Y}&=\hat{f}_k(x)\\&=\tfrac{1}{2}d_0^{(k)}T_0(x)+d_1^{(k)}T_1(x)+\cdots+d_k^{(k)}T_k(x),\end{aligned} \tag{8.12}$$

say. The appropriate recurrence relationships for carrying out these computations are [by substituting (8.11) in (8.8)]

$$c_j^{(r+1)}=c_{j+1}^{(r)}+c_{|j-1|}^{(r)}-2a_{r+1}c_j^{(r)}-b_rc_j^{(r-1)} \tag{8.13}$$

and, by substituting (8.11) and (8.12) in the equation

$$\hat{f}_{k+1}(x) = \hat{f}_k(x) + \hat{\gamma}_{k+1}\phi_{k+1}(x),$$

$$d_j^{(r)} = d_j^{(r-1)} + \hat{\gamma}_r c_j^{(r)}, \tag{8.14}$$

where $j = 0, 1, \ldots, r+1$ and $c_j^{(r)} = d_j^{(r)} = 0$ for $j > r$.

Although the above modification takes about two to three times as long as Forsythe's method, the computing time is generally small in either case. Therefore, because time is not generally the decisive factor, the modification is recommended by Clenshaw and Hayes [1965: p. 168] as it presents a convenient output in concise form. The $c_j^{(r)}$, for example, carry more information than the a_r and b_r. Hayes [1969] also shows that the recurrence relation (8.8) can operate entirely in terms of the coefficients $c_j^{(r)}$ and certain of the quantities $\Sigma_i \phi_r(x_i) T_s(x_i)$. If these numbers are stored then we need not store either the x_i or the $\phi_r(x_i)$. Another useful feature of Clenshaw's modification, pointed out by Hayes [1970a: p. 52], is that the coefficients $c_j^{(k)}$ (for increasing j and fixed k) behave in a very similar manner to RSS_k (for increasing k); they decrease steadily, except possibly at the start, and then settle down to constant values. This feature, illustrated by Examples A and B in Section 8 of Hayes [1970a], provides additional "evidence" for determining the degree of the polynomial fit.

When the coefficients $d_j^{(k)}$ in (8.12) have been computed, $\hat{f}$ can be evaluated at any desired value of x by a procedure given by Clenshaw [1955]. In this we first compute the auxilliary numbers $g_k, g_{k-1}, \ldots, g_0$ from the recurrence relation

$$g_i = 2xg_{i+1} - g_{i+2} + d_i^{(k)}$$

starting with $g_{k+1} = g_{k+2} = 0$. The required value of $\hat{f}$ is then given by

$$\hat{f}_k(x) = \tfrac{1}{2}(g_0 - g_2). \tag{8.15}$$

An error analysis of Clenshaw's modification is given by Clenshaw and Hayes [1965: p. 169]. In particular they give a method for estimating the numerical error in each $\hat{\gamma}_j$; this error can then be used to deduce an error estimate for $d_j^{(r)}$ in Equation (8.12), using (8.14) and the computed values of $c_j^{(r)}$.

8.2.3 *Weighted Least Squares*

It is sometimes desirable to carry out a weighted least squares fit, particularly if a residual plot of the unweighted fit suggests a changing variance.

For instance, we may transform Y (e.g., take $\log Y$) to achieve a good polynomial fit, but in the process affect the variance. Now weighted least squares consists of minimizing

$$\sum_{i=1}^{n} w_i\left(Y_i-\beta_0-\beta_1 x_i-\cdots-\beta_k x_i^k\right)^2$$

where $w_i>0$ $(i=1,2,\ldots,n)$. This problem can be solved once again by using orthogonal polynomials $\tilde{\phi}_r(x)$, say, which satisfy the condition

$$\sum_{i=1}^{n} w_i\tilde{\phi}_r(x_i)\tilde{\phi}_s(x_i)=0, \qquad r\neq s.$$

In this case the fitted polynomial is

$$\tilde{f}_k(x)=\tilde{\gamma}_0+\tilde{\gamma}_1\tilde{\phi}_1(x)+\cdots+\tilde{\gamma}_k\tilde{\phi}_k(x),$$

where

$$\tilde{\gamma}_r=\frac{\sum_i w_i Y_i\tilde{\phi}_r(x_i)}{\sum_i w_i\tilde{\phi}_r^2(x_i)} \qquad (r=0,1,\ldots,k), \tag{8.16}$$

and the variance–covariance matrix of the $\tilde{\gamma}_r$ is a diagonal matrix with elements $\sigma^2\{\sum_i w_i\phi_r^2(x_i)\}^{-1}$. The polynomials can be generated by a relation analogous to (8.8), namely,

$$\tilde{\phi}_{r+1}(x)=2(x-\tilde{a}_{r+1})\tilde{\phi}_r(x)-\tilde{b}_r\tilde{\phi}_{r-1}(x)$$

starting with $\tilde{\phi}_0(x)=1$ and $\tilde{\phi}_1(x)=2(x-\tilde{a}_1)$, where

$$\tilde{a}_{r+1}=\frac{\sum_i w_i x_i\tilde{\phi}_r^2(x_i)}{\sum_i w_i\tilde{\phi}_r^2(x_i)}$$

and

$$\tilde{b}_r=\frac{\sum_i w_i\tilde{\phi}_r^2(x_i)}{\sum_i w_i\tilde{\phi}_{r-1}^2(x_i)}.$$

The rest of the theory, with regard to the use of Chebyshev polynomials, goes through as for the unweighted case; the only changes are in the coefficients $\tilde{a}_r$, $\tilde{b}_r$, and $\tilde{\gamma}_r$.

8.2.4 Application of Constraints

A frequent requirement in curve fitting is for the fitting function $f(x)$, and possibly its derivatives also, to take specified values at certain values of x. For example, the function may be required to pass through the origin or to join smoothly on to a straight line at some point, or we may wish to fit the data in two adjoining ranges separately, forcing continuity up to some order of derivative at the joint. To satisfy such requirements Clenshaw and Hayes [1965] consider setting

$$f(x) = \mu(x) + \nu(x)g(x), \tag{8.17}$$

where $\mu(x)$ is a simple function (usually a polynomial) chosen to satisfy the constraints, $\nu(x)$ is an associated "zeroizing" polynomial chosen to ensure that $f(x)$ satisfies the specifications, and $g(x)$ is a polynomial to be determined by least squares. For example, if we specify $f(0)=1, f(1)=f'(1)=0$, we may take

$$\mu(x) = (1-x)^2 \tag{8.18}$$

and

$$\nu(x) = x(1-x)^2.$$

Having chosen $\mu(x)$ and $\nu(x)$ we now fit a polynomial $G(x)$, of the form $\nu(x)g(x)$, to the modified data $Y_i' = Y_i - \mu(x_i)$. This can be done readily using Forsythe's iterative method [Equation (8.8)], but starting with $\phi_0(x) = \nu(x)$ instead of $\phi_0(x)=1$; $\nu(x)$ will then be a factor of each $\phi_r(x)$, and hence of the fitted curve $\hat{f}(x) = \mu(x) + \hat{G}(x)$.

If, in general, we wish to fit a polynomial for $x \geqslant x_0$ satisfying the end point conditions $f(x_0) = m_0, f'(x_0) = m_1, \ldots, f^{(s-1)}(x_0) = m_{s-1}$, then we choose

$$\mu(x) = \sum_{j=0}^{s-1} \frac{m_j(x-x_0)^j}{j!} \qquad \text{and} \qquad \nu(x) = (x-x_0)^s.$$

Further details are given in Cadwell and Williams [1961].

Clenshaw and Hayes [1965] note that nonpolynomial behavior can be introduced into the fit by a suitable choice of $\mu(x)$. For example, if it is required that $f(0)=1$, $f(1)=0$, and $f'(1)=\infty$, they suggest

$$\mu(x)=\gamma(1-x)^{v}+(1-\gamma)(1-x) \qquad (0<v<1)$$

and

$$\nu(x)=x(1-x).$$

(It is assumed that satisfactory values of γ and v can be chosen by plotting $\log Y_i$ against $\log(1-x_i)$ for values of x_i near unity; for x near 1 the dominant term in $\mu(x)$ is $\gamma(1-x)^{v}$.) In general, if the forced values are finite then both $\mu(x)$ and $\nu(x)$ can be polynomials.

If Clenshaw's modification is to be used in the fitting process, it is necessary that the polynomial $\nu(x)$ be first expressed in Chebyshev-series form. When $\mu(x)$ is a polynomial, as in (8.18), it is best to convert this to Chebyshev-series form also so that the final fit $\hat{f}(x)$ can be as a single Chebyshev series (Hayes [1974]).

Equation (8.17) can also be estimated by fitting a polynomial $g(x)$ to the "scaled" Y_i values

$$Y_i''=\{\nu(x_i)\}^{-1}[Y_i-\mu(x_i)]$$

instead of to the Y_i. We now minimize

$$\sum_{i=1}^{n} \nu^2(x_i)[Y_i''-g(x_i)]^2$$

which is equivalent to a weighted least squares fit with weights $w_i=\nu^2(x_i)$. The methods of the preceding section now apply and, if $\hat{g}(x)$ is the fitted polynomial, we have

$$\hat{f}(x)=\mu(x)+\nu(x)\hat{g}(x).$$

In conclusion we mention one other form of constraint. If the polynomial is constrained to be nonnegative, nondecreasing, or convex, then the quadratic programming type method of Hudson [1969] can be used for fitting the polynomial.

8.2.5 *Equally Spaced x-Values*

Suppose that the x values are equally spaced so that they can be transformed to

$$x_i=i-\tfrac{1}{2}(n+1), \qquad (i=1,2,\ldots,n). \tag{8.19}$$

Then we have the following system of orthogonal polynomials (generally ascribed to Chebyshev):

$$\phi_0(x)=1$$

$$\phi_1(x)=\lambda_1 x$$

$$\phi_2(x)=\lambda_2\left(x^2-\tfrac{1}{12}(n^2-1)\right)$$

$$\phi_3(x)=\lambda_3\left(x^3-\tfrac{1}{20}(3n^2-7)x\right)$$

$$\phi_4(x)=\lambda_4\left(x^4-\tfrac{1}{14}(3n^2-13)x+\tfrac{3}{560}(n^2-1)(n^2-9)\right), \text{ etc.,}$$

where the λ_r are chosen so that the values $\phi_r(x_i)$ are all positive and negative integers. These polynomials are extensively tabulated in Pearson and Hartley [1970] for $n=1(1)52$ and $r=1(1)6$ $(r \leqslant n-1)$; a section of their table is given in Table 8.1. To illustrate the use of this table suppose that $n=3$. Then $x_i=-1,0,1$, $\phi_0(x)=1$, $\phi_1(x)=\lambda_1 x=x$, $\phi_2(x)=\lambda_2(x-\frac{2}{3})=3x^2-2$, and the fitted polynomial is

$$\hat{f}(x)=\hat{\beta}_0+\hat{\beta}_1 x+\hat{\beta}_2(3x^2-2),$$

where

$$\hat{\beta}_0=\overline{Y}$$

$$\hat{\beta}_1=\frac{\sum_i \phi_1(x_i)Y_i}{\sum \phi_1^2(x_i)}=\tfrac{1}{2}\{(-1)Y_1+(0)Y_2+(1)Y_3\}=\tfrac{1}{2}(Y_3-Y_1)$$

Table 8.1 Values of the Orthogonal Polynomials, $\phi_r(x)$, for the Equally Spaced x-Data of Equation (8.19)

	$n=3$		$n=4$			$n=5$			
	ϕ_1	ϕ_2	ϕ_1	ϕ_2	ϕ_3	ϕ_1	ϕ_2	ϕ_3	ϕ_4
	-1	1	-3	1	-1	-2	2	-1	1
	0	-2	-1	-1	3	-1	-1	2	-4
	1	1	1	-1	-3	0	-2	0	6
			3	1	1	1	-1	-2	-4
						2	2	1	1
$\sum_i \phi_r^2(x_i)$	2	6	20	4	20	10	14	10	70
λ_r	1	3	2	1	$\frac{10}{3}$	1	1	$\frac{5}{6}$	$\frac{35}{12}$

and

$$\hat{\beta}_2 = \tfrac{1}{6}\{Y_1 - 2Y_2 + Y_3\}.$$

Also the residual sum of squares is given by [Equation (8.7)]

$$\begin{aligned}\text{RSS}_3 &= \sum_{i=1}^{3}\left(Y_i - \overline{Y}\right)^2 - \hat{\beta}_1^2 \sum_i \phi_1^2(x_i) - \hat{\beta}_2^2 \sum_i \phi_2^2(x_i)\\ &= \sum\left(Y_i - \overline{Y}\right)^2 - 2\hat{\beta}_1^2 - 6\hat{\beta}_2^2.\end{aligned}$$

A helpful numerical example is given in Draper and Smith [1966: p. 155].

The theory of this section can be used for fitting polynomials up to degree 6 by hand. However, its main application is in the theory of experimental design where various sums of squares are sometimes split into linear, quadratic, etc. components.

A simple method for generating the orthogonal polynomials iteratively when $x = 0, 1, \ldots, n-1$, due to Fisher and Yates [1957], is described by Jennrich and Sampson [1971].

8.3 PIECEWISE POLYNOMIAL FITTING

8.3.1 Unsatisfactory Fit

Sometimes a polynomial fit is unsatisfactory even when orthogonal polynomials up to, say, degree 20 are fitted. This lack of fit is usually revealed in several ways. One symptom is the failure of RSS_k to settle down to a constant value as k increases; the residual sum of squares may, for example, just continue to decrease slowly. Another symptom is the behavior of the residuals: a residual plot of e_i (or d_i) versus x_i will continue to exhibit a systematic pattern instead of a random one (see, for example, Hayes [1970a: Section 8, Example E]). In the worst cases there will be waves in the fitted curve which eventually become oscillations between adjacent data points usually near the ends of the range. These difficulties most frequently arise when the behavior of the underlying function is very different in one part of the range from another. It may, for example, be rapidly varying in one region and slowly varying in another. Hayes [1970a] suggests two methods for dealing with such cases.

First, if the "misbehavior" occurs at one end of the range it is often possible to remove the problem by a suitable transformation, taking an origin close to the misbehavior. Although this is a matter of trial and error, Hayes points out that we improve with practice! If we wish to stretch out the part of the x-axis where the function is rapidly changing and contract the part where it is slowly varying we can transform x. For example, a log transformation is often appropriate to cases that appear to have a vertical asymptote, and a fractional power to cases that appear to have an infinite derivative at a finite value of the function. This problem can sometimes be handled by introducing nonpolynomial terms into the fit as described in Section 8.2.4. It is sometimes helpful to transform Y, though this may necessitate a weighted least squares analysis due to a nonconstant variance; a useful collection of graphs showing the different types of functions is given in Daniel and Wood [1971: pp. 20–24].

One other way of dealing with difficult cases is by dividing the range into segments and fitting different curves in different segments. The problem of choosing suitable dividing points is not an easy one and, once again, it is likely to be a process of trial and error. However, having chosen a particular subdivision, we first fit a polynomial to one of the two end segments using the data in that segment plus a few points outside (to ensure that we get better values for the derivatives at the join with the next segment). We then compute the values, at this join, of the polynomial and of as many derivatives as we wish to make continuous (usually first derivative continuity is sufficient). The next step is to use the method of Section 8.2.4 to force these values into the fit for the next subrange, again using a few points beyond the next join. For example, if the join point is $x=1.4$ and we estimate $f(1.4)=2.3$ and $f'(1.4)=1.6$, then we can use

$$\mu(x)=2.3+1.6(x-1.4) \qquad \text{and} \qquad \nu(x)=(x-1.4)^2.$$

Using the above method we proceed, segment by segment, from one end of the curve to the other. We finally examine the overall fit and introduce further joins where the fit is unsatisfactory. An example where piecewise fitting was used is given by Hayes [1970a: Section 8, Example F].

8.3.2 *Use of Spline Functions*

The above piecewise fitting method suffers from a major defect: undue weight is given to the initial segments. This means that roundoff errors and the effects of a poor fit are passed on cumulatively via the join points (called knots) so that the last segment may be distorted because of erroneous constraints at one end (see, for example, Payne [1970: Figs. 3 and 4]). What we need, therefore, is a least squares method for simultaneously fitting all the segments, subject to the constraints of the continuity of the function and certain derivatives at the knots. This particular problem of piecewise approximation has led to the theory of spline functions, a theory pioneered by Schoenberg [1946].

A *spline function* $s(x)$ of order k (degree $k-1$), with knots $\lambda_1,\lambda_2,\ldots,\lambda_h$ (where $\lambda_1<\lambda_2<\cdots<\lambda_h$) and domain $[a,b]$ $(-\infty\leqslant a<\lambda_1, \lambda_h<b\leqslant\infty)$, is a function with the following properties:

(1) In each of the intervals

$$a<x\leqslant\lambda_1; \qquad \lambda_{j-1}\leqslant x\leqslant\lambda_j \; (j=2,3,\ldots,h); \qquad \lambda_h\leqslant x<b$$

$s(x)$ is a polynomial of degree $k-1$ at most.

(2) $s(x)$ and its derivatives up to order $(k-2)$ are continuous. (When a and b are finite, which is the usual case in practice, some authors call $\lambda_0=a$ and $\lambda_{h+1}=b$ knots also.)

The cubic spline $(k=4)$ is a satisfactory function for fitting data, and second-derivative continuity is usually adequate for most practical problems. A full discussion on the least squares fitting of cubic splines when the knots are known is given by Poirier [1973]. He also discusses the question of hypothesis testing for linear and quadratic segments, and hypothesis testing for structural changes. Further references on the properties of cubic splines are Hayes [1970b: Chapters 4, 6, 8, 9] and Hayes [1974]; the latter reference discusses the question of choosing the knots when they are unknown (see also Wold [1974: pp. 2–3] for somc helpful suggestions).

Any cubic spline with knots λ_i has a unique representation in the form

$$s(x)=\sum_{j=0}^{3}\alpha_j x^j+\sum_{i=1}^{h}\beta_i(x-\lambda_i)_+^3, \tag{8.20}$$

where

$$\begin{aligned} A_+ &= A, \quad \text{when } A\geqslant 0\\ &= 0, \quad \text{when } A\leqslant 0. \end{aligned}$$

This representation contains $h+4$ basis functions (four power terms and h one-sided cubics), the smallest number by which the general cubic spline with h knots can be represented. However, from a computational vantage point, a much better representation is provided by the so-called *B*-splines or fundamental splines. A *cubic* B-*spline* is a cubic spline with the characteristic that it is nonzero only over four adjacent intervals between knots. Specifically, we define the cubic *B*-spline $M_i(x)$ to be a cubic spline with knots $\lambda_1,\lambda_2,\ldots,\lambda_h$ which is zero everywhere except in the range $\lambda_{i-4}<x<\lambda_i$ (some authors use $\lambda_{i-2}<x<\lambda_{i+2}$). It transpires that $M_i(x)$ [see Equation (8.21)] has the same sign (conveniently taken to be positive) throughout the range $\lambda_{i-4}<x<\lambda_i$ and has a single local maximum. (A summary of the properties of *B*-splines is given by Curry and Schoenberg [1966].)

To define the full set of B-splines we introduce eight additional knots $\lambda_{-3},\lambda_{-2},\lambda_{-1},\lambda_0,\lambda_{h+1},\lambda_{h+2},\lambda_{h+3}$, and λ_{h+4}. These must satisfy

$$\lambda_{-3}<\lambda_{-2}<\lambda_{-1}<\lambda_0\leqslant a$$

and

$$b\leqslant\lambda_{h+1}<\lambda_{h+2}<\lambda_{h+3}<\lambda_{h+4},$$

but are otherwise arbitrary. It is usual to take $\lambda_0=a$ and $\lambda_{h+1}=b$; the other λ_i are arbitrary and can be chosen for numerical convenience. Except for possible differences in rounding error, they do not affect the value of the least squares spline in $[a,b]$. With this augmented set of knots we can define $h+4$ fundamental splines

$$M_i(x)=\sum_{j=i-4}^{i}\left\{\frac{(x-\lambda_j)_+^3}{\prod_{\substack{m=i-4\\ m\neq j}}^{i}(\lambda_j-\lambda_m)}\right\},\qquad (i=1,2,\ldots,h+4),\qquad (8.21)$$

and the general cubic spline with knots $\lambda_1,\lambda_2,\ldots,\lambda_h$ now has a unique representation in the range $[a,b]$ of the form

$$s(x)=\sum_{i=1}^{h+4}\gamma_i M_i(x).$$

The above discussion on fundamental splines comes from Hayes [1974], and his paper should be consulted for references and details of least squares fitting. For further applications of spline functions to data analysis see Wold[1974].

8.3.3 *Multiphase Polynomial Regression*

In Section 7.6 we discussed two-phase linear regression, that is, a linear model which allowed the possibility of a change in slope. A generalization of this might be described as *multiphase polynomial regression*, in which we have a polynomial that changes its form at one or more points due to (possible) physical changes in the underlying process. Robison [1964] considered the case of just two polynomials with a known changeover (join) point, thus generalizing the theory of Section 7.6.1 (due to Sprent [1961]). Using cubic splines Poirier [1973] generalized Robison's work to the case of any number of (cubic) polynomials with known join points.

When the join points are unknown, then the problem is much more difficult. If there are good a priori reasons for postulating one or more possible structural changes, then possibly the methods of Hudson [1966], McGee and Carleton [1970], and Gallant and Fuller [1973] can be used. However, if it is just a question of fitting the best piecewise polynomial with an unknown number of pieces, the methods of the previous sections would seem appropriate. The question of choosing suitable knots in fitting a cubic spline is discussed by Hayes [1974].

8.4 OPTIMAL ALLOCATION OF POINTS

In designing an experiment for a kth-degree polynomial fit, the experimenter is faced with an allocation problem; given that x is scaled to be in $[-1,1]$, at what values of x should the n values of Y be observed? Obviously the answer to this question depends on our purpose for fitting the polynomial. For example, are we interested in estimation (with regard to all the polynomial coefficients, or just a subset of coefficients), interpolation (e.g., calibration curves), or extrapolation (prediction)? To meet such questions a variety of optimality criteria have been suggested, the most studied being D-optimality (Section 3.11). Hoel [1958: pp. 1137–1138] discovered that a D-optimal design is also one which minimizes

$$\max_{-1 \leqslant x \leqslant 1} \operatorname{var}\left[\sum_{r=0}^{k} \hat{\beta}_r x^r\right], \qquad (=v(x), \text{ say}). \tag{8.22}$$

A design with this property is called a *minimax* (or G-optimal) design, and the equivalence of the D-optimal and minimax criteria was proved by Kiefer and Wolfowitz [1960] (the so-called equivalence theorem: see St. John and Draper [1975] for a helpful review).

A first step toward finding the D-optimal design was made by De La Garza [1954]. He showed that corresponding to any allocation of n x values there exists another allocation at just $k+1$ points x_i $(i=0,1,\ldots,k)$ with the same $\mathbf{X}'\mathbf{X}$ matrix. The optimality problem was then solved by Hoel [1958] and Guest [1958] (see also Kiefer and Wolfowitz [1959] for a more general treatment) who showed that for D-optimality np_i observations $(i=0,1,\ldots,k)$ should be made at x_i, where $p_i=1/(1+k)$ (that is, equal allocation) and the x_i are the $k+1$ zeros of $(1-x^2)P_k'(x)$, $P_k(x)$ being the kth-degree Legendre polynomial. In this case $v(x_i)=\sigma^2(k+1)/n$, a unique property that can be used to test for optimality. The values of x_i are listed by Kussmaul [1969] for $k=1(1)5$.

Sometimes a researcher is interested in just a subset of the coefficients

$\beta_0, \beta_1, \ldots, \beta_k$, and D-optimality has been defined for this particular problem (cf. St. John and Draper [1975] for a general review). If we are specifically interested in just β_k then most reasonable design criteria reduce to finding the design that minimizes var$[\hat{\beta}_k]$. This problem was solved by Kiefer and Wolfowitz [1959], and the optimal solution is

$$p_i = \frac{1}{2k}, \qquad i = 0, k$$

$$= \frac{1}{k}, \qquad i = 1, 2, \ldots, k-1$$

at the points

$$x_i = -\cos\left(\frac{i\pi}{k}\right), \qquad (i = 0, 1, \ldots, k). \tag{8.23}$$

(These are called Chebyshev points because $T_k(x_i) = 1$.) When studying optimality it is convenient to allow the allocations np_i to be (theoretically) nonintegral; such designs are called approximate or continuous designs. In practice the np_i may not be integers so that our working design is generally an integer approximation to the optimal continuous design. The cases $k = 1, 2$ are considered in further detail by Atkinson [1972].

If we are likely to be interested in any of the individual $\hat{\beta}_r$ then we can define optimality in terms of

$$\max_{0 \leqslant r \leqslant k} \text{var}\left[\hat{\beta}_r\right].$$

Elvring [1959] called a design with this property the minimax s.p. (single parameter) design, and Studden [1968] gave the optimal solution. Murty [1971] highlighted a useful sufficient condition for detecting optimality in the above sense, and gave a number of examples.

The extrapolation problem has also received considerable attention in the literature. If we wish to find the design which minimizes the prediction var$[\sum_{r=0}^{k} \hat{\beta}_r x^r]$ at x, where $|x| > 1$, then the solution is once again (8.23) but with the allocation (Hoel and Levine [1964])

$$p_i = \frac{|L_i(x)|}{\sum_{i=0}^{k} |L_i(x)|} \qquad (i = 0, 1, \ldots, k),$$

where

$$L_i(x) = \frac{(x - x_0) \cdots (x - x_{i-1})(x - x_{i+1}) \cdots (x - x_k)}{(x_i - x_0) \cdots (x_i - x_{i-1})(x_i - x_{i+1}) \cdots (x_i - x_k)}$$

is the Lagrange polynomial. This work has been extended by Kiefer and Wolfowitz [1965] (see Karlin and Studden [1966], Studden [1968], and, in particular, Herzberg and Cox [1972]).

The above solutions to the design problem have one serious drawback. We assumed that k is known, and Y is observed at only $k+1$ different values of x so that the (weighted) least squares fit using the mean of Y at each x_i is exact, with no residual for examining underlying assumptions. Since in practice k is not generally known, an optimal solution may not be satisfactory as it makes no provision for examining the adequacy of the fit. To get round this problem Box and Draper [1959, 1963] (see also Kupper [1973], Kupper and Meydrech [1973], and Thompson [1973]) defined optimality in terms of minimizing the integrated mean square error of prediction; they focused their attention on the squared bias term because this term seemed to be the major contributing factor. Karson et al. [1969] and Cote et al. [1973] took a similar viewpoint and considered designs which, in the first instance, minimized the integrated squared bias. Stigler [1971], however, criticized the bias criteria and suggested modifications of the usual D-optimal and minimax designs. He called his modification a C-restricted design as he imposed the condition that, if the additional term $\beta_{k+1}x^{k+1}$ is fitted,

$$\text{var}\left[\hat{\beta}_{k+1}\right] \leqslant \frac{\sigma^2 C}{n}, \tag{8.24}$$

where C is prechosen. The choice of C reflects a compromise between two conflicting goals: precise inferences about β_{k+1}, and precise inferences about the kth-degree polynomial. On the one hand, C should be chosen sufficiently small so that practically significant departures from the model can be detected with a specified precision (for example, $H:\beta_{k+1}=0$ has a specified power); on the other hand, large values of C will yield more efficient designs for the kth-degree polynomial fit. Unfortunately, finding the C-restricted optimal solution is not easy, and only the solution for $k=1$ is given by Stigler.

Several other approaches to the problem of optimal design when k is unknown have been suggested. For example, Hoel [1968] and Kussmaul [1969] recommend using a D-optimal design for the highest degree that the experimenter is prepared to entertain. Atwood [1971], however, assumes that the polynomial to be fitted is of a given degree (say, s) plus a "small" polynomial of given higher degree k, and suggests using a weighted combination of D-optimal designs for degrees s and k. His emphasis is on robustness rather than on optimality.

The problem of finding optimal designs for estimating the slope of a polynomial regression has been considered by Murty and Studden [1972];

the case of a second-degree polynomial is investigated by Ott and Mendenhall [1972].

8.5 POLYNOMIAL REGRESSION IN SEVERAL VARIABLES

8.5.1 Surface fitting

The problem of fitting a second-degree polynomial in two variables, namely,

$$f(x_1,x_2)=\beta_0+\beta_1x_1+\beta_2x_2+\beta_{11}x_1^2+\beta_{12}x_1x_2+\beta_{22}x_2^2, \qquad (8.25)$$

has received considerable attention in the literature. It transpires that all the theory of orthogonal polynomials can be generalized to handle the two-dimensional case. Once again Chebyshev polynomials play a fundamental role and the reader is referred to Hayes [1974] for details.

8.5.2 Response Surfaces

One of the most important applications of polynomial regression in several variables is in the study of response surfaces. We illustrate some of the basic features of response surface methodology by considering the simple case of just two regressors.

Suppose that the "response" (yield) η from a given experiment is an unknown function of two variables x_1 (temperature) and x_2 (concentration), namely, $\eta=g(x_1,x_2)$. It is assumed that this three-dimensional surface is "well-behaved," in particular is smooth with a single well-defined peak. The response η is measured with error so that we actually observe $Y=\eta+\varepsilon$, where $E[\varepsilon]=0$ and $\text{var}[\varepsilon]=\sigma^2$. The basic problem of response theory then is to estimate the coordinates, (x_{01},x_{02},η_0) say, of the summit.

One method of doing this is to use a sequence of experiments and a steepest ascent technique to "climb" up the surface. For points away from the summit the surface is relatively linear so that it can be represented locally by a plane, namely,

$$E[Y]=\beta_0+\beta_1x_1+\beta_2x_2. \qquad (8.26)$$

To estimate the coefficients β_i we can, for example, use a so-called 2^2 design in which we observe Y at the four vertices of a small rectangle, with center P_1, in the (x_1,x_2) plane (Fig. 8.1). Suppose that Y_{rs} is the observed Y at (x_{r1},x_{s2}), where x_{r1} $(r=1,2)$ are the two chosen values of x_1 and x_{s2}

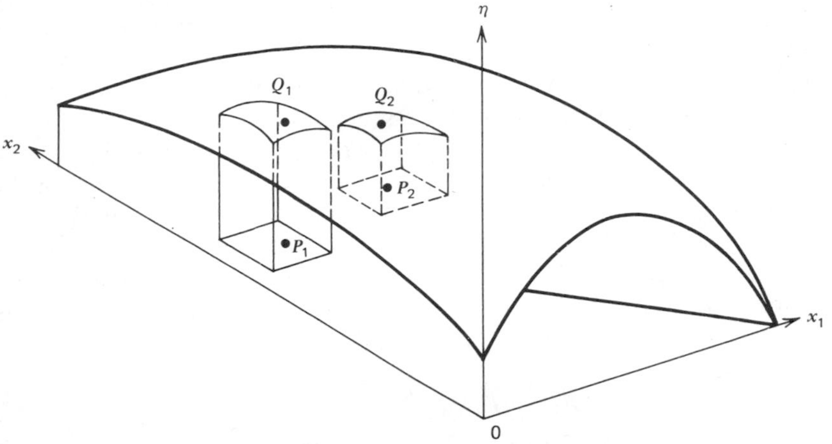

Fig. 8.1 A response surface.

$(s=1,2)$ are the two chosen values of x_2. Then we can fit the model

$$Y_{rs}=\beta_0+\beta_1x_{r1}+\beta_2x_{s2}+\varepsilon_{rs}, \tag{8.27}$$

where $r=1,2$ and $s=1,2$, and obtain the fitted plane

$$\hat{Y}=\phi(x_1,x_2)=\hat{\beta}_0+\hat{\beta}_1x_1+\hat{\beta}_2x_2. \tag{8.28}$$

If Q_1 is the point on this plane vertically above P_1 we can then use the fitted plane, which approximates to the surface in the neighborhood of Q_1, to help us climb up the surface to a higher point Q_2 and thus obtain a higher yield Y. For example, if $\hat{\beta}_1$ and $\hat{\beta}_2$ are both positive in (8.28) we would increase x_1 and x_2. However, the most efficient way of climbing up the surface is to choose the direction of steepest slope. To find this so-called path of steepest ascent we now consider the following problem. Suppose we wish to maximize $\phi(d_1,d_2)-\phi(0,0)$ subject to $d_1^2+d_2^2=r^2$. Using a Lagrange multiplier λ we have

$$\frac{\partial\phi}{\partial d_i}+2\lambda d_i=0 \qquad (i=1,2),$$

or, setting ϕ equal to the right side of (8.28), $d_i\propto\hat{\beta}_i$ for a maximum. Therefore regarding Q_1 as origin, the (x_1,x_2) coordinates of the next experimental observation should be $(k\hat{\beta}_1,k\hat{\beta}_2)$ for some $k>0$. By steadily increasing k we can go on measuring Y until we reach a point P_2 in the (x_1,x_2) plane at which the increase in Y due to a change in k becomes very small, or possibly negative. A new 2^2 experiment is then carried out on another small rectangle using P_2 as center, and another plane (8.28) is fitted. The path of steepest ascent is redetermined and, once again we proceed in this direction until there is little change in Y, at, say, P_3. In this way we climb the surface toward the summit.

As we approach the summit, $\hat{\beta}_1$ and $\hat{\beta}_2$ get smaller so that progress by the method of steepest ascent becomes more difficult; the curvature of the surface also begins to have a significant effect on the yield. When we are in the region of the summit we can then fit a quadratic polynomial of the form (8.25) using, say, a 3^2 design; that is, we use three appropriate values for each of x_1 and x_2 and observe Y at the 9 design points. By shifting the origin and rotating the axes, the fitted surface

$$y=\hat{\beta}_0+\hat{\beta}_1x_1+\hat{\beta}_2x_2+\hat{\beta}_{11}x_1^2+\hat{\beta}_{12}x_1x_2+\hat{\beta}_2x_2^2 \tag{8.29}$$

can be expressed in the canonical form

$$y-c_3=\lambda_1(x_1-c_1)^2+\lambda_2(x_2-c_2)^2 \qquad (\lambda_1,\lambda_2>0), \tag{8.30}$$

where (c_1,c_2,c_3) is an estimate of the summit point (x_{01},x_{02},η_0). The triple (c_1,c_2,c_3) can be most readily found by differentiating (8.29) partially with respect to x_1 and x_2, and solving the resulting pair of equations for x_1 and x_2; c_3 is the value of y in (8.29) at the solution (c_1,c_2).

The above rather sketchy description of response surface methodology leaves a number of questions unanswered such as the following.

(1) In the above discussion we used a 2^2 design to carry out a planar fit (called a first order design) and a 3^2 design for the quadratic fit (called a second order design). This raises the question, what is the best design to use in each case?

(2) How do we know when to change from a first-order to a second-order design?

(3) How do we select the values of k in $(k\hat{\beta}_1,k\hat{\beta}_2)$?

(4) What happens if, in our climb, we run into a stationary point which is not the maximum, or a slowly rising ridge? [Such a situation is indicated when one or other of the λ_i in Equation (8.30) is negative.]

Although there is not space to consider these and other important practical questions, a few comments about first-order designs are perhaps appropriate. We saw in Section 3.5 (Lemma) that designs with orthogonal structure have certain optimal properties. In particular the 2^2 design lies in this category, and is D-optimal (Box and Draper [1971]) if we scale the values of x_1 and x_2 so that they take the values ± 1. To examine this orthogonal structure it is convenient to represent the two levels of x_1 symbolically by 1 and a, and the two levels of x_2 by 1 and b so that the four possible combinations $(1,1),(1,a),(1,b),(a,b)$ can be represented symbolically (by multiplying the levels together in each pair) as 1, a, b, and ab, with observed Y values Y_1, Y_a, Y_b, and Y_{ab}, respectively. In this notation (8.27) becomes

$$\begin{bmatrix} Y_1 \\ Y_a \\ Y_b \\ Y_{ab} \end{bmatrix} = \begin{bmatrix} 1 & -1 & -1 \\ 1 & 1 & -1 \\ 1 & -1 & 1 \\ 1 & 1 & 1 \end{bmatrix} \begin{bmatrix} \beta_0 \\ \beta_1 \\ \beta_2 \end{bmatrix} + \begin{bmatrix} \varepsilon_1 \\ \varepsilon_a \\ \varepsilon_b \\ \varepsilon_{ab} \end{bmatrix}, \tag{8.31}$$

or $\mathbf{Y} = \mathbf{X}\boldsymbol{\beta} + \boldsymbol{\varepsilon}$, where the columns of $\mathbf{X}$ are mutually orthogonal and satisfy the conditions of the Lemma in Section 3.5. Then

$$\hat{\boldsymbol{\beta}} = (\mathbf{X}'\mathbf{X})^{-1}\mathbf{X}'\mathbf{Y} = \tfrac{1}{4} \begin{bmatrix} 1 & 1 & 1 & 1 \\ -1 & 1 & -1 & 1 \\ -1 & -1 & 1 & 1 \end{bmatrix} \begin{bmatrix} Y_1 \\ Y_a \\ Y_b \\ Y_{ab} \end{bmatrix}$$

so that

$$\begin{aligned} \hat{\beta}_0 &= \bar{Y}, \\ \hat{\beta}_1 &= \tfrac{1}{4}(-Y_1 + Y_a - Y_b - Y_{ab}) \\ &= \tfrac{1}{2}\left[\tfrac{1}{2}(Y_a + Y_{ab}) - \tfrac{1}{2}(Y_1 + Y_b)\right] \\ &= \tfrac{1}{2}(\textit{average effect of first factor at upper level} \\ &\qquad - \textit{average effect of first factor at lower level}) \end{aligned}$$

and

$$\begin{aligned} \hat{\beta}_2 &= \tfrac{1}{4}(-Y_1 - Y_a + Y_b + Y_{ab}) \\ &= \tfrac{1}{2}\left[\tfrac{1}{2}(Y_b + Y_{ab}) - \tfrac{1}{2}(Y_1 + Y_a)\right]. \end{aligned}$$

If we use "factor" terminology and call x_1 and x_2 factors A and B, then $\hat{\beta}_1$ and $\hat{\beta}_2$ can be identified as estimates of what we might call the *main effects* of A and B, respectively.

The general question of optimal designs for response surface applications is discussed further by Box and Draper [1971, 1975], Atkinson [1972], Thompson [1973], and Mitchell [1974b]. Further references on response surface methodology are Davies [1960], Hill and Hunter [1966: a review article], John [1971: Chapter 10], Guttman et al. [1971: 435ff], and Myers [1971].

In addition to the method of steepest ascent another technique known as evolutionary operation (EVOP), proposed by Box [1957] and Box and Draper [1969], is recommended for industrial use. However, this technique does not seem to be as widely used as it might be, and the reader is referred to Hahn and Dershowitz [1974] and Lowe [1974] for some helpful comments.

MISCELLANEOUS EXERCISES 8

1. Using the method of orthogonal polynomials described in Section 8.2.5, fit a third-degree equation to the following data:

y (index):	9.8	11.0	13.2	15.1	16.0
x (year):	1950	1951	1952	1953	1954

   Test the hypothesis that a second-degree equation is adequate.

2. Verify Equations (8.13) and (8.14) in Section 8.2.2.
3. Show that the least squares estimates of β_1 and β_2 for the model (8.31) in Section 8.5.2 are still unbiased even when the true model includes an interaction term β_{12}, that is,

$$E[Y]=\beta_0+\beta_1x_1+\beta_2x_2+\beta_{12}x_1x_2.$$

   Find the least squares estimate of β_{12}.

4. Suppose that the regression curve

$$E[Y]=\beta_0+\beta_1x+\beta_2x^2$$

   has a local maximum at $x=x_m$ where x_m is near the origin. If Y is observed at n points x_i $(i=1,2,\ldots,n)$ in $[-a,a], \bar{x}=0$, and the usual normality assumptions hold, outline a method for finding a confidence interval for x_m.
   [*Hint*: Use the method of Section 7.2.2.]

   Williams [1959: p. 110]

CHAPTER 9

Analysis of Variance

9.1 ONE-WAY CLASSIFICATION

9.1.1 Representation as a Regression Model

In Section 4.1.3, Example 4.2, we showed how the general regression theory could be applied to the problem of comparing the means of two normal populations when the population variances are equal. We now extend this theory to the case of comparing I normal populations ($I \geqslant 2$).

Let Y_{ij} be the jth sample observation ($j = 1, 2, \ldots, J$) on the ith normal population $N(\mu_i, \sigma^2)$ ($i = 1, 2, \ldots, I$), so that we have the following array of data:

		Sample mean
Population 1:	$Y_{11}, Y_{12}, \ldots, Y_{1J}$	$\overline{Y}_{1\cdot}$
Population 2:	$Y_{21}, Y_{22}, \ldots, Y_{2J}$	$\overline{Y}_{2\cdot}$
⋯	⋯	⋯
Population I:	$Y_{I1}, Y_{I2}, \ldots, Y_{IJ}$	$\overline{Y}_{I\cdot}$

In order to apply the general regression theory once again, we combine the above information into the single model

$$Y_{ij} = \mu_i + \varepsilon_{ij} \qquad (i = 1, 2, \ldots, I; \quad j = 1, 2, \ldots, J),$$

where the ε_{ij} are i.i.d. (independently and identically distributed) as

$N(0,\sigma^2)$. Using vector notation, this model takes the form:

$$\begin{bmatrix} Y_{11} \\ Y_{12} \\ \cdots \\ Y_{1J} \\ \hline Y_{21} \\ Y_{22} \\ \cdots \\ Y_{2J} \\ \hline \cdots \\ \hline Y_{I1} \\ Y_{I2} \\ \cdots \\ Y_{IJ} \end{bmatrix} = \begin{bmatrix} 1 & 0 & 0 & \cdots & 0 \\ 1 & 0 & 0 & \cdots & 0 \\ \cdots & \cdots & \cdots & \cdots & \cdots \\ 1 & 0 & 0 & \cdots & 0 \\ \hline 0 & 1 & 0 & \cdots & 0 \\ 0 & 1 & 0 & \cdots & 0 \\ \cdots & \cdots & \cdots & \cdots & \cdots \\ 0 & 1 & 0 & \cdots & 0 \\ \hline \cdots & \cdots & \cdots & \cdots & \cdots \\ \hline 0 & 0 & 0 & \cdots & 1 \\ 0 & 0 & 0 & \cdots & 1 \\ \cdots & \cdots & \cdots & \cdots & \cdots \\ 0 & 0 & 0 & \cdots & 1 \end{bmatrix} \begin{bmatrix} \mu_1 \\ \mu_2 \\ \cdots \\ \mu_I \end{bmatrix} + \begin{bmatrix} \varepsilon_{11} \\ \varepsilon_{12} \\ \cdots \\ \varepsilon_{1J} \\ \hline \varepsilon_{21} \\ \varepsilon_{22} \\ \cdots \\ \varepsilon_{2J} \\ \hline \cdots \\ \hline \varepsilon_{I1} \\ \varepsilon_{I2} \\ \cdots \\ \varepsilon_{IJ} \end{bmatrix} \tag{9.1}$$

or

$$\mathbf{Y} = \mathbf{X}\boldsymbol{\mu} + \boldsymbol{\varepsilon}, \tag{9.2}$$

where $\boldsymbol{\mu} = [(\mu_i)]$, $\boldsymbol{\varepsilon} \sim N_n(\mathbf{0}, \sigma^2\mathbf{I}_n)$, and $n = IJ$. Since the columns of $\mathbf{X}$ are linearly independent, (9.2) is a special case of the (full rank) linear regression model of Section 3.1 with $p = I$ and $\boldsymbol{\beta} = \boldsymbol{\mu}$ (β_0 is not included).

The null hypothesis of interest is $H: \mu_1 = \mu_2 = \cdots = \mu_I$ $(=\mu$, say), or $\mu_1 - \mu_I = \mu_2 - \mu_I = \cdots = \mu_{I-1} - \mu_I$. In matrix form this is

$$\begin{bmatrix} 1 & 0 & 0 & \cdots & 0 & -1 \\ 0 & 1 & 0 & \cdots & 0 & -1 \\ \cdots & \cdots & \cdots & \cdots & \cdots & \cdots \\ 0 & 0 & 0 & \cdots & 1 & -1 \end{bmatrix} \begin{bmatrix} \mu_1 \\ \mu_2 \\ \cdots \\ \mu_I \end{bmatrix} = \mathbf{0}, \tag{9.3}$$

or $\mathbf{A}\boldsymbol{\mu} = \mathbf{0}$. Here the rows of $\mathbf{A}$ are obviously linearly independent so that $\mathbf{A}$ is $q \times p$ of rank q, where $q = I - 1$. Thus H is a linear hypothesis so that the theory of Chapter 4 applies and an F-test of H is given by the statistic

$$F = \frac{(\text{RSS}_H - \text{RSS})/q}{\text{RSS}/(n-p)}. \tag{9.4}$$

To find RSS we minimize $\boldsymbol{\varepsilon}'\boldsymbol{\varepsilon} = \sum_i \sum_j (Y_{ij} - \mu_i)^2$ with respect to the μ_i. Taking partial derivatives with respect to the μ_i we have

$$-2\sum_j (Y_{ij} - \hat{\mu}_i) = 0 \qquad \text{or} \qquad \hat{\mu}_i = \overline{Y}_{i\cdot},$$

and

$$\text{RSS} = \sum_i \sum_j (Y_{ij} - \hat{\mu}_i)^2 = \sum_i \sum_j \left(Y_{ij} - \overline{Y}_{i\cdot}\right)^2$$

To find RSS_H we minimize $\boldsymbol{\varepsilon}'\boldsymbol{\varepsilon}$ subject to the constraints given by H, the simplest way being to use H to reduce the number of free parameters to just one, μ, say. Hence minimizing $\Sigma\Sigma(Y_{ij} - \mu)^2$ with respect to μ, we obtain $\hat{\mu}_H = \overline{Y}_{\cdot\cdot}$, where $\overline{Y}_{\cdot\cdot} = \Sigma\Sigma Y_{ij}/IJ$ (the "grand mean"), and

$$\text{RSS}_H = \Sigma\Sigma(Y_{ij} - \hat{\mu}_H)^2 = \Sigma\Sigma\left(Y_{ij} - \overline{Y}_{\cdot\cdot}\right)^2.$$

There is one further algebraic simplification:

$$\begin{aligned}\Sigma\Sigma\left(Y_{ij} - \overline{Y}_{\cdot\cdot}\right)^2 &= \Sigma\Sigma\left(Y_{ij} - \overline{Y}_{i\cdot} + \overline{Y}_{i\cdot} - \overline{Y}_{\cdot\cdot}\right)^2 \\ &= \Sigma\Sigma\left(Y_{ij} - \overline{Y}_{i\cdot}\right)^2 + \Sigma\Sigma\left(\overline{Y}_{i\cdot} - \overline{Y}_{\cdot\cdot}\right)^2,\end{aligned}$$

since the cross-product term is zero, and hence

$$\text{RSS}_H - \text{RSS} = \sum_i \sum_j \left(\overline{Y}_{i\cdot} - \overline{Y}_{\cdot\cdot}\right)^2 \left(= J \sum_{i=1}^{I} \left(\overline{Y}_{i\cdot} - \overline{Y}_{\cdot\cdot}\right)^2 \right).$$

Thus (9.4) becomes

$$F = \frac{J\Sigma\left(\overline{Y}_{i\cdot} - \overline{Y}_{\cdot\cdot}\right)^2/(I-1)}{\Sigma\Sigma\left(Y_{ij} - \overline{Y}_{i\cdot}\right)^2/(IJ-I)} \left(= \frac{S_H^2}{S^2}, \text{ say} \right), \tag{9.5}$$

which has the $F_{I-1, IJ-I}$ distribution when H is true.

Finally we note that (9.1) can also be expressed in the form

$$Z_r = \mu_1 d_{r1} + \mu_2 d_{r2} + \cdots + \mu_I d_{rI} + \varepsilon_r \qquad (r = 1, 2, \ldots, n),$$

where $\mathbf{Z} = \mathbf{Y}$ and d_{ri} is the rth observation on the ith dummy categorical regressor d_i ($i = 1, 2, \ldots, I$); that is, $d_i = +1$ when Z_r is an observation from the ith population and zero otherwise.

9.1.2 Computations

It is standard practice to set out the various sums of squares in the form of a table (Table 9.1). Frequently the lines below "corrected total" (which is short for "total sum of squares corrected for the mean") are omitted from the table. The terminology used for the column labeled "source" varies in the literature. Instead of "between populations" one encounters "between groups" or "treatments." The "error" sum of squares, variously called "within groups," "within populations," or "residual" sum of squares, provides a pooled estimate of σ^2.

Table 9.1 Analysis of Variance Table for a One-Way Classification

Source	Sum of Squares (SS)	Degrees of Freedom (df)	$\frac{\text{SS}}{\text{df}}$
Between populations	$J\sum_i(\bar{Y}_{i\cdot}-\bar{Y}_{\cdot\cdot})^2$	$I-1$	S_H^2
Error	$\sum_i\sum_j(Y_{ij}-\bar{Y}_{i\cdot})^2$	$IJ-I$	S^2
Corrected total	$\sum_i\sum_j(Y_{ij}-\bar{Y}_{\cdot\cdot})^2$	$IJ-1$	
Mean	$IJ\bar{Y}_{\cdot\cdot}^2$	1	
Total	$\sum_i\sum_j Y_{ij}^2$	IJ	

If the calculations are performed on a desk calculator the following formulae are useful:

$$\begin{aligned}\sum_i\sum_j(\bar{Y}_{i\cdot}-\bar{Y}_{\cdot\cdot})^2 &= J\sum_i(\bar{Y}_{i\cdot}-\bar{Y}_{\cdot\cdot})^2\\ &= J\left[\sum_i \bar{Y}_{i\cdot}^2 - I\bar{Y}_{\cdot\cdot}^2\right]\\ &= \sum_i\frac{(Y_{i\cdot})^2}{J}-\frac{(Y_{\cdot\cdot})^2}{IJ},\end{aligned}\tag{9.6}$$

where $Y_{i\cdot}=\Sigma_j Y_{ij}$ and $Y_{\cdot\cdot}=\Sigma_i\Sigma_j Y_{ij}$, and

$$\sum_i\sum_j(Y_{ij}-\bar{Y}_{i\cdot})^2=\sum_i\sum_j Y_{ij}^2-\sum_i\frac{(Y_{i\cdot})^2}{J}.\tag{9.7}$$

Formulas (9.6) and (9.7) require only the computation of totals and their squares so that the amount of rounding error due to dividing is kept to a minimum. However, in each case, one must take care when taking the difference of two terms, particularly when the two terms are nearly equal. If too few decimal places are carried over the difference could have a very low accuracy. For this reason computer programs should work with the differences $\bar{Y}_{i\cdot} - \bar{Y}_{\cdot\cdot}$ and $Y_{ij} - \bar{Y}_{i\cdot}$ directly (or else use the methods of Youngs and Cramer [1971] and Ling [1974]), rather than use (9.6) and (9.7).

It is perhaps worth noting that simple rules for remembering the above two formulas can be readily devised. The expression $(\bar{Y}_{i\cdot} - \bar{Y}_{\cdot\cdot})^2$ indicates two terms: the first term is preceded by Σ_i and the second term is negative. Also the denominators J and IJ represent the numbers of observations within each set of parentheses squared, that is, in $(Y_{i\cdot})^2$ and $(Y_{\cdot\cdot})^2$, respectively. For example, applying this "rule" to $\Sigma_i\Sigma_j(Y_{ij} - \bar{Y}_{i\cdot} - \bar{Y}_{\cdot j} + \bar{Y}_{\cdot\cdot})^2$ we have, noting where the signs and the subscripts occur,

$$\sum_i \sum_j Y_{ij}^2 - \sum_i \frac{(Y_{i\cdot})^2}{J} - \sum_j \frac{(Y_{\cdot j})^2}{I} + \frac{(Y_{\cdot\cdot})^2}{IJ}.$$

Experiments involving a one-way classification are usually called one-factor experiments. For example, we may wish to compare the action of six different kinds of drugs, or one drug at six different concentrations; here drug is the *factor*, and we have six *levels* of the factor.

9.1.3 Expected Values

We know from the general theory of Theorem 4.1 that

$$\begin{aligned} E\left[\Sigma\Sigma\left(Y_{ij} - \bar{Y}_{i\cdot}\right)^2\right] &= E[\text{RSS}] \\ &= (n-p)\sigma^2 \\ &= (IJ - I)\sigma^2 \end{aligned}$$

and

$$\begin{aligned} E\left[\Sigma\Sigma\left(\bar{Y}_{i\cdot} - \bar{Y}_{\cdot\cdot}\right)^2\right] &= E[\text{RSS}_H - \text{RSS}] \\ &= q\sigma^2 + (\mathbf{A}\boldsymbol{\beta} - \mathbf{c})'[\mathbf{A}(\mathbf{X}'\mathbf{X})^{-1}\mathbf{A}']^{-1}(\mathbf{A}\boldsymbol{\beta} - \mathbf{c}) \\ &= q\sigma^2 + (\text{RSS}_H - \text{RSS})_{\mathbf{Y} = \mathcal{E}[\mathbf{Y}]} \\ &= (I-1)\sigma^2 + \Sigma\Sigma(\mu_i - \bar{\mu}_{\cdot})^2. \end{aligned} \qquad (9.8)$$

Here we have used our knowledge of the degrees of freedom, obtained by studying the various ranks associated with the underlying regression model in Section 9.1.1, to derive the expected values. Alternatively, in some experimental designs where the degrees of freedom are hard to find directly from the ranks, the degrees of freedom can be found by using the general formula

$$E[\mathbf{Y}'\mathbf{A}\mathbf{Y}] = \sigma^2 \operatorname{tr}\mathbf{A} + \boldsymbol{\theta}'\mathbf{A}\boldsymbol{\theta},$$

where $\boldsymbol{\theta} = \mathcal{E}[\mathbf{Y}]$ and $\operatorname{tr}\mathbf{A}$ is the sum of the coefficients of Y^2 terms. The latter coefficients can be identified by using (9.6) and (9.7); from (9.6), for example, we have

$$\begin{aligned}\operatorname{tr}\mathbf{A} &= \sum_i \frac{1}{J}(J \times 1) - \frac{1}{IJ}(IJ \times 1)\\ &= I - 1.\end{aligned}$$

9.1.4 *Reparametrization of the Model*

Having established above that the general regression theory for testing a linear hypothesis can be used, we next obtained expressions for RSS and $\text{RSS}_H - \text{RSS}$. We now consider an alternative method for deriving these sums of squares by reparametrising the model. Let

$$\begin{aligned}\mu_i &= \bar{\mu}_. + (\mu_i - \bar{\mu}_.)\\ &= \mu + \alpha_i, \end{aligned} \tag{9.9}$$

say, where $\alpha_. = \sum_{i=1}^I \alpha_i = 0$; then H is the hypothesis that $\alpha_1 = \alpha_2 = \cdots \alpha_{I-1} = 0$ (the constraint $\alpha_. = 0$ implies that $\alpha_I = 0$). Now consider a similar decomposition of the ε_{ij}, namely,

$$\bar{\varepsilon}_{i.} = \bar{\varepsilon}_{..} + \bar{\varepsilon}_{i.} - \bar{\varepsilon}_{..},$$

so that

$$\varepsilon_{ij} = \bar{\varepsilon}_{..} + (\bar{\varepsilon}_{i.} - \bar{\varepsilon}_{..}) + (\bar{\varepsilon}_{ij} - \bar{\varepsilon}_{i.}). \tag{9.10}$$

Then squaring and summing on i and j, we find that cross-product terms vanish, e.g.,

$$\begin{aligned}\sum_i \sum_j (\bar{\varepsilon}_{i.} - \bar{\varepsilon}_{..})(\varepsilon_{ij} - \bar{\varepsilon}_{i.}) &= \sum_i (\bar{\varepsilon}_{i.} - \bar{\varepsilon}_{..}) \sum_j (\varepsilon_{ij} - \bar{\varepsilon}_{i.})\\ &= 0. \end{aligned} \tag{9.11}$$

Hence

$$\Sigma\Sigma\varepsilon_{ij}^2=\Sigma\Sigma\bar{\varepsilon}_{..}^2+\Sigma\Sigma(\bar{\varepsilon}_{i.}-\bar{\varepsilon}_{..})^2+\Sigma\Sigma(\varepsilon_{ij}-\bar{\varepsilon}_{i.})^2, \tag{9.12}$$

or, substituting $\varepsilon_{ij}=Y_{ij}-\mu-\alpha_i$,

$$\Sigma\Sigma(Y_{ij}-\mu-\alpha_i)^2=\Sigma\Sigma(\overline{Y}_{..}-\mu)^2+\Sigma\Sigma(\overline{Y}_{i.}-\overline{Y}_{..}-\alpha_i)^2+\Sigma\Sigma(Y_{ij}-\overline{Y}_{i.})^2. \tag{9.13}$$

With this decomposition of $\boldsymbol{\varepsilon}'\boldsymbol{\varepsilon}$ we can find RSS, RSS_H, and RSS_H-RSS by inspection. For instance, the right side of (9.13) is minimized, subject to $\alpha_.=0$, when μ and α_i are equal to $\overline{Y}_{..}$ ($=\hat{\mu}$, say) and $\overline{Y}_{i.}-\overline{Y}_{..}$ ($=\hat{\alpha}_i$, say), respectively; thus

$$\begin{aligned}\text{RSS}&=\Sigma\Sigma(Y_{ij}-\hat{\mu}-\hat{\alpha}_i)^2\\&=\Sigma\Sigma(Y_{ij}-\overline{Y}_{i.})^2.\end{aligned}$$

Similarly, when H is true, (9.13) becomes

$$\Sigma\Sigma(Y_{ij}-\mu)^2=\Sigma\Sigma(\overline{Y}_{..}-\mu)^2+\Sigma\Sigma(\overline{Y}_{i.}-\overline{Y}_{..})^2+\Sigma\Sigma(Y_{ij}-\overline{Y}_{i.})^2,$$

so that the right side is minimized when $\mu=\hat{\mu}$; thus

$$\begin{aligned}\text{RSS}_H&=\Sigma\Sigma(Y_{ij}-\hat{\mu})^2\\&=\Sigma\Sigma(\overline{Y}_{i.}-\overline{Y}_{..})^2+\Sigma\Sigma(Y_{ij}-\overline{Y}_{i.})^2\end{aligned}$$

and

$$\text{RSS}_H-\text{RSS}=\Sigma\Sigma(\overline{Y}_{i.}-\overline{Y}_{..})^2=J\sum_i\hat{\alpha}_i^2. \tag{9.14}$$

Although the above reparametrization does not really simplify the derivation of the F-statistic for testing H, we will find that it does lead to simplifications in the two-way and higher-way classifications discussed later.

9.1.5 Geometrical Considerations

It is of academic interest to look more closely at the geometry underlying the decomposition (9.12). In the same way that $\boldsymbol{\varepsilon}$ represents the string of elements ε_{ij}, let $\mathbf{U}=[(U_{ij})]$, $\mathbf{V}=[(V_{ij})]$, and $\mathbf{W}=[(W_{ij})]$, where $U_{ij}=\bar{\varepsilon}_{..}$, $V_{ij}=\bar{\varepsilon}_{i.}-\bar{\varepsilon}_{..}$ and $W_{ij}=\varepsilon_{ij}-\bar{\varepsilon}_{i.}$. Then, from (9.10), we have

$$\boldsymbol{\varepsilon}=\mathbf{U}+\mathbf{V}+\mathbf{W}, \tag{9.15}$$

and $\mathbf{U}$, $\mathbf{V}$, and $\mathbf{W}$ are mutually orthogonal; for example, from (9.11) we have $\mathbf{V}'\mathbf{W}=0$. Thus

$$\begin{aligned}\boldsymbol{\varepsilon}'\boldsymbol{\varepsilon}&=(\mathbf{U}+\mathbf{V}+\mathbf{W})'(\mathbf{U}+\mathbf{V}+\mathbf{W})\\&=\mathbf{U}'\mathbf{U}+\mathbf{V}'\mathbf{V}+\mathbf{W}'\mathbf{W},\end{aligned} \tag{9.16}$$

which is Equation (9.12). If $\mathbf{U}=\mathbf{A}\boldsymbol{\varepsilon}$, $\mathbf{V}=\mathbf{B}\boldsymbol{\varepsilon}$, and $\mathbf{W}=\mathbf{C}\boldsymbol{\varepsilon}$, then it can be shown that $\mathbf{A}$, $\mathbf{B}$, and $\mathbf{C}$ are all symmetric and idempotent, and $\mathbf{AB}=\mathbf{BC}=\mathbf{CA}=\mathbf{O}$. Hence, by Theorems 2.7 (Corollary 1) and 2.8, $\mathbf{U}'\mathbf{U}/\sigma^2$ ($=\boldsymbol{\varepsilon}'\mathbf{A}'\mathbf{A}\boldsymbol{\varepsilon}/\sigma^2=\boldsymbol{\varepsilon}'\mathbf{A}^2\boldsymbol{\varepsilon}/\sigma^2=\boldsymbol{\varepsilon}'\mathbf{A}\boldsymbol{\varepsilon}/\sigma^2$), $\mathbf{V}'\mathbf{V}/\sigma^2$, and $\mathbf{W}'\mathbf{W}/\sigma^2$ are mutually independent and distributed as chi-square with degrees of freedom given by the traces of $\mathbf{A}$, $\mathbf{B}$, and $\mathbf{C}$, respectively. Thus when H is true,

$$\begin{aligned}\mathbf{V}'\mathbf{V}&=\Sigma\Sigma(\bar{\varepsilon}_{i.}-\bar{\varepsilon}_{..})^2\\&=\Sigma\Sigma(\bar{Y}_{i.}-\bar{Y}_{..}-\alpha_i)^2\\&=\Sigma\Sigma(\bar{Y}_{i.}-\bar{Y}_{..})^2\end{aligned}$$

and

$$F=\frac{\mathbf{V}'\mathbf{V}/\operatorname{tr}\mathbf{B}}{\mathbf{W}'\mathbf{W}/\operatorname{tr}\mathbf{C}},$$

showing that F of (9.5) has an F-distribution when H is true.

It is of interest to note that mutual independence of the quadratic forms can be proved directly as follows. By the repeated application of Theorem 1.5 (for one-dimensional vectors, that is, for scalars), we have

$$\begin{aligned}\operatorname{cov}[V_{ij},W_{rs}]&=\operatorname{cov}[\bar{\varepsilon}_{i.}-\bar{\varepsilon}_{..},\varepsilon_{rs}-\bar{\varepsilon}_{r.}]\\&=\operatorname{cov}[\bar{\varepsilon}_{i.},\varepsilon_{rs}]-\operatorname{cov}[\bar{\varepsilon}_{i.},\bar{\varepsilon}_{r.}]-\operatorname{cov}[\bar{\varepsilon}_{..},\varepsilon_{rs}]+\operatorname{cov}[\bar{\varepsilon}_{..},\bar{\varepsilon}_{r.}]\\&=\frac{1}{J}\delta_{ir}\operatorname{var}[\varepsilon_{is}]-\delta_{ir}\operatorname{var}[\bar{\varepsilon}_{i.}]-\frac{1}{IJ}\operatorname{var}[\varepsilon_{rs}]+\frac{1}{I}\operatorname{var}[\bar{\varepsilon}_{r.}]\\&=\frac{1}{J}\delta_{ir}\sigma^2-\delta_{ir}\frac{\sigma^2}{J}-\frac{1}{IJ}\sigma^2+\frac{1}{I}\cdot\frac{\sigma^2}{J}=0,\end{aligned}$$

and $\mathcal{C}[\mathbf{V},\mathbf{W}]=\mathbf{O}$. Similarly it can be shown that $\mathcal{C}[\mathbf{U},\mathbf{V}]$ and $\mathcal{C}[\mathbf{U},\mathbf{W}]$ are also zero so that by Corollary 1 of Theorem 2.7 (Section 2.3), $\mathbf{U}'\mathbf{U}$, $\mathbf{V}'\mathbf{V}$, and $\mathbf{W}'\mathbf{W}$ are mutually independent.

9.1.6 *Identifiability Constraints*

We note that the reparametrized model

$$Y_{ij}=\mu+\alpha_i+\varepsilon_{ij} \tag{9.17}$$

is a less than full rank model since the first column, $\mathbf{1}_n$, of the associated design matrix $\mathbf{X}$ is the sum of the other columns [cf. (3.40) of Section 3.8.1]. The constraint $\Sigma_i\alpha_i=0$, which arose from the definition of α_i [Equation (9.9)], is necessary and sufficient for the identifiability of the α_i. However, we could equally have begun with the model (9.17) and imposed an identifiability constraint of the form $\Sigma_i d_i\alpha_i=0$. By adding the row $(0,d_1,d_2,\ldots,d_I)$ to the matrix $\mathbf{X}$ in (9.1) we see that this row is linearly independent of the rows of $\mathbf{X}$, and the columns of the augmented matrix are linearly independent, provided $\Sigma_i d_i\neq 0$. Hence any linear combination of the α_i which is not a contrast can be used for identifiability (Theorem 3.9, Section 3.8.1). For general d_i the least squares estimates of μ and α_i can be found directly from $\hat{\mu}+\hat{\alpha}_i=\hat{\mu}_i=\overline{Y}_{i\cdot}$; thus $\hat{\mu}=\Sigma_i d_i\overline{Y}_{i\cdot}/\Sigma_i d_i$, etc. However, as we have seen above, the choice of $d_i=1$ $(i=1,2,\ldots,I)$ leads to a simple analysis.

9.1.7 *Confidence Intervals*

If H is rejected by the F-test then the next step is to find out how the μ_i differ from one another. In particular we are usually interested in such differences as $\mu_r-\mu_s$, $\mu_1-\frac{1}{2}(\mu_2+\mu_3)$, $\frac{1}{2}(\mu_1+\mu_2)-\frac{1}{3}(\mu_3+\mu_4+\mu_5)$, etc. These linear combinations are called contrasts in the μ_i as they are of the form $\Sigma_i c_i\mu_i$ where $\Sigma_i c_i=0$. Since $\Sigma c_i\overline{Y}_{i\cdot}$ $(=\Sigma c_i\hat{\mu}_i=\mathbf{c}'\hat{\boldsymbol{\beta}}$, say) has variance $\Sigma c_i^2\sigma^2/J$ we can readily show, using the argument of Section 4.1.5, that

$$\sum_i c_i\overline{Y}_{i\cdot}\pm t_{IJ-I}^{(1/2)\alpha}S\left[\frac{\sum_i c_i^2}{J}\right]^{1/2} \tag{9.18}$$

is a two-sided $100(1-\alpha)\%$ confidence interval for $\theta=\Sigma c_i\mu_i$, where $S^2=\Sigma\Sigma(Y_{ij}-\overline{Y}_{i\cdot})^2/(IJ-I)$. We note that θ is significantly different from zero if the above interval does not contain zero.

If we are interested in several contrasts, chosen prior to seeing the data, then we are faced with the problems of simultaneous interval estimation discussed in Chapter 5. Three methods are described there and, in general, the maximum modulus t intervals provide the shortest intervals. However, if we are interested in all possible contrasts then Scheffé's method is appropriate. Setting $\boldsymbol{\phi}=\mathbf{A}\boldsymbol{\mu}$, where $\mathbf{A}$ is given by Equation (9.3), then the set of all linear combinations $\mathbf{h}'\boldsymbol{\phi}=\sum_{i=1}^{I-1}h_i\phi_i$ is the set of all contrasts $\sum_{i=1}^{I}c_i\mu_i$ [Equation (5.16)]. Hence from the last paragraph of Section 5.1.1 (with $\mathbf{c}=\mathbf{0}$) we see that the F-statistic for H will be significant if and only if one or more of the Scheffé intervals

$$\sum c_i\overline{Y}_{i\cdot}\pm\left[(I-1)F_{I-1,IJ-I}^{\alpha}\right]^{1/2}S\left(\frac{\sum c_i^2}{J}\right)^{1/2}$$

do not contain zero (that is, one or more contrasts $\sum_i c_i\mu_i$ is nonzero). Therefore, if F is significant we can search through the likely contrasts to see which ones are responsible for significance (though such a search through a potentially infinite number of contrasts may not always be straightforward).

If we are interested in just the differences $\mu_r-\mu_s$, then it is possible to construct a set of confidence intervals for all such differences with an overall confidence of $100(1-\alpha)\%$. These intervals take the form

$$\overline{Y}_{r\cdot}-\overline{Y}_{s\cdot}\pm\frac{q_{I,IJ-I}^{\alpha}S}{\sqrt{J}} \tag{9.19}$$

or

$$\overline{Y}_{r\cdot}-\overline{Y}_{s\cdot}\pm d,$$

say, where $q_{k,\nu}^{\alpha}$ is the upper $100\alpha\%$ point of the Studentized range distribution with parameters k and ν (that is, the distribution of the range of k independent $N(0,1)$ random variables divided by $(V/\nu)^{1/2}$, where V is independently distributed as χ_ν^2). This method, due to Tukey, is described in detail by Scheffé [1959: p. 73] and Miller [1966: p. 37]; tables are given, for example, in Pearson and Hartley [1966: pp. 191–193]. Since any difference $\overline{Y}_{r\cdot}-\overline{Y}_{s\cdot}$ greater than d suggests that $\mu_r-\mu_s\neq 0$, we can sort the μ_i into groups which contain means that are not significantly different from one another. For example, if $d=10.4$ and the ranked sample means are given as follows:

$$\begin{array}{cccccc} i: & 5 & 1 & 3 & 2 & 4 \\ \overline{Y}_{i\cdot}: & 25.4 & 32.6 & 39.2 & 40.8 & 52.1 \\ \cline{2-3} \\ \cline{3-5} \end{array}$$

then $\mu_5 < \mu_3, \mu_2, \mu_4$; $\mu_1 < \mu_4$; $\mu_3 < \mu_4$; $\mu_2 < \mu_4$; and the appropriate groups of means are (μ_5, μ_1), (μ_1, μ_3, μ_2), and μ_4 (it is common practice to underline the groups as above).

Another method for sorting the μ_i into such groups is Duncan's multiple range test (Miller [1966: p. 81]). Although very popular with research workers, this technique has not gained universal acceptance by statisticians; in particular the variable significance level associated with the procedure has caused much controversy (O'Neill and Wetherill [1971: pp. 226–227]).

There are two other approaches to this problem of sorting means into groups which show considerable promise. Since we are essentially involved with decision making, a natural approach is to set up the problem in a decision theory framework; such a formulation is given by Waller and Duncan [1969]. On the other hand, the problem could be regarded as one of dividing up a set of sample points (representing the sample means) into nonsignificant clusters; a potentially useful cluster analysis technique for doing this is described by Scott and Knott [1974].

In a paper by Carmer and Swanson [1973] a number of pairwise procedures like the ones mentioned above were compared using simulation. The methods that fared best for the particular experiments carried out were the so-called Least Significant Difference (LSD) method based on simple pairwise t-comparisons

$$\overline{Y}_{r\cdot} - \overline{Y}_{s\cdot} \pm t_{IJ-I}^{(1/2)\alpha} S \left(\frac{2}{J}\right)^{1/2} \tag{9.20}$$

(which are only made if F is significant), and the Bayesian method of Waller and Duncan [1969].

9.1.8 *Underlying Assumptions*

In Section 6.3.2 we saw that quadratically balanced F-tests are robust with regard to departures from the assumption of normality. Now it is easy to test for quadratic balance once we have derived the F-statistic; we simply check both the numerator and the denominator to see if, in each case, the coefficient of Y_{rs}^2 is the same for all r, s. Because this holds true in (9.5) we see that the one-way classification with equal numbers of observations per mean enjoys this robustness.

Scheffé [1959: chapter 10] has also discussed the question of robustness in some detail and demonstrated that the F-statistic is robust with respect to unequal population variances but not with respect to the presence of intraclass correlation. In the latter case Bhargava and Srivastava [1973] show how to modify the Tukey and Scheffé multiple confidence intervals to allow for the presence of a constant correlation: basically their method amounts to replacing S^2 by

$$\frac{\Sigma\Sigma\left(Y_{ij}-\bar{Y}_{i\cdot}-\bar{Y}_{\cdot j}+\bar{Y}_{\cdot\cdot}\right)^2}{(I-1)(J-1)}$$

and adjusting the degrees of freedom accordingly.

A number of procedures are available for testing for equal population variances. In particular there is Scheffé's [1959: pp. 83–87] approximate F-test (due to Box [1953]), which is a good example of the use of weighted least squares; an approximate F-test based on absolute deviations (Levene [1960], Draper and Hunter [1969]); and two other tests, called the chi-square and jackknife tests, due to Layard [1973] (these have been modified slightly by Brown and Forsythe [1974]). In a simulation study Layard [1973] demonstrated that the chi-square and jackknife tests are satisfactory if the emphasis is on robustness and power; on the other hand, the Box test is less powerful but more robust with respect to the significance level. If correlation is present as well then the procedures suggested by Han [1968] may be useful.

When the population variances are unequal we can use Spjøtvoll's [1972b] modification of Scheffé's multiple confidence intervals. If we are interested in just a single confidence interval for, say, $\Sigma a_i\mu_i$, then the robust method of Scott and Smith [1971] is available (see the end of Section 9.1.9).

9.1.9 *Unequal Numbers of Observations per Mean*

If we now have J_i observations on the ith normal population then the only change in the theory of Section 9.1.1 is that $n=\Sigma_{i=1}^{I}J_i$. The F-statistic for testing H is now

$$F=\frac{\sum_{i=1}^{I}\sum_{j=1}^{J_i}\left(\bar{Y}_{i\cdot}-\bar{Y}_{\cdot\cdot}\right)^2/(I-1)}{\sum_{i=1}^{I}\sum_{j=1}^{J_i}\left(Y_{ij}-\bar{Y}_{i\cdot}\right)^2\Big/\left(\sum_i J_i-I\right)}. \tag{9.21}$$

In Section 9.1.2 J and IJ are replaced by J_i and $\Sigma_i J_i$, respectively, in Equations (9.6) and (9.7); the "rules" for remembering these equations still apply. However, we run into some difficulties when we use the reparametrization suggested in Section 9.1.4. The decomposition (9.10) is still orthogonal provided we define $\bar{\varepsilon}_{\cdot\cdot}=\Sigma_i\Sigma_j\varepsilon_{ij}/\Sigma_i J_i$; for example, (9.11) is still valid, and

$$\begin{aligned}\Sigma\Sigma\bar{\varepsilon}_{..}(\bar{\varepsilon}_{i.}-\bar{\varepsilon}_{..}) &= \bar{\varepsilon}_{..}\sum_i J_i(\bar{\varepsilon}_{i.}-\bar{\varepsilon}_{..})\\ &= \bar{\varepsilon}_{..}\left(\sum_i\sum_j \varepsilon_{ij}-\sum_i\sum_j \varepsilon_{ij}\right)\\ &= 0,\end{aligned}$$

so that (9.12) still holds. However, (9.13) is no longer true as, for example,

$$\begin{aligned}\bar{\varepsilon}_{..} &= \overline{Y}_{..}-\left(\mu+\frac{\sum_i J_i\alpha_i}{\sum_i J_i}\right)\\ &\neq \overline{Y}_{..}-\mu. \qquad (9.22)\end{aligned}$$

However, we can get round this difficulty by simply choosing a different identifiability constraint, namely, $\Sigma_i J_i\alpha_i=0$; in this case (9.13) still holds. (We recall from Section 9.1.6 that any constraint $\Sigma d_i\alpha_i=0$ which is not a contrast can be used.)

Although the reparametrization is mainly of academic interest as far as the one-way classification is concerned, it does at least provide a suitable method for handling higher-way classifications, as we shall see later. In practice the difficulties encountered above with the unequal numbers case are typical of higher-way classifications with unequal numbers per mean.

We note that the multiple confidence intervals of Scheffé and Tukey can still be used (Spjøtvoll and Stoline [1973]); the latter are recommended if primary interest is in a pairwise comparison of the means and the J_i are not too different. However, the F-statistic for testing H is no longer robust with respect to nonnormality (because F is no longer quadratically balanced) or unequal population variances. When the population variances are unequal we can use Spjøtvoll's [1972b] modification of Scheffé's multiple confidence intervals. If we are interested in just a single confidence interval for, say, $\Sigma a_i\mu_i$, then we can use the approximately $N(0,1)$ statistic (Scott and Smith [1971])

$$Z=\frac{\Sigma a_i\overline{Y}_{i.}-\Sigma a_i\mu_i}{\left\{\sum_i a_i^2\tilde{S}_i^2/J_i\right\}^{1/2}} \qquad (\text{each } J_i>3)$$

where $\tilde{S}_i^2=\Sigma_{j=1}^{J_i}(Y_{ij}-\overline{Y}_{i.})^2/(J_i-3)$.

9.2 TWO-WAY CLASSIFICATION

9.2.1 *Representation as a Regression Model*

Consider an experiment in which two factors A and B are allowed to vary. Suppose there are I levels of A and J levels of B, and let Y_{ijk} be the kth experimental observation ($k=1,2,\ldots,K$: $K>1$) on the combination of the ith level of A with the jth level of B. We assume that the Y_{ijk} are independently distributed as $N(\mu_{ij},\sigma^2)$ so that

$$Y_{ijk}=\mu_{ij}+\varepsilon_{ijk} \qquad (i=1,2,\ldots,I;\quad J=1,2,\ldots,J;\quad k=1,2,\ldots,K), \quad (9.23)$$

where the ε_{ijk} are i.i.d. $N(0,\sigma^2)$. By writing

$$\mathbf{Y}'=(Y_{111},Y_{112},\ldots,Y_{11K},Y_{121},Y_{122},\ldots,Y_{IJ1},Y_{IJ2},\ldots,Y_{IJK}),$$

$$\boldsymbol{\mu}'=(\mu_{11},\mu_{12},\ldots,\mu_{1J},\mu_{21},\mu_{22},\ldots,\mu_{2J},\ldots,\mu_{I1},\mu_{I2},\ldots,\mu_{IJ}),$$

etc., and using the same approach which led to Equation (9.1), we find that (9.23) can be expressed in the regression form

$$\mathbf{Y}=\mathbf{X}\boldsymbol{\mu}+\boldsymbol{\varepsilon},$$

where $\boldsymbol{\varepsilon}$ is $N_n(\mathbf{0},\sigma^2\mathbf{I}_n)$, $n=IJK$, and $\mathbf{X}$ is $n\times IJ$ of rank IJ. Minimizing $\sum\sum\sum(Y_{ijk}-\mu_{ij})^2$ with respect to μ_{ij} we find that $\text{RSS}=\sum\sum\sum(Y_{ijk}-\overline{Y}_{ij\cdot})^2$ with $n-p=IJK-IJ$ degrees of freedom.

The next question to consider is, what hypotheses are of interest? In the first place we would like to know whether the two factors interact in any way, that is, whether the effect of factor A at level i, say, depends on the level of factor B. If there were no such interaction we would expect the difference in means $\mu_{i_1j}-\mu_{i_2j}$ to depend only on i_1 and i_2 and not on j. Mathematically this means that

$$\begin{aligned}\mu_{i_1j}-\mu_{i_2j}&=\phi(i_1,i_2), \quad \text{say}\\ &=\sum_{j=1}^{J}\frac{\phi(i_1,i_2)}{J}\\ &=\overline{\mu}_{i_1\cdot}-\overline{\mu}_{i_2\cdot} \qquad (\text{for all } i_1,i_2),\end{aligned}$$

or

$$\mu_{i_1j}-\overline{\mu}_{i_1\cdot}=\mu_{i_2j}-\overline{\mu}_{i_2\cdot}.$$

This last equation implies that $\mu_{ij}-\bar{\mu}_{i\cdot}$ does not depend on i so that

$$\begin{aligned}\mu_{ij}-\bar{\mu}_{i\cdot} &= \Phi(j), \quad \text{say}\\ &= \sum_{i=1}^{I}\frac{\Phi(j)}{I}\\ &= \bar{\mu}_{\cdot j}-\bar{\mu}_{\cdot\cdot}\end{aligned}$$

or

$$\mu_{ij}-\bar{\mu}_{i\cdot}-\bar{\mu}_{\cdot j}+\bar{\mu}_{\cdot\cdot}=0 \qquad \text{for all } i,j. \tag{9.24}$$

We note that this expression is symmetric in i and j so that we would arrive at the same result if we assumed the difference $\mu_{ij_1}-\mu_{ij_2}$ to depend only on j_1 and j_2 and not on i.

The linear combination

$$(\alpha\beta)_{ij}=\mu_{ij}-\bar{\mu}_{i\cdot}-\bar{\mu}_{\cdot j}+\bar{\mu}_{\cdot\cdot} \qquad (=\mathbf{a}_{ij}'\boldsymbol{\mu}, \quad \text{say})$$

is called the interaction between the ith level of A and the jth level of B. The hypothesis of no interaction is then

$$H_{AB}:(\alpha\beta)_{ij}=0 \qquad (i=1,2,\ldots,I;\quad j=1,2,\ldots,J).$$

If there is no interaction then we can treat the two factors separately and test, for example, the hypothesis that all the levels of A have the same effect; that is given j, μ_{ij} does not vary with i, or equivalently

$$\begin{aligned}\mu_{ij} &= \theta(j)\\ &= \bar{\mu}_{\cdot j}.\end{aligned} \tag{9.25}$$

Combining $\mu_{ij}-\bar{\mu}_{\cdot j}=0$ with $(\alpha\beta)_{ij}=0$ we have the hypothesis $H_A:\alpha_i=0$ $(i=1,2,\ldots,I)$, where

$$\alpha_i=\bar{\mu}_{i\cdot}-\bar{\mu}_{\cdot\cdot}. \tag{9.26}$$

Similarly a test for no difference in the levels of B is $H_B:\beta_j=0$ $(j=1,2,\ldots,J)$, where

$$\beta_j=\bar{\mu}_{\cdot j}-\bar{\mu}_{\cdot\cdot}. \tag{9.27}$$

Here α_i is called the *main* ith *effect* of A, and β_j is called the *main* jth *effect* of B. We note that there is little point in testing H_A or H_B if H_{AB} has been rejected. In this case H_A, for example, is no longer equivalent to $\mu_{ij}-\bar{\mu}_{\cdot j}=0$ (all i,j).

9.2.2 Test Statistics

All three hypotheses H_A, H_B, and H_{AB} can be expressed in the form $\mathbf{C}\boldsymbol{\mu}=\mathbf{0}$ so that the general regression theory can be applied to each hypothesis. However, there still remain two problems: first, we have to find the rank of $\mathbf{C}$ in each case and, second, we have to minimize $\|\mathbf{Y}-\mathbf{X}\boldsymbol{\mu}\|^2$ subject to $\mathbf{C}\boldsymbol{\mu}=\mathbf{0}$ to find RSS_H.

The ranks can be found heuristically as follows. Since $(\alpha\beta)_{\cdot j}=0$ ($j=1,2,\ldots,J$) and $(\alpha\beta)_{i\cdot}=0$ ($i=1,2,\ldots,I$), the number of mathematically independent equations of the form $(\alpha\beta)_{ij}=0$ is $IJ-I-J+1$ [$+1$ as $(\alpha\beta)_{\cdot\cdot}=0$ can be derived from each set]. If $\mathbf{C}\boldsymbol{\mu}=\mathbf{0}$ represents this reduced set of $IJ-I-J+1$ $(=(I-1)(J-1))$ equations, then $\mathbf{C}$ will be $(I-1)(J-1)\times IJ$, of rank $(I-1)(J-1)$. Similarly, since $\alpha_{\cdot}=0$ and $\beta_{\cdot}=0$, H_A is represented by $(I-1)$ linearly independent equations and H_B by $(J-1)$ independent equations.

To find RSS_H for each hypothesis we consider the following reparametrization:

$$\begin{aligned}\mu_{ij}&=\bar{\mu}_{\cdot\cdot}+(\bar{\mu}_{i\cdot}-\bar{\mu}_{\cdot\cdot})+(\bar{\mu}_{\cdot j}-\bar{\mu}_{\cdot\cdot})+(\mu_{ij}-\bar{\mu}_{i\cdot}-\bar{\mu}_{\cdot j}+\bar{\mu}_{\cdot\cdot})\\&=\mu+\alpha_i+\beta_j+(\alpha\beta)_{ij},\end{aligned}\tag{9.28}$$

with the corresponding decomposition of ε_{ijk}, namely,

$$\begin{aligned}\varepsilon_{ijk}=\bar{\varepsilon}_{\cdots}&+(\bar{\varepsilon}_{i\cdot\cdot}-\bar{\varepsilon}_{\cdots})+(\bar{\varepsilon}_{\cdot j\cdot}-\bar{\varepsilon}_{\cdots})\\&+(\bar{\varepsilon}_{ij\cdot}-\bar{\varepsilon}_{i\cdot\cdot}-\bar{\varepsilon}_{\cdot j\cdot}+\bar{\varepsilon}_{\cdots})+(\varepsilon_{ijk}-\bar{\varepsilon}_{ij\cdot}).\end{aligned}$$

Squaring, and summing on i, j, and k, we find, as in Section 9.1.4 that the cross-product terms vanish and

$$\sum\sum\sum\varepsilon_{ijk}^2=\sum\sum\sum\bar{\varepsilon}_{\cdots}^2+\sum\sum\sum(\bar{\varepsilon}_{i\cdot\cdot}-\bar{\varepsilon}_{\cdots})^2+\cdots+\sum\sum\sum(\varepsilon_{ijk}-\bar{\varepsilon}_{ij\cdot})^2.\tag{9.29}$$

Setting $\varepsilon_{ijk}=Y_{ijk}-\mu-\alpha_i-\beta_j-(\alpha\beta)_{ij}$, and using $\alpha_{\cdot}=0$, etc., we obtain

$$\begin{aligned}&\sum\sum\sum(Y_{ijk}-\mu-\alpha_i-\beta_j-(\alpha\beta)_{ij})^2\\&\quad=\sum\sum\sum(\bar{Y}_{\cdots}-\mu)^2+\sum\sum\sum(\bar{Y}_{i\cdot\cdot}-\bar{Y}_{\cdots}-\alpha_i)^2+\sum\sum\sum(\bar{Y}_{\cdot j\cdot}-\bar{Y}_{\cdots}-\beta_j)^2\\&\quad\quad+\sum\sum\sum(\bar{Y}_{ij\cdot}-\bar{Y}_{i\cdot\cdot}-\bar{Y}_{\cdot j\cdot}+\bar{Y}_{\cdots}-(\alpha\beta)_{ij})^2+\sum\sum\sum(Y_{ijk}-\bar{Y}_{ij\cdot})^2.\end{aligned}\tag{9.30}$$

By inspection, the right side of (9.30) is minimized (subject to $\alpha_. = 0$ etc.) when the unknown parameters take the values

$$\hat{\mu} = \overline{Y}_{...}, \qquad \hat{\alpha}_i = \overline{Y}_{i..} - \overline{Y}_{...}, \qquad \hat{\beta}_j = \overline{Y}_{.j.} - \overline{Y}_{...},$$

and

$$(\widehat{\alpha\beta})_{ij} = \overline{Y}_{ij.} - \overline{Y}_{i..} - \overline{Y}_{.j.} + \overline{Y}_{...}.$$

Hence

$$\text{RSS} = \sum\sum\sum\left(Y_{ijk} - \overline{Y}_{ij.}\right)^2$$

as before.

To find $\text{RSS}_{H_{AB}}$ we must minimize (9.30) subject to $(\alpha\beta)_{ij} = 0$ for all i,j. By inspection this minimum occurs at $\hat{\mu}$, $\hat{\alpha}_i$, and $\hat{\beta}_j$ so that

$$\text{RSS}_{H_{AB}} = \sum\sum\sum\left(\overline{Y}_{ij.} - \overline{Y}_{i..} - \overline{Y}_{.j.} + \overline{Y}_{...}\right)^2 + \sum\sum\sum\left(Y_{ijk} - \overline{Y}_{ij.}\right)^2$$

and

$$\begin{aligned}\text{RSS}_{H_{AB}} - \text{RSS} &= \sum\sum\sum\left(\overline{Y}_{ij.} - \overline{Y}_{i..} - \overline{Y}_{.j.} + \overline{Y}_{...}\right)^2 \\ &= K\sum_i \sum_j (\widehat{\alpha\beta})_{ij}^2.\end{aligned}$$

The F-statistic for testing H_{AB} is therefore

$$F = \frac{K\sum_i \sum_j (\widehat{\alpha\beta})_{ij}^2/(I-1)(J-1)}{\text{RSS}/(IJK - IJ)} = \frac{S_{AB}^2}{S^2}, \tag{9.31}$$

say, which has an F-distribution with $(I-1)(J-1)$ and $IJK - IJ$ degrees of freedom, respectively, when H_{AB} is true.

Test statistics for H_A and H_B are obtained in a similar fashion. Setting $\alpha_i = 0$, for example, in (9.30) we find that the minimum of (9.30) now occurs at $\hat{\mu}$, $\hat{\beta}_j$, and $(\widehat{\alpha\beta})_{ij}$. Thus

$$\text{RSS}_{H_A} = \sum\sum\sum\left(\overline{Y}_{i..} - \overline{Y}_{...}\right)^2 + \sum\sum\sum\left(Y_{ijk} - \overline{Y}_{ij.}\right)^2,$$

$$\begin{aligned}\text{RSS}_{H_A} - \text{RSS} &= \sum\sum\sum\left(\overline{Y}_{i..} - \overline{Y}_{...}\right)^2 \\ &= JK\sum_i \hat{\alpha}_i^2,\end{aligned}$$

and

$$F = \frac{JK\sum_i \hat{\alpha}_i^2/(I-1)}{\text{RSS}/(IJK-IJ)} = \frac{S_A^2}{S^2}, \tag{9.32}$$

say, is the F-statistic for testing H_A. The corresponding statistic for H_B is

$$F = \frac{IK\sum_j \hat{\beta}_j^2/(J-1)}{\text{RSS}/(IJK-IJ)} = \frac{S_B^2}{S^2}. \tag{9.33}$$

Although the number of degrees of freedom for each RSS_H was found heuristically, it can also be obtained from the coefficient of σ^2 in $E[\text{RSS}_H]$ (these expected values are given in the next section).

9.2.3 *Analysis of Variance Table*

As in the one-way classification the various sums of squares are normally set out in the form of a table (Table 9.2). The various sums of squares in Table 9.2 add up to $\sum\sum\sum Y_{ijk}^2$; this follows from the fact that (9.29) is an *identity* in the ε_{ijk} and therefore also holds for the Y_{ijk}. Since

$$\sum\sum\sum\left(\overline{Y}_{i\cdot\cdot} - \overline{Y}_{\cdots}\right)^2 = JK\sum_i \hat{\alpha}_i^2,$$

Table 9.2 Analysis of Variance for a Two-Way Classification with K $(K>1)$ Observations per Population Mean

Source	Sum of Squares (SS)	Degrees of Freedom (df)	$\frac{\text{SS}}{\text{df}}$
A main effects	$JK\sum_i \hat{\alpha}_i^2$	$I-1$	S_A^2
B main effects	$IK\sum_j \hat{\beta}_j^2$	$J-1$	S_B^2
AB interactions	$K\sum_i\sum_j \widehat{(\alpha\beta)}_{ij}^2$	$(I-1)(J-1)$	S_{AB}^2
Error	$\sum_i\sum_j\sum_k (Y_{ijk}-\overline{Y}_{ij\cdot})^2$	$IJ(K-1)$	S^2
Corrected total	$\sum_i\sum_j\sum_k (Y_{ijk}-\overline{Y}_{\cdots})^2$	$IJK-1$	
Mean	$IJK\overline{Y}_{\cdots}^2$	1	
Total	$\sum_i\sum_j\sum_k Y_{ijk}^2$	IJK	

this sum of squares is called the "sum of squares due to the A main effects," though some textbooks use the term "row sum of squares." Similar designations apply to the next two sums of squares in Table 9.2 as they are $IK\Sigma_j \hat{\beta}_j^2$ and $K\Sigma\Sigma_{ij} \widehat{(\alpha\beta)}_{ij}^2$, respectively. The sum of squares labeled "error" gives a pooled estimate of σ^2 based on all IJ normal populations; this term is also called the "within populations" or "residual" sum of squares. In spite of the ordering in the table we test H_{AB} first as the definitions for the A and B main effects make more sense when the AB interactions are zero. If the interactions are zero then the experiment is actually equivalent to two one-factor experiments—one for A and one for B.

Using the method for expanding quadratic forms suggested in Section 9.1.2, we have

$$\Sigma\Sigma\Sigma\left(\bar{Y}_{i..} - \bar{Y}_{...}\right)^2 = \sum_i \frac{(Y_{i..})^2}{JK} - \frac{(Y_{...})^2}{IJK},$$

$$\Sigma\Sigma\Sigma\left(\bar{Y}_{.j.} - \bar{Y}_{...}\right)^2 = \sum_j \frac{(Y_{.j.})^2}{IK} - \frac{(Y_{...})^2}{IJK},$$

$$\Sigma\Sigma\Sigma\left(\bar{Y}_{ij.} - \bar{Y}_{i..} - \bar{Y}_{.j.} + \bar{Y}_{...}\right)^2 = \sum_i \sum_j \frac{(Y_{ij.})^2}{K} - \sum_i \frac{(Y_{i..})^2}{JK} - \sum_j \frac{(Y_{.j.})^2}{IK} + \frac{(Y_{...})^2}{IJK},$$

and

$$\Sigma\Sigma\Sigma\left(Y_{ijk} - \bar{Y}_{ij.}\right)^2 = \Sigma\Sigma\Sigma Y_{ijk}^2 - \sum_i \sum_j \frac{(Y_{ij.})^2}{K}.$$

By adding up the coefficients of all the squared terms Y_{ijk}^2 the trace of each quadratic form can be found by inspection. Hence (Section 9.1.3):

$$E\left[(I-1)S_A^2\right] = \sigma^2(I-1) + JK\sum_i \alpha_i^2$$

$$E\left[(J-1)S_B^2\right] = \sigma^2(J-1) + IK\sum_j \beta_j^2$$

$$E\left[(I-1)(J-1)S_{AB}^2\right] = \sigma^2(I-1)(J-1) + K\sum_i \sum_j (\alpha\beta)_{ij}^2$$

and

$$E\left[S^2\right] = \sigma^2.$$

9.2.4 Confidence Intervals

When discussing the one-way classification with equal numbers per mean, methods were given in Section 9.1.7 for finding confidence intervals for contrasts $\Sigma_i c_i \mu_i$. Since $\Sigma c_i \alpha_i = \Sigma c_i(\mu_i - \bar{\mu}) = \Sigma c_i \mu_i$, we see that the methods given there apply to contrasts in the α_i. In a similar fashion we can construct confidence intervals for contrasts in the main effects $\{\alpha_i\}, \{\beta_j\}$, and interactions $\{(\alpha\beta)_{ij}\}$ for the two-way classification. For example, if $H_A : \boldsymbol{\phi} = \mathbf{A}\boldsymbol{\mu} = \mathbf{0}$ $[\boldsymbol{\phi}' = (\alpha_1, \alpha_2, \ldots, \alpha_{I-1})]$ is rejected, we can use Scheffé's method, given below, for finding out which contrasts are responsible.

Following Section 9.1.7 we note that

$$\begin{aligned}\sum_{i=1}^{I-1} h_i \alpha_i &= \sum_{i=1}^{I-1} h_i(\bar{\mu}_{i\cdot} - \bar{\mu}_{\cdot\cdot}) \\ &= \sum_{i=1}^{I-1}\left(h_i - \frac{1}{I}\sum_{i=1}^{I-1} h_i\right)\bar{\mu}_{i\cdot} + \left(-\frac{1}{I}\sum_{i=1}^{I-1} h_i\right)\bar{\mu}_{I\cdot} \\ &= \sum_{i=1}^{I} c_i \bar{\mu}_{i\cdot} \\ &= \sum_{i=1}^{I} c_i \alpha_i, \end{aligned} \tag{9.34}$$

where $\Sigma c_i = 0$. Conversely, by writing $\alpha_I = -\Sigma_{i=1}^{I-1}\alpha_i$ we see that (9.34) is expressible in the form $\Sigma_{i=1}^{I-1} h_i \alpha_i$. These two statements imply that the set of all linear combinations $\mathbf{h}'\boldsymbol{\phi}$ is the set of all contrasts in $\alpha_1, \alpha_2, \ldots, \alpha_I$. Now

$$\sum_i c_i \hat{\alpha}_i = \Sigma c_i(\bar{Y}_{i\cdot\cdot} - \bar{Y}_{\cdot\cdot\cdot}) = \Sigma c_i \bar{Y}_{i\cdot\cdot},$$

so that

$$\operatorname{var}[\Sigma c_i \hat{\alpha}_i] = \frac{\sigma^2 \sum_i c_i^2}{(JK)}$$

and

$$1 - \alpha = \operatorname{pr}\left[\Sigma c_i \alpha_i \in \Sigma c_i \bar{Y}_{i\cdot\cdot} \pm \{(I-1)F_{I-1,\nu}^{\alpha}\}^{1/2} S\left\{\frac{\Sigma c_i^2}{(JK)}\right\}^{1/2}, \text{ all contrasts}\right]$$

where

$$S^2 = \sum_i \sum_j \sum_k (Y_{ijk} - \bar{Y}_{ij\cdot})^2/\nu \qquad \text{and} \qquad \nu = IJK - IJ.$$

If H_B is rejected then the same procedure can be carried out for contrasts $\sum_{j=1}^{J} c_j \beta_j$. We note that Tukey's method can be used for examining the differences $\alpha_r - \alpha_s$ $(=\bar{\mu}_{r\cdot} - \bar{\mu}_{s\cdot})$ and $\beta_r - \beta_s$ $(=\bar{\mu}_{\cdot r} - \bar{\mu}_{\cdot s})$ for all r, s.

If H_{AB} is rejected then there is little point in testing either H_A or H_B. In this case we can examine all the contrasts $\sum\sum_{ij} c_{ij}(\alpha\beta)_{ij}$ $(=\sum\sum c_{ij}\mu_{ij})$, where $\sum_i c_{ij} = 0$ (all j) and $\sum_j c_{ij} = 0$ (all i), using Scheffé's method. However, Tukey's method does not apply here as the covariances of the $\widehat{(\alpha\beta)}_{ij}$ are not equal (cf. Scheffé [1959: p. 110]). The case when $c_{ij} = a_i b_j$ is discussed in detail by Gabriel et al. [1973] and three methods of simultaneous inference are described by Bradu and Gabriel [1974].

9.2.5 *Unequal Numbers of Observations per Mean*

In Section 9.2.3 we saw that the various test sums of squares in the analysis of variance table are additive, and represent an orthogonal decomposition of the vector $\mathbf{Y} = [(Y_{ijk})]$. Also from (9.30) it readily follows that the numerator sum of squares for testing H_A against the full model is the same as the numerator for testing H_A *given* H_{AB} is true. In fact if we use the expression $(\hat{\mathbf{Y}} - \hat{\mathbf{Y}}_H)'(\hat{\mathbf{Y}} - \hat{\mathbf{Y}}_H)$ [(3.65), Section 3.9.1] for the numerator sums of squares, this result follows immediately: the two sums of squares are $\sum\sum\{\hat{\mu} + \hat{\alpha}_i + \hat{\beta}_j + \widehat{(\alpha\beta)}_{ij} - (\mu + \hat{\beta}_j + \widehat{(\alpha\beta)}_{ij}\}^2$ and $\sum\sum\{\hat{\mu} + \hat{\alpha}_i + \hat{\beta}_j - (\hat{\mu} + \hat{\beta}_j)\}^2$, respectively, and they are both equal to $\sum\sum\hat{\alpha}_i^2$.

However, suppose that we now have K_{ij} observations on the population $N(\mu_{ij}, \sigma^2)$. If we still use the same model as for the balanced (equal numbers) case of Section 9.2.1, namely,

$$Y_{ijk} = \mu + \alpha_i + \beta_j + (\alpha\beta)_{ij} + \varepsilon_{ijk} \qquad (k = 1, 2, \ldots, K_{ij})$$

with identifiability constraints $\alpha_\cdot = \beta_\cdot = 0$, etc., then we no longer obtain a simple decomposition like (9.30). This means that the various test sums of squares are no longer additive, and the numerator sum of squares for testing H_A will now depend on whether we can assume H_{AB} to be true or not. This situation is usually described as a "nonorthogonal" analysis of variance because the test sums of squares do not come from orthogonal vectors as in the balanced case. Unfortunately, there seems to be some confusion in the literature as to the proper method for analyzing such a model (see Francis [1973] and, in particular, Nelder [1974]), so that a few comments are perhaps appropriate.

The first step is to test H_{AB} against the full model and the reader is referred to Scheffé [1959: Section 4.4] for further details of this test. If H_{AB} is accepted then the next step is to test H_A and H_B for the model

$$Y_{ijk} = \mu + \alpha_i + \beta_j + \varepsilon_{ijk} \tag{9.35}$$

in which H_{AB} is assumed to be true. As already mentioned, testing H_A against (9.35) leads to a different numerator sum of squares from that used in testing H_A against the full model because of the lack of orthogonality. Contrary to Scheffé [1959: pp. 116–118], the latter procedure is not appropriate; if the interactions $(\alpha\beta)_{ij}$ are not zero then $H_A: \alpha_i = \bar{\mu}_{i\cdot} - \bar{\mu}_{\cdot\cdot} = 0$ (all i) is not equivalent to $\mu_{ij} - \bar{\mu}_{i\cdot} = 0$ (all i,j), the hypothesis originally intended (cf. Section 9.2.1). There is not much point in testing whether the average effects of A are zero when some of the individual effects $\mu_{ij} - \bar{\mu}_{i\cdot}$ are not zero.

Because (9.30) no longer holds, the various test sums of squares are not so readily computed. However, the method of Section 11.5.4 is particularly appropriate in this case. In applying the algorithm we note that various other identifiability constraints can be used. In one particular case, namely, $K_{ij} = K_{i\cdot}K_{\cdot j}/K_{\cdot\cdot}$ (all i,j), we can obtain a simple orthogonal decomposition like (9.30) if we choose the identifiability constraints

$$\sum_i v_i\alpha_i = \sum_j w_j\beta_j = \sum_i v_i(\alpha\beta)_{ij} = \sum_j w_j(\alpha\beta)_{ij} = 0,$$

where $v_i = K_{i\cdot}$ and $w_j = K_{\cdot j}$ (Scheffé [1959: p. 119], Seber [1966: p. 47]).

9.3 HIGHER-WAY CLASSIFICATIONS WITH EQUAL NUMBERS PER MEAN

9.3.1 *Definition of Interactions and Main Effects*

The extension of the preceding theory to higher-way classifications with equal numbers of observations per mean is fairly straightforward and it is demonstrated briefly by considering the three-way classification:

$$Y_{ijkm} = \mu_{ijk} + \varepsilon_{ijkm}, \tag{9.36}$$

where $i = 1,2,\ldots,I$; $j = 1,2,\ldots,J$; $k = 1,2,\ldots,K$; $m = 1,2,\ldots,M$, and the ε_{ijkm} are i.i.d. $N(0,\sigma^2)$. Here we have three factors: A at I levels, B at J levels, C at K levels; and there are M $(M > 1)$ observations per population mean for each of the IJK means. In addition to the (first-order) interactions between A and B, B and C, A and C, we now have the possibility of a (second-order) interaction between all three factors. If, however, the factors only interact in pairs so that, for example, the AB interactions are not affected by C, then an AB interaction would be the same for all levels of C.

Mathematically this means that

$$\mu_{ijk} - \bar{\mu}_{i\cdot k} - \bar{\mu}_{\cdot jk} + \bar{\mu}_{\cdot\cdot k} = \psi(i,j)$$

$$= \sum_{k=1}^{K} \psi(i,j)/K$$

$$= \bar{\mu}_{ij\cdot} - \bar{\mu}_{i\cdot\cdot} - \bar{\mu}_{\cdot j\cdot} + \bar{\mu}_{\cdot\cdot\cdot}$$

or

$$(\alpha\beta\gamma)_{ijk} = \mu_{ijk} - \bar{\mu}_{ij\cdot} - \bar{\mu}_{\cdot jk} - \bar{\mu}_{i\cdot k} + \bar{\mu}_{i\cdot\cdot} + \bar{\mu}_{\cdot j\cdot} + \bar{\mu}_{\cdot\cdot k} - \bar{\mu}_{\cdot\cdot\cdot}$$
$$= 0.$$

(A numerical example demonstrating this is given in Exercise 8 at the end of this chapter.) Since $(\alpha\beta\gamma)_{ijk}$ is symmetric in i,j,k we see that we would have arrived at the same result if we considered the BC interaction at different levels of A, or the AC interaction at different levels of B. It therefore seems appropriate to define $(\alpha\beta\gamma)_{ijk}$ as the second-order interaction between the ith level of A, the jth level of B, and the kth level of C. We refer to these interactions as simply ABC interactions.

Our two-factor concepts of Section 9.2 can be carried over to this situation by considering a two-way table for each level of C. For example the interaction of the ith level of A with the jth level of B, given that C is at level k, is

$$\mu_{ijk} - \bar{\mu}_{i\cdot k} - \bar{\mu}_{\cdot jk} + \bar{\mu}_{\cdot\cdot k}. \tag{9.37}$$

The average of these over the levels of C, namely,

$$(\alpha\beta)_{ij} = \bar{\mu}_{ij\cdot} - \bar{\mu}_{i\cdot\cdot} - \bar{\mu}_{\cdot j\cdot} + \bar{\mu}_{\cdot\cdot\cdot},$$

we call the interaction of the ith level of A with the jth level of B. We similarly define the BC and AC interactions to be

$$(\beta\gamma)_{jk} = \bar{\mu}_{\cdot jk} - \bar{\mu}_{\cdot j\cdot} - \bar{\mu}_{\cdot\cdot k} + \bar{\mu}_{\cdot\cdot\cdot}$$

and

$$(\alpha\gamma)_{ik} = \bar{\mu}_{i\cdot k} - \bar{\mu}_{i\cdot\cdot} - \bar{\mu}_{\cdot\cdot k} + \bar{\mu}_{\cdot\cdot\cdot}.$$

By analogy with Section 9.2.1 we also define the following main effects:

A main effects: $\alpha_i = \bar{\mu}_{i\cdot\cdot} - \bar{\mu}_{\cdot\cdot\cdot}$,

B main effects: $\beta_j = \bar{\mu}_{\cdot j\cdot} - \bar{\mu}_{\cdot\cdot\cdot}$, and

C main effects: $\gamma_k = \bar{\mu}_{\cdot\cdot k} - \bar{\mu}_{\cdot\cdot\cdot}$.

9.3.2 Hypothesis Testing

With the above definitions, and defining $\mu = \bar{\mu}_{...}$, we have the reparametrization:

$$\mu_{ijk} = \mu + \alpha_i + \beta_j + \gamma_k + (\alpha\beta)_{ij} + (\beta\gamma)_{jk} + (\alpha\gamma)_{ik} + (\alpha\beta\gamma)_{ijk} \tag{9.38}$$

where

$$\alpha_{.} = \beta_{.} = \gamma_{.} = 0,$$

$$(\alpha\beta)_{i\cdot} = (\alpha\beta)_{\cdot j} = (\beta\gamma)_{j\cdot} = (\beta\gamma)_{\cdot k} = (\alpha\gamma)_{i\cdot} = (\alpha\gamma)_{\cdot k} = 0$$

and

$$(\alpha\beta\gamma)_{ij\cdot} = (\alpha\beta\gamma)_{\cdot jk} = (\alpha\beta\gamma)_{i\cdot k} = 0, \tag{9.39}$$

these conditions holding for all values of the subscripts i, j, and k.

The appropriate order for hypothesis testing is as follows: second-order interactions zero ($H_{ABC}: (\alpha\beta\gamma)_{ijk} = 0$, all i,j,k); first-order interactions zero ($H_{AB}: (\alpha\beta)_{ij} = 0$, all i,j; $H_{BC}: (\beta\gamma)_{jk} = 0$, all j,k; $H_{AC}: (\alpha\gamma)_{ik} = 0$, all i,k); and main effects zero ($H_A: \alpha_i = 0$, all i; $H_B: \beta_j = 0$, all j; $H_C: \gamma_k = 0$, all k). When H_{ABC} is true the three-factor experiment becomes equivalent to three independent two-factor experiments, one for each pair of factors, and the first-order interactions are readily interpreted. For example, (9.37) is now the same for all k so it is equal to the average over k [which is $(\alpha\beta)_{ij}$]. Similarly when H_{AB} is also true then the three-factor experiment becomes equivalent to two independent one-factor experiments for A and B, respectively, and the main effects α_i and β_j have a simple interpretation, e.g., $\alpha_i = \bar{\mu}_{i..} - \bar{\mu}_{...} = \bar{\mu}_{ij\cdot} - \bar{\mu}_{\cdot j\cdot} = \mu_{ijk} - \bar{\mu}_{\cdot jk}$.

As in the two-way classification the general regression theory can be applied here. For example, writing

$$\mathbf{Y}' = (Y_{1111}, Y_{1112}, \ldots, Y_{IJKM})$$

(9.36) can be expressed in the form $\mathbf{Y} = \mathbf{X}\boldsymbol{\mu} + \boldsymbol{\varepsilon}$, where $\mathbf{X}$ is $n \times p$ of rank p, $n = IJKM$, and $p = IJK$. Minimizing $\Sigma\Sigma\Sigma\Sigma(Y_{ijkm} - \mu_{ijk})^2$ with respect to μ_{ijk} we obtain

$$\text{RSS} = \Sigma\Sigma\Sigma\Sigma\left(Y_{ijkm} - \bar{Y}_{ijk\cdot}\right)^2 \tag{9.40}$$

with $(n-p)$ degrees of freedom. To find RSS_H for each hypothesis we split up ε_{ijkm} in a manner suggested by (9.38), namely:

$$
\begin{aligned}
\varepsilon_{ijkm} = \bar{\varepsilon}_{....} &+ (\bar{\varepsilon}_{i...} - \bar{\varepsilon}_{....}) + (\bar{\varepsilon}_{.j..} - \bar{\varepsilon}_{....}) \\
&+ (\bar{\varepsilon}_{..k.} - \bar{\varepsilon}_{....}) + (\bar{\varepsilon}_{ij..} - \bar{\varepsilon}_{i...} - \bar{\varepsilon}_{.j..} + \bar{\varepsilon}_{....}) \\
&+ (\bar{\varepsilon}_{i.k.} - \bar{\varepsilon}_{i...} - \bar{\varepsilon}_{..k.} + \bar{\varepsilon}_{....}) \\
&+ (\bar{\varepsilon}_{.jk.} - \bar{\varepsilon}_{.j..} - \bar{\varepsilon}_{..k.} + \bar{\varepsilon}_{....}) \\
&+ (\bar{\varepsilon}_{ijk.} - \bar{\varepsilon}_{ij..} - \bar{\varepsilon}_{.jk.} - \bar{\varepsilon}_{i.k.} + \bar{\varepsilon}_{i...} + \bar{\varepsilon}_{.j..} + \bar{\varepsilon}_{..k.} - \bar{\varepsilon}_{....}) \\
&+ (\varepsilon_{ijkm} - \bar{\varepsilon}_{ijk.}).
\end{aligned}
$$

Squaring and summing on i, j, k, and m, we find that the cross-product terms vanish so that

$$\Sigma\Sigma\Sigma\Sigma\, \varepsilon_{ijkm}^2 = \Sigma\Sigma\Sigma\Sigma\, \bar{\varepsilon}_{....}^2 + \cdots + \Sigma\Sigma\Sigma\Sigma (\varepsilon_{ijkm} - \bar{\varepsilon}_{ijk.})^2.$$

Setting $\varepsilon_{ijkm} = Y_{ijkm} - \mu_{ijk}$, and using Equations (9.38) and (9.39), we find that

$$
\begin{aligned}
&\Sigma\Sigma\Sigma\Sigma (Y_{ijkm} - \mu - \alpha_i - \cdots - (\alpha\beta\gamma)_{ijk})^2 \\
&\qquad = \Sigma\Sigma\Sigma\Sigma (\bar{Y}_{....} - \mu)^2 + \Sigma\Sigma\Sigma\Sigma (\bar{Y}_{i...} - \bar{Y}_{....} - \alpha_i)^2 \\
&\qquad\quad + \cdots + \Sigma\Sigma\Sigma\Sigma (Y_{ijkm} - \bar{Y}_{ijk.})^2. \qquad (9.41)
\end{aligned}
$$

By inspection the left side of (9.41) is minimized when the unknown parameters take the values

$$
\begin{aligned}
\hat{\mu} &= \bar{Y}_{....}, \\
\hat{\alpha}_i &= \bar{Y}_{i...} - \bar{Y}_{....}, \quad \hat{\beta}_j = \bar{Y}_{.j..} - \bar{Y}_{....}, \quad \hat{\gamma}_k = \bar{Y}_{..k.} - \bar{Y}_{....}, \\
(\widehat{\alpha\beta})_{ij} &= \bar{Y}_{ij..} - \bar{Y}_{i...} - \bar{Y}_{.j..} + \bar{Y}_{....}, \text{ etc.,} \\
(\widehat{\alpha\beta\gamma})_{ijk} &= \bar{Y}_{ijk.} - \bar{Y}_{ij..} - \bar{Y}_{.jk.} - \bar{Y}_{i.k.} + \bar{Y}_{i...} + \bar{Y}_{.j..} + \bar{Y}_{..k.} - \bar{Y}_{....},
\end{aligned}
$$

and the minimum value is, of course, RSS of (9.40). Testing any particular hypothesis is now very straightforward. For example, if we wish to test H_A we set $\alpha_i = 0$ in (9.41) and minimize with respect to the other parameters.

We see, by inspection, that the minimum occurs at the same values of the remaining parameters so that

$$\text{RSS}_{H_A} = \Sigma\Sigma\Sigma\Sigma\left(\overline{Y}_{i\cdots} - \overline{Y}_{\cdots\cdot}\right)^2 + \Sigma\Sigma\Sigma\Sigma\left(Y_{ijkm} - \overline{Y}_{ijk\cdot}\right)^2.$$

Hence

$$\begin{aligned}\text{RSS}_{H_A} - \text{RSS} &= \Sigma\Sigma\Sigma\Sigma\left(\overline{Y}_{i\cdots} - \overline{Y}_{\cdots\cdot}\right)^2 \\ &= JKM\sum_i \hat{\alpha}_i^2,\end{aligned}$$

with $(I-1)$ degrees of freedom, and the appropriate F-statistic is

$$F = \frac{JKM\sum_i \hat{\alpha}^2/(I-1)}{\text{RSS}/(IJK(M-1))} = \frac{S_A^2}{S^2},$$

say. This statistic has an F-distribution with $I-1$ and $IJK(M-1)$ degrees of freedom when H_A is true.

The various quadratic forms, together with their degrees of freedom, are listed in Table 9.3. The number of degrees of freedom can be obtained using the intuitive approach given in Section 9.2.2, or else it can be found directly from the trace of the quadratic form. For example, using the rule of thumb method for expanding quadratic forms we have

$$\begin{aligned}&\Sigma\Sigma\Sigma\Sigma\left(\overline{Y}_{ij\cdot\cdot} - \overline{Y}_{i\cdots} - \overline{Y}_{\cdot j\cdot\cdot} + \overline{Y}_{\cdots\cdot}\right)^2 \\ &\qquad = \sum_i\sum_j \frac{(Y_{ij\cdot\cdot})^2}{KM} - \sum_i \frac{(Y_{i\cdots})^2}{JKM} - \sum_j \frac{(Y_{\cdot j\cdot\cdot})^2}{IKM} + \frac{(Y_{\cdots\cdot})^2}{IJKM},\end{aligned}$$

and the trace of the symmetric matrix underlying this quadratic is the sum of the coefficients of the terms Y_{ijkm}^2, namely,

$$\begin{aligned}&\frac{1}{KM}\sum_i\sum_j KM - \frac{1}{JKM}\sum_i JKM - \frac{1}{IKM}\sum_j IKM + \frac{1}{IJKM}\cdot IJKM \\ &\qquad = IJ - I - J + 1 \\ &\qquad = (I-1)(J-1).\end{aligned}$$

Table 9.3 Analysis of Variance of a Three-Way Classification with M Observations per Population Mean

Source	Sum of Squares (SS)	Degrees of Freedom (df)	$\frac{\text{SS}}{\text{df}}$
A main effects	$JKM\sum_i \hat{\alpha}_i^2$	$I-1$	S_A^2
B main effects	$IKM\sum_j \hat{\beta}_j^2$	$J-1$	S_B^2
C main effects	$IJM\sum_k \hat{\gamma}_k^2$	$K-1$	S_C^2
AB interactions	$KM\sum_i \sum_j \widehat{(\alpha\beta)}_{ij}^2$	$(I-1)(J-1)$	S_{AB}^2
BC interactions	$IM\sum_j \sum_k \widehat{(\beta\gamma)}_{jk}^2$	$(J-1)(K-1)$	S_{BC}^2
AC interactions	$JM\sum_i \sum_k \widehat{(\alpha\gamma)}_{ik}^2$	$(I-1)(K-1)$	S_{AC}^2
ABC interactions	$M\sum_i \sum_j \sum_k \widehat{(\alpha\beta\gamma)}_{ijk}^2$	$(I-1)(J-1)(K-1)$	S_{ABC}^2
Error	$\sum_i \sum_j \sum_k \sum_m (Y_{ijkm} - \overline{Y}_{ijk\cdot})^2$	$IJKM - IJK$	S^2
Corrected total	$\sum_i \sum_j \sum_k \sum_m (Y_{ijkm} - \overline{Y}_{\cdots\cdot})^2$	$IJKM-1$	
Mean	$IJKM\overline{Y}_{\cdots\cdot}^2$	1	
Total	$\sum_i \sum_j \sum_k \sum_m Y_{ijkm}^2$	$IJKM$	

9.4 CLASSIFICATIONS WITH ONE OBSERVATION PER MEAN

9.4.1 *Derivation of Test Statistics*

Suppose that in a two-way classification there is only one observation per mean so that the model becomes

$$Y_{ij} = \mu_{ij} + \varepsilon_{ij} \qquad (i=1,2,\ldots,I;\, j=1,2,\ldots,J), \tag{9.42}$$

where the ε_{ij} are i.i.d. $N(0,\sigma^2)$. We now have IJ observations but $IJ+1$ unknown parameters ($\{\mu_{ij}\}$ and σ^2) so that we cannot estimate all the parameters without imposing at least one constraint to reduce the number of "free" parameters. However, typically, such data come from a randomized block design with I treatments and J blocks. Because the treatments are randomized within each block we would expect the interaction between treatment and block number to be small. A reasonable assumption is therefore

$$(\alpha\beta)_{ij} = \mu_{ij} - \bar{\mu}_{i\cdot} - \bar{\mu}_{\cdot j} + \bar{\mu}_{\cdot\cdot} = 0 \qquad \text{for all } i,j, \tag{9.43}$$

or $\mathbf{C}\boldsymbol{\mu} = \mathbf{0}$, where $\mathbf{C}$ is an $IJ \times IJ$ matrix with $(I-1)(J-1)$ linearly independent rows (cf. Section 9.2.2). Writing (9.42) in the form $\mathbf{Y} = \mathbf{X}\boldsymbol{\mu} + \boldsymbol{\varepsilon}$ where $\mathbf{X} = \mathbf{I}_n$ $(n = IJ)$, we can test the hypotheses $H_A: \alpha_i = \bar{\mu}_{i\cdot} - \bar{\mu}_{\cdot\cdot} = 0$ (or $\mathbf{A}\boldsymbol{\mu} = \mathbf{0}$) and $H_B: \beta_j = \bar{\mu}_{\cdot j} - \bar{\mu}_{\cdot\cdot} = 0$ (or $\mathbf{B}\boldsymbol{\mu} = \mathbf{0}$) subject to the constraints $\mathbf{C}\boldsymbol{\mu} = \mathbf{0}$ using the general theory of Section 4.6. For example, to test H_A we use

$$F = \frac{(\text{RSS}_{H_A} - \text{RSS})/q}{\text{RSS}/[n-(n-k)]}$$

where $q = \text{rank}\,\mathbf{A} = I-1$ and $k = \text{rank}\,\mathbf{C} = (I-1)(J-1)$. The residual sums of squares RSS_{H_A} and RSS can be found directly by using the method of Section 9.2.2. Thus, by analogy with $\mu_{ij} = \mu + \alpha_i + \beta_j + (\alpha\beta)_{ij}$, we have the decomposition

$$\varepsilon_{ij} = \bar{\varepsilon}_{\cdot\cdot} + (\bar{\varepsilon}_{i\cdot} - \bar{\varepsilon}_{\cdot\cdot}) + (\bar{\varepsilon}_{\cdot j} - \bar{\varepsilon}_{\cdot\cdot}) + (\bar{\varepsilon}_{ij} - \bar{\varepsilon}_{i\cdot} - \bar{\varepsilon}_{\cdot j} + \bar{\varepsilon}_{\cdot\cdot}),$$

which leads to

$$\sum\sum \varepsilon_{ij}^2 = \sum\sum \bar{\varepsilon}_{\cdot\cdot}^2 + \sum\sum(\bar{\varepsilon}_{i\cdot} - \bar{\varepsilon}_{\cdot\cdot})^2 + \sum\sum(\bar{\varepsilon}_{\cdot j} - \bar{\varepsilon}_{\cdot\cdot})^2 + \sum\sum(\varepsilon_{ij} - \bar{\varepsilon}_{i\cdot} - \bar{\varepsilon}_{\cdot j} + \bar{\varepsilon}_{\cdot\cdot})^2.$$

Applying the constraints (9.43) to the model we have

$$\varepsilon_{ij} = Y_{ij} - \mu - \alpha_i - \beta_j,$$

and hence

$$\sum\sum(Y_{ij} - \mu - \alpha_i - \beta_j)^2 = \sum\sum(\bar{Y}_{\cdot\cdot} - \mu)^2 + \sum\sum(\bar{Y}_{i\cdot} - \bar{Y}_{\cdot\cdot} - \alpha_i)^2$$

$$+ \sum\sum(\bar{Y}_{\cdot j} - \bar{Y}_{\cdot\cdot} - \beta_j)^2 + \sum\sum(Y_{ij} - \bar{Y}_{i\cdot} - \bar{Y}_{\cdot j} + \bar{Y}_{\cdot\cdot})^2. \tag{9.44}$$

The left side of (9.44) is minimized when $\mu = \bar{Y}_{\cdot\cdot}$ $(= \hat{\mu})$, $\alpha_i = \bar{Y}_{i\cdot} - \bar{Y}_{\cdot\cdot}$ $(= \hat{\alpha}_i)$ and $\beta_j = \bar{Y}_{\cdot j} - \bar{Y}_{\cdot\cdot}$ $(= \hat{\beta}_j)$, so that

$$\begin{aligned}\text{RSS} &= \sum\sum(Y_{ij} - \bar{Y}_{i\cdot} - \bar{Y}_{\cdot j} + \bar{Y}_{\cdot\cdot})^2\\ &= \sum\sum(\widehat{\alpha\beta})_{ij}^2,\end{aligned}$$

say. Since the interactions are zero we see from the above equation that the interaction sum of squares takes over the role of the error sum of squares, and an unbiased estimate of σ^2 is RSS$/(I-1)(J-1)$. However, it should be pointed out that an estimate of σ^2 can be found by making much weaker assumptions; not all the interactions need be zero (Johnson and Graybill [1972a, b]).

Setting $\alpha_i = 0$ we see, by inspection, that the left side of (9.44) is minimized when $\mu = \hat{\mu}$ and $\beta_j = \hat{\beta}_j$ so that

$$\text{RSS}_{H_A} = \Sigma\Sigma\left(\overline{Y}_{i\cdot} - \overline{Y}_{\cdot\cdot}\right)^2 + \Sigma\Sigma\left(Y_{ij} - \overline{Y}_{i\cdot} - \overline{Y}_{\cdot j} + \overline{Y}_{\cdot\cdot}\right)^2,$$

and

$$\text{RSS}_{H_A} - \text{RSS} = \Sigma\Sigma\left(\overline{Y}_{i\cdot} - \overline{Y}_{\cdot\cdot}\right)^2.$$

Hence the F-statistic for testing H_A is

$$F = \frac{\Sigma\Sigma\left(\overline{Y}_{i\cdot} - \overline{Y}_{\cdot\cdot}\right)^2/(I-1)}{\Sigma\Sigma\left(Y_{ij} - \overline{Y}_{i\cdot} - \overline{Y}_{\cdot j} + \overline{Y}_{\cdot\cdot}\right)^2/(I-1)(J-1)}$$

$$= \frac{J\sum_i \hat{\alpha}_i^2/(I-1)}{\Sigma\Sigma(\widehat{\alpha\beta})_{ij}^2/(I-1)(J-1)}. \tag{9.45}$$

The test statistic for H_B follows by interchanging i and j, namely,

$$F = \frac{I\sum_j \hat{\beta}_j^2/(J-1)}{\Sigma\Sigma(\widehat{\alpha\beta})_{ij}^2/(I-1)(J-1)}. \tag{9.46}$$

The whole procedure can be summarized as in Table 9.4.

The extension of the above theory to higher-way classifications is straightforward. For example, in the three-way analysis of variance described by Table 9.3 we simply set $M = 1$ and use the ABC interaction sum of squares for our error sum of squares.

Table 9.4 Analysis of Variance for a Two-Way Classification with One Observation per Population Mean

Source	Sum of Squares (SS)	Degrees of Freedom (df)	$\frac{\text{SS}}{\text{df}}$
A main effects (treatments)	$J\sum_i \hat{\alpha}_i^2$	$I-1$	S_A^2
B main effects (blocks)	$I\sum_j \hat{\beta}_j^2$	$J-1$	S_B^2
Error	$\sum_i \sum_j \widehat{(\alpha\beta)}_{ij}^2$	$(I-1)(J-1)$	S^2
Corrected total	$\sum_i \sum_j (Y_{ij}-\overline{Y}_{..})^2$	$IJ-1$	
Mean	$IJ\overline{Y}_{..}^2$	1	
Total	$\sum_i \sum_j Y_{ij}^2$	IJ	

9.4.2 *Underlying Assumptions*

The effect of departures from the assumption of zero interactions on (9.45) is discussed by Scheffé [1959: p. 134]. This assumption (9.43) is frequently called the assumption of additivity as $\mu_{ij}=\mu+\alpha_i+\beta_j$ and the main effects α_i and β_j are "additive." Because we cannot estimate all the parameters in (9.42) we cannot test (9.43) against a general class of alternatives $(\alpha\beta)_{ij}\neq 0$ [for at least one pair (i,j)]. However, we can test (9.43) against a suitably restricted class of alternatives and several such classes have been considered. For example, if we assume $(\alpha\beta)_{ij}=G\alpha_i\beta_j$, then Tukey's [1949] well-known test for additivity is equivalent to testing the null hypothesis $H_G: G=0$ against the alternative $G\neq 0$ (Scheffé [1959: pp. 129–137]). Tukey's test statistic is

$$F=\frac{\text{SS}_G}{(\text{RSS}-\text{SS}_G)/[(I-1)(J-1)-1]}, \tag{9.47}$$

where

$$\text{SS}_G=\frac{\left(\sum_i \sum_j \hat{\alpha}_i \hat{\beta}_j Y_{ij}\right)^2}{\sum_i \hat{\alpha}_i^2 \sum_j \hat{\beta}_j^2}$$

and

$$\text{RSS} = \sum_i \sum_j (\widehat{\alpha\beta})_{ij}^2,$$

Then F has an F-distribution with 1 and $IJ - I - J$ degrees of freedom, respectively, when the underlying model is

$$Y_{ij} = \mu + \alpha_i + \beta_j + \varepsilon_{ij}.$$

It is instructive to derive (9.47) from the following lemma due to Scheffé [1959: p. 144, Ex. 4.19].

LEMMA Suppose that $\mathbf{Y} \sim N_n(\mathbf{X}\boldsymbol{\beta}, \sigma^2\mathbf{I}_n)$, where $\mathbf{X}$ is $n \times p$ of rank p, and define $\hat{\boldsymbol{\theta}} = \mathbf{X}\hat{\boldsymbol{\beta}}$, where $\hat{\boldsymbol{\beta}}$ is the least squares estimate of $\boldsymbol{\beta}$. Let $\mathbf{Z} = \mathbf{f}(\hat{\boldsymbol{\theta}})$ be a continuous function of $\hat{\boldsymbol{\theta}}$ (chosen before the outcome of $\mathbf{Y}$ is inspected) and let $\hat{\boldsymbol{\phi}}$ be the same linear function of $\mathbf{Z}$ that $\hat{\boldsymbol{\theta}}$ is of $\mathbf{Y}$. Define $R = \|\mathbf{Y} - \hat{\boldsymbol{\theta}}\|^2$ and

$$R_1 = \frac{\mathbf{Z}'(\mathbf{Y} - \hat{\boldsymbol{\theta}})}{\{(\mathbf{Z} - \hat{\boldsymbol{\phi}})'(\mathbf{Z} - \hat{\boldsymbol{\phi}})\}^{1/2}}.$$

Then

$$F_0 = \frac{R_1^2}{(R - R_1^2)/(n-p-1)} \sim F_{1,n-p-1}.$$

Proof. $\hat{\boldsymbol{\theta}} = \mathbf{X}(\mathbf{X}'\mathbf{X})^{-1}\mathbf{X}'\mathbf{Y} = \mathbf{P}\mathbf{Y}$ so that $\hat{\boldsymbol{\phi}} = \mathbf{P}\mathbf{Z}$ and

$$R_1 = \frac{\mathbf{Z}'(\mathbf{I}_n - \mathbf{P})\mathbf{Y}}{\{\mathbf{Z}'(\mathbf{I}_n - \mathbf{P})\mathbf{Z}\}^{1/2}} \qquad (= \mathbf{Z}'(\mathbf{I}_n - \mathbf{P})\mathbf{Y}/c_Z, \text{ say}).$$

Consider the distributions of R and R_1 conditional on $\mathbf{Z} = \mathbf{z}$. Since R is independent of $\hat{\boldsymbol{\beta}}$ [Theorem 3.5 (iii), Section 3.4], and therefore of $\mathbf{Z}$, the conditional distribution of R/σ^2 is the same as the unconditional distribution, namely, χ^2_{n-p} [Theorem 3.5 (iv)]. Also $R_1 = \mathbf{z}'(\mathbf{I}_n - \mathbf{P})\mathbf{Y}/c_z$ is a linear combination of the Y_i so that

$$E[R_1] = \frac{\mathbf{z}'(\mathbf{I}_n - \mathbf{P})\mathbf{X}\boldsymbol{\beta}}{c_z} = 0$$

and

$$\text{var}[R_1] = \frac{\mathbf{z}'(\mathbf{I}_n - \mathbf{P})\mathcal{D}[\mathbf{Y}](\mathbf{I}_n - \mathbf{P})'\mathbf{z}}{c_z^2} = \sigma^2$$

imply that $R_1 \sim N(0, \sigma^2)$. Now setting $\mathbf{u} = (\mathbf{I}_n - \mathbf{P})\mathbf{Y}$ and $\mathbf{v} = (\mathbf{I}_n - \mathbf{P})\mathbf{z}$, and invoking the Cauchy–Schwartz inequality (*A4.11*), we have

$$R - R_1^2 = \mathbf{Y}'(\mathbf{I}_n - \mathbf{P})\mathbf{Y} - \frac{\{\mathbf{z}'(\mathbf{I}_n - \mathbf{P})\mathbf{Y}\}^2}{\mathbf{z}'(\mathbf{I}_n - \mathbf{P})\mathbf{z}}$$

$$= \mathbf{u}'\mathbf{u} - \frac{(\mathbf{u}'\mathbf{v})^2}{\mathbf{v}'\mathbf{v}}$$

$$= \frac{(\mathbf{u}'\mathbf{u})(\mathbf{v}'\mathbf{v}) - (\mathbf{u}'\mathbf{v})^2}{\mathbf{v}'\mathbf{v}}$$

$$\geqslant 0.$$

Then since $R/\sigma^2 \sim \chi^2_{n-p}$ and $R_1^2/\sigma^2 \sim \chi_1^2$ we have, by Theorem 2.9 (Section 2.4), that $(R - R_1^2)/\sigma^2$ and R_1^2/σ^2 are independently distributed as χ^2_{n-p-1} and χ_1^2, respectively. Thus $F_0 \sim F_{1,n-p-1}$ and, because the F-distribution does not depend on $\mathbf{z}$, it is also the unconditional distribution of F_0. □

To apply the above lemma to (9.47) we simply define $\mathbf{Z} = \mathbf{f}(\hat{\boldsymbol{\theta}})$ by $Z_{ij} = \hat{\theta}_{ij}^2$, where $\hat{\theta}_{ij} = \hat{\mu} + \hat{\alpha}_i + \hat{\beta}_j$. Then

$$\|\mathbf{Y} - \hat{\boldsymbol{\theta}}\|^2 = \text{RSS} = \Sigma\Sigma\left(Y_{ij} - \overline{Y}_{i\cdot} - \overline{Y}_{\cdot j} + \overline{Y}_{\cdot\cdot}\right)^2$$

and

$$R_1^2 = \frac{\left\{\left[(\mathbf{I}_n - \mathbf{P})\mathbf{Z}\right]'\mathbf{Y}\right\}^2}{\mathbf{Z}'(\mathbf{I}_n - \mathbf{P})\mathbf{Z}}$$

$$= \frac{\left\{\Sigma\Sigma\left(Z_{ij} - \overline{Z}_{i\cdot} - \overline{Z}_{\cdot j} + \overline{Z}_{\cdot\cdot}\right)Y_{ij}\right\}^2}{\Sigma\Sigma\left(Z_{ij} - \overline{Z}_{i\cdot} - \overline{Z}_{\cdot j} + \overline{Z}_{\cdot\cdot}\right)^2}.$$

Using $\hat{\alpha}_\cdot = \hat{\beta}_\cdot = 0$ we have, after some algebra, that

$$Z_{ij} - \overline{Z}_{i\cdot} - \overline{Z}_{\cdot j} + \overline{Z}_{\cdot\cdot} = 2\hat{\alpha}_i\hat{\beta}_j$$

so that

$$R_1^2 = \frac{\left\{\Sigma\Sigma\hat{\alpha}_i\hat{\beta}_j Y_{ij}\right\}^2}{\Sigma\Sigma\hat{\alpha}_i^2\hat{\beta}_j^2}, \tag{9.48}$$

and we have derived (9.47). A similar method can be used for deriving a test for interaction for other experimental designs which assume additivity, for example, the Latin square.

Tukey's test [which was originally proposed without specifying any particular form for $(\alpha\beta)_{ij}$] seems to have reasonably good power for the alternatives $G \neq 0$ (Ghosh and Sharma [1963]), and the effect of nonnormality on the test is examined empirically by Yates [1972]. Several generalizations of the procedure have been proposed (cf. Johnson and Graybill [1972a, b] for references) and all these tests would appear to have reasonably good power when $(\alpha\beta)_{ij}$ is a function of the α_i or β_j. Johnson and Graybill [1972b] also proposed a test for interaction which would have a reasonable power when the underlying model is

$$Y_{ij} = \mu + \alpha_i + \beta_j + \lambda\gamma_i\delta_j + \varepsilon_{ij}, \tag{9.49}$$

where $\alpha_. = \beta_. = \gamma_. = \delta_. = 0$ and $\Sigma_i \gamma_i^2 = \Sigma_j \delta_j^2 = 1$.

Residual plots based on the residuals $(\widehat{\alpha\beta})_{ij}$ must be interpreted with caution. Any irregularities could be due to either departures from the usual normality assumptions or to the presence of nonzero interactions (that is, $[E(\widehat{\alpha\beta})_{ij}] \neq 0$ for some i, j). Because the F-tests (9.45) and (9.46) are quadratically balanced we would expect the test statistics to be robust with regard to nonnormality. Any heterogeneity of variance, or error correlations within blocks could be tested using, for example, the methods of Han [1969].

9.4.3 *An Alternative Approach*

If we assume additivity for the randomized block design, we have the model

$$Y_{ij} = \mu + \alpha_i + \beta_j + \varepsilon_{ij}.$$

This is a less than full rank model and can therefore be investigated using any of the approaches in Section 3.8. For example, we can impose identifiability constraints of the form $\alpha_. = \beta_. = 0$ and then use the algorithm described in Section 11.5.4 for generating residual sums of squares. This algorithm would be useful in the analysis of incomplete block designs, where each block does not contain all the treatments.

9.5 DESIGNS WITH SIMPLE BLOCK STRUCTURE

In addition to the cross-classification designs considered above there are also the so-called hierarchical or nested designs. For example, suppose we have I cities, J factories within each city, and a sample of size K is taken from each factory, giving the model $Y_{ijk}=\theta_{ijk}+\varepsilon_{ijk}$ $(i=1,2,\ldots,I;\ j=1,2,\ldots,J;\ k=1,2,\ldots,K)$. Then the appropriate reparametrization of the model is

$$\theta_{ijk}=\bar{\theta}_{\cdots}+\left(\bar{\theta}_{i\cdot\cdot}-\bar{\theta}_{\cdots}\right)+\left(\bar{\theta}_{ij\cdot}-\bar{\theta}_{i\cdot\cdot}\right)+\left(\theta_{ijk}-\bar{\theta}_{ij\cdot}\right) \tag{9.50}$$

or, since $\theta_{ijk}=\mu_{ij}$ $(k=1,2,\ldots,K)$,

$$\mu_{ij}=\mu+\alpha_i+\beta_{ij} \tag{9.51}$$

with identifiability constraints $\alpha_{\cdot}=0$ and $\beta_{i\cdot}=0$ (all i). The hypotheses of interest are $H_1:\beta_{ij}=0$ (no variation within each city) and $H_2:\alpha_i=0$ (no variation between cities), and the appropriate decomposition of ε_{ijk} is

$$\varepsilon_{ijk}=\bar{\varepsilon}_{\cdots}+\left(\bar{\varepsilon}_{i\cdot\cdot}-\bar{\varepsilon}_{\cdots}\right)+\left(\bar{\varepsilon}_{ij\cdot}-\bar{\varepsilon}_{i\cdot\cdot}\right)+\left(\varepsilon_{ijk}-\bar{\varepsilon}_{ij\cdot}\right). \tag{9.52}$$

Once again we have an orthogonal decomposition of $\boldsymbol{\varepsilon}$, and F-statistics are readily obtained for testing H_1 and H_2; the details are left as an exercise (Exercise 3 at the end of this chapter).

Many of the designs currently used are a mixture of both crossing and nesting. When every nesting classification used has equal numbers of subunits nested in each unit, then the experimental units are said to have a *simple block structure* and the following elegant theory (due to Nelder [1965a, b]) is available for handling such designs.

Now all simple block structures can be built using two basic operations, nesting (denoted by $\rightarrow$) and crossing (denoted by $\times$). The two simplest structures are written as $B_1\rightarrow B_2$ (one-way classification with equal numbers per mean) and $B_1\times B_2$ (two-way classification with one observation per mean). Elements in either of these expressions may themselves be expressions of this type; for example, $B_1\rightarrow(B_2\rightarrow B_3)$ (the hierarchical design described above), $B_1\rightarrow(B_2\times B_3)$, $(B_1\times B_2)\rightarrow B_3$ (two-way classification with equal numbers per mean) and $(B_1\times B_2)\times B_3$ (three-way classification with one observation per mean). We note that

$$B_1\rightarrow(B_2\rightarrow B_3)=(B_1\rightarrow B_2)\rightarrow B_3,\quad \text{and}$$

$$B_1\times(B_2\times B_3)=(B_1\times B_2)\times B_3.$$

Suppose that we change our notation slightly and let $Y_{i_1i_2\cdots i_r}$ represent an observation from such a design with r "blocks" $B_1, B_2,\ldots,B_r$, for example, Y_{ijk} now becomes $Y_{i_1i_2i_3}$. Let n_k be the number of units in B_k so that $i_k=1,2,\ldots,n_k$ $(k=1,2,\ldots,r)$. The first step in the analysis of variance of the design is to set up a degrees of freedom (df) identity. For instance, for

$B_1 \to B_2$ we have (cf. Table 9.1 in Section 9.1.2)

$$\begin{aligned} n_1 n_2 &\equiv 1 + (n_1 - 1) + (n_1 n_2 - n_1) \\ &= 1 + \quad \nu_1 \quad + \quad n_1 \nu_2 \end{aligned}$$

where $\nu_i = n_i - 1$. The randomized block design, $B_1 \times B_2$, leads to the identity (cf. Table 9.4 in Section 9.4.1)

$$\begin{aligned} n_1 n_2 &\equiv 1 + (n_1 - 1) + (n_2 - 1) + (n_1 - 1)(n_2 - 1) \\ &= 1 + \quad \nu_1 \quad + \quad \nu_2 \quad + \quad \nu_1 \nu_2. \end{aligned}$$

The identities for more complex structures can be built up by the use of the following nesting and crossing functions

$$N(n_1, n_2) = 1 + \nu_1 + n_1 \nu_2$$

and

$$C(n_1, n_2) = 1 + \nu_1 + \nu_2 + \nu_1 \nu_2.$$

The rules for using these functions are as follows: (1) If on substitution for n_i a term (such as ν_i) becomes zero, it is ignored; and any unity occurring in a product is suppressed, for example,

$$N(1, n_2) = 1 + 0 + 1 \cdot \nu_2 = 1 + \nu_2$$

and

$$C(n_1, n_2) = N(1, n_1) N(1, n_2). \tag{9.53}$$

(2) The arguments in N and C may themselves be N and C functions, in which case ν retains its meaning as $n - 1$ where n may now be an N or C expression. However, the n_1 that occurs in the expansion of $N(n_1, n_2)$ is to be thought of as the algebraic sum of all the terms in the expansion of n_1 and is denoted by $\bar{N}$ or $\bar{C}$.

EXAMPLE 9.1 $(B_1 \to B_2) \to B_3$.

$$\begin{aligned} n_1 n_2 n_3 &\equiv N(N(n_1, n_2), n_3) \\ &= 1 + [N(n_1, n_2) - 1] + \bar{N}(n_1, n_2) \nu_3 \\ &= 1 + (\nu_1 + n_1 \nu_2) + n_1 n_2 \nu_3 \\ &= 1 + (n_1 - 1) + (n_1 n_2 - n_1) + (n_1 n_2 n_3 - n_1 n_2) \end{aligned} \tag{9.54}$$

EXAMPLE 9.2 $(B_1 \times B_2) \rightarrow B_3$.

$$\begin{aligned} n_1 n_2 n_3 &\equiv N(C(n_1, n_2), n_3) \\ &= 1 + [C(n_1, n_2) - 1] + \bar{C}(n_1, n_2)\nu_3 \\ &= 1 + (\nu_1 + \nu_2 + \nu_1\nu_2) + n_1 n_2 \nu_3 \\ &= 1 + (n_1 - 1) + (n_2 - 1) + (n_1 n_2 - n_1 - n_2 + 1) \\ &\quad + (n_1 n_2 n_3 - n_1 n_2). \end{aligned} \tag{9.55}$$

EXAMPLE 9.3 $B_1 \times (B_2 \rightarrow B_3)$.

$$\begin{aligned} n_1 n_2 n_3 &\equiv C(n_1, N(n_2, n_3)) \\ &= 1 + \nu_1 + [N(n_2, n_3) - 1] + \nu_1[N(n_2, n_3) - 1] \\ &= 1 + \nu_1 + (\nu_2 + n_2\nu_3) + \nu_1(\nu_2 + n_2\nu_3) \\ &= 1 + \nu_1 + \nu_2 + \nu_1\nu_2 + n_2\nu_3 + \nu_1 n_2 \nu_3 \\ &= 1 + (n_1 - 1) + (n_2 - 1) + (n_1 n_2 - n_1 - n_2 + 1) \\ &\quad + (n_2 n_3 - n_2) + (n_1 n_2 n_3 - n_1 n_2 - n_2 n_3 + n_2) \end{aligned} \tag{9.56}$$

Alternatively, using (9.53), we have

$$\begin{aligned} n_1 n_2 n_3 &\equiv N(1, n_1) N(1, N(n_2, n_3)) \\ &= (1 + \nu_1)\{1 + [N(n_2, n_3) - 1]\} \\ &= (1 + \nu_1)(1 + \nu_2 + n_2\nu_3) \\ &= 1 + \nu_1 + \nu_2 + \nu_1\nu_2 + n_2\nu_3 + \nu_1 n_2 \nu_3. \end{aligned}$$

Once the identity has been established it is then a simple matter to obtain the appropriate reparametrisation of $\theta_{i_1 i_2 \cdots i_r} = E[Y_{i_1 i_2 \cdots i_r}]$ and the corresponding orthogonal decomposition of $\boldsymbol{\varepsilon} = [(\varepsilon_{i_1 i_2 \cdots i_r})]$. For example, corresponding to Equation (9.54) we have Equations (9.50), (9.51), and (9.52): we associate with each part of each term in (9.54) a mean of the vector $\boldsymbol{\theta}$ with the same sign and averaged over all those indices which are absent. In the case of Example 9.3 above we have

$$\begin{aligned}\theta_{ijk} &= \bar{\theta}_{...} + (\bar{\theta}_{i..} - \bar{\theta}_{...}) + (\bar{\theta}_{.j.} - \bar{\theta}_{...}) \\ &\quad + (\bar{\theta}_{ij.} - \bar{\theta}_{i..} - \bar{\theta}_{.j.} + \bar{\theta}_{...}) + (\bar{\theta}_{.jk} - \bar{\theta}_{.j.}) \\ &\quad + (\theta_{ijk} - \bar{\theta}_{ij.} - \bar{\theta}_{.jk} + \bar{\theta}_{.j.}) \qquad (9.57) \\ &= \mu + \alpha_i + \beta_j + (\alpha\beta)_{ij} + \beta_{jk} + (\alpha\gamma)_{ijk}, \qquad (9.58)\end{aligned}$$

say, with the usual identifiability constraints $\alpha_. = \beta_. = (\alpha\beta)_{i.} = (\alpha\beta)_{.j} = \beta_{j.} = (\alpha\gamma)_{.jk} = (\alpha\gamma)_{ij.} = 0$ (all i,j,k). All the parameters in (9.58), except possibly μ, correspond to hypotheses of interest, and the least squares estimates of these parameters are simply the corresponding terms in the same decomposition of Y_{ijk}, e.g., $\hat{\beta}_{jk} = \bar{Y}_{.jk} - \bar{Y}_{.j.}$. This simple method of finding the least squares estimates can be verified by using the same decomposition for ε_{ijk}. Such a decomposition leads to

$$\begin{aligned}\Sigma\Sigma\Sigma\varepsilon_{ijk}^2 &= \Sigma\Sigma\Sigma\bar{\varepsilon}_{...}^2 + \Sigma\Sigma\Sigma(\bar{\varepsilon}_{i..} - \bar{\varepsilon}_{...})^2 + \Sigma\Sigma\Sigma(\bar{\varepsilon}_{.j.} - \bar{\varepsilon}_{...})^2 \\ &\quad + \Sigma\Sigma\Sigma(\bar{\varepsilon}_{ij.} - \bar{\varepsilon}_{i..} - \bar{\varepsilon}_{.j.} + \bar{\varepsilon}_{...})^2 + \Sigma\Sigma\Sigma(\bar{\varepsilon}_{.jk} - \bar{\varepsilon}_{.j.})^2 \\ &\quad + \Sigma\Sigma\Sigma(\varepsilon_{ijk} - \bar{\varepsilon}_{ij.} - \bar{\varepsilon}_{.jk} + \bar{\varepsilon}_{.j.})^2 \qquad (9.59)\end{aligned}$$

which, on substituting $\varepsilon_{ijk} = Y_{ijk} - \mu - \cdots - (\alpha\gamma)_{ijk}$ and using the identifiability constraints, gives the least squares estimates by inspection. If we assume that $(\alpha\gamma)_{ijk} = 0$ (all i,j,k), we can test whether members of the sets of parameters $\{\alpha_i\}$, $\{\beta_j\}$, $\{(\alpha\beta)_{ij}\}$, and $\{\beta_{ij}\}$ are zero using F-statistics with $\Sigma\Sigma\Sigma(Y_{ijk} - \bar{Y}_{ij.} - \bar{Y}_{.jk} + \bar{Y}_{.j.})^2$ in the denominator. The various numerator sums of squares are given by the terms of (9.59) expressed as an identity in Y_{ijk}, and the associated degrees of freedom are given by the identity (9.56). The analysis of variance table follows from the decomposition of $\Sigma\Sigma\Sigma Y_{ijk}^2$.

The above simple rules form the basis of a general computer system, called GENSTAT, that can handle a wide range of linear models. References and further details concerning the system are given in Wilkinson and Rogers [1973].

MISCELLANEOUS EXERCISES 9

1. If the ε_{ij} $(i=1,2,\ldots,I;\ j=1,2,\ldots,J)$ are independently distributed as $N(0,\sigma^2)$, prove that

$$\sum_i \sum_j (\bar{\varepsilon}_{i\cdot} - \bar{\varepsilon}_{\cdot\cdot})^2 \quad \text{and} \quad \sum_i \sum_j (\varepsilon_{ij} - \bar{\varepsilon}_{i\cdot} - \bar{\varepsilon}_{\cdot j} + \bar{\varepsilon}_{\cdot\cdot})^2$$

are statistically independent.

2. Let $Y_{ij} = \mu + \alpha_i + \varepsilon_{ij}$ $(i=1,2,\ldots,I;\ j=1,2,\ldots,J)$, where $\sum_i d_i \alpha_i = 0$ $(\sum_i d_i \neq 0)$ and $E[\varepsilon_{ij}] = 0$ for all i,j. Using the method of Lagrange multipliers find the least squares estimates of μ and α_i.
[*Hint*: Show that the Lagrange multiplier is zero.]

3. Let $Y_{ijk} = \mu_{ij} + \varepsilon_{ijk}$, where

$$\begin{aligned} \mu_{ij} &= \bar{\mu}_{\cdot\cdot} + (\bar{\mu}_{i\cdot} - \bar{\mu}_{\cdot\cdot}) + (\mu_{ij} - \bar{\mu}_{i\cdot}) \\ &= \mu + \quad \alpha_i \quad + \quad \beta_{ij}, \end{aligned}$$

say, $i=1,2,\ldots,I;\ j=1,2,\ldots,J;\ k=1,2,\ldots,K$, and the ε_{ijk} are independently distributed as $N(0,\sigma^2)$.
   (a) Find the least squares estimates of μ, α_i, and β_{ij}, and show that they are mutually statistically independent.
   (b) Obtain a test statistic for testing the hypotheses $H_1: \beta_{ij} = 0$ (all i,j) and $H_2: \alpha_i = 0$ (all i).

4. Let $Y_{ij} = \mu_i + \varepsilon_{ij}$ $(i=1,2,\ldots,I;\ j=1,2,\ldots,J)$, where the ε_{ij} are independently distributed as $N(0,\sigma^2)$.
   (a) When $I=4$ obtain an F-statistic for testing the hypothesis that $\mu_1 = 2\mu_2 = 3\mu_3$.
   (b) When $I=2$, show that the F-statistic for testing $\mu_1 = \mu_2$ is the square of the usual t-test for testing the hypothesis that the means of two normally distributed populations are equal, given that their variances are equal.

5. Suppose that we have the model

$$Y_{ijk} = \mu + \alpha_i + \beta_j + \gamma_k + \varepsilon_{ijk},$$

where $i=1,2,\ldots,I;\ j=1,2,\ldots,J;\ k=1,2,\ldots,K$, $\sum_i \alpha_i = \sum_j \beta_j = \sum_k \gamma_k = 0$, and the ε_{ijk} are independently distributed as $N(0,\sigma^2)$.
   (a) Express μ, α_i, β_j, and γ_k in terms of the parameters $\mu_{ijk} = E[Y_{ijk}]$.
   (b) Obtain a test statistic for testing the hypothesis $H: \alpha_i = 0$ (all i).

(c) Prove that

$$\frac{\sum_i \sum_j \sum_k (\bar{\varepsilon}_{ij\cdot} - \bar{\varepsilon}_{i\cdot\cdot} - \bar{\varepsilon}_{\cdot j\cdot} + \bar{\varepsilon}_{\cdots})^2}{\sigma^2} \sim \chi^2_{(I-1)(J-1)}.$$

[*Hint*: Split up $\sum_i \sum_j \sum_k (\bar{\varepsilon}_{ij\cdot} - \bar{\varepsilon}_{i\cdot\cdot})^2$ into two sums of squares.]

6. Given the population means μ_{ij} $(i=1,2,\ldots,I;\ j=1,2,\ldots,J)$, let

$$A_i = \sum_j v_j \mu_{ij} \qquad \left(\sum_j v_j = 1\right)$$

$$B_j = \sum_i u_i \mu_{ij} \qquad \left(\sum_i u_i = 1\right)$$

and

$$\mu = \sum_i u_i A_i = \sum_j v_j B_j = \sum_i \sum_j u_i v_j \mu_{ij}.$$

Define $\alpha_i = A_i - \mu$, $\beta_j = B_j - \mu$, and

$$(\alpha\beta)_{ij} = \mu_{ij} - A_i - B_j + \mu.$$

(a) Show that $\sum_i u_i \alpha_i = \sum_j v_j \beta_j = 0$, $\sum_i u_i (\alpha\beta)_{ij} = 0$ (all j), and $\sum_j v_j (\alpha\beta)_{ij} = 0$ (all i).

(b) Conversely, given

$$\mu_{ij} = \mu + \alpha_i + \beta_j + (\alpha\beta)_{ij},$$

show that the parameters in the above equation are uniquely determined by the constraints in (a).

(c) Prove that if the interactions $\{(\alpha\beta)_{ij}\}$ are all zero for some system of weights $\{u_i\}$ and $\{v_j\}$, then they are zero for every system of weights. In that case show that every contrast in the $\{\alpha_i\}$, or $\{\beta_j\}$, has a value that does not depend on the system of weights.

Scheffé [1959: Section 4.1]

7. Let $Y_{ijk} = \mu + \alpha_i + \beta_j + (\alpha\beta)_{ij} + \varepsilon_{ijk}$, where $i=1,2,\ldots,I$; $j=1,2,\ldots,J$; $k=1,2,\ldots,K_{ij}$; and the ε_{ijk} are independently distributed as $N(0,\sigma^2)$. Given $K_{ij} = K_{i\cdot}K_{\cdot j}/K_{\cdot\cdot}$ for all i,j, find a test statistic for testing the hypothesis $H:(\alpha\beta)_{ij} = 0$ (all i,j).
[*Hint*: By Exercise 6, the validity of H does not depend on the weights used in the identifiability constraints $\sum_i u_i \alpha_i = \sum_j v_j \beta_j = 0$. We can therefore use $u_i = K_{i\cdot}/K_{\cdot\cdot}$ and $v_j = K_{\cdot j}/K_{\cdot\cdot}$ and find the least squares estimates of α_i and β_j when H is true.]

Scheffé [1959: p. 119]

8. A three-factor experiment has population means μ_{ijk} $(i=1,2,3;\ j=1,2,3;\ k=1,2)$ given by the following tables:

C_1	B_1	B_2	B_3	Mean
A_1	5	6	10	7
A_2	7	7	1	5
A_3	6	5	7	6
Mean	6	6	6	6

C_2	B_1	B_2	B_3	Mean
A_1	9	7	14	10
A_2	9	6	3	6
A_3	9	5	10	8
Mean	9	6	9	8

Show that the ABC interactions are zero.

9. Consider the linear model $Y_{ijk}=\mu_{ijk}+\varepsilon_{ijk}$, where $i=1,2,\ldots,I$; $j=1,2,\ldots,J$; $k=1,2,\ldots,K$; and the ε_{ijk} are independently distributed as $N(0,\sigma^2)$. Let

$$\mu_{ijk}=\bar{\mu}_{\cdots}+(\bar{\mu}_{i\cdot\cdot}-\bar{\mu}_{\cdots})+(\bar{\mu}_{ij\cdot}-\bar{\mu}_{i\cdot\cdot})+(\bar{\mu}_{\cdot\cdot k}-\bar{\mu}_{\cdots})+\Delta_{ijk}$$

$$=\mu+\alpha_i+\beta_{ij}+\gamma_k,$$

say, where $\Delta_{ijk}=0$ (all i,j,k).

(a) Find the least squares estimates of μ, α_i, β_{ij}, and γ_k.

(b) Obtain an F-statistic for testing $H:\alpha_i=0$ (all i).

10. Use Nelder's method of Section 9.5 to obtain an analysis of variance table for the following designs:

(a) $B_1\times(B_2\times B_3)$;

(b) $(B_1\rightarrow B_2)\times B_3$.

CHAPTER 10

Analysis of Covariance and Missing Observations

10.1 ANALYSIS OF COVARIANCE

10.1.1 Least Squares Estimation

In an experimental situation a particular "factor" may be involved either quantitatively or qualitatively. For example, suppose we investigate the effect of temperature (t) and concentration (c) of a certain reagent on the yield Y of a process and we fit the regression model

$$\begin{aligned} E[Y_i] &= \beta_0 + \beta_1 t_i + \beta_2 c_i + \beta_{11} t_i^2 + \beta_{12} t_i c_i + \beta_{22} c_i^2 \\ &= \gamma_0 + \gamma_1 z_{i1} + \gamma_2 z_{i2} + \gamma_3 z_{i3} + \gamma_4 z_{i4} + \gamma_5 z_{i5}, \text{ say} \qquad (i = 1, 2, \ldots, n). \end{aligned}$$

Here temperature and concentration enter quantitatively and our model takes the form $\mathscr{E}[\mathbf{Y}] = \mathbf{Z}\boldsymbol{\gamma}$, where $\mathbf{Z}$ is an $n \times 6$ data matrix. However, in experimental designs we frequently have factors that are strictly qualitative for example, geographical location, type of fertilizer, variety of grain, method of treatment, type of drug, etc. For instance, we may wish to compare the effects of three different drugs on people by measuring some response Y. If Y_{ij} is the response from the jth patient taking the ith drug, then a one-way analysis of variance (one factor at three levels) can be carried out using the model $E[Y_{ij}] = \mu_i \quad (i = 1, 2, 3; j = 1, 2, \ldots, J)$ or $\mathscr{E}[\mathbf{Y}] = \mathbf{X}\boldsymbol{\beta}$, where the qualitative nature of the factor is indicated by the elements of the design matrix $\mathbf{X}$ being zero or one [Equation (9.1)]. Of course quantitative factors like temperature and concentration can also be treated qualitatively so that the above one-way model could refer to, say, three different temperatures or temperature groupings.

Generally speaking we talk about *analysis of variance* when all the factors are treated qualitatively, and refer to *regression analysis* when all the factors are treated quantitatively. If we have a mixture of the two, say, $\mathcal{E}[\mathbf{Y}]=\mathbf{X}\boldsymbol{\beta}+\mathbf{Z}\boldsymbol{\gamma}$, in which some factors are treated qualitatively and some are treated quantitatively, we use the term *analysis of covariance* (Scheffé [1959]). For example, we may find that the effect of a drug may depend on the age of the patient so that one model might be

$$E[Y_{ij}]=\mu_i+\gamma_{i1}z_{ij}+\gamma_{i2}z_{ij}^2,$$

where z_{ij} is the age of the jth patient taking drug i. This model can be expressed in the form

$$\mathcal{E}[\mathbf{Y}]=\mathbf{X}\boldsymbol{\beta}+\mathbf{Z}\boldsymbol{\gamma}$$

where

$$\mathbf{Z}\boldsymbol{\gamma}=\begin{bmatrix} z_{11} & 0 & 0 & z_{11}^2 & 0 & 0 \\ z_{12} & 0 & 0 & z_{12}^2 & 0 & 0 \\ \cdots & \cdots & \cdots & \cdots & \cdots & \cdots \\ z_{1J} & 0 & 0 & z_{1J}^2 & 0 & 0 \\ 0 & z_{21} & 0 & 0 & z_{21}^2 & 0 \\ 0 & z_{22} & 0 & 0 & z_{22}^2 & 0 \\ \cdots & \cdots & \cdots & \cdots & \cdots & \cdots \\ 0 & z_{2J} & 0 & 0 & z_{2J}^2 & 0 \\ 0 & 0 & z_{31} & 0 & 0 & z_{31}^2 \\ 0 & 0 & z_{32} & 0 & 0 & z_{32}^2 \\ \cdots & \cdots & \cdots & \cdots & \cdots & \cdots \\ 0 & 0 & z_{3J} & 0 & 0 & z_{3J}^2 \end{bmatrix}\begin{bmatrix}\gamma_{11}\\ \gamma_{12}\\ \gamma_{13}\\ \gamma_{21}\\ \gamma_{22}\\ \gamma_{23}\end{bmatrix}.$$

If there is no interaction between age and type of drug, that is, the effect of age is the same for each drug, then the model can be simplified to

$$E[Y_{ij}]=\mu_i+\gamma_1 z_{ij}+\gamma_2 z_{ij}^2$$

or

$$\mathbf{Z}\boldsymbol{\gamma}=\begin{bmatrix} z_{11} & z_{11}^2 \\ z_{12} & z_{12}^2 \\ \cdots & \cdots \\ z_{3J} & z_{3J}^2 \end{bmatrix}\begin{pmatrix}\gamma_1\\ \gamma_2\end{pmatrix}.$$

In addition to age there may be a body weight effect which also does not interact with drug type. A suitable model might then be

$$E[Y_{ij}] = \mu_i + \gamma_1 z_{ij} + \gamma_2 z_{ij}^2 + \gamma_3 w_{ij}, \tag{10.1}$$

where w_{ij} is the weight of the jth patient taking the ith drug; if the drugs change the weight then w_{ij} could refer to the initial body weight. The three quantities *age*, $(age)^2$, and *weight* are commonly called *concomitant* variables and frequently they are random variables rather than variables controlled by the experimenter. This means that the methods considered in this chapter refer strictly to conditional models; for example, the left side of (10.1) should be $E[Y_{ij}|Z_{ij}=z_{ij}, W_{ij}=w_{ij}]$, and the assumptions about the usual "error" term ε_{ij} will also be conditional on the observed values of Z_{ij} and W_{ij}.

If, in the above experiment, age and weight are likely to have a considerable effect on the drug action, and we are particularly interested in this effect, then it would be more appropriate to design a three-way layout with three factors treated qualitatively: drug, age, and weight. Since analysis of covariance models do not generally have the same robust properties enjoyed by analysis of variance models (see Section 10.1.3), the former should be used with discretion. Some helpful comments on the choice of appropriate models are given by Cochran [1957, 1969].

Consider the general analysis of covariance model

$$G: \mathcal{E}[\mathbf{Y}] = \mathbf{X}\boldsymbol{\beta} + \mathbf{Z}\boldsymbol{\gamma} = (\mathbf{X}, \mathbf{Z})\begin{pmatrix} \boldsymbol{\beta} \\ \boldsymbol{\gamma} \end{pmatrix} = \mathbf{W}\boldsymbol{\delta}, \tag{10.2}$$

say, where $\mathbf{X}$ is $n \times p$, $\mathbf{Z}$ is $n \times t$ of rank t, and the columns of $\mathbf{Z}$ are linearly independent of the columns of $\mathbf{X}$. Although G can be analyzed as one large regression model, the common approach is to utilize what we know about the analysis of variance model $\mathcal{E}[\mathbf{Y}] = \mathbf{X}\boldsymbol{\beta}$ and use the method of two-step least squares described in Section 3.7.3. This method applies even if $\mathbf{X}$ has less than full rank or there are constraints on $\boldsymbol{\beta}$ (Section 3.8.3). Before giving two examples we wish to emphasize the assumption that the variables in $\mathbf{Z}$ are not affected by the "treatments" in $\mathbf{X}$. For example, as pointed out above, if a particular drug causes a weight change then w_{ij} should refer to the *initial* weight which is, of course, unaffected by the drug.

EXAMPLE 10.1 (randomized block design) Consider the model

$$G: E[Y_{ij}] = \mu + \alpha_i + \beta_j + \gamma z_{ij} \qquad (i = 1, 2, \ldots, I; \quad j = 1, 2, \ldots, J),$$

where $\Sigma_i \alpha_i = \Sigma_j \beta_j = 0$. For example, in an agricultural experiment carried out in a sandy area, the I treatments may refer to different fertilizers, the J blocks to strips of land, and z_{ij} may be a measure of the sand content in the (i,j)th plot. Even though the analysis of variance model $E[Y_{ij}]=\mu+\alpha_i+\beta_j$ is not of full rank, the method of two-step least squares can still be used to find the least squares estimates of the unknown parameters and the residual sum of squares for G. The steps for the two-step least squares are as follows (cf. Section 3.7.3):

(1) Find the least squares estimates $\hat{\mu}=\bar{Y}_{..}$, $\hat{\alpha}_i=\bar{Y}_{i.}-\bar{Y}_{..}$, and $\hat{\beta}_j=\bar{Y}_{.j}-\bar{Y}_{..}$, and the residual sum of squares

$$\text{RSS}=\sum_i\sum_j\left(Y_{ij}-\bar{Y}_{i.}-\bar{Y}_{.j}+\bar{Y}_{..}\right)^2 \qquad (=\mathbf{Y}'\mathbf{R}\mathbf{Y})$$

for the model $E[Y_{ij}]=\mu+\alpha_i+\beta_j$.

(2) Replace Y_{ij} by $Y_{ij}-\gamma z_{ij}$ in RSS, and minimize the resulting expression with respect to γ. Thus

$$\begin{aligned} r&=\sum\sum\left[Y_{ij}-\bar{Y}_{i.}-\bar{Y}_{.j}+\bar{Y}_{..}-\gamma\left(z_{ij}-\bar{z}_{i.}-\bar{z}_{.j}+\bar{z}_{..}\right)\right]^2\\ &=R_{yy}-2\gamma R_{yz}+\gamma^2R_{zz}, \end{aligned}$$

say, where

$$R_{yz}=\sum\sum\left(Y_{ij}-\bar{Y}_{i.}-\bar{Y}_{.j}+\bar{Y}_{..}\right)\left(z_{ij}-\bar{z}_{i.}-\bar{z}_{.j}+\bar{z}_{..}\right),$$

etc., and $dr/d\gamma=0$ gives us

$$-2R_{yz}+2\gamma R_{zz}=0. \tag{10.3}$$

Hence the least squares estimate of γ in G is $\hat{\gamma}_G=R_{yz}/R_{zz}$.

(3) The residual sum of squares for G is

$$\begin{aligned} r_{\min}&=R_{yy}-2\hat{\gamma}_GR_{yz}+\hat{\gamma}_G^2R_{zz}\\ &=R_{yy}-\hat{\gamma}_G^2R_{zz} \qquad (10.4)\\ &=R_{yy}-\frac{R_{yz}^2}{R_{zz}}. \qquad (10.5) \end{aligned}$$

(4) Replacing Y_{ij} by $Y_{ij}-\hat{\gamma}_Gz_{ij}$ in $\hat{\alpha}_i$ etc, we obtain

$$\hat{\alpha}_{i,G}=\bar{Y}_{i.}-\bar{Y}_{..}-\hat{\gamma}_G(\bar{z}_{i.}-\bar{z}_{..}),\ \text{etc.} \tag{10.6}$$

(5) The coefficient of 2γ in (10.3) is R_{zz} so that

$$\text{var}[\hat{\gamma}_G] = \frac{\sigma^2}{R_{zz}}. \tag{10.7}$$

This last result can also be obtained directly (cf. Exercise 1 at the end of this chapter).

EXAMPLE 10.2 (one-way classification) To analyze the model

$$G: E[Y_{ij}] = \mu_i + \gamma z_{ij} \qquad (i=1,2,\ldots,I; j=1,2,\ldots,J),$$

we apply the method of two-step least squares to $E[Y_{ij}] = \mu_i$. For the latter model $\hat{\mu}_i = \overline{Y}_{i.}$ and $\text{RSS} = \Sigma\Sigma(Y_{ij} - \overline{Y}_{i.})^2$, and it transpires that the procedure is identical to that given in Example 10.1, the only difference being that $R_{yz} = \Sigma\Sigma(Y_{ij} - \overline{Y}_{i.})(z_{ij} - \bar{z}_{i.})$, etc., and $\hat{\mu}_{i,G} = \overline{Y}_{i.} - \hat{\gamma}_G \bar{z}_{i.}$.

10.1.2 *Hypothesis Testing*

We now demonstrate the basic ideas of hypothesis testing in analysis of covariance models by considering several examples.

EXAMPLE 10.3 (testing regression lines for parallelism) Let

$$G: Y_{ij} - \mu_i + \gamma_i z_{ij} + \varepsilon_{ij} \qquad (i=1,2,\ldots,I; j=1,2,\ldots,J),$$

where the ε_{ij} are independently and identically distributed as $N(0,\sigma^2)$, be I regression lines with J observations per line, and suppose we wish to test $H: \gamma_1 = \gamma_2 = \cdots = \gamma_I$ ($=\gamma$, say). Now for the model $E[Y_{ij}] = \mu_i$ we have $\text{RSS} = \Sigma\Sigma(Y_{ij} - \overline{Y}_{i.})^2$, and replacing Y_{ij} by $Y_{ij} - \gamma_i z_{ij}$, we have

$$\begin{aligned} r &= \sum_i \sum_j \left[Y_{ij} - \overline{Y}_{i.} - \gamma_i (z_{ij} - \bar{z}_{i.}) \right]^2 \\ &= \sum_i R_{yyi} - 2\sum_i \gamma_i R_{yzi} + \sum_i \gamma_i^2 R_{zzi}, \end{aligned}$$

where

$$R_{yzi} = \sum_j \left(Y_{ij} - \overline{Y}_{i.} \right)\left(z_{ij} - \bar{z}_{i.} \right),$$

etc. Then $\partial r / \partial \gamma_i = 0$ gives us

$$\hat{\gamma}_{i,G} = \frac{R_{yzi}}{R_{zzi}}$$

and

$$\begin{aligned}\mathrm{RSS}_G &= r_{\min}\\ &= \sum_i R_{yyi} - \sum_i \left(\hat{\gamma}_{i,G}^2 R_{zzi}\right)\\ &= R_{yy} - \sum_i \left(\frac{R_{yzi}^2}{R_{zzi}}\right).\end{aligned}$$

The number of degrees of freedom associated with RSS_G will be $IJ-2I$ since the matrix $(\mathbf{X},\mathbf{Z})$ underlying G is obviously $IJ\times 2I$ of rank $2I$.

When H is true, G reduces to the model discussed in Example 10.2, so that from Equation (10.5)

$$\mathrm{RSS}_H = R_{yy} - \frac{R_{yz}^2}{R_{zz}}.$$

Since H takes the form

$$\left[\;\overbrace{\mathbf{O}}^{I}\;\middle|\;\overbrace{\begin{matrix}1 & 0 & \cdot\cdot & 0 & -1\\ 0 & 1 & \cdot\cdot & 0 & -1\\ \cdot & \cdot & \cdot & \cdot & \cdot\\ 0 & 0 & \cdots & 1 & -1\end{matrix}}^{I}\;\right]\left(\frac{\boldsymbol{\mu}}{\boldsymbol{\gamma}}\right)=\mathbf{0}$$

or $\mathbf{A}\boldsymbol{\delta}=\mathbf{0}$, where $\mathbf{A}$ is $I-1\times 2I$ of rank $I-1$, the F-statistic for testing H is

$$\begin{aligned}F &= \frac{(\mathrm{RSS}_H-\mathrm{RSS}_G)/(I-1)}{\mathrm{RSS}_G/(IJ-2I)}\\ &= \frac{\left\{\left(\sum_i R_{yzi}^2/R_{zzi}\right)-R_{yz}^2/R_{zz}\right\}/(I-1)}{\left\{R_{yy}-\sum_i R_{yzi}^2/R_{zzi}\right\}/(IJ-2I)}.\end{aligned}$$

Although the notation is different, this result is the same as that given in Section 7.5.2.

EXAMPLE 10.4 (randomized block design) We consider testing two different hypotheses for the randomized block model of Example 10.1.

(i) Test $H_\gamma : \gamma = 0$.

From Example 10.1 [Equation (10.5)] we have

$$\mathrm{RSS}_G = R_{yy} - \frac{R_{yz}^2}{R_{zz}}$$

with $(I-1)(J-1)-1$ degrees of freedom (since we have added one more parameter γ to the usual model; Section 9.4.1). When $\gamma=0$ we are back to the usual model with no concomitant variables so that $\mathrm{RSS}_{H_\gamma} = R_{yy}$. Arguing as in Example 10.3, there is one degree of freedom associated with H_γ so that F-statistic for testing this hypothesis is

$$F = \frac{\mathrm{RSS}_{H_\gamma} - \mathrm{RSS}_G}{\mathrm{RSS}_G / \{(I-1)(J-1)-1\}}$$

$$= \frac{R_{yz}^2 / R_{zz}}{\left\{R_{yy} - \left(R_{yz}^2 / R_{zz}\right)\right\} / \{(I-1)(J-1)-1\}}.$$

If this test is significant, that is, the concomitant variable z cannot be ignored, we may wish to find a confidence interval for γ. Applying the general regression theory to (10.2) we have that $\hat{\boldsymbol{\delta}} \sim N_{p+t}(\boldsymbol{\delta}, \sigma^2(\mathbf{W}'\mathbf{W})^{-1})$ and $\hat{\boldsymbol{\delta}}$ is independent of RSS_G $(=[(I-1)(J-1)-1]S^2)$. Since in the above example $\mathrm{var}[\hat{\gamma}_G] = \sigma^2/R_{zz}$ [Equation (10.7)], we have $\hat{\gamma}_G \sim N(\gamma, \sigma^2/R_{zz})$ (being an element of $\hat{\boldsymbol{\delta}}$) and

$$T = \frac{\hat{\gamma}_G - \gamma}{S/\sqrt{R_{zz}}}$$

has a t-distribution with $(I-1)(J-1)-1$ degrees of freedom.

(ii) Test $H: \alpha_i = 0$ for all i.

Once again we apply the method of two-step least squares to the model $E[Y_{ij}] = \mu + \beta_j$ to find RSS_H, where

$$H: E\left[Y_{ij}\right] = \mu + \beta_j + \gamma z_{ij}.$$

Beginning with

$$\sum_i \sum_j \left(Y_{ij} - \hat{\mu} - \hat{\beta}_j\right)^2 = \sum\sum\left[Y_{ij} - \bar{Y}_{..} - \left(\bar{Y}_{.j} - \bar{Y}_{..}\right)\right]^2$$

$$= \sum\sum\left(Y_{ij} - \bar{Y}_{.j}\right)^2,$$

we replace Y_{ij} by $Y_{ij}-\gamma z_{ij}$ to get

$$r=\sum\sum\left[Y_{ij}-\overline{Y}_{.j}-\gamma(z_{ij}-\bar{z}_{.j})\right]^2$$

$$=S_{yy}-2\gamma S_{yz}+\gamma^2 S_{zz},$$

where

$$S_{yz}=\sum\sum\left(Y_{ij}-\overline{Y}_{.j}\right)(z_{ij}-\bar{z}_{.j}),$$

etc. Taking $dr/d\gamma=0$ gives us $\hat{\gamma}_H=S_{yz}/S_{zz}$ and

$$\begin{aligned}\text{RSS}_H&=r_{\min}\\&=S_{yy}-2\hat{\gamma}_H S_{yz}+\hat{\gamma}_H^2 S_{zz}\\&=S_{yy}-\frac{S_{yz}^2}{S_{zz}}.\end{aligned}$$

Finally, the F-statistic for testing H is

$$F=\frac{(\text{RSS}_H-\text{RSS}_G)/(I-1)}{\text{RSS}_G/\{(I-1)(J-1)-1\}}$$

$$=\frac{\left\{S_{yy}-S_{yz}^2/S_{zz}-\left(R_{yy}-R_{yz}^2/R_{zz}\right)\right\}/(I-1)}{\left\{R_{yy}-R_{yz}^2/R_{zz}\right\}/\{(I-1)(J-1)-1\}}.$$

The number of degrees of freedom for the numerator follows from arguments like those given in Section 9.2.2.

If F is significant we may wish to examine all contrasts $\psi=\sum_i c_i\alpha_i(\sum_i c_i=0)$ in the α_i's using Scheffé's multiple-comparison method (Section 9.2.4). This can be done using the estimate [cf. Equation (10.6)]

$$\begin{aligned}\hat{\psi}_G&=\sum_i c_i\hat{\alpha}_{i,G}\\&=\sum_i c_i\left\{\left(\overline{Y}_{i.}-\overline{Y}_{..}\right)-\hat{\gamma}_G\left(\bar{z}_{i.}-\bar{z}_{..}\right)\right\}\\&=\sum_i c_i\hat{\alpha}_i-\hat{\gamma}_G\sum_i c_i\bar{z}_{i.}.\end{aligned}$$

Since, from

$$\operatorname{cov}\left[\bar{Y}_{r.}-\bar{Y}_{..}, Y_{ij}-\bar{Y}_{i.}-\bar{Y}_{.j}+\bar{Y}_{..}\right]=0$$

for all r,i,j, we have (see also Equation 3.34)

$$\operatorname{cov}\left[\hat{\alpha}_r, \hat{\gamma}_G\right]=0 \qquad (r=1,2,\ldots,I),$$

and

$$\begin{aligned}
\operatorname{var}\left[\hat{\psi}_G\right] &= \operatorname{var}\left[\sum_i c_i\hat{\alpha}_i\right]+\left(\sum_i c_i\bar{z}_{i.}\right)^2 \operatorname{var}\left[\hat{\gamma}_G\right]\\
&= \operatorname{var}\left[\sum c_i \bar{Y}_{i.}\right]+\frac{\left(\sum c_i\bar{z}_{i.}\right)^2\sigma^2}{R_{zz}}\\
&= \sigma^2\left\{\frac{\sum c_i^2}{J}+\frac{\left(\sum c_i\bar{z}_{i.}\right)^2}{R_{zz}}\right\}\\
&= \sigma^2 v,
\end{aligned}$$

say. Hence a set of multiple confidence intervals is given by

$$1-\alpha=\operatorname{pr}\left[\psi_G \in \hat{\psi}_G \pm \left\{(I-1)F^{\alpha}_{I-1,(I-1)(J-1)-1} S^{\prime 2} v\right\}^{1/2}, \text{ for all contrasts}\right].$$

The various sums of squares and cross-products that are required for an analysis of covariance are usually set out in a table like Table 10.1. Here

$$A_{yz}=\sum\sum\left(\bar{Y}_{i.}-\bar{Y}_{..}\right)\left(\bar{z}_{i.}-\bar{z}_{..}\right),$$

$$B_{yz}=\sum\sum\left(\bar{Y}_{.j}-\bar{Y}_{..}\right)\left(\bar{z}_{.j}-\bar{z}_{..}\right),$$

$$R_{yz}=\sum\sum\left(Y_{ij}-\bar{Y}_{i.}-\bar{Y}_{.j}+\bar{Y}_{..}\right)\left(z_{ij}-\bar{z}_{i.}-\bar{z}_{.j}+\bar{z}_{..}\right)$$

and

$$T_{yz}=\sum\sum\left(Y_{ij}-\bar{Y}_{..}\right)\left(z_{ij}-\bar{z}_{..}\right).$$

We also note that S_{yz} defined above is given by

$$\begin{aligned}
S_{yz} &= \sum\sum\left(Y_{ij}-\bar{Y}_{.j}\right)\left(z_{ij}-\bar{z}_{.j}\right)\\
&= A_{yz}+R_{yz}.
\end{aligned}$$

Table 10.1 Sums of Squares and Cross-Products for an Analysis of Covariance of a Randomized Block Design

Source	(y,y)	(y,z)	(z,z)
A main effects (treatments)	A_{yy}	A_{yz}	A_{zz}
B main effects (blocks)	B_{yy}	B_{yz}	B_{zz}
Error	R_{yy}	R_{yz}	R_{zz}
Corrected total	T_{yy}	T_{yz}	T_{zz}

Rules similar to those used for expanding sums of squares can be readily generalized to deal with sums of cross-products. For example,

$$A_{yz} = \sum_i \frac{(Y_{i.}z_{i.})}{J} - \frac{(Y_{..}z_{..})}{IJ}$$

and

$$R_{yz} = \sum_i \sum_j Y_{ij}z_{ij} - \sum_i \frac{(Y_{i.}z_{i.})}{I} - \sum_j \frac{(Y_{.j}z_{.j})}{J} + \frac{(Y_{..}z_{..})}{IJ}.$$

10.1.3 *Underlying Assumptions*

The general comments of Chapter 6 apply to analysis of covariance models because the latter are special cases of the general linear regression model. For example, in the light of Section 6.3.1 we would expect to find that in balanced design models (which, without concomitant variables, are robust to nonnormality) it is the extent of the "nonnormality" in the concomitant variables that determines the sensitivity of any F-test to nonnormality in the Y observations. This fact is demonstrated, for example, by Atiqullah [1964] who examines the effect on nonnormality on F, the F-statistic for testing $H: \mu_i = \mu$ $(i = 1, 2, \ldots, I)$ in the following one-way layout with r concomitant variables:

$$E[Y_{ij}] = \mu_i + \gamma_1(z_{ij1} - \bar{z}_{..1}) + \cdots + \gamma_r(z_{ijr} - \bar{z}_{..r}), \tag{10.8}$$

where $i = 1, 2, \ldots, I$ and $j = 1, 2, \ldots, J$. He does this using the method of Section 6.3.2 to find approximate expressions for the mean and variance of $Z = \frac{1}{2}\log F$; these expressions are set out in detail for the special case $(r = 1)$

$$E[Y_{ij}] = \mu_i + \gamma(z_{ij} - \bar{z}_{..}). \tag{10.9}$$

Atiqullah also examines what happens to $\frac{1}{2}\log F$ for the above model when the z_{ij} are independently normal with common variance σ_z^2, but the underlying model is actually

$$E[Y_{ij}] = \mu_i + \gamma_i(z_{ij} - \bar{z}_{..}) \tag{10.10}$$

or

$$E[Y_{ij}] = \mu_i + \gamma(z_{ij} - \bar{z}_{..}) + \delta(z_{ij} - \bar{z}_{..})^2. \tag{10.11}$$

In the case of (10.10) he shows that, for large I, the effect of variable γ_i is not likely to be serious if

$$\frac{2\sigma_z^2\sigma_\gamma^2}{\sigma^2} = O(I^{-2}),$$

where

$$\sigma_\gamma^2 = \frac{\sum_{i=1}^{I}(\gamma_i - \bar{\gamma})^2}{I}.$$

However, the effect of the quadratic term in (10.11) is much more serious, though the effect is mitigated when $\delta = O(I^{-1})$.

The adequacy of any particular model can be examined using the general residual plotting methods of Section 6.6. However, additional plots are available for examining assumptions specifically related to the concomitant variables. For example, in model (10.9), the assumption that the regression of Y on z is linear and independent of the "treatments" (that is, independent of i) can be examined by plotting Y_{ij} versus z_{ij} for each i ($J \geqslant 3$). If all the plots are linear and have similar slopes then the assumption is reasonable; a test for equal slopes can also be made (Section 10.1.2, Example 10.3). However, this method is not possible when $J<3$, as in unreplicated experimental designs such as randomized blocks and Latin squares. In this case Snee [1971] recommends plotting the usual Y-residual versus the z-residual (computed in exactly the same way) for each treatment to check for equal slope; a combined plot of the Y- versus z-residuals will provide a check on linearity. For example, in the randomized block design $E[Y_{ij}] = \mu + \alpha_i + \beta_j$, the Y- and z-residuals are $Y_{ij} - \bar{Y}_{i.} - \bar{Y}_{.j} + \bar{Y}_{..}$ and $z_{ij} - \bar{z}_{i.} - \bar{z}_{.j} + \bar{z}_{..}$, respectively. The method of plotting is demonstrated by Snee using a Graeco-Latin square design.

In analysis of covariance models it is generally more difficult to determine the values of the concomitant variables in advance as other features of the underlying design are usually more important (e.g., balance, with equal numbers of observations per cell mean). This means that the concomitant variables are generally random variables so that any analysis is conditional on their observed values. If, in addition, the concomitant variables are measured with error then the method of analysis proposed by DeGracie and Fuller [1972] can be used.

10.2 MISSING OBSERVATIONS

10.2.1 Estimation Theory

In using experimental designs the situation sometimes arises when one or more observations is "missing," for example, a plant dies, a test tube breaks, or a patient withdraws from the treatment. When this happens the design loses its symmetry balance), so that the usual analysis of variance calculations on the remaining data are no longer as straightforward. One common method for handling this situation is to find appropriate estimates of the missing observations so that an analysis of variance carried out on the now "complete" data is equivalent to the analysis of the data actually available. By working with complete data the design matrix $\mathbf{X}$ is readily specified and the symmetry of the design can be exploited to the full, as far as computations are concerned.

Suppose that in the general model $\mathbf{Y}=\mathbf{X}\boldsymbol{\beta}+\boldsymbol{\varepsilon}$, where $\mathbf{X}$ is $n\times p$, m of the n observations are missing. Then, by relabeling the Y_i if necessary, we have

$$\mathbf{Y}=\begin{pmatrix}\mathbf{Y}_1\\ \mathbf{Y}_2\end{pmatrix}=\begin{pmatrix}\mathbf{X}_1\\ \mathbf{X}_2\end{pmatrix}\boldsymbol{\beta}+\boldsymbol{\varepsilon},$$

where only the $n-m$ observations $\mathbf{Y}_1$ are available or observed. Let

$$\begin{aligned}\|\mathbf{Y}-\mathbf{X}\boldsymbol{\beta}\|^2&=\|\mathbf{Y}_1-\mathbf{X}_1\boldsymbol{\beta}\|^2+\|\mathbf{Y}_2-\mathbf{X}_2\boldsymbol{\beta}\|^2\\ &=Q_1(\boldsymbol{\beta})+Q_2(\mathbf{Y}_2,\boldsymbol{\beta}),\end{aligned}$$

say, and let $\hat{\boldsymbol{\beta}}$ be a least squares estimate of $\boldsymbol{\beta}$; that is, $Q_1(\boldsymbol{\beta})$ is minimized at $\boldsymbol{\beta}=\hat{\boldsymbol{\beta}}$, where $\hat{\boldsymbol{\beta}}$ is a solution of the normal equations $\mathbf{X}_1'\mathbf{X}_1\boldsymbol{\beta}=\mathbf{X}_1'\mathbf{Y}_1$. Then $\|\mathbf{Y}-\mathbf{X}\boldsymbol{\beta}\|^2$ will be minimized with respect to both $\mathbf{Y}_2$ and $\boldsymbol{\beta}$ if we set $\boldsymbol{\beta}=\hat{\boldsymbol{\beta}}$ and $\mathbf{Y}_2=\hat{\mathbf{Y}}_2=\mathbf{X}_2\hat{\boldsymbol{\beta}}$ (that is, $Q_2=0$). Since $Q_2(\hat{\mathbf{Y}}_2,\hat{\boldsymbol{\beta}})=0$ we have the useful relationship

$$Q_1(\hat{\boldsymbol{\beta}})=\min_{\boldsymbol{\beta},\mathbf{Y}_2}\|\mathbf{Y}-\mathbf{X}\boldsymbol{\beta}\|^2 \tag{10.12}$$

$$=\min_{\boldsymbol{\beta}}\left(\min_{\mathbf{Y}_2}\|\mathbf{Y}-\mathbf{X}\boldsymbol{\beta}\|^2\right). \tag{10.13}$$

Here $\hat{\mathbf{Y}}_2$ is called the least squares estimate of $\mathbf{Y}_2$.

We note that the above arguments, given in various guises by many authors (for example, Wilkinson [1958a, b; 1960] and the Letters to the Editor in *The American Statistician* [1972 (4)]), hold irrespective of whether $\mathbf{X}$ has full rank p or not. More general "coordinate-free" arguments which allow for other representations of $\boldsymbol{\theta} = \mathcal{E}[\mathbf{Y}]$ (e.g., $\mathbf{A}\boldsymbol{\theta} = \mathbf{0}$ instead of $\boldsymbol{\theta} = \mathbf{X}\boldsymbol{\beta}$) have been given by Kruskal [1960] and Seber [1966: Chapter 8].

Suppose now that $\mathbf{X}_1$ is $n \times p$ of rank p; then

$$\hat{\mathbf{Y}}_2 = \mathbf{X}_2\hat{\boldsymbol{\beta}} = \mathbf{X}_2(\mathbf{X}_1'\mathbf{X}_1)^{-1}\mathbf{X}_1'\mathbf{Y}_1,$$

which is simply the value we would "predict" for $\mathbf{Y}_2$ at the set of x-values given by $\mathbf{X}_2$ (Cramer [1972]). However, a more useful formula that utilizes the full design matrix $\mathbf{X}$ (which is also of rank p since it contains the p linearly independent rows of $\mathbf{X}_1$) can be found by adding $\mathbf{X}_2'\hat{\mathbf{Y}}_2 = \mathbf{X}_2'\mathbf{X}_2\hat{\boldsymbol{\beta}}$ and $\mathbf{X}_1'\mathbf{Y}_1 = \mathbf{X}_1'\mathbf{X}_1\hat{\boldsymbol{\beta}}$, namely,

$$\mathbf{X}'\hat{\mathbf{Y}} = \mathbf{X}'\mathbf{X}\hat{\boldsymbol{\beta}}. \tag{10.14}$$

Multiplying this equation by $\mathbf{X}_2(\mathbf{X}'\mathbf{X})^{-1}$, we have

$$\mathbf{X}_2(\mathbf{X}'\mathbf{X})^{-1}\mathbf{X}'\begin{pmatrix}\mathbf{Y}_1\\ \hat{\mathbf{Y}}_2\end{pmatrix} = \mathbf{X}_2\hat{\boldsymbol{\beta}} = \hat{\mathbf{Y}}_2 \tag{10.15}$$

or

$$\left(\mathbf{I}_m - \mathbf{X}_2(\mathbf{X}'\mathbf{X})^{-1}\mathbf{X}_2'\right)\hat{\mathbf{Y}}_2 = \mathbf{X}_2(\mathbf{X}'\mathbf{X})^{-1}\mathbf{X}'\mathbf{Y}_1. \tag{10.16}$$

Equations (10.14) and $\hat{\mathbf{Y}}_2 = \mathbf{X}\hat{\boldsymbol{\beta}}$ tell us that

$$Q_1(\hat{\boldsymbol{\beta}}) = \min_{\mathbf{Y}_2}\left(\min_{\boldsymbol{\beta}} \|\mathbf{Y} - \mathbf{X}\boldsymbol{\beta}\|^2\right), \tag{10.17}$$

so that we can find $\hat{\mathbf{Y}}_2$ by first minimizing $\|\mathbf{Y} - \mathbf{X}\boldsymbol{\beta}\|^2$ with respect to $\boldsymbol{\beta}$, and then minimizing the residual sum of squares with respect to the missing observations $\mathbf{Y}_2$, a result due to Yates [1933]. The usefulness of Equation (10.14) will become apparent in the following examples.

EXAMPLE 10.5 (randomized block design) Let

$$E[Y_{ij}] = \theta_{ij} = \mu + \alpha_i + \beta_j \qquad (i = 1, 2, \ldots, I;\ j = 1, 2, \ldots, J)$$

represent the randomized block design and suppose that the last observation Y_{IJ} is missing. Then, from $\hat{\boldsymbol{\theta}}=\mathbf{X}\hat{\boldsymbol{\beta}}$ and $\hat{\mathbf{Y}}_2=\mathbf{X}_2\hat{\boldsymbol{\beta}}$ $(=\hat{\boldsymbol{\theta}}_2$, say),

$$\begin{aligned}\hat{Y}_{IJ}&=\hat{\theta}_{IJ}\\&=\hat{\mu}+\hat{\alpha}_I+\hat{\beta}_J\\&=\overline{Y}_{..}+(\overline{Y}_{I.}-\overline{Y}_{..})+(\overline{Y}_{.J}-\overline{Y}_{..})\\&=\overline{Y}_{I.}+\overline{Y}_{.J}-\overline{Y}_{..}\\&=\frac{1}{J}(Y_{I*}+\hat{Y}_{IJ})+\frac{1}{I}(Y_{*J}+\hat{Y}_{IJ})-\frac{1}{IJ}(Y_{**}+\hat{Y}_{IJ}),\end{aligned}$$

where

$$Y_{I*}=\sum_{j=1}^{J-1}Y_{Ij},\ Y_{*J}=\sum_{i=1}^{I-1}Y_{iJ}\qquad\text{and}\qquad Y_{**}=\left(\sum_{i=1}^{I}\sum_{j=1}^{J}Y_{ij}\right)-Y_{IJ}.$$

Solving for $\hat{Y}_{IJ}$ we have

$$\hat{Y}_{IJ}=\frac{IY_{I*}+JY_{*J}-Y_{**}}{(I-1)(J-1)}. \tag{10.18}$$

From Equation (10.17) the residual sum of squares for the *available* data is $\Sigma\Sigma(Y_{ij}-\overline{Y}_{i.}-\overline{Y}_{.j}+\overline{Y}_{..})^2$ where Y_{IJ} is replaced by $\hat{Y}_{IJ}$. We note that the residual $\hat{Y}_{IJ}-\overline{Y}_{I.}-\overline{Y}_{.J}+\overline{Y}_{..}$ $(=\hat{Y}_{IJ}-\hat{\theta}_{IJ})$ is zero.

EXAMPLE 10.6 (one-way classification) Let

$$E[Y_{ij}]=\mu_i\qquad(i=1,2,\ldots,I;\ j=1,2,\ldots,J)$$

and suppose that Y_{IJ} is missing. Then

$$\begin{aligned}\hat{Y}_{IJ}&=\hat{\mu}_I\\&=\overline{Y}_{I.}\\&=\frac{1}{J}(Y_{I*}+\hat{Y}_{IJ})\end{aligned}$$

and

$$\hat{Y}_{IJ}=\frac{Y_{I*}}{J-1},$$

the average of the remaining $J-1$ observations on the mean μ_I.

Applications of the above technique to Latin and Youden squares are demonstrated by Jaech [1966], and further formulas for estimating a single observation are given for most of the standard designs in Cochran and Cox [1957].

10.2.2 Hypothesis Testing

Given the general model $\mathbf{Y}=\mathbf{X}\boldsymbol{\beta}+\boldsymbol{\varepsilon}$, where $\mathbf{X}$ is $n\times p$ of rank $r(\leqslant p)$, suppose we wish to test the (testable) hypothesis $H:\mathbf{A}\boldsymbol{\beta}=\mathbf{0}$, where $\mathbf{A}$ is $q\times p$ of rank q. If there are no missing observations then the F-statistic for testing H would be

$$F=\frac{(\mathrm{RSS}_H-\mathrm{RSS})/q}{\mathrm{RSS}/(n-r)}$$

$$=\frac{(\|\mathbf{Y}-\mathbf{X}\hat{\boldsymbol{\beta}}_H\|^2-\|\mathbf{Y}-\mathbf{X}\hat{\boldsymbol{\beta}}\|^2)/q}{(\|\mathbf{Y}-\mathbf{X}\hat{\boldsymbol{\beta}}\|^2)/(n-r)}.$$

However, if $\mathbf{Y}_2$ is missing it can be estimated by $\mathbf{u}=\hat{\mathbf{Y}}_2$, where $\mathbf{u}$ is obtained by minimizing RSS with respect to $\mathbf{Y}_2$ (or, equivalently, by solving $\hat{\mathbf{Y}}_2-\mathbf{X}_2\hat{\boldsymbol{\beta}}=\mathbf{0}$ as in Examples 10.5 and 10.6 above); RSS_u, the minimum value of RSS, is then the correct residual sum of squares for the original model. On the other hand, $\mathrm{RSS}_{H,u}$, the value of RSS_H at $\mathbf{Y}_2=\mathbf{u}$, is not the correct residual sum of squares under H so that $\mathrm{RSS}_{H,u}-\mathrm{RSS}_u$ is not the correct hypothesis sum of squares for testing H. What we require is an estimate of $\mathbf{Y}_2$ which is specific to the model H. Thus solving $\hat{\mathbf{Y}}_{2,H}=\mathbf{X}_2\hat{\boldsymbol{\beta}}_H$ for $\mathbf{v}=\hat{\mathbf{Y}}_{2,H}$ we obtain $\mathrm{RSS}_{H,v}$, and the appropriate F-statistic for testing H is now

$$F=\frac{(\mathrm{RSS}_{H,v}-\mathrm{RSS}_u)/q}{\mathrm{RSS}_u/(n-r-m)}.$$

Here m degrees of freedom are "lost" in the denominator as m missing observations $\mathbf{Y}_2$ are estimated (the $(n-m)\times p$ matrix $\mathbf{X}_1$ is assumed to have the same rank as $\mathbf{X}$, namely, r).

Since $\mathrm{RSS}_{H,v}$ can also be obtained by minimizing RSS_H with respect to $\mathbf{Y}_2$, we must have

$$\mathrm{RSS}_{H,v}\leqslant \mathrm{RSS}_{H,u}$$

with the probability of equality [that is, $\mathrm{pr}(\mathbf{u}=\mathbf{v})$] being zero. Therefore if we use

$$F_c=\frac{(\mathrm{RSS}_{H,u}-\mathrm{RSS}_u)/q}{\mathrm{RSS}_u/(n-r-m)}$$

and treat this as an F-variable with q and $n-r-m$ degrees of freedom then we have a *conservative* test because the numerator is too large (Yates [1933]).

Instead of computing $\text{RSS}_{H,v}$ a common procedure is to simply correct the numerator of F_c by subtracting the correction

$$b=\text{RSS}_{H,u}-\text{RSS}_{H,v}$$

and using the following lemma. (We note that b need be computed only if F_c is significant).

LEMMA Writing $\mathbf{v}=\mathbf{X}_2\hat{\boldsymbol{\beta}}_H=\mathbf{A}_1\mathbf{Y}_1+\mathbf{A}_2\mathbf{v}$, we have

$$b=(\mathbf{u}-\mathbf{v})'(\mathbf{I}_m-\mathbf{A}_2)(\mathbf{u}-\mathbf{v}).$$

(This result was proved by Wilkinson [1958b]; a more general proof is given in Seber [1966: p. 70]).

Proof. If $\mathbf{X}\hat{\boldsymbol{\beta}}_H=\mathbf{P}_H\mathbf{Y}$ then, from Equation (4.5),

$$\mathbf{P}_H^2=\mathbf{P}_H,\ (\mathbf{I}_n-\mathbf{P}_H)^2=\mathbf{I}_n-\mathbf{P}_H$$

and

$$\begin{aligned}\|\mathbf{Y}-\mathbf{X}\hat{\boldsymbol{\beta}}_H\|^2&=\mathbf{Y}'(\mathbf{I}_n-\mathbf{P}_H)'(\mathbf{I}_n-\mathbf{P}_H)\mathbf{Y}\\&=\mathbf{Y}'(\mathbf{I}_n-\mathbf{P}_H)\mathbf{Y}.\end{aligned}$$

Let $\mathbf{Y}_u'=(\mathbf{Y}_1',\mathbf{u}')$, $\mathbf{Y}_v'=(\mathbf{Y}_1',\mathbf{v}')$, and partition $\mathbf{P}_H$ in the form

$$\mathbf{P}_H=\begin{pmatrix}\mathbf{P}_{11} & \mathbf{P}_{12}\\ \mathbf{P}_{21} & \mathbf{P}_{22}\end{pmatrix}\begin{matrix}\}n-m\\ \}m\end{matrix}.$$

Now from

$$\begin{aligned}\mathbf{v}&=\mathbf{X}_2\hat{\boldsymbol{\beta}}_H\\&=(\mathbf{0},\mathbf{I}_m)\mathbf{X}\hat{\boldsymbol{\beta}}_H\\&=(\mathbf{0},\mathbf{I}_m)\mathbf{P}_H\mathbf{Y}_v\\&=(\mathbf{P}_{21},\mathbf{P}_{22})\mathbf{Y}_v\\&=\mathbf{P}_{21}\mathbf{Y}_1+\mathbf{P}_{22}\mathbf{v},\end{aligned}\tag{10.19}$$

we have $\mathbf{A}_1 = \mathbf{P}_{21}$ and $\mathbf{A}_2 = \mathbf{P}_{22}$. Also

$$\begin{aligned}(\mathbf{Y}_u - \mathbf{Y}_v)'(\mathbf{I}_n - \mathbf{P}_H)\mathbf{Y}_v &= (\mathbf{0}', (\mathbf{u}-\mathbf{v})')\begin{pmatrix} \mathbf{I}_{n-m} - \mathbf{P}_{11}, & -\mathbf{P}_{12} \\ -\mathbf{P}_{21}, & \mathbf{I}_m - \mathbf{P}_{22} \end{pmatrix}\begin{pmatrix} \mathbf{Y}_1 \\ \mathbf{v} \end{pmatrix} \\ &= (\mathbf{u}-\mathbf{v})'\left[-\mathbf{P}_{21}\mathbf{Y}_1 + (\mathbf{I}_m - \mathbf{P}_{22})\mathbf{v} \right] \\ &= 0 \qquad [\text{by } (10.19)]\end{aligned}$$

so that

$$\begin{aligned} b &= \text{RSS}_{H,u} - \text{RSS}_{H,v} \\ &= \mathbf{Y}_u'(\mathbf{I}_n - \mathbf{P}_H)\mathbf{Y}_u - \mathbf{Y}_v'(\mathbf{I}_n - \mathbf{P}_H)\mathbf{Y}_v \\ &= (\mathbf{Y}_u - \mathbf{Y}_v)'(\mathbf{I}_n - \mathbf{P}_H)(\mathbf{Y}_u - \mathbf{Y}_v) + 2(\mathbf{Y}_u - \mathbf{Y}_v)'(\mathbf{I}_n - \mathbf{P}_H)\mathbf{Y}_v \\ &= (\mathbf{u}-\mathbf{v})'(\mathbf{I}_m - \mathbf{P}_{22})(\mathbf{u}-\mathbf{v}) \end{aligned}$$

□

EXAMPLE 10.7 Suppose we wish to test $H: \alpha_i = 0$ for all i for the randomized block design of Example 10.5. Applying the technique used there to $H: E[Y_{ij}] = \mu + \beta_j$, we have

$$\begin{aligned} v - \hat{Y}_{IJ,H} & \\ &= \hat{\mu}_H + \hat{\beta}_{J,H} \\ &= \overline{Y}_{..} + (\overline{Y}_{.J} - \overline{Y}_{..}) \\ &= \overline{Y}_{.J} \\ &= \frac{Y_{*J} + v}{I} \end{aligned} \tag{10.20}$$

so that

$$v\left(1 - \frac{1}{I}\right) = \frac{Y_{*J}}{I},$$

or

$$v = \frac{Y_{*J}}{I-1}.$$

Therefore, applying the above lemma [Equation (10.19)] to Equation (10.20), we have $\mathbf{I}_m - \mathbf{P}_{22} = 1 - 1/I$ so that

$$b = (u - v)^2\left(1 - \frac{1}{I}\right)$$

and

$$\begin{aligned} \text{RSS}_{H,v} - \text{RSS}_u &= \text{RSS}_{H,u} - \text{RSS}_u - b \\ &= \left\{ \sum_i \sum_j \left(\bar{Y}_{i.} - \bar{Y}_{..}\right)^2 \right\}_{Y_{IJ} = u} - b. \end{aligned}$$

The F-statistic for testing H is now

$$F = \left\{ \frac{\left[\sum\sum\left(\bar{Y}_{i.} - \bar{Y}_{..}\right)^2 - b\right]/(I-1)}{\sum\sum\left(Y_{ij} - \bar{Y}_{i.} - \bar{Y}_{.j} + \bar{Y}_{..}\right)^2/[(I-1)(J-1)-1]} \right\}_{Y_{IJ} = u}.$$

Another, slightly more complicated, method for finding b in the above example is given by Kshirsagar [1971].

10.2.3 Analysis of Covariance Method

When more than one observation is missing Yates [1933] has suggested using the formula for one missing observation iteratively, starting with guessed values for all but one missing observation; the iterations are continued until all the residuals corresponding to the missing observations are negligible. However, this method is not suitable for, say, an all-purpose computer program because each design requires its own formulas for u, v, and b.

A more general iterative method has been described by Hartley [1956]. If there is one missing observation, we can find the appropriate estimate of that observation by analyzing the data three times using equally spaced values (such as $-1, 0, 1$) for the missing observation. The estimate is then given by a simple formula involving only the three values used and the three residual sums of squares obtained. If there is more than one missing observation the method is used iteratively.

Another general iterative method, described by Healy and Westmacott [1956], requires only a subroutine for finding residuals in the individual cells. Starting with guessed values for the missing observations (for example, cell, row, or column means for the data available), we perform the analysis for the complete model and then subtract the residual from the guessed value for each missing observation. This difference is the new estimate for the missing observation and we continue the process until all the residuals for missing observations are negligible. Although this procedure is iterative, even for one missing value, it converges fairly rapidly to the least squares solution. However, a modification of the procedure, given by Pearce [1965: Section 7.3], Pearce and Jeffers [1971], and Preece [1971], increases the speed of convergence; in the case of one missing value only two steps at most are required. This modification can also be applied to the case of mixed-up values (Preece and Gower [1974]). Another iterative method for factorial designs, which becomes noniterative with a single missing value, has been proposed by Shearer [1973].

An alternative noniterative procedure for handling missing data was suggested by Bartlett [1937b: p. 151]. The data are augmented with arbitrary values (e.g., zeros) for the m missing observations, and the influence of these values on the analysis of the augmented data is removed by an analysis of covariance on m dummy concomitant variables. The ith dummy variable takes the value 1 in the position corresponding to the ith missing observation, and the value zero elsewhere. We see below that when missing observations are set equal to zero, the least squares estimate of the ith missing observation is the negative of the regression coefficient of the ith dummy variable. This method has been described in a number of places (for example, Tocher [1952], Wilkinson [1960], and Seber [1966]) and it is illustrated by Example 10.8. A noniterative method for missing observations in crossed classifications is described by Haseman and Gaylor [1973].

EXAMPLE 10.8 Suppose that the observation Y_{IJ} is missing in the randomized block design described above in Example 10.5. Then we assume the model

$$E[Y_{ij}] = \mu + \alpha_i + \beta_j + \gamma z_{ij} \tag{10.21}$$

where $Y_{IJ} = c$ (an arbitrary constant) and $z_{ij} = \delta_{iI}\delta_{jJ}$. The least squares estimate $\hat{\gamma}$, say, of γ is readily found by the method of two-step least squares described in Example 10.1 (Section 10.1.1). Thus

$$\hat{\gamma} = \frac{R_{yz}}{R_{zz}}$$

where

$$R_{yz} = \sum\sum\left(Y_{ij} - \overline{Y}_{i.} - \overline{Y}_{.j} + \overline{Y}_{..}\right)(z_{ij} - \bar{z}_{i.} - \bar{z}_{.j} + \bar{z}_{..})$$

$$= \sum\sum\left(Y_{ij} - \overline{Y}_{i.} - \overline{Y}_{.j} + \overline{Y}_{..}\right)z_{ij}$$

$$= Y_{IJ} - \overline{Y}_{I.} - \overline{Y}_{.J} + \overline{Y}_{..}$$

$$= c\left(1 - \frac{1}{J} - \frac{1}{I} + \frac{1}{IJ}\right) - \frac{Y_{I*}}{J} - \frac{Y_{*J}}{I} + \frac{Y_{**}}{IJ}$$

and

$$R_{zz} = \sum\sum\left(z_{ij} - \bar{z}_{i.} - \bar{z}_{.j} + \bar{z}_{..}\right)^2$$

$$= \sum\sum\left(\delta_{iI}\delta_{jJ} - \frac{1}{J}\delta_{iI} - \frac{1}{I}\delta_{jJ} + \frac{1}{IJ}\right)^2$$

$$= \sum\sum\left(\delta_{iI} - \frac{1}{I}\right)^2\left(\delta_{ij} - \frac{1}{J}\right)^2$$

$$= \frac{(I-1)(J-1)}{IJ}.$$

If $c = 0$ then, from Equation (10.18), we see that

$$-\hat{\gamma} = \frac{IY_{I*} + JY_{*J} - Y_{**}}{(I-1)(J-1)} = \hat{Y}_{IJ}.$$

The above result is to be expected because the two-step least squares method applied to (10.21) amounts to simply replacing Y_{ij} by $Y_{ij} - \gamma z_{ij}$, that is, replacing Y_{IJ} by $-\gamma$, and minimizing the residual sum of squares with respect to γ. This is identical to Yates's method of minimizing R_{yy} with respect to Y_{IJ}. We note, in passing, that [cf. Equation (10.4)]

$$\{R_{yy}\}_{Y_{IJ} = \hat{Y}_{IJ}} = R_{yy} - \hat{\gamma}^2 R_{zz}$$

$$= \{R_{yy}\}_{Y_{IJ}=0} - \frac{\hat{Y}_{IJ}^2\,(I-1)(J-1)}{IJ}.$$

Having demonstrated the analysis of covariance method for a simple case, we now provide a general theory and a computational procedure for handling m missing observations. Consider the analysis of covariance model

$$G: \mathbf{Y}_0 = \mathbf{X}\boldsymbol{\beta} + \mathbf{Z}\boldsymbol{\gamma} + \boldsymbol{\varepsilon}$$

where $\mathbf{Y}_0 = (\mathbf{Y}', \mathbf{0}')'$ and

$$\mathbf{Z} = \begin{pmatrix} \mathbf{O} \\ \mathbf{I}_m \end{pmatrix} \begin{matrix} \}n-m \\ \}m \end{matrix}$$

$$= (\mathbf{z}_1, \mathbf{z}_2, \ldots, \mathbf{z}_m).$$

If $\mathbf{X}$ has full rank and $\mathbf{P} = \mathbf{X}(\mathbf{X}'\mathbf{X})^{-1}\mathbf{X}'$, then the least squares estimate of γ is the solution of [cf. (3.32) in Section 3.7]

$$(\mathbf{Z}'(\mathbf{I}_n - \mathbf{P})\mathbf{Z})\hat{\gamma}_G = \mathbf{Z}'(\mathbf{I}_n - \mathbf{P})\mathbf{Y}_0$$

$$= -\mathbf{Z}'\mathbf{P}\mathbf{Y}_0 \qquad (\text{since } \mathbf{Z}'\mathbf{Y}_0 = \mathbf{0}), \tag{10.22}$$

that is, of

$$(\mathbf{I}_m - \mathbf{X}_2(\mathbf{X}'\mathbf{X})^{-1}\mathbf{X}_2')\hat{\gamma}_G = -\mathbf{X}_2(\mathbf{X}'\mathbf{X})^{-1}\mathbf{X}_1'\mathbf{Y}_1. \tag{10.23}$$

From Equation (10.16) we note that $\hat{\mathbf{Y}}_2 = -\hat{\gamma}_G$, which, once again, is to be expected because of the equivalence of the method of two-step least squares and Yates's method.

This analysis of covariance approach to missing observations does not seem to be used very much because, on the surface, it requires a general analysis of covariance computer program with m concomitant variables. However, Rubin [1972] has pointed out that this is not the case; all we need is a routine for finding residuals and a subroutine for inverting an $m \times m$ matrix. To see this we note, first of all, that $(\mathbf{I}_n - \mathbf{P})\mathbf{z}_j$ is the residual vector that one would get if $\mathbf{z}_j$ is used as the observation vector $\mathbf{Y}$. Because $\mathbf{z}_j$ is the vector with its $(n-m+j)$th element unity and the other elements zero,

$$(\mathbf{Z}'(\mathbf{I}_n - \mathbf{P})\mathbf{Z})_{ij} = \mathbf{z}_i'(\mathbf{I}_n - \mathbf{P})\mathbf{z}_j$$

is the $(n-m+i)$th element of $(\mathbf{I}_n - \mathbf{P})\mathbf{z}_j$, namely, the residual corresponding to the ith missing observation when *all* the observations, including the missing ones, are assigned the value 0 except the jth missing observation which is assigned the value one. Similarly

$$(\mathbf{Z}'(\mathbf{I}_n - \mathbf{P})\mathbf{Y}_0)_i = \mathbf{z}_i'(\mathbf{I}_n - \mathbf{P})\mathbf{Y}_0$$

can be identified as the residual corresponding to the ith missing observation when all the missing observations (and not the elements of $\mathbf{Y}_1$) are assigned the value zero. Thus from Equation (10.22)

$$\begin{aligned}\hat{\mathbf{Y}}_2 &= -\hat{\gamma}_G \\ &= -\left[\mathbf{Z}'(\mathbf{I}_n-\mathbf{P})\mathbf{Z}\right]^{-1}\mathbf{Z}'(\mathbf{I}_n-\mathbf{P})\mathbf{Y}_0\end{aligned}$$

or, in terms of Rubin's notation (with column instead of row vectors), $\mathbf{X}=-\mathbf{R}^{-1}\boldsymbol{\rho}$. If $\mathbf{Z}'(\mathbf{I}_n-\mathbf{P})\mathbf{Z}$ is singular we require a generalized inverse (Wilkinson [1958a]); from the general theory of Section 10.2.1 any solution of Equation (10.16), that is, of $\hat{\mathbf{Y}}_2=\mathbf{X}_2\hat{\boldsymbol{\beta}}$, will do. In practice the singular case arises when sections, such as blocks, of a design are missing.

10.2.4 Missing Observations in Analysis of Covariance Models

Suppose that an analysis of covariance is required on a set of Y observations and the corresponding measurements of r concomitant variables $w_1, w_2, \ldots, w_r$. If some Y observations are missing, then Wilkinson [1957, 1958a] gives the following simple procedure:

(1) Discard all measurements of $w_1, w_2, \ldots, w_r$ that correspond to the missing Y values.
(2) Fit a set of missing Y values as though there were no concomitant variables.
(3) With exactly parallel calculations, fit sets of missing values for $w_1, w_2, \ldots, w_r$ to replace those discarded.
(4) Carry out the analysis of covariance on the completed data for Y and $w_1, w_2, \ldots, w_r$.

We note, in conclusion, that Rubin's technique described above can be used for estimating the missing Y and w_i values.

MISCELLANEOUS EXERCISES 10

1. Prove Equation (10.7) directly.
2. Let $Y_{ij}=\mu_i+\gamma_1 z_{ij}+\gamma_2 w_{ij}+\varepsilon_{ij}$, where $i=1,2,\ldots,I$; $j=1,2,\ldots,J$; and the ε_{ij} are independently distributed as $N(0,\sigma^2)$.
   (a) Derive the least squares estimate of γ_1 and show that it is an unbiased estimate of γ_1.
   (b) Find the variance–covariance matrix of the least squares estimates $\hat{\gamma}_i$ $(i=1,2)$ of γ_i.
   (c) Under what conditions are $\hat{\gamma}_1$ and $\hat{\gamma}_2$ statistically independent?

3. Let $Y_{ijk}=\mu_{ij}+\gamma_{ij}z_{ijk}+\varepsilon_{ijk}$, where $i=1,2,\ldots,I$; $j=1,2,\ldots,J$; $k=1,2,\ldots,K$; and the ε_{ijk} are independently distributed as $N(0,\sigma^2)$.
   (a) Obtain a test statistic for testing the hypothesis
   $$H:\gamma_{ij}=\gamma \qquad (\text{all } i,j).$$
   (b) Assuming H to be true, derive a $100(1-\alpha)\%$ confidence interval for γ.
4. Let $Y_{ijk}=\mu+\alpha_i+\beta_{ij}+\gamma z_{ijk}+\varepsilon_{ijk}$, where the ε_{ijk} are as in Exercise 3, $\alpha_{.}=0$, and $\beta_{i.}=0$ $(i=1,2,\ldots,I)$. Derive test statistics for the following hypotheses:
   $$\text{(a) } \gamma=0, \qquad \text{(b) } \alpha_i=0 \text{ (all } i).$$
5. Consider the model $Y_{ij}=\mu_i+\gamma_i x_j+\varepsilon_{ij}$, where $i=1,2$; $j=1,2,\ldots,J$; and the ε_{ij} are independently distributed as $N(0,\sigma^2)$. Using the analysis of covariance method, derive a F-statistic for testing the hypothesis that $\gamma_1=\gamma_2$. Show that this statistic is the square of the usual t-statistic for testing whether two lines are parallel.
6. Given the one-way classification
   $$Y_{ij}=\mu_i+\varepsilon_{ij} \qquad (i=1,2,\ldots,I;\, j=1,2,\ldots,J_i)$$
   where Y_{11} is missing, find u and b such that the usual F-statistic for testing $H:\mu_1=\mu_2=\cdots=\mu_I$ can be expressed in the form
   $$F=\left\{\frac{\left(\sum\sum(\bar{Y}_{i.}-\bar{Y}_{..})^2-b\right)/(I-1)}{\sum\sum(Y_{ij}-\bar{Y}_{i.})^2/\left(\sum_i J_i-I-1\right)}\right\}_{Y_{11}=u}$$
7. Show how to obtain a set of confidence intervals for all contrasts $\sum_i c_i\alpha_i(\sum_i c_i=0)$ in the one-way model $Y_{ij}=\mu+\alpha_i+\gamma z_{ij}+\varepsilon_{ij}$ $(i=1,2,\ldots,I;\ j=1,2,\ldots,J_i)$ where $\sum_i J_i\alpha_i=0$, such that the overall confidence for the set is $100(1-\alpha)\%$.
8. If $c\neq 0$ in Example 10.8 (Section 10.2.3), show that $\hat{Y}_{IJ}=c-\hat{\gamma}$.
9. In Examples 10.5 and 10.6 (Section 10.2.1) verify that the estimate of the missing observation can be found by minimizing the residual sum of squares with respect to the missing observation.
10. Obtain estimates of Y_{123}, which is missing, for the models given in Exercises 5 and 9 at the end of Chapter 9.

CHAPTER 11

Computational Techniques for Fitting a Specified Regression

11.1 INTRODUCTION

Given the regression model $\mathbf{Y}=\mathbf{X}\boldsymbol{\beta}+\boldsymbol{\varepsilon}$, where $\mathbf{X}$ is $n\times p$, we now consider algorithms for carrying out the following steps:

(1) Solve the normal equations $\mathbf{X}'\mathbf{X}\hat{\boldsymbol{\beta}}=\mathbf{X}'\mathbf{Y}$.
(2) Calculate $\mathfrak{D}[\hat{\boldsymbol{\beta}}]$.
(3) Calculate the residual $\mathbf{e}=\mathbf{Y}-\mathbf{X}\hat{\boldsymbol{\beta}}$.
(4) Calculate the residual sum of squares $\text{RSS}=\mathbf{e}'\mathbf{e}$. We also give algorithms for handling the case when $\mathbf{X}$ is not of full rank. Later we consider the following procedures.
(5) Update the regression model (that is, add or remove a row of $\mathbf{X}$).
(6) Add or remove a regressor (that is, add or remove a column of $\mathbf{X}$).
(7) Calculate an F-statistic for a general linear hypothesis.

In the section below we consider steps (1) to (4) only. For notational convenience we set $\mathbf{B}=\mathbf{X}'\mathbf{X}$, $\mathbf{c}=\mathbf{X}'\mathbf{Y}$, and $\mathbf{x}=\hat{\boldsymbol{\beta}}$ (the latter is included to avoid using $\hat{\beta}_0$ for the first element of $\hat{\boldsymbol{\beta}}$); the normal equations can now be written

$$\mathbf{B}\mathbf{x}=\mathbf{c}. \tag{11.1}$$

If $\mathbf{X}$ has rank p, $\mathbf{B}$ is nonsingular (in fact positive definite), and (11.1) has a unique solution $\mathbf{B}^{-1}\mathbf{c}$, that is, $\hat{\boldsymbol{\beta}}=(\mathbf{X}'\mathbf{X})^{-1}\mathbf{X}'\mathbf{Y}$. Some of the more common algorithms for finding this solution are described below. They all consist of reducing the problem to that of solving a triangular system of equations as such systems can be solved very accurately (Wilkinson [1965, 1967]). These algorithms are compared in Section 11.4.

11.2 FULL RANK CASE

11.2.1 Gaussian Elimination

This method consists of reducing **B** to an upper triangular matrix,

$$\mathbf{V}=\begin{bmatrix} v_{11} & v_{12} & v_{13} & \cdots & v_{1p} \\ 0 & v_{22} & v_{23} & \cdots & v_{2p} \\ \cdots & \cdots & \cdots & \cdots & \cdots \\ 0 & 0 & 0 & \cdots & v_{pp} \end{bmatrix}, \tag{11.2}$$

with positive diagonal elements (positive since **B** is positive definite). The reduction is achieved by a series of nonsingular elementary row transformations in which multiples of each row of **B** are successively subtracted from the rows below to give zeros below the diagonal (Fox [1964: Chapter 3], Wilkinson [1965, 1967]). Since the product of nonsingular transformations is also nonsingular we essentially find a $p \times p$ nonsingular matrix **K** (which turns out to be lower triangular) such that

$$\mathbf{KB}=\mathbf{V}, \tag{11.3}$$

so that our normal equations are now equivalent to

$$\begin{aligned}\mathbf{Vx}&=\mathbf{KBx}\\ &=\mathbf{Kc}\\ &=\mathbf{d},\end{aligned}$$

say. If we carry out these transformations on the augmented matrix $(\mathbf{B}:\mathbf{c})$ we end up with $(\mathbf{V}:\mathbf{d})$, and the elements of $\mathbf{x}$ are readily found by *back-substitution*, namely,

$$x_p=\frac{d_p}{v_{pp}},$$

$$x_{p-1}=\frac{d_{p-1}-x_p v_{p-1,p}}{v_{p-1,p-1}},$$

$$x_{p-2}=\frac{d_{p-2}-x_p v_{p-2,p}-x_{p-1}v_{p-2,p-1}}{v_{p-2,p-2}},$$

etc. This solution of $\mathbf{Vx}=\mathbf{d}$ is, for obvious reasons, commonly called the *backward solution* of the Gaussian elimination procedure.

It is always possible to reduce **B** to **V** row by row, beginning at row 1 and ending at row p, provided the leading principal minor determinants of orders 1 to $(p-1)$ of $\mathbf{B}$ are all nonzero (Wilkinson [1965: p. 204]); this condition is satisfied as $\mathbf{B}$ is positive definite (*A4.7*). Also, since $\mathbf{B}$ is positive definite, the usual "pivoting" is not needed.

If we proceed still further with elementary row transformations and reduce $\mathbf{B}$ to the identity matrix $\mathbf{I}_p$, then the same transformations performed on $(\mathbf{I}_p:\mathbf{c})$ will lead to $(\mathbf{B}^{-1}:\mathbf{x})$, thus giving $\hat{\boldsymbol{\beta}}$ and $\mathcal{D}[\hat{\boldsymbol{\beta}}]=\sigma^2\mathbf{B}^{-1}$. This method is frequently called Jordan elimination.

If we wish to carry out residual plots then there is no short method for finding $\mathbf{e}$—we simply calculate $\mathbf{Y}-\mathbf{X}\hat{\boldsymbol{\beta}}$, and RSS then follows. However, if we are interested in just RSS $(=\mathbf{Y}'\mathbf{Y}-\mathbf{Y}'\mathbf{X}(\mathbf{X}'\mathbf{X})^{-1}\mathbf{X}'\mathbf{Y})$, then we can find it directly by suitably augmenting the matrix $\mathbf{B}$. We simply apply Gaussian elimination to the first p columns of the $(p+1)\times(p+1)$ augmented matrix

$$\begin{pmatrix} \mathbf{X}'\mathbf{X}, & \mathbf{X}'\mathbf{Y} \\ \mathbf{Y}'\mathbf{X}, & \mathbf{Y}'\mathbf{Y} \end{pmatrix} = \begin{pmatrix} \mathbf{B} & \mathbf{c} \\ \mathbf{c}' & \mathbf{Y}'\mathbf{Y} \end{pmatrix}$$

and obtain (see Exercise 2 at the end of the chapter)

$$\begin{pmatrix} \mathbf{V} & \mathbf{d} \\ \mathbf{0}' & \text{RSS} \end{pmatrix}. \tag{11.4}$$

It seems that Gaussian elimination was popularized by Doolittle [1878] so that the procedure is sometimes called the Doolittle method (Dwyer [1941]).

11.2.2 Cholesky Decomposition (Square Root) Method

Since $\mathbf{B}$ is positive definite it can be expressed *uniquely* in the form (*A4.10*)

$$\mathbf{B}=\mathbf{U}'\mathbf{U}, \tag{11.5}$$

where $\mathbf{U}$ is a real upper triangular matrix with positive diagonal elements. This factorization of $\mathbf{B}$ is called the Cholesky decomposition, and its application to regression (under the title of the square root method) seems to have been popularized by Dwyer [1945]. Some writers prefer to use lower triangular matrices so that $\mathbf{B}$ is written in the form $\mathbf{L}\mathbf{L}'$, where $\mathbf{L}=\mathbf{U}'$.

By equating corresponding elements in (11.5) we find that $\mathbf{U}$ can be computed, row by row, from the following expressions:

$$u_{11}=b_{11}^{1/2}, \qquad u_{1j}=\frac{b_{1j}}{u_{11}} \qquad (j=2,3,\ldots,p) \tag{11.6}$$

and, for $i=2,3,\ldots,p$,

$$u_{ii}=\left(b_{ii}-\sum_{k=1}^{i-1}u_{ki}^2\right)^{1/2}, \tag{11.7}$$

$$u_{ij}=\frac{b_{ij}-\sum_{k=1}^{i-1}u_{ki}u_{kj}}{u_{ii}} \qquad (j=i+1,\ldots,p). \tag{11.7}$$

This algorithm has the advantage that if the ith row of $\mathbf{U}$ is being computed, then it is only necessary to have available the ith row of $\mathbf{B}$ and the $(i-1)$ previously computed rows of $\mathbf{U}$. This is of considerable advantage when $\mathbf{B}$ is so large that it is necessary to store it in auxiliary storage and it is not known a priori in what order the rows are to be recalled.

Once $\mathbf{U}$ has been calculated, it is then a simple matter to solve $\mathbf{U}'\mathbf{U}\mathbf{x}=\mathbf{c}$ by solving the triangular systems $\mathbf{U}'\mathbf{z}=\mathbf{c}$ for $\mathbf{z}$, and $\mathbf{U}\mathbf{x}=\mathbf{z}$ for $\mathbf{x}$. Also

$$\mathbf{z}'\mathbf{z}=\mathbf{x}'\mathbf{U}'\mathbf{U}\mathbf{x}=\hat{\boldsymbol{\beta}}'\mathbf{X}'\mathbf{X}\hat{\boldsymbol{\beta}} \tag{11.8}$$

so that from Equation (3.9) we have

$$\text{RSS}=\mathbf{Y}'\mathbf{Y}-\mathbf{z}'\mathbf{z}.$$

However, if we work with the augmented matrix $(\mathbf{X}:\mathbf{Y})$ instead of $\mathbf{X}$, we find that $\mathbf{z}$ and δ $(=\sqrt{\text{RSS}}\,)$ can be generated at the same time as $\mathbf{U}$ (see Exercise 4 at the end of this chapter).

The inverse of $\mathbf{B}$ may be obtained by solving the p equations $\mathbf{B}\mathbf{x}=\boldsymbol{\alpha}_i$ $(i=1,2,\ldots,p)$, where $\boldsymbol{\alpha}_i$ is the unit vector with 1 in the ith position and zeros elsewhere; the solution corresponding to $\boldsymbol{\alpha}_i$ is the ith column of $\mathbf{B}^{-1}$. However, since $\mathbf{U}$ is upper triangular its inverse, $\mathbf{T}$, say (which is also upper triangular), is readily calculated, and $\mathbf{B}^{-1}$ can be computed in the form

$$\mathbf{B}^{-1}=\mathbf{U}^{-1}(\mathbf{U}')^{-1}=\mathbf{T}\mathbf{T}'. \tag{11.9}$$

Here $\mathbf{T}$ is given by

$$t_{ii}=\frac{1}{u_{ii}}$$

$$t_{ij}=-\frac{\sum_{k=i+1}^{j}u_{ik}t_{kj}}{u_{ii}} \qquad (j=i+1,\ldots,p)$$

for $i=1,2,\ldots,p$, and

$$(\mathbf{B}^{-1})_{rs}=\sum_{k=s}^{p} t_{rk}t'_{ks}$$

$$=\sum_{k=s}^{p} t_{rk}t_{sk} \qquad (s=r,r+1,\ldots,p),$$

which is the product of the rth and sth rows of $\mathbf{T}$. We only need to calculate the upper triangle of elements as $\mathbf{B}^{-1}$ is symmetric. Martin et al. [1965] point out that the t_{rs} is no longer required once $(\mathbf{B}^{-1})_{rs}$ has been computed; this fact may be useful when there is a shortage of storage space.

The inverse of $\mathbf{B}$ can also be found directly from $\mathbf{U}$ by solving

$$\mathbf{U}\mathbf{B}^{-1}=(\mathbf{U}')^{-1} \tag{11.10}$$

for the columns of $\mathbf{B}^{-1}$, beginning with the last column (Fox and Hayes [1951], Plackett [1960: p. 4]); see Exercise 5 at the end of the chapter. In commenting on this method Golub [1969: p. 378] states that the number of operations is roughly the same as for the preceding method using $\mathbf{T}\mathbf{T}'$.

An alternative factorization of $\mathbf{B}$ which involves no more multiplications, but avoids the square roots of (11.6) and (11.7), has been given by Martin et al. [1965]. Let $\mathbf{D}=\text{diag}(u_{11},u_{22},\ldots,u_{pp})$, and let

$$\tilde{\mathbf{U}}=\mathbf{D}^{-1}\mathbf{U}=\begin{bmatrix} 1 & \tilde{u}_{12} & \tilde{u}_{13} & \cdots & \tilde{u}_{1p} \\ 0 & 1 & \tilde{u}_{23} & \cdots & \tilde{u}_{2p} \\ \cdots & \cdots & \cdots & \cdots & \cdots \\ 0 & 0 & 0 & \cdots & 1 \end{bmatrix},$$

where $\tilde{\mathbf{U}}$ is a unit upper triangular matrix. Then

$$\mathbf{B}=\mathbf{U}'\mathbf{U}=\tilde{\mathbf{U}}'\mathbf{D}^2\tilde{\mathbf{U}}=\tilde{\mathbf{U}}'\mathbf{D}_1\tilde{\mathbf{U}}, \tag{11.11}$$

say, where $\mathbf{D}_1=\mathbf{D}^2$ is a positive diagonal matrix. The normal equations, which now take the form $\tilde{\mathbf{U}}'\mathbf{D}_1\tilde{\mathbf{U}}\mathbf{x}=\mathbf{c}$, are solved by successively solving

$$\tilde{\mathbf{U}}'\boldsymbol{\phi}=\mathbf{c} \quad \text{for} \quad \boldsymbol{\phi}, \qquad \text{and} \qquad \tilde{\mathbf{U}}\mathbf{x}=\mathbf{D}_1^{-1}\boldsymbol{\phi} \quad \text{for} \quad \mathbf{x}. \tag{11.12}$$

Martin et al. show that this solution for $\mathbf{x}$ involves the same number of multiplications as for the Cholesky decomposition but only half as many divisions. Also $\mathbf{B}^{-1}$ can be readily computed in the form

$$\mathbf{B}^{-1}=\tilde{\mathbf{U}}^{-1}\mathbf{D}_1^{-1}(\tilde{\mathbf{U}}')^{-1}. \tag{11.13}$$

11.2.3 Triangular Decomposition

From (11.10) we have the unique factorization

$$\mathbf{B}=\tilde{\mathbf{U}}'(\mathbf{D}_1\tilde{\mathbf{U}})$$

$$=\tilde{\mathbf{L}}\mathbf{V}, \qquad (11.14)$$

say, where $\tilde{\mathbf{L}}$ $(=\tilde{\mathbf{U}}')$ is a unit lower triangular matrix, and $\mathbf{V}$ is an upper triangular matrix. The elements of $\tilde{\mathbf{L}}$ and $\mathbf{V}$ may be determined in p steps; in the rth step we determine the rth row of $\mathbf{V}$ and then the rth column of $\tilde{\mathbf{L}}$ (Wilkinson [1967, 1974]). Thus

$$v_{rj}=b_{rj}-\sum_{k=1}^{r-1}\tilde{l}_{rk}v_{kj} \qquad (j=r,r+1,\dots,p)$$

and

$$\tilde{l}_{ir}=\frac{b_{ir}-\sum_{k=1}^{r-1}\tilde{l}_{ik}v_{kr}}{v_{rr}} \qquad (i=r+1,\dots,p).$$

We solve the normal equations $\tilde{\mathbf{L}}\mathbf{V}\mathbf{x}=\mathbf{c}$ by solving $\tilde{\mathbf{L}}\mathbf{d}=\mathbf{c}$ for $\mathbf{d}$, and $\mathbf{V}\mathbf{x}=\mathbf{d}$ for $\mathbf{x}$ (called the forward and backward solutions, respectively).

The factorization (11.14) forms the basis of a computational technique called the abbreviated Doolittle procedure. Details of this procedure for desk calculation are given, for example, by Dwyer [1941, 1944], Anderson and Bancroft [1952], and Graybill [1961: p. 151]. From $(\mathbf{B}:\mathbf{c})$ we calculate the array (for $p=3$, say)

v_{11}	v_{12}	v_{13}	d_1
1	$\tilde{u}_{12}$	$\tilde{u}_{13}$	$\tilde{d}_1$
	v_{22}	v_{23}	d_2
	1	$\tilde{u}_{23}$	$\tilde{d}_2$
		v_{33}	d_3
		1	$\tilde{d}_3$

where $\tilde{u}_{ij}=v_{ij}/v_{ii}$ $(=\tilde{l}_{ji})$ and $\tilde{d}_i=d_i/v_{ii}$. Here $\mathbf{x}$ is obtained by backward substitution in $\tilde{\mathbf{U}}\mathbf{x}=\tilde{\mathbf{d}}$ (or $\mathbf{V}\mathbf{x}=\mathbf{d}$). In practice this method is basically a version of Gaussian elimination; from (11.3) it transpires that $\mathbf{B}=\mathbf{K}^{-1}\mathbf{V}$ and $\mathbf{K}^{-1}=\tilde{\mathbf{L}}$ (Fox [1964], Wilkinson [1973]).

From (11.11) an alternative factorization of $\mathbf{B}$ is

$$\mathbf{B} = (\mathbf{D}_1\tilde{\mathbf{U}})'\tilde{\mathbf{U}}$$

$$= \mathbf{L}\tilde{\mathbf{U}},$$

say, and this forms the basis of a computational technique known as Crout's method (Fox [1964]). Details for desk computation are given by Graybill [1969: p. 290–294], though he associates his method with the abbreviated Doolittle procedure (the two methods are essentially the same; only the format is different).

Wilkinson [1967, 1974] discusses in detail the above methods of factorizing $\mathbf{B}$ and gives an error analysis for the solution $\mathbf{x}$.

11.2.4 Orthogonal-Triangular Decomposition

Using the Gram–Schmidt orthogonalization process it is possible to find an orthonormal set of vectors $\mathbf{q}_i$ $(i = 1, 2, \ldots, p)$ which forms a basis for the space spanned by the columns $\mathbf{x}_i$ of $\mathbf{X}$. Algebraically this means that there exist u_{ij} such that

$$\begin{aligned}
\mathbf{q}_1 &= \frac{1}{u_{11}} \cdot \mathbf{x}_1, \\
\mathbf{q}_2 &= \frac{1}{u_{22}} \cdot \mathbf{x}_2 - \frac{u_{12}}{u_{22}} \cdot \mathbf{q}_1, \\
&\cdots\cdots\cdots\cdots\cdots\cdots\cdots\cdots \\
\mathbf{q}_p &= \frac{1}{u_{pp}} \cdot \mathbf{x}_p - \frac{u_{1p}}{u_{pp}} \cdot \mathbf{q}_1 - \cdots - \frac{u_{p-1,p}}{u_{pp}} \cdot \mathbf{q}_{p-1} \qquad (u_{ii} > 0)
\end{aligned}$$

or

$$\begin{aligned}
\mathbf{x}_1 &= u_{11}\mathbf{q}_1 \\
\mathbf{x}_2 &= u_{12}\mathbf{q}_1 + u_{22}\mathbf{q}_2 \\
&\cdots\cdots\cdots\cdots\cdots\cdots \\
\mathbf{x}_p &= u_{1p}\mathbf{q}_1 + u_{2p}\mathbf{q}_2 + \cdots + u_{pp}\mathbf{q}_p.
\end{aligned}$$

Thus

$$\begin{aligned}
\mathbf{X} &= (\mathbf{x}_1, \mathbf{x}_2, \ldots, \mathbf{x}_p) \\
&= (\mathbf{q}_1, \mathbf{q}_2, \ldots, \mathbf{q}_p)\mathbf{U} \\
&= \mathbf{Q}_p\mathbf{U}, \qquad (11.15)
\end{aligned}$$

say, where $\mathbf{U} = [(u_{ij})]$ is a $p \times p$ upper triangular matrix, and $\mathbf{Q}_p$ is $n \times p$ with orthonormal columns ($\mathbf{Q}_p'\mathbf{Q}_p = \mathbf{I}_p$). Since

$$\mathbf{B} = \mathbf{X}'\mathbf{X} = \mathbf{U}'\mathbf{Q}_p'\mathbf{Q}_p\mathbf{U} = \mathbf{U}'\mathbf{U},$$

$\mathbf{U'U}$ is the Cholesky decomposition of $\mathbf{B}$ so that $\mathbf{U}$ is unique if its diagonal elements are chosen to be positive; in this case $\mathbf{Q}_p$ $(=\mathbf{XU}^{-1})$ is also unique.

The normal equations are now

$$\mathbf{U'Ux}=\mathbf{X'Y}=\mathbf{U'Q}_p'\mathbf{Y}=\mathbf{U'z},$$

say, or, because $\mathbf{U}$ is nonsingular,

$$\mathbf{Ux}=\mathbf{z}. \tag{11.16}$$

Once we have found

$$\mathbf{Q}_p'(\mathbf{X}:\mathbf{Y})=(\mathbf{U}:\mathbf{z})$$

we can readily solve the triangular system of Equations (11.16). The residual vector is then given by

$$\begin{aligned}\mathbf{e}&=\mathbf{Y}-\mathbf{X}\hat{\boldsymbol{\beta}}\\&=\mathbf{Y}-\mathbf{Xx}\\&=\mathbf{Y}-\mathbf{Q}_p\mathbf{Ux}\\&=\mathbf{Y}-\mathbf{Q}_p\mathbf{z}\end{aligned} \tag{11.17}$$

and, from Equation (11.8),

$$\mathbf{e'e}=\mathbf{Y'Y}-\mathbf{z'z}.$$

An alternative decomposition is available if we consider the orthogonal matrix

$$\begin{aligned}\mathbf{Q}&=(\mathbf{Q}_p:\mathbf{q}_{p+1},\mathbf{q}_{p+2},\ldots,\mathbf{q}_n)\\&=(\mathbf{Q}_p:\mathbf{Q}_{n-p}),\end{aligned}$$

say, which is obtained by adding $(n-p)$ further orthonormal columns to $\mathbf{Q}_p$ to make up a full set of n orthonormal vectors for n-dimensional Euclidean space. Then

$$\mathbf{X}=\mathbf{Q}\left(\frac{\mathbf{U}}{\mathbf{O}}\right)\}\ (n-p)\times p'$$

(sometimes written $\mathbf{X}=\mathbf{QR}$—hence the name "QR algorithm") and

$$\mathbf{Q'X}=\begin{pmatrix}\mathbf{U}\\\mathbf{O}\end{pmatrix}. \tag{11.18}$$

Once again we find

$$\mathbf{Q}'(\mathbf{X}:\mathbf{Y}) = \begin{pmatrix} \mathbf{Q}'_p \\ \mathbf{Q}'_{n-p} \end{pmatrix} (\mathbf{X}:\mathbf{Y})$$

$$= \begin{pmatrix} \mathbf{U}:\mathbf{z} \\ \mathbf{O}:\mathbf{t} \end{pmatrix}, \tag{11.19}$$

say, and solve $\mathbf{Ux}=\mathbf{z}$. Using Equation (11.17), the residuals are given by

$$\mathbf{e} = \mathbf{Q}\begin{pmatrix} \mathbf{z} \\ \mathbf{t} \end{pmatrix} - \mathbf{Q}_p\mathbf{z}$$

$$= \mathbf{Q}_{n-p}\mathbf{t}$$

$$(= \mathbf{Q}_{n-p}\mathbf{Q}'_{n-p}\mathbf{Y}), \tag{11.20}$$

though it has been observed by Gentleman that this method of computing $\mathbf{e}$ may be numerically unstable. If $\mathbf{X}$ is stored (as is the case in iterative refinement—Section 11.6), $\mathbf{e}$ is best calculated from $\mathbf{Y}-\mathbf{Xx}$. However, the provision of $\mathbf{t}$ in (11.19) is useful in two respects. First,

$$\mathbf{e}'\mathbf{e} = \mathbf{t}'\mathbf{Q}'_{n-p}\mathbf{Q}_{n-p}\mathbf{t} = \mathbf{t}'\mathbf{t}, \tag{11.21}$$

and second, since $\mathcal{E}[\mathbf{t}] = \mathbf{Q}'_{n-p}\mathbf{X}\boldsymbol{\beta} = \mathbf{0}$ and $\mathcal{D}[\mathbf{t}] = \sigma^2\mathbf{Q}'_{n-p}\mathbf{Q}_{n-p} = \sigma^2\mathbf{I}_{n-p}$,

$$\mathbf{t} \sim N_{n-p}(\mathbf{0}, \sigma^2\mathbf{I}_{n-p}). \tag{11.22}$$

The t_i, which are independently and identically distributed as $N(0,\sigma^2)$ under the usual least squares assumptions, can be used to test for departures from these assumptions (Section 6.6.5).

The basis of the procedure underlying (11.19) is to carry out an orthogonal transformation (that is, a rotation) on $(\mathbf{X}:\mathbf{Y})$. This can be done in a stepwise fashion using a series of "elementary" rotations such as the Householder or Givens transformations described below. The matrix $\mathbf{Q}'$ can then be found, if necessary, by applying the transformations to the identity matrix $\mathbf{I}_n$ or, in the case of Householder transformations, by storing and multiplying the elementary rotations; $\mathbf{Q}_p$ is most readily found from $\mathbf{Q}_p = \mathbf{XU}^{-1}$ (we saw in Section 11.2.3 that the inversion of a triangular matrix is straightforward).

We now discuss in detail numerical methods for finding the matrices $\mathbf{U}$ and $\mathbf{Q}$ (or $\mathbf{Q}_p$) given above.

a GRAM–SCHMIDT ORTHOGONALIZATION

Square roots can be avoided by using the following decomposition:

$$\begin{aligned}\mathbf{X} &= \mathbf{Q}_p\mathbf{U}\\ &= \mathbf{Q}_p\mathbf{D}\tilde{\mathbf{U}}\\ &= \mathbf{R}_p\tilde{\mathbf{U}},\end{aligned} \tag{11.23}$$

where $\mathbf{D}=\mathrm{diag}(u_{11},u_{22},\ldots,u_{pp})$, $\tilde{\mathbf{U}}=[(\tilde{u}_{ij})]$ is an upper triangular matrix with unit diagonal elements, and $\mathbf{R}_p$ is an $n\times p$ matrix with orthogonal columns such that

$$\mathbf{R}_p'\mathbf{R}_p=\mathbf{D}'\mathbf{Q}_p'\mathbf{Q}_p\mathbf{D}=\mathbf{D}^2=\mathbf{D}_1,$$

say. Using an orthogonalizing process we can therefore transform $\mathbf{X}$ to $\mathbf{R}_p$, where the columns of $\mathbf{R}_p$ are not normalized, and solve the normal equations using the method of (11.12). If we use $(\mathbf{X}:\mathbf{Y})$ instead of $\mathbf{X}$, then the additional element of $\mathbf{D}_1$ is $\mathbf{e}'\mathbf{e}$ (see Exercise 6 at the end of the chapter).

We can also carry out the same square-root free process on the augmented matrix

$$\mathbf{X}_{\mathrm{AUG}}=\begin{pmatrix}\mathbf{X} & -\mathbf{Y}\\ \mathbf{I}_p & \mathbf{0}\end{pmatrix}$$

instead of $\mathbf{X}$. The result (Jordan [1968]) is an augmented $\mathbf{R}_p$ matrix, namely,

$$\mathbf{R}_{\mathrm{AUG}}=\begin{pmatrix}\mathbf{R}_p & \mathbf{e}\\ \tilde{\mathbf{U}}^{-1} & -\mathbf{x}\end{pmatrix},$$

so that the least squares solution $\mathbf{x}$ and the residual $\mathbf{e}$ are generated at the same time. The matrix $(\mathbf{X}'\mathbf{X})^{-1}$ can be obtained from $\tilde{\mathbf{U}}^{-1}\mathbf{D}_1^{-1}(\tilde{\mathbf{U}}')^{-1}$ [cf. Equation (11.13)].

There are two basic algorithms for transforming $\mathbf{X}$ to $\mathbf{R}_p$, or $\mathbf{X}_{\mathrm{AUG}}$ to $\mathbf{R}_{\mathrm{AUG}}$: the Classical Gram–Schmidt Algorithm (CGSA) and the Modified Gram–Schmidt Algorithm (MGSA). The CGSA for $\mathbf{X}$ may be described geometrically as follows: at the kth stage make the kth column vector orthogonal to each of the $k-1$ previously orthogonalized column vectors, and do this for column vectors subscripted $k=2,3,\ldots,p$. The MGSA has the following geometrical description: at the kth stage make the column vectors subscripted $k, k+1, \ldots, p$ orthogonal to the $(k-1)$th column vector, and do this for columns subscripted $k=2,3,\ldots,p$. Further details of these algorithms are given by Björck [1967a, b; 1968], Jordan [1968], Golub [1969], and, in particular, Clayton [1971], and Farebrother [1974]. Experimental comparisons of the two algorithms given by Rice [1966], Jordan [1968], and Wampler [1970: pp. 556–557], and the theoretical analysis of Björck [1967a], indicate that the MGSA is more accurate and more stable than the CGSA. If $\mathbf{X}$ is at all "ill-conditioned" (Section 11.4) then, using CGSA, the computed columns of $\mathbf{R}_p$ or $\mathbf{R}_{\text{AUG}}$ soon lose their orthogonality. This means that CGSA should never be used without reorthogonalization; this greatly increases the amount of computation. Reorthogonalization is not needed when using the MGSA.

b HOUSEHOLDER TRANSFORMATIONS

A square matrix of the form $\mathbf{H}=\mathbf{I}_n-2\mathbf{v}\mathbf{v}'$, where $\mathbf{v}'\mathbf{v}=1$, is defined to be a Householder transformation. Here $\mathbf{H}=\mathbf{H}'$ and $\mathbf{H}'\mathbf{H}=\mathbf{I}_n$ so that $\mathbf{H}$ is symmetric and orthogonal. We are particularly interested in such transformations of the form

$$\mathbf{H}^{(i)}=\mathbf{I}_n-2\mathbf{v}^{(i)}\mathbf{v}^{(i)\prime}, \tag{11.24}$$

where $\mathbf{v}^{(i)}=(0,0,\ldots,0,v_i^{(i)},v_{i+1}^{(i)},\ldots,v_n^{(i)})'$ and $\mathbf{v}^{(i)\prime}\mathbf{v}^{(i)}=1$. This transformation can also be written in the form

$$\mathbf{H}^{(i)}=\begin{pmatrix}\mathbf{I}_{n-i} & \mathbf{O}\\ \mathbf{O} & \mathbf{K}_i\end{pmatrix} \tag{11.25}$$

where $\mathbf{K}_i$ is an $i\times i$ Householder transformation. We now indicate briefly how the matrix $\mathbf{Q}'$ of Equation (11.18) can be written as a product of p Householder transformations of the form (11.24).

With $\mathbf{X}=(\mathbf{x}_1,\mathbf{x}_2,\ldots,\mathbf{x}_p)$, the first column $\mathbf{x}_1$ can be transformed by a suitable $\mathbf{H}^{(1)}$ to a vector with each element zero except the first. Thus

$$\mathbf{H}^{(1)}\mathbf{X}=(u_{11}\boldsymbol{\alpha}_1,\mathbf{H}^{(1)}\mathbf{x}_2,\ldots,\mathbf{H}^{(1)}\mathbf{x}_p), \qquad u_{11}\neq 0,$$

where $\boldsymbol{\alpha}_j$ $(j=1,2,\ldots,n)$ is an $n\times 1$ vector with the jth element unity and the other elements zero; geometrically this orthogonal transformation amounts to rotating the columns of $\mathbf{X}$ until $\mathbf{x}_1$ lies along the "first" axis (represented by $\boldsymbol{\alpha}_1$). The next step is to rotate the remaining columns about the first axis, using a suitable $\mathbf{H}^{(2)}$, until $\mathbf{x}_2$ lies in the plane formed by the first and second axes; $\mathbf{x}_2$ is now a linear combination of $\boldsymbol{\alpha}_1$ and $\boldsymbol{\alpha}_2$. Then, as $\mathbf{v}^{(2)\prime}\boldsymbol{\alpha}_1=0$, $\mathbf{H}^{(2)}\boldsymbol{\alpha}_1=\boldsymbol{\alpha}_1$ and hence

$$\mathbf{H}^{(2)}\mathbf{H}^{(1)}\mathbf{X}=(u_{11}\boldsymbol{\alpha}_1,u_{12}\boldsymbol{\alpha}_1+u_{22}\boldsymbol{\alpha}_2,\mathbf{H}^{(2)}\mathbf{H}^{(1)}\mathbf{x}_3,\ldots,\mathbf{H}^{(2)}\mathbf{H}^{(1)}\mathbf{x}_p).$$

This process can be continued so that at the jth step we have

$$\mathbf{H}^{(j)}\mathbf{H}^{(j-1)}\cdots\mathbf{H}^{(1)}\mathbf{x}_j = u_{1j}\boldsymbol{\alpha}_1 + u_{2j}\boldsymbol{\alpha}_2 + \cdots + u_{jj}\boldsymbol{\alpha}_j.$$

Finally,

$$\begin{aligned}\mathbf{Q}'\mathbf{X} &= \mathbf{H}^{(p)}\mathbf{H}^{(p-1)}\cdots\mathbf{H}^{(1)}\mathbf{X}\\ &= (u_{11}\boldsymbol{\alpha}_1, u_{12}\boldsymbol{\alpha}_1 + u_{22}\boldsymbol{\alpha}_2, \ldots, u_{1p}\boldsymbol{\alpha}_1 + u_{2p}\boldsymbol{\alpha}_2 + \cdots + u_{pp}\boldsymbol{\alpha}_p)\\ &= \left[\begin{array}{cccc} u_{11} & u_{12} & \cdots & u_{1p}\\ 0 & u_{22} & \cdots & u_{2p}\\ \cdots & \cdots & \cdots & \cdots\\ 0 & \cdots & \cdots & u_{pp}\\ \hline \multicolumn{4}{c}{\mathbf{O}} \end{array}\right],\end{aligned}$$

and $\mathbf{Q}'$ is orthogonal as the product of orthogonal matrices is also orthogonal.

Details of the above algorithm are given by Golub [1965] and Businger and Golub [1965] who also give a program in Algol 60. These authors also give two slight modifications for improving the numerical efficiency. First, each Householder transformation requires the computation of two square roots (see Golub and Styan [1973: p. 255] for details); however, if we write $\mathbf{H} = \mathbf{I} - \mathbf{v}(\mathbf{v}'\mathbf{v})^{-1}\mathbf{v}'$ (which is a projection matrix since $\mathbf{H}^2 = \mathbf{H}$), only one square root per transformation is required. Second, there is a slight numerical advantage to be gained by allowing the columns of $\mathbf{X}$ to be chosen in the following order: at the jth stage ($j = 1, 2, \ldots, p$) choose that column of $\mathbf{X}$ out of the remaining $p - j + 1$ possibilities which maximizes the sum of squares of its last $n - j + 1$ elements. This procedure, called pivoting, amounts to maximizing the next diagonal element of $\mathbf{U}$ at each stage, and algebraically it is equivalent to finding $\mathbf{Q}'$ such that

$$\mathbf{Q}'\mathbf{X}\boldsymbol{\Pi} = \begin{pmatrix}\mathbf{U}\\ \mathbf{O}\end{pmatrix} \tag{11.26}$$

where $\boldsymbol{\Pi}$ is a $p \times p$ permutation matrix which permutes the columns of $\mathbf{X}$ into the appropriate order. Because $\boldsymbol{\Pi}^{-1} = \boldsymbol{\Pi}'$, the normal equations become, after substitution for $\mathbf{X}$,

$$\boldsymbol{\Pi}\mathbf{U}'\mathbf{U}\boldsymbol{\Pi}'\hat{\boldsymbol{\beta}} = \boldsymbol{\Pi}\mathbf{U}'\mathbf{Q}_p'\mathbf{Y}$$

or, because $\boldsymbol{\Pi}\mathbf{U}'$ is nonsingular,

$$\mathbf{U}(\boldsymbol{\Pi}'\hat{\boldsymbol{\beta}}) = \mathbf{Q}_p'\mathbf{Y} = \mathbf{z}.$$

We then solve $\mathbf{U}\mathbf{x} = \mathbf{z}$ for $\mathbf{x}$ and hence obtain $\hat{\boldsymbol{\beta}} = \boldsymbol{\Pi}\mathbf{x}$.

How does the Householder method compare with the Gram–Schmidt process? It was mentioned above that the CGSA produces columns of $\mathbf{Q}_p$ which may be far from orthogonal, particularly if $\mathbf{X}$ is ill-conditioned. These columns must be reorthogonalized if we wish to produce a matrix of "comparable orthogonality" as the $\mathbf{Q}$ produced by Householder transformations (Wilkinson [1965: p. 244]). The MGSA has the advantage of being relatively easy to program and, experimentally (Jordan [1968], Wampler [1970]), it seems to be slightly more accurate than the Householder procedure. However, it requires slightly more computation.

C GIVENS TRANSFORMATIONS

Another type of orthogonal transformation is the Givens transformation

$$
\begin{array}{cc}
 & \begin{array}{cccccccccccc} & & & & \begin{array}{c}\text{column}\\ j\end{array} & & & & \begin{array}{c}\text{column}\\ k\end{array} & & & \end{array} \\
\mathbf{G}= & \left[\begin{array}{cccccccccccc}
1 &&&&&&&&&&& \\
& 1 &&&&&&&&&& \\
&& \ddots &&&&&&&&& \\
&&& 1 &&&&&&&& \\
&&&& \cos\theta &&&& \sin\theta &&& \\
&&&&& 1 &&&&&& \\
&&&&&& \ddots &&&&& \\
&&&&&&& 1 &&&& \\
&&&& -\sin\theta &&&& \cos\theta &&& \\
&&&&&&&&& 1 && \\
&&&&&&&&&& \ddots & \\
&&&&&&&&&&& 1
\end{array}\right]
\end{array}
$$

where j and k are arbitrary, and all the elements of $\mathbf{G}$ not shown are zeros. Basically $\mathbf{G}$ consists of the planar rotation

$$
\begin{pmatrix} \cos\theta & \sin\theta \\ -\sin\theta & \cos\theta \end{pmatrix}
$$

with ones on the remaining diagonal positions so that when $\mathbf{G}$ is applied to $\mathbf{X}$, it simply rotates each of the two-dimensional column vectors formed by the jth and kth rows through an angle θ, leaving the remaining rows unchanged. In particular we can choose θ so that two rows

$$
\begin{array}{ll}
\text{row } j\text{:} & 0,\ldots,0,r_i,r_{i+1},\ldots,r_m,\ldots,r_p \\
\text{row } k\text{:} & 0,\ldots,0,s_i,s_{i+1},\ldots,s_m,\ldots,s_p
\end{array}
$$

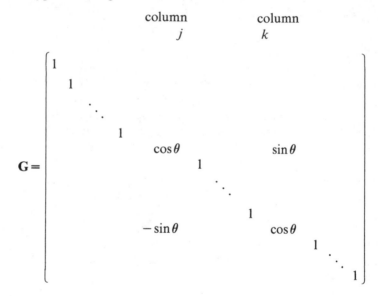

become

$$0, \ldots, 0, r_i', r_{i+1}', \ldots, r_m', \ldots, r_p'$$
$$0, \ldots, 0, 0, s_{i+1}', \ldots, s_m', \ldots, s_p',$$

where

$$r_i' = \sqrt{(r_i^2 + s_i^2)}$$

$$c = \frac{r_i}{r_i'} \qquad (=\cos\theta) \tag{11.27}$$

$$s = \frac{s_i}{r_i'} \qquad (=\sin\theta)$$

$$r_m' = cr_m + ss_m \tag{11.28}$$

and

$$s_m' = -sr_m + cs_m. \tag{11.29}$$

When $r_i = 0$ and $s_i > 0$, $\theta = \frac{1}{2}\pi$, and **G** simply interchanges the rows, with a change in sign, namely,

$$\begin{matrix} 0, \ldots, 0, s_i, s_{i+1}, \ldots, s_p \\ 0, \ldots, 0, 0, -r_{i+1}, \ldots, -r_p. \end{matrix} \tag{11.30}$$

The sign of the first row of (11.30) is also changed if $s_i < 0$. If $r_i = s_i = 0$, $\theta = 0$ and there is no change.

We can therefore annihilate (reduce to zero) the elements of **X** below the diagonal by a series of Givens transformations; thus $\mathbf{Q}'$ can be written in the form

$$\mathbf{Q}' = \mathbf{G}_t \mathbf{G}_{t-1} \cdots \mathbf{G}_2 \mathbf{G}_1.$$

This reduction can be done in two ways (Wilkinson [1965: pp. 239–240]):

(1) Transform the first row successively with the second, third, fourth, etc., row so as to annihilate the last $n-1$ elements of the first column; transform the second row successively with the third, fourth, etc., row so as to annihilate the last $n-2$ elements in the second column; in general, transform the jth row successively with the $(j+1)$th, $(j+2)$th, $\ldots$, nth row so as to annihilate the last $n-j$ elements in the jth column.

(2) At the kth step $(k = 1, 2, \ldots, n-1)$ transform the $(k+1)$th row with each of the rows above, beginning with the first, so that the first $(k+1)$ rows of **X** are reduced to upper triangular form.

Method (1) requires storing all of $\mathbf{X}$ in the computer, whereas method (2) can be applied to each row of $\mathbf{X}$ as it is fed in.

The transformations (11.28) and (11.29) are also applied to the elements of $\mathbf{Y}$ to get $\mathbf{z}$ and $\mathbf{t}$ of (11.19). For example, if rows j and k are transformed, then

$$Y_j' = cY_j + sY_k \qquad \text{and} \qquad Y_k' = -sY_j + cY_k.$$

Once we have reduced $(\mathbf{X}:\mathbf{Y})$ to the form (11.19) we can, if we wish, carry out further rotations on pairs of elements of $\mathbf{t}$ and systematically reduce $\mathbf{t}$ to $(\delta, 0, \ldots, 0)'$, where all the elements except the first are now zero. In this case the residual sum of squares, $\mathbf{e}'\mathbf{e}$, is simply δ^2, and $\mathbf{e}$ is δ times the first column of $\mathbf{Q}_{n-p}$ [cf. Equation (11.20)]. We can always ensure that $\delta > 0$ by changing the sign, if necessary, with a further "rotation" of $\theta = \pi$.

In comparing the Householder and Givens transformations, the reduction of $\mathbf{X}$ to upper triangular form requires about $np^2 - \frac{1}{2}p^3$ multiplications and p square roots for the former, and about $2np^2$ multiplications and np square roots for the latter; for this reason the Householder method has been generally preferred. However, the Givens transformation has two distinct advantages: $\mathbf{X}$ can be processed one row at a time [method (2) above], and zeros already present in $\mathbf{X}$ are readily exploited to reduce arithmetic cost. As pointed out by Gentleman [1973] (whose paper forms the basis for this section), the importance of these advantages is threefold. First, in regression and particularly factorial designs, $\mathbf{X}$ may be too large to keep on high-speed store, and therefore must be generated or fetched as required. Since there are generally many more rows than columns, processing by rows minimizes the high-speed store and/or transfers to backing store required. Also it is often more natural to generate or fetch $\mathbf{X}$ by rows. For example, in regression problems each row corresponds to an observation on the model and we may wish to update $(\mathbf{X}:\mathbf{Y})$ by adding more observations (rows); in factorial designs it is usually more convenient to generate rows than columns (Fowlkes [1969]). Second, for experimental designs, $\mathbf{X}$ frequently has a large number of zeros. Using Givens transformations and merely exploiting zeros in obvious ways (but not tailoring the computation to the particular matrix structure) Gentleman reports that reductions in computing cost of up to 70% have been measured for the analysis of some conventional unbalanced factorial designs. Third, even after a model has been analyzed and $\mathbf{X}$ and $\mathbf{Y}$ have been reduced to $\mathbf{U}$ and $\mathbf{Q}'\mathbf{Y}$, it is often necessary to include further observations or impose constraints (Section 11.5.4). Irrespective of how the reduction was done, whether by Householder or Givens transformations, the new matrix can be retriangularized by Givens transformations, taking advantage of the structure already present (Section 11.8).

Gentleman also presents a modification of the above method, similar to that given for the Gram–Schmidt method above, which avoids the computation of square roots and takes only three quarters as many multiplications. This modification, which is described briefly in Section 11.3.4, competes very favorably with the Householder method (Gentleman [1974a, b]).

11.3 WEIGHTED LEAST SQUARES

11.3.1 The Normal Equations

If we wish to carry out a weighted least squares with (positive) weights w_i, then our normal equations are now

$$\mathbf{X}'\mathbf{W}\mathbf{X}\boldsymbol{\beta}^* = \mathbf{X}'\mathbf{W}\mathbf{Y},$$

where $\mathbf{W} = \operatorname{diag}[w_1, w_2, \ldots, w_n]$ $(= \mathbf{V}^{-1}$ of Section 3.6). By writing these equations in the form

$$(\mathbf{W}^{1/2}\mathbf{X})'(\mathbf{W}^{1/2}\mathbf{X})\,\boldsymbol{\beta}^* = (\mathbf{W}^{1/2}\mathbf{X})'(\mathbf{W}^{1/2}\mathbf{Y}),$$

we see that all the methods of Section 11.2 can be used if we now work with $\mathbf{W}^{1/2}\mathbf{X}$ and $\mathbf{W}^{1/2}\mathbf{Y}$; that is, we multiply the ith row of $(\mathbf{X}:\mathbf{Y})$ by $\sqrt{w_i}$. However, it is of interest to note that square roots can be avoided in the Cholesky, Gram–Schmidt, and Givens methods, as we see below.

11.3.2 Cholesky Method

From

$$\mathbf{W}^{1/2}\mathbf{X} = \mathbf{W}^{1/2}\mathbf{Q}_p\mathbf{U}$$

we have

$$\begin{aligned}\mathbf{X}'\mathbf{W}\mathbf{X} &= \mathbf{U}'\mathbf{Q}_p'\mathbf{W}\mathbf{Q}_p\mathbf{U}\\ &= \mathbf{U}'\mathbf{W}\mathbf{U},\end{aligned}$$

since the introduction of the diagonal matrix $\mathbf{W}$ affects only the norms of the columns of $\mathbf{Q}_p$ and not their mutual orthogonality. Therefore, proceeding as in Equation (11.11) we have the decomposition

$$\begin{aligned}\mathbf{X}'\mathbf{W}\mathbf{X} &= \tilde{\mathbf{U}}'\mathbf{D}\mathbf{W}\mathbf{D}\tilde{\mathbf{U}}\\ &= \tilde{\mathbf{U}}'\mathbf{D}_2\tilde{\mathbf{U}},\end{aligned}$$

where

$$\mathbf{D}_2 = \mathbf{DWD} = \mathbf{WD}^2 = \mathbf{WD}_1$$

is a diagonal matrix with positive diagonal elements. To find $\boldsymbol{\beta}^*$ we proceed as follows. Solve

$$\tilde{\mathbf{U}}'\boldsymbol{\phi} = \mathbf{X}'(\mathbf{WY}) \quad \text{for} \quad \boldsymbol{\phi},$$

and

$$\tilde{\mathbf{U}}\boldsymbol{\beta}^* = \mathbf{D}_2^{-1}\boldsymbol{\phi} \quad \text{for} \quad \boldsymbol{\beta}^*. \tag{11.31}$$

The above algebra indicates that the square root-free technique of Martin et al. [1965], mentioned briefly in the last paragraph of Section 11.2.2, can be easily adapted to deal with weighted least squares: we simply work with $\mathbf{WD}_1$ and $\mathbf{WY}$ instead of $\mathbf{D}_1$ and $\mathbf{Y}$ in (11.12).

11.3.3 Gram–Schmidt Method

The algorithm described in Section 11.2.4 can be easily modified to deal with weighted least squares. In the process of reducing $\mathbf{X}$ to $\tilde{\mathbf{U}}$ the elements of the diagonal matrix $\mathbf{D}_1$ $(=\mathbf{R}_p'\mathbf{R}_p)$ can be stored. Thus the method of Section 11.3.2, and in particular (11.31), can be used here also if we use $\mathbf{WX}$ instead of $\mathbf{X}$; $\mathbf{D}_2$ is stored instead of $\mathbf{D}_1$.

11.3.4 Gentleman's Method

Consider first of all the problem of finding $\tilde{\mathbf{U}}$ and $\mathbf{D}_1$ such that $\mathbf{U} = \mathbf{D}_1^{1/2}\tilde{\mathbf{U}}$. Using a similar notation to that given in Section 11.2.4c we consider transforming a row of the product $\mathbf{D}_1^{1/2}\tilde{\mathbf{U}}$ with a scaled row of $\mathbf{X}$:

$$\begin{array}{l} 0,\ldots,0,\sqrt{d}\,,\sqrt{d}\,\tilde{u}_{i+1},\ldots,\sqrt{d}\,\tilde{u}_m,\ldots,\sqrt{d}\,\tilde{u}_p \\ 0,\ldots,0,\sqrt{\delta}\,v_i,\sqrt{\delta}\,v_{i+1},\ldots,\sqrt{\delta}\,v_m,\ldots,\sqrt{\delta}\,v_p \end{array} \tag{11.32}$$

Using Equations (11.33), the transformed equations become (Gentleman [1973, 1974a]

$$\begin{array}{l} 0,\ldots,0,\sqrt{d'}\,,\ldots,\sqrt{d'}\,\tilde{u}_m',\ldots,\sqrt{d'}\,\tilde{u}_p' \\ 0,\ldots,0,0,\ldots,\sqrt{\delta'}\,v_m',\ldots,\sqrt{\delta'}\,v_p', \end{array}$$

where

$$d' = d + \delta v_i^2,$$

$$\bar{c} = \frac{d}{d + \delta v_i^2} = \frac{d}{d'}$$

$$\delta' = \frac{d\delta}{d + \delta v_i^2} = \bar{c}\delta$$

$$\bar{s} = \frac{\delta v_i}{d + \delta v_i^2} = \frac{\delta v_i}{d'}, \tag{11.33}$$

$$v'_m = v_m - v_i \tilde{u}_m \qquad (\tilde{u}_i = 1)$$

and

$$\tilde{u}'_m = \bar{c}\tilde{u}_m + \bar{s}v_m.$$

This means that the transformed rows can be expressed as a row of a new $\mathbf{D}_1^{1/2}\tilde{\mathbf{U}}$ and a new scaled row of $\mathbf{X}$, the scaling factor having possibly changed. Storing $\mathbf{D}_1$ and $\tilde{\mathbf{U}}$ takes no more space than storing $\tilde{\mathbf{U}}$, and the above updating formulas not only avoid square roots but also take only half as many multiplications (Gentleman [1974a: p. 452]). In the unweighted case we always set $\delta = 1$; in the weighted least squares situation we set δ equal to the weight w attached to the particular row of $\mathbf{X}$ and thus obtain $\mathbf{D}_2$ $(=\mathbf{D}_1\mathbf{W})$ instead of $\mathbf{D}_1$.

If the above method is applied to $(\mathbf{X}:\mathbf{Y})$ we obtain

$$\begin{pmatrix} \tilde{\mathbf{U}} : \tilde{\mathbf{z}} \\ \mathbf{O} : \tilde{\mathbf{t}} \end{pmatrix},$$

where $\tilde{\mathbf{U}}\boldsymbol{\beta}^* = \tilde{\mathbf{z}}$ and $\text{RSS} = \tilde{\mathbf{t}}'\tilde{\mathbf{t}}$.

11.4 COMPARISON OF METHODS

A set of linear equations $\mathbf{Bx} = \mathbf{c}$ is said to be *ill-conditioned* if small errors or variations in the elements of $\mathbf{B}$ and $\mathbf{c}$ have a large effect on the *exact* solution $\mathbf{x}$. For example, the difference $\delta\mathbf{x}$ between the solution of $\mathbf{Bx} = \mathbf{c}$ and that of

$$(\mathbf{B} + \delta\mathbf{B})(\mathbf{x} + \delta\mathbf{x}) = \mathbf{c} + \delta\mathbf{c}$$

can be expressed as

$$\delta\mathbf{x}=(\mathbf{B}+\delta\mathbf{B})^{-1}(\delta\mathbf{c}-\delta\mathbf{B}\mathbf{x}),$$

and its value depends critically on the inverse matrix. If $\mathbf{B}$ is near singularity, that is, small changes in its elements can cause singularity, then $\delta\mathbf{x}$ could be very large.

In the case of the normal least squares equation $\mathbf{B}$ $(=\mathbf{X}'\mathbf{X})$ and $\mathbf{c}$ $(=\mathbf{X}'\mathbf{Y})$ will contain roundoff errors because they must be computed from $\mathbf{X}$ and $\mathbf{Y}$. Even if $\mathbf{B}$ could be computed exactly, it would not necessarily be stored exactly in the computer; all numbers are stored in binary mode, and a decimal number like 0.1 is a nonterminating binary fraction, just as $\frac{1}{7}$ is a nonterminating decimal fraction. This means that if $\mathbf{X}$ is ill-conditioned, that is, "small" changes in the elements of $\mathbf{X}$ can cause "large" changes in $(\mathbf{X}'\mathbf{X})^{-1}$ and $\hat{\boldsymbol{\beta}}=(\mathbf{X}'\mathbf{X})^{-1}\mathbf{X}'\mathbf{Y}$, then any errors in the formation of $\mathbf{X}'\mathbf{X}$ could have a serious effect on the stability and accuracy of the solution. The situation is made even worse by the fact that $\mathbf{x}$ cannot be computed exactly; roundoff errors accumulate in the solving process and these could tip the balance and make the problem unstable if $\mathbf{X}'\mathbf{X}$ is near singularity. The problem of ill-conditioning is particularly serious in polynomial regression (see Section 8.1.1).

A measure of the ill-conditioning of a matrix $\mathbf{X}$ is its *condition number* $\mathcal{K}[\mathbf{X}]$ which we define as the ratio of the largest to the smallest nonzero singular value of $\mathbf{X}$ (other definitions are used). The singular values of $\mathbf{X}$ are the positive square roots of the eigenvalues of $\mathbf{X}'\mathbf{X}$. We note the following facts about $\mathcal{K}[\mathbf{X}]$:

(1) $\mathcal{K}[\mathbf{X}'\mathbf{X}]=(\mathcal{K}[\mathbf{X}])^2$.
(2) Since $\mathbf{X}'\mathbf{X}=\mathbf{U}'\mathbf{U}$, $\mathcal{K}[\mathbf{U}]=\mathcal{K}[\mathbf{X}]$.

Since $\mathcal{K}>1$ we see from (1) that $\mathbf{X}'\mathbf{X}$ is more ill-conditioned than $\mathbf{X}$ so that unless $\mathcal{K}[\mathbf{X}]$ is of moderate magnitude and $\mathbf{X}'\mathbf{X}$ can be formed accurately, it is better not to form $\mathbf{X}'\mathbf{X}$ at all. Since $\mathcal{K}[\mathbf{X}]$ is unknown before the computing begins, and noting (2), it is safer then to work with $\mathbf{X}$ and obtain $\mathbf{U}$ directly using the orthogonal decomposition methods of Section 11.2.4; however, it is not quite as simple as this because the effect of $\mathcal{K}^2[\mathbf{X}]$ cannot be avoided entirely (Golub [1969: p. 385], Wilkinson [1974]). Of the three approaches given there it seems that the Householder method is currently (1974) most popular among numerical analysts; it is slightly faster than the MGSA, about twice as fast as Givens' method, and has a similar accuracy to MGSA and Givens. However, the square root-free modification of Givens' method is highly competitive as it costs about the same as the Householder method, is easily adapted to weighted least squares, and has all the advantages outlined in Section 11.2.4c. It is also readily adapted to the less than full rank case (Section 11.5.4). Error analyses of all the methods are given by Wilkinson [1967, 1974], Golub [1969: pp. 382–385] and Gentleman [1973].

Various modifications are given in Section 11.5 to allow for the possibility of a singular $\mathbf{X}'\mathbf{X}$. For example, by using pivoting, that is, allowing column interchanges for $\mathbf{X}$ so as to maximize the next diagonal element of $\mathbf{U}$ at each stage, all the methods (except Givens') in Section 11.2 can be generalized to allow for singular $\mathbf{X}'\mathbf{X}$. However, the Givens method can be used, without pivoting, if identifiability constraints are imposed (Section 11.5.4). The singular value decomposition, which seems to be at least as accurate as the other methods, is particularly useful when the rank of $\mathbf{X}$ is unknown or when $\mathbf{X}$ is of full rank but is ill-conditioned in an unpredictable way as, for example, in ridge regression (Chambers [1971]). However, it is about two to four times as expensive as the Householder algorithm, and is not as adaptable.

The Householder method is readily generalized to deal with the problem of computing an F-statistic for testing a general linear hypothesis (Section 11.10). It is also readily adapted to allow for adding a row to $(\mathbf{X}:\mathbf{Y})$ (Section 11.8) or adding a column to $\mathbf{X}$ (Section 11.9); rows and columns of $\mathbf{X}$ are best removed using Givens transformations. The MGSA also provides a simple method for adding a column to $\mathbf{X}$ but it is not so easy to add rows. The Givens method is a good method for adding rows, but adding columns is more difficult. Clearly a combination of the Householder and Givens methods is appropriate.

In practice it is not uncommon for one regressor to be highly correlated with a linear combination of the other regressors so that the columns of $\mathbf{X}$ will be close to being linearly dependent. This means that $\mathbf{X}'\mathbf{X}$ will be near singularity, the smallest eigenvalue will be small, and $\mathcal{K}[\mathbf{X}]$ will be large. For example, Longley's [1967] test data has $\mathcal{K} \approx 4.8 \times 10^9$, whereas Wampler's [1970] polynomial models have a value of the order of 10^6. In fact polynomial regression models are notoriously ill-conditioned when the degree is higher than about 5 or 6, particularly if the values of x are equally spaced (see Section 8.1.1). We would therefore expect any reasonable program to either produce a correctly rounded solution $\hat{\boldsymbol{\beta}}$ or indicate that $\mathbf{X}'\mathbf{X}$ is too ill-conditioned for this to be achieved without working to higher precision (and may even be singular). Pivoting so that u_{rr} is maximized at each stage indicates when $\mathbf{X}$ modified by rounding errors has less than full rank; at some step u_{rr} is less than some tolerance level. A higher degree of accuracy in $\hat{\boldsymbol{\beta}}$ and $\mathbf{e}$ can be obtained by iterative refinement (Section 11.6), though this does not work if the model is too ill-conditioned. If there is no perceptible improvement in $\hat{\boldsymbol{\beta}}$ after one cycle of iteration then the problem is highly ill-conditioned. In this case if one cannot work to a higher precision then it may be appropriate to use the more accurate singular-value decomposition method (Section 11.5.5), or treat the model as though it were singular. The programme of Björck and Golub [1967] using Householder transformations fulfills the above requirements; it has a suitable "failure" exit when **X** modified by rounding errors has less than full rank or when there is no perceptible improvement with iterative refinement.

As a partial safeguard against computational difficulties it is recommended that the means be subtracted from **X** and **Y** as in Section 11.7.1 before computing the regression. This so-called "centering" of the data reduces the condition number of **X** (Golub and Styan [1974]). Some authors also recommend normalizing the columns of **X** and **Y** to unit length as in Section 11.7.2; this is particularly recommended for the singular value decomposition method.

In concluding this comparison of least squares algorithms some mention should be made of one other algorithm due to Bauer [1965]. His algorithm, which uses a decomposition of **X** like (11.23) in Section 11.2.4, uses a Gaussian elimination scheme but with a suitably weighted combination of rows used for elimination instead of a selected row. Excellent Algol programs for this algorithm and for most of the other methods described in Section 11.2 are detailed in Wilkinson and Reinsch [1971]. A numerical comparison of programs using the algorithms in 11.2 is given by Wampler [1970]. For a general reference on the use of computers for data analysis, the reader is referred to Muller [1970].

11.5 LESS THAN FULL RANK CASE

If **X** has rank r $(r<p)$, then $\mathbf{B}=\mathbf{X}'\mathbf{X}$ is singular (in fact positive semidefinite), and the normal equations $\mathbf{B}\boldsymbol{x}=\mathbf{c}$ no longer have a unique solution. Five computational methods for handling this problem are given below.

11.5.1 Cholesky Method for a Generalized Inverse of $X'X$

If **B** is positive semidefinite it can still be expressed in the form $\mathbf{B}=\mathbf{U}'\mathbf{U}$, where **U** is a real upper triangular matrix. Although the rows of **U** can still be obtained from Equations (11.6) and (11.7), we find that $p-r$ of the diagonal elements u_{ii} are now zero; the other elements of **U** in the corresponding rows are then taken to be zero. Now any solution of $\mathbf{U}\mathbf{x}=\mathbf{z}$, where **z** is a solution of $\mathbf{U}'\mathbf{z}=\mathbf{c}$, is a solution of $\mathbf{U}'\mathbf{U}\mathbf{x}=\mathbf{c}$. Hence if $\mathbf{U}^-$ is a generalized inverse of **U** (cf. Section 3.8.1c), then

$$\begin{aligned}\mathbf{x} &= \mathbf{U}^{-}\mathbf{z} \\ &= \mathbf{U}^{-}(\mathbf{U}')^{-}\mathbf{c} \\ &= \mathbf{U}^{-}(\mathbf{U}^{-})'\mathbf{c}, \\ &= \mathbf{B}^{-}\mathbf{c}, \end{aligned} \tag{11.34}$$

say. For example, if the columns of **B** are permuted so that the first r columns are linearly independent, then, for the permuted matrix ($\tilde{\mathbf{B}}$, say), we have

$$\mathbf{U} = \left[\begin{array}{ccccc|ccc} u_{11} & u_{12} & \cdots & u_{1r} & & u_{1,r+1} & \cdots & u_{1p} \\ 0 & u_{22} & \cdots & u_{2r} & & u_{2,r+1} & \cdots & u_{2p} \\ \cdots & \cdots & \cdots & \cdots & & \cdots & \cdots & \cdots \\ & \cdots & \cdots & u_{rr} & & u_{r,r+1} & \cdots & u_{rp} \\ \hline & & \mathbf{O} & & & & \mathbf{O} & \end{array}\right] \}\ p-r\times r$$

$$= \left(\begin{array}{c|c} \mathbf{U}_{11} & \mathbf{U}_{12} \\ \hline \mathbf{O} & \mathbf{O} \end{array}\right), \tag{11.35}$$

where $u_{ii} > 0$ $(i = 1, 2, \ldots, r)$. Then

$$\mathbf{U}^{-} = \begin{pmatrix} \mathbf{U}_{11}^{-1} & \mathbf{O} \\ \mathbf{O} & \mathbf{O} \end{pmatrix}$$

is a generalized inverse of **U**, and

$$\tilde{\mathbf{B}}^{-} = \left(\begin{array}{c|c} \mathbf{U}_{11}^{-1}(\mathbf{U}_{11}^{-1})' & \mathbf{O} \\ \hline \mathbf{O} & \mathbf{O} \end{array}\right),$$

with its columns suitably permuted, will be a generalized inverse of **B**.

Healy [1968a] points out that rounding errors in the formation of **B** may mean that **B** is not exactly singular; even when it is, further rounding errors are involved in forming the elements of **U**. This means that the appropriate u_{ii} will not vanish and the generalized inverse mechanism must be called into play whenever a u_{ii}^2 is "unduly small"—say, less than some small fraction of u_{ii}. Two such algorithms for finding **U** and $\mathbf{B}^{-}$ are given by Healy [1968b] (see also Farebrother and Berry [1974]).

11.5.2 Abbreviated Doolittle Method

In Section 11.2.3 it was shown that the various methods such as the abbreviated Doolittle procedure, which used a triangular decomposition, are closely related to the Cholesky method. It is therefore not surprising to find that a method similar to that in Section 11.5.1 is available for adapting the abbreviated Doolittle procedure to the less than full rank case. Details of this adaptation are given by Rohde and Harvey [1965]; they use a **QR** decomposition of **X** to verify their procedure algebraically.

11.5.3 Generalized Inverse of X Using Householder Transformations

When **X** is not of full rank we still have the decomposition [Equation (11.18)]

$$\mathbf{Q}'\mathbf{X} = \begin{pmatrix} \mathbf{U} \\ \mathbf{O} \end{pmatrix},$$

but now $(r-p)$ diagonal elements of **U** will be zero, the remaining diagonal elements being positive. However, if we allow column interchanges for **X** so that the next diagonal element of **U** at each stage is maximized (that is, pivoting is used), then **X** can be reduced by r Householder transformations to the above triangular form, but with **U** taking the form of (11.35). Hence

$$\mathbf{Q}'\mathbf{X}\boldsymbol{\Pi} = \begin{pmatrix} \mathbf{U}_{11} & \mathbf{U}_{12} \\ \mathbf{O} & \mathbf{O} \end{pmatrix}$$

for some orthogonal **Q** and a $p \times p$ permutation matrix **Π**. Now consider the matrix (Golub and Styan [1973])

$$\mathbf{X}^* = \boldsymbol{\Pi}\begin{pmatrix} \mathbf{U}_{11}^{-1} & \mathbf{O} \\ \mathbf{O} & \mathbf{O} \end{pmatrix}\mathbf{Q}';$$

then, because **Π** is also orthogonal,

$$\begin{aligned} \mathbf{X}\mathbf{X}^* &= \mathbf{Q}\begin{pmatrix} \mathbf{U}_{11} & \mathbf{U}_{12} \\ \mathbf{O} & \mathbf{O} \end{pmatrix}\boldsymbol{\Pi}'\boldsymbol{\Pi}\begin{pmatrix} \mathbf{U}_{11}^{-1} & \mathbf{O} \\ \mathbf{O} & \mathbf{O} \end{pmatrix}\mathbf{Q}' \\ &= \mathbf{Q}\begin{pmatrix} \mathbf{I}_r & \mathbf{O} \\ \mathbf{O} & \mathbf{O} \end{pmatrix}\mathbf{Q}' \end{aligned} \qquad (11.36)$$

is symmetric, $\mathbf{XX}^*\mathbf{X} = \mathbf{X}$, and $\mathbf{X}^*\mathbf{XX}^* = \mathbf{X}^*$. Hence $\mathbf{X}^*$ is a generalized inverse of $\mathbf{X}$ satisfying conditions (a), (b), and (c) of Section 3.8.1c, and therefore $\mathbf{X}^*\mathbf{Y}$ is a solution of the normal equations.

To find the residuals we partition $\mathbf{Q}=(\mathbf{Q}_r,\mathbf{Q}_{n-r})$, where $\mathbf{Q}_r$ is $n\times r$, and consider

$$\mathbf{Q}'\mathbf{Y}=\begin{pmatrix}\mathbf{Q}_r'\mathbf{Y}\\ \mathbf{Q}_{n-r}'\mathbf{Y}\end{pmatrix}=\begin{pmatrix}\mathbf{z}\\ \mathbf{t}\end{pmatrix},$$

say, where $\mathbf{z}$ is now $r\times 1$ instead of $p\times 1$. Since, by (11.36), $\mathbf{X}\mathbf{X}^*=\mathbf{Q}_r\mathbf{Q}_r'$,

$$\begin{aligned}\mathbf{e}&=\mathbf{Y}-\mathbf{X}\hat{\boldsymbol{\beta}}\\ &=\mathbf{Y}-\mathbf{X}\mathbf{X}^*\mathbf{Y}\\ &=(\mathbf{I}_n-\mathbf{X}\mathbf{X}^*)\mathbf{Y}\\ &=(\mathbf{I}_n-\mathbf{Q}_r\mathbf{Q}_r')\mathbf{Q}\begin{pmatrix}\mathbf{z}\\ \mathbf{t}\end{pmatrix}\\ &=\mathbf{Q}_{n-r}\mathbf{t}\qquad(\text{since}\quad \mathbf{Q}_r'\mathbf{Q}_{n-r}=\mathbf{O})\\ &=\mathbf{Q}_{n-r}\mathbf{Q}_{n-r}'\mathbf{Y}\end{aligned}$$

and

$$\mathbf{e}'\mathbf{e}=\mathbf{t}'\mathbf{Q}_{n-r}'\mathbf{Q}_{n-r}\mathbf{t}=\mathbf{t}'\mathbf{t}.$$

11.5.4 Orthogonal Decomposition with Identifiability Constraints

In Section 3.8.1 we saw that if the equations $\mathbf{H}\boldsymbol{\beta}=\mathbf{0}$ (where $\mathbf{H}$ is $s\times p$ of rank $p-r$) are identifiability constraints, and

$$\mathbf{G}=\begin{pmatrix}\mathbf{X}\\ \mathbf{H}\end{pmatrix},$$

then the normal equations take the form:

$$\mathbf{G}'\mathbf{G}\hat{\boldsymbol{\beta}}=\mathbf{X}'\mathbf{Y}=\mathbf{G}'\begin{pmatrix}\mathbf{Y}\\ \mathbf{0}\end{pmatrix}.$$

Since (by Theorem 3.9 of Section 3.8.1) $\mathbf{G}$ is $(n+t)\times p$ of rank p, an orthogonal decomposition method can be applied to $(\mathbf{X}:\mathbf{Y})$ augmented by the rows $(\mathbf{H}:\mathbf{0})$.

This method of solving the normal equations has a number of advantages. First, s does not have to equal $p-r$; any redundant constraints will be automatically taken care of (a useful feature in some analysis of variance problems). Second, the order in which the regressors are brought into the model is fixed, being determined by the order of the columns of $\mathbf{X}$. This means that any submodel determined by, say, the first p_1 columns of $\mathbf{G}$ can be tested against the full model. After the first p_1 columns have been reduced to upper triangular form, the sum of squares of the corresponding transformed $\mathbf{Y}$ elements is the regression sum of squares ($=\mathbf{Y}'\mathbf{Y}-\text{RSS}_{p_1}$) for the submodel fitted thus far. We can therefore test the submodel using the F statistic

$$F=\frac{n-r}{p-p_1}\cdot\frac{\text{RSS}_{p_1}-\text{RSS}_p}{\text{RSS}_p}$$

$$=\frac{n-r}{p-p_1}\cdot\frac{\text{change in regression sum of squares}}{\text{RSS}_p}.$$

A good illustration of this is analysis of covariance (Chapter 10); the concomitant variables are ordered first. For example, consider the one-way classification

$$E[Y_{ij}]=\mu+\alpha_i+\gamma z_{ij},$$

with a single concomitant variable z, and an identifiability constraint $\Sigma_i \alpha_i=0$. (For ease of exposition we assume $i=1,2$ and $j=1,2$.) Then $\mathbf{G}\boldsymbol{\beta}$ takes the form

$$\mathbf{G}\boldsymbol{\beta}=\begin{bmatrix} z_{11} & 1 & 1 & 0 \\ z_{12} & 1 & 1 & 0 \\ z_{21} & 1 & 0 & 1 \\ z_{22} & 1 & 0 & 1 \\ 0 & 0 & 1 & 1 \end{bmatrix}\begin{bmatrix} \gamma \\ \mu \\ \alpha_1 \\ \alpha_2 \end{bmatrix}$$

and $H:\alpha_1=\alpha_2=0$ can be tested by recording RSS_2 after the first two columns of $\mathbf{G}$ have been transformed to upper triangular form.

In testing submodels there is one possible difficulty; the identifiability constraints for the full model may not be appropriate for the submodel. For example, suppose we have

$$\mathbf{G}\boldsymbol{\beta}=\begin{pmatrix} \mathbf{X}_1, & \mathbf{X}_2 \\ \mathbf{H}_1, & \mathbf{H}_2 \end{pmatrix}\begin{pmatrix} \boldsymbol{\beta}_1 \\ \boldsymbol{\beta}_2 \end{pmatrix},$$

where $\mathbf{X}_1$ is the design matrix for the submodel. Then although the equations $\mathbf{H}\boldsymbol{\beta}=\mathbf{0}$ may be identifiability constraints for $\mathbf{X}\boldsymbol{\beta}$, the equations $\mathbf{H}_1\boldsymbol{\beta}_1=\mathbf{0}$ may not be identifiability constraints for $\mathbf{X}_1\boldsymbol{\beta}_1$ (for example, the rows of $\mathbf{H}_1$ may not be linearly independent of the rows of $\mathbf{X}_1$, as required by Theorem 3.9). Fortunately this problem does not generally arise in analysis of variance and covariance problems because we usually have

$$\mathbf{G}=\begin{bmatrix} \mathbf{X}_1 & \mathbf{X}_2 \\ \mathbf{H}_1 & \mathbf{O} \\ \mathbf{O} & \mathbf{H}_2 \end{bmatrix}.$$

In the randomized block design

$$E[Y_{ij}]=\mu+\alpha_i+\beta_j,$$

for example, we can use the identifiability constraints $\Sigma_i\alpha_i=\Sigma_j\beta_j=0$.

If identifiability constraints cannot be specified a priori, then the following method of Gentleman [1974a, b] can be used. When applying an orthogonal decomposition algorithm, any rank deficiency theoretically appears as a zero row in $(\mathbf{U}:\mathbf{z})$. In practice, through roundoff error, this will show up as a row with a very small diagonal element; the rows below will also be computed but they will be nonsense. Suppose then that the $(i+1)$th row is approximately zero so that the $(i+1)$th column of $\mathbf{X}$ appears to be linearly dependent on the first i columns. We can then overcome this difficulty by forcing the $(i+1)$th element of $\boldsymbol{\beta}$ to be zero; that is, we add the identifiability constraint $(\mathbf{h}':0)$, where $\mathbf{h}$ has its $(i+1)$th element equal to unity and its remaining elements equal to zero. Retriangularization of the augmented system, starting with the $(i+1)$th column, will then produce a satisfactory decomposition (provided, of course, there are no further linear dependencies; otherwise, we repeat the process and add further identifiability constraints). The nature of the linear dependence can also be found. Suppose that $\mathbf{x}_{i+1}$, the $(i+1)$th column of $\mathbf{X}$, takes the form $\mathbf{x}_{i+1}=\mathbf{X}_1\mathbf{a}$, where $\mathbf{X}_1$ consists of the first i columns of $\mathbf{X}$. Then by regarding the column above the (theoretically) zero $(i+1)$th diagonal element as the right side in a system of equations equal to the triangle to the left of this column, we can solve these equations and obtain $\mathbf{a}$. (The validity of this procedure follows from the fact that the equation $\mathbf{x}_{i+1}-\mathbf{X}_1\mathbf{a}=\mathbf{0}$ corresponds to a model in which the residual sum of squares is zero; that is, it is of the form $\mathbf{Y}-\mathbf{X}\hat{\boldsymbol{\beta}}=\mathbf{0}$).

11.5.5 Singular Value Decomposition

In Section 3.8.1c we saw that that $\mathbf{X}^+\mathbf{Y}$ is a solution of the least squares equations, where $\mathbf{X}^+$ is the Moore–Penrose inverse of $\mathbf{X}$. Now it can be shown that an $n\times p$ matrix $\mathbf{X}$ can be expressed in the form (*A.10*)

$$\mathbf{X}=\mathbf{P}\boldsymbol{\Sigma}\mathbf{Q}', \tag{11.37}$$

where $\mathbf{P}$ is an $n \times p$ matrix consisting of p orthonormalized eigenvectors associated with the p largest eigenvalues of $\mathbf{XX}'$ (thus $\mathbf{P}'\mathbf{P}=\mathbf{I}_p$), $\mathbf{Q}$ is a $p \times p$ orthogonal matrix consisting of the orthonormalized eigenvectors of $\mathbf{X}'\mathbf{X}$, and $\mathbf{\Sigma}=\text{diag}(\sigma_1,\sigma_2,\ldots,\sigma_p)$ is a $p \times p$ diagonal matrix. Here $\sigma_1 \geqslant \sigma_2 \geqslant \cdots \geqslant \sigma_p \geqslant 0$ and these diagonal elements of $\mathbf{\Sigma}$ (called the singular values of $\mathbf{X}$) are the square roots of the eigenvalues of $\mathbf{X}'\mathbf{X}$. These eigenvalues are nonnegative as $\mathbf{X}'\mathbf{X}$ is positive semidefinite; also $\sigma_{r+1}=\sigma_{r+2}=\cdots=\sigma_p=0$ as $\mathbf{X}$ has rank r. This decomposition of $\mathbf{X}$ is called the *singular-value decomposition*, and it can be shown that

$$\mathbf{X}^+ = \mathbf{Q}\mathbf{\Sigma}^+\mathbf{P}', \tag{11.38}$$

where $\mathbf{\Sigma}^+ = \text{diag}(\sigma_1^{-1},\sigma_2^{-1},\ldots,\sigma_r^{-1},0,\ldots,0)$.

To find $\hat{\boldsymbol{\beta}}=\mathbf{Q}\mathbf{\Sigma}^+\mathbf{P}'\mathbf{Y}$ it is only necessary to compute $\mathbf{Q}$, $\mathbf{\Sigma}$, and $\mathbf{P}'\mathbf{Y}$. Details of an algorithm for doing this are given by Golub [1969] and Golub and Reinsch [1970]. It provides an accurate method when $\mathbf{X}$ has full rank, but is highly ill-conditioned. It is also useful when r is unknown; a σ_i is deemed to be zero if its value is less than a certain tolerance.

11.6 IMPROVING THE SOLUTION BY ITERATIVE REFINEMENT

If $\mathbf{e}=\mathbf{Y}-\mathbf{Xx}$ is the *exact* residual vector, then $\mathbf{e}+\mathbf{Xx}=\mathbf{Y}$ and $\mathbf{X}'\mathbf{e}=\mathbf{0}$. Hence

$$\begin{pmatrix} \mathbf{I}_n & \mathbf{X} \\ \mathbf{X}' & \mathbf{O} \end{pmatrix}\begin{pmatrix} \mathbf{e} \\ \mathbf{x} \end{pmatrix}=\begin{pmatrix} \mathbf{Y} \\ \mathbf{0} \end{pmatrix}$$

or

$$\mathbf{Mv}=\mathbf{m}, \tag{11.39}$$

say. Since $\mathbf{X}'\mathbf{X}=\mathbf{U}'\mathbf{U}$, $\mathbf{M}$ can be expressed in the form

$$\mathbf{M}=\begin{pmatrix} \mathbf{I}_n & \mathbf{O} \\ \mathbf{X}' & \mathbf{U}' \end{pmatrix}\begin{pmatrix} \mathbf{I}_n & \mathbf{X} \\ \mathbf{O} & -\mathbf{U} \end{pmatrix}$$

$$=\mathbf{L}_M\mathbf{U}_M,$$

say, where $\mathbf{L}_M$ and $\mathbf{U}_M$ are lower and upper triangular matrices, respectively.

Suppose $\mathbf{x}^{(0)}$ is the numerical solution of the normal equations. Let $\mathbf{e}^{(0)}$ be the calculated residual and let

$$\mathbf{v}^{(0)} = \begin{pmatrix} \mathbf{e}^{(0)} \\ \mathbf{x}^{(0)} \end{pmatrix}.$$

Then, because of rounding errors in finding $\mathbf{U}$, $\mathbf{z}$, and solving $\mathbf{Ux}=\mathbf{z}$, $\mathbf{Mv}^{(0)}$ will not be exactly equal to $\mathbf{m}$ in (11.39), and there will be a nonzero difference

$$\mathbf{h}^{(0)} = \mathbf{m} - \mathbf{Mv}^{(0)}.$$

If, therefore, $\mathbf{v} = \mathbf{v}^{(0)} + \boldsymbol{\delta}$, say, $\boldsymbol{\delta}$ satisfies

$$\mathbf{M}\boldsymbol{\delta} = \mathbf{Mv} - \mathbf{Mv}^{(0)} = \mathbf{h}^{(0)},$$

and we can use the *same* $\mathbf{M}$ (that is, use the same $\mathbf{U}$) to obtain an approximate solution for $\boldsymbol{\delta}$, say, $\boldsymbol{\delta}^{(0)}$. Under certain reasonable conditions we would expect the new solution

$$\mathbf{v}^{(1)} = \mathbf{v}^{(0)} + \boldsymbol{\delta}^{(0)}$$

of (11.39) to be more accurate than $\mathbf{v}^{(0)}$. Because $\boldsymbol{\delta}^{(0)}$ is only an approximation we can repeat the process. Thus, for $\mathbf{v}^{(0)}$ given, the algorithm may be summarized as follows:

(1) Compute $\mathbf{h}^{(k)} = \mathbf{m} - \mathbf{Mv}^{(k)}$ using $\mathbf{M}$ of (11.39).
(2) Solve $\mathbf{M}\boldsymbol{\delta}^{(k)} = \mathbf{h}^{(k)}$ by solving the triangular systems $\mathbf{L}_M \mathbf{w}^{(k)} = \mathbf{h}^{(k)}$ and $\mathbf{U}_M \boldsymbol{\delta}^{(k)} = \mathbf{w}^{(k)}$.
(3) Compute $\mathbf{v}^{(k+1)} = \mathbf{v}^{(k)} + \boldsymbol{\delta}^{(k)}$.

The process continues until $\|\boldsymbol{\delta}^{(k)}\| / \|\mathbf{v}^{(k)}\| < \varepsilon$, where ε is a predetermined constant, or until some other criterion is satisfied.

This method for improving the numerical solution of a set of linear equations, in this case (11.39), is called iterative refinement. The above adaptation to the least squares problem is given by Björck and Golub [1967], and is discussed briefly by Golub [1969: p. 385]. It can be used for improving the numerical solution of the normal equations obtained by either the Cholesky or orthogonal decomposition methods; all we need is $\mathbf{U}$, and $\mathbf{X}$ saved.

Iterative refinement has been discussed in detail by various authors, for example, Golub and Wilkinson [1966], Moler [1967], Wilkinson [1967], and Fletcher [1975]. It appears that, provided $\mathbf{v}^{(0)}$ has some correct digits (that is, $\mathbf{X}$ is not too ill-conditioned), $\mathbf{v}^{(k)}$ will converge to the exact solution $\mathbf{v}$

(within "working accuracy") if the $\mathbf{h}^{(k)}$ are calculated using double precision. Since elements of the vector $\mathbf{v}^{(k)}$ are single precision numbers, this double precision may be achieved by accumulating inner products (without intermediate rounding) when computing $\mathbf{h}^{(k)}$; provided the subtraction $\mathbf{m}-\mathbf{M}\mathbf{v}^{(k)}$ is carried out before rounding, the remaining steps may be carried out in single precision. On some computers the logical design is such that the accumulation of inner products is scarcely faster than true double-precision arithmetic. However, on others, the accumulation of inner products is essentially as fast as ordinary single precision. If inner products cannot be accumulated then $\mathbf{h}^{(k)}$ *must* be computed using true double precision arithmetic.

Even if a greater accuracy is not required, iterative refinement should perhaps be carried out for at least one iteration as a check on the conditioning of $\mathbf{X}$. If there is no perceptible improvement then $\mathbf{X}$ is ill-conditioned and the program should have a suitable "failure" exit such as that given by Björck and Golub [1967]; this exit is used by Wampler [1970].

11.7 CENTERING AND SCALING THE DATA

11.7.1 Centering the Data

Since the elements of the first column of $\mathbf{X}$ are generally chosen to be unity, the normal equations are $(p=4)$

$$\begin{bmatrix} n & \Sigma x_{i1} & \Sigma x_{i2} & \Sigma x_{i3} \\ \Sigma x_{i1} & \Sigma x_{i1}^2 & \Sigma x_{i1}x_{i2} & \Sigma x_{i1}x_{i3} \\ \Sigma x_{i2} & \Sigma x_{i2}x_{i1} & \Sigma x_{i2}^2 & \Sigma x_{i2}x_{i3} \\ \Sigma x_{i3} & \Sigma x_{i3}x_{i1} & \Sigma x_{i3}x_{i2} & \Sigma x_{i3}^3 \end{bmatrix} \begin{bmatrix} \hat\beta_0 \\ \hat\beta_1 \\ \hat\beta_2 \\ \hat\beta_3 \end{bmatrix} = \begin{bmatrix} \Sigma Y_i \\ \Sigma x_{i1} Y_i \\ \Sigma x_{i2} Y_i \\ \Sigma x_{i3} Y_i \end{bmatrix}$$

where $i=1,2,\ldots,n$. Dividing the first equation by n and subtracting suitable multiples of the first equation from the others, the remaining elements in the first column of $\mathbf{X}'\mathbf{X}$ can be reduced to zero; that is, we have

$$\begin{bmatrix} 1 & \bar{x}_1 & \bar{x}_2 & \bar{x}_3 \\ 0 & \Sigma \tilde{x}_{i1}^2 & \Sigma \tilde{x}_{i1}\tilde{x}_{i2} & \Sigma \tilde{x}_{i1}\tilde{x}_{i3} \\ 0 & \Sigma \tilde{x}_{i2}\tilde{x}_{i1} & \Sigma \tilde{x}_{i2}^2 & \Sigma \tilde{x}_{i2}\tilde{x}_{i3} \\ 0 & \Sigma \tilde{x}_{i3}\tilde{x}_{i1} & \Sigma \tilde{x}_{i3}\tilde{x}_{i2} & \Sigma \tilde{x}_{i3}^2 \end{bmatrix} \begin{bmatrix} \hat\beta_0 \\ \hat\beta_1 \\ \hat\beta_2 \\ \hat\beta_3 \end{bmatrix} = \begin{bmatrix} \bar{Y} \\ \Sigma \tilde{x}_{i1} \tilde{Y}_i \\ \Sigma \tilde{x}_{i2} \tilde{Y}_i \\ \Sigma \tilde{x}_{i3} \tilde{Y}_i \end{bmatrix}, \qquad (11.40)$$

where $\tilde{x}_{ij} = x_{ij} - \bar{x}_j$, $\bar{x}_j = \Sigma_i x_{ij}/n$, and $\tilde{Y}_i = Y_i - \bar{Y}$. We have also made use of the formulas

$$\sum_i \tilde{x}_{ij}\tilde{x}_{ik} = \sum_i (x_{ij} - \bar{x}_j)(x_{ik} - \bar{x}_k) \tag{11.41}$$

$$= \sum_i x_{ij}x_{ik} - \frac{\left(\sum_i x_{ij}\right)\left(\sum_i x_{ik}\right)}{n}, \tag{11.42}$$

$$\sum_i \tilde{x}_{ij}\tilde{Y}_i = \sum_i x_{ij}Y_i - \frac{\left(\sum_i x_{ij}\right)\left(\sum_i Y_i\right)}{n}.$$

Hence setting $\mathbf{b}' = (\hat{\beta}_1, \hat{\beta}_2, \hat{\beta}_3)$, $\tilde{\mathbf{X}} = [(\tilde{x}_{ij})]$, and $\tilde{\mathbf{Y}} = [(\tilde{Y}_i)]$, we have

$$\hat{\beta}_0 = \bar{Y} - b_1\bar{x}_1 - b_2\bar{x}_2 - b_3\bar{x}_3,$$

and $\mathbf{b}$ is the solution of

$$\tilde{\mathbf{X}}'\tilde{\mathbf{X}}\mathbf{b} = \tilde{\mathbf{X}}'\tilde{\mathbf{Y}}. \tag{11.43}$$

In practical terms it means that we can find $\mathbf{b}$ by subtracting off averages and working with the "centered" data $Y_i - \bar{Y}$ and $x_{ij} - \bar{x}_j$. This approach is also suggested intuitively by the fitted model

$$\hat{Y}_i = \hat{\beta}_0 + b_1x_{i1} + b_2x_{i2} + b_3x_{i3}$$

or, substituting for $\hat{\beta}_0$,

$$\hat{Y}_i - \bar{Y} = b_1(x_{i1} - \bar{x}_1) + b_2(x_{i2} - \bar{x}_2) + b_3(x_{i3} - \bar{x}_3)$$

$$= b_1\tilde{x}_{i1} + b_2\tilde{x}_{i2} + b_3\tilde{x}_{i3}.$$

We note that the residuals, the residual sum of squares, and the square of the multiple correlation coefficient R^2 can also be expressed in terms of the centered data:

$$\begin{aligned} e_i &= Y_i - \hat{Y}_i \\ &= Y_i - \bar{Y} - b_1(x_{i1} - \bar{x}_1) - \cdots - b_3(x_{i3} - \bar{x}_3) \\ &= \tilde{Y}_i - b_1\tilde{x}_{i1} - b_2\tilde{x}_{i2} - b_3\tilde{x}_{i3}, \end{aligned}$$

$$\begin{aligned} \text{RSS} &= \mathbf{e}'\mathbf{e} \\ &= (\tilde{\mathbf{Y}} - \tilde{\mathbf{X}}\mathbf{b})'(\tilde{\mathbf{Y}} - \tilde{\mathbf{X}}\mathbf{b}) \\ &= \tilde{\mathbf{Y}}'\tilde{\mathbf{Y}} - \mathbf{b}'\tilde{\mathbf{X}}'\tilde{\mathbf{Y}} \qquad [\text{by (11.43)}] \qquad (11.44) \\ &= \tilde{\mathbf{Y}}'\tilde{\mathbf{Y}} - \tilde{\mathbf{Y}}'\tilde{\mathbf{X}}(\tilde{\mathbf{X}}'\tilde{\mathbf{X}})^{-1}\tilde{\mathbf{X}}'\tilde{\mathbf{Y}}, \qquad (11.45) \end{aligned}$$

and, from Equation (4.30) in Section 4.2,

$$\begin{aligned} 1 - R^2 &= \frac{\mathbf{e}'\mathbf{e}}{\Sigma(Y_i - \bar{Y})^2} \\ &= \mathbf{e}'\mathbf{e}/\tilde{\mathbf{Y}}'\tilde{\mathbf{Y}} \qquad (11.46) \end{aligned}$$

or [by (11.44)]

$$R^2 = \frac{\mathbf{b}'\tilde{\mathbf{X}}'\tilde{\mathbf{Y}}}{\tilde{\mathbf{Y}}'\tilde{\mathbf{Y}}}.$$

Although the above theory is developed for the case $p=4$, it is obviously true in general.

Suppose we wish to test the hypothesis $H: \beta_j = 0$. Then the F-statistic is [Equation (4.13) in Section 4.1.3]

$$F = \frac{b_j^2}{S^2 d_{jj}}, \qquad (11.47)$$

where $S = \text{RSS}/(n-p)$ and $\sigma^2 d_{jj} = \text{var}[b_j]$. To find d_{jj} we note first of all that

$$\sum_i \tilde{x}_{ij}\tilde{Y}_i = \sum_i \tilde{x}_{ij} Y_i$$

for all j. Then

$$\tilde{\mathbf{X}}'\tilde{\mathbf{Y}}=\tilde{\mathbf{X}}'\mathbf{Y},$$

$$\mathbf{b}=(\tilde{\mathbf{X}}'\tilde{\mathbf{X}})^{-1}\tilde{\mathbf{X}}'\mathbf{Y},$$

and hence

$$\mathcal{D}[\mathbf{b}]=\sigma^2(\tilde{\mathbf{X}}'\tilde{\mathbf{X}})^{-1}.$$

This means that d_{jj} is the jth diagonal element of the above inverse matrix, a result proved directly in Example 4.3, Section 4.1.3; there $\mathbf{V}=\tilde{\mathbf{X}}'\tilde{\mathbf{X}}$. Now by relabeling, if necessary, we can assume that $j=1$. Let

$$\tilde{\mathbf{X}}'\tilde{\mathbf{X}}=\begin{pmatrix} a_{11} & \mathbf{a}_{12}' \\ \mathbf{a}_{21} & \mathbf{A}_{22} \end{pmatrix},$$

then, by $A7$,

$$(\tilde{\mathbf{X}}'\tilde{\mathbf{X}})^{-1}=\begin{pmatrix} (a_{11}-\mathbf{a}_{12}'\mathbf{A}_{22}^{-1}\mathbf{a}_{21})^{-1} & ,* \\ * & ,* \end{pmatrix}$$

and $d_{11}^{-1}=a_{11}-\mathbf{a}_{12}'\mathbf{A}_{22}^{-1}\mathbf{a}_{21}$. Comparing this with Equation (11.45), namely,

$$\mathbf{e}'\mathbf{e}=\tilde{\mathbf{Y}}'\tilde{\mathbf{Y}}-(\tilde{\mathbf{X}}'\tilde{\mathbf{Y}})'(\tilde{\mathbf{X}}'\tilde{\mathbf{X}})^{-1}(\tilde{\mathbf{X}}'\tilde{\mathbf{Y}}),$$

we see that d_{11}^{-1} is the residual sum of squares that we would obtain from regressing $\tilde{x}_1$ on $\tilde{x}_2,\tilde{x}_3,\ldots,\tilde{x}_{p-1}$. In particular, from (11.46), we have

$$d_{11}^{-1}=(1-R_1^2)\sum_i \tilde{x}_{i1}^2=(1-R_1^2)s_1^2,$$

say, where R_1^2 is the square of the multiple correlation coefficient obtained by regressing $\tilde{x}_1$ on the other regressors. Therefore, in general,

$$d_{jj}^{-1}=(1-R_j^2)s_j^2, \tag{11.48}$$

where R_j^2 is the square of the multiple correlation coefficient obtained by regressing $\tilde{x}_j$ on the other regressors, and $s_j^2=\Sigma_i\tilde{x}_{ij}^2$.

It is interesting to note that the reduction of the first column of $\mathbf{X}'\mathbf{X}$ to $(1,0,0,0)'$ in (11.40) is simply the first step in Gaussian elimination. Because elimination (in the form of the abbreviated Doolittle procedure) was widely used on desk calculators, it is not surprising that subtracting means has become standard statistical practice. For computing $\tilde{\mathbf{X}}'\tilde{\mathbf{X}}$ the early computer programs tended to use the formula (11.42), which is generally considered to be more appropriate to desk computation, rather than (11.41) (Longley [1967]); however, see Ling [1974: p. 866] for some interesting results. Other methods for computing these quantities are described by Youngs and Cramer [1971: pp. 664–665] and Ling [1974] (see Exercise 13 at the end of the chapter). However, the lesson to be learned today from the numerical analysts is that we should work directly with $\tilde{\mathbf{X}}$ using the orthogonal decomposition methods of Section 11.2.4, and perhaps avoid computing $\tilde{\mathbf{X}}'\tilde{\mathbf{X}}$ altogether.

11.7.2 Scaling (Equilibration)

Some authors suggest that $\tilde{\mathbf{Y}}$ and the columns of $\tilde{\mathbf{X}}$ should be scaled so that they have unit norms; this means working with

$$\mathbf{Z}=\frac{\tilde{\mathbf{Y}}}{(\tilde{\mathbf{Y}}'\tilde{\mathbf{Y}})^{1/2}} \qquad \left(=\frac{\tilde{\mathbf{Y}}}{s_Y}, \text{ say}\right)$$

and the matrix $\mathbf{W}=[(w_{ij})]$, where

$$w_{ij}=\frac{\tilde{x}_{ij}}{\left(\sum_i \tilde{x}_{ij}^2\right)^{1/2}} \qquad \left(=\frac{\tilde{x}_{ij}}{s_j}, \text{ say}\right).$$

Although there is little theoretical support for this scaling, we consider its effect in detail because it forms the basis of several well-known procedures for stepwise regression (Chapter 12).

We note that

$$(\mathbf{W}'\mathbf{W})_{jk}=\frac{\sum_i \tilde{x}_{ij}\tilde{x}_{ik}}{s_j s_k} \qquad (=r_{jk}, \text{ say}),$$

the simple correlation between the regressors x_j and x_k, and

$$(\mathbf{W}'\mathbf{Z})_j=\frac{\sum_i \tilde{x}_{ij}\tilde{Y}_i}{s_j s_Y} \qquad (=r_{jY}, \text{ say}),$$

the simple correlation between x_j and Y. Therefore the equations $\mathbf{W}'\mathbf{W}\mathbf{a}=\mathbf{W}'\mathbf{Z}$ take the form

$$\mathbf{R}_{xx}\mathbf{a}=\mathbf{r}_{xy},$$

where

$$\mathbf{R}_{xx}=\begin{bmatrix} 1 & r_{12} & r_{13} & \cdots & r_{1,p-1} \\ r_{21} & 1 & r_{23} & \cdots & r_{2,p-1} \\ \cdots & \cdots & \cdots & \cdots & \cdots \\ r_{p-1,1} & \cdots & \cdots & \cdots & 1 \end{bmatrix} \quad \text{and} \quad \mathbf{r}_{xy}=\begin{bmatrix} r_{1Y} \\ r_{2Y} \\ \cdots \\ r_{p-1,Y} \end{bmatrix}.$$

Here $r_{jk}=r_{kj}$, and $\mathbf{R}_{xx}$ is called the *correlation matrix*. Applying the methods of Section 11.2.4 to $(\mathbf{W}:\mathbf{Z})$ instead of $(\tilde{\mathbf{X}}:\tilde{\mathbf{Y}})$ leads to the solution

$$\mathbf{a}=\mathbf{R}_{xx}^{-1}\mathbf{r}_{xy}.$$

We note that $\mathbf{R}_{xx}^{-1}$ can be recovered from the resulting upper triangular matrix [Equation (11.9) in Section 11.2.2], and the residual sum of squares for this model is

$$\frac{\mathbf{e}'\mathbf{e}}{s_Y^2}=1-R^2. \tag{11.49}$$

The elements of $\mathbf{b}$ are obtained from the relation $b_j=a_j s_Y/s_j$, and the diagonal elements d_{jj} of $(\tilde{\mathbf{X}}'\tilde{\mathbf{X}})^{-1}$ are obtained from

$$d_{jj}=\frac{(\mathbf{R}_{xx}^{-1})_{jj}}{s_j^2}. \tag{11.50}$$

The F-statistic for testing the hypothesis $H:\beta_j=0$ is given by [cf. (11.47)]

$$\begin{aligned} F&=\frac{b_j^2(n-p)}{\mathbf{e}'\mathbf{e}d_{jj}} \\ &=\frac{a_j^2(n-p)s_Y^2}{\mathbf{e}'\mathbf{e}(\mathbf{R}_{xx}^{-1})_{jj}} \\ &=\frac{a_j^2(n-p)}{(1-R^2)(\mathbf{R}_{xx}^{-1})_{jj}} \\ &=\frac{a_j^2(n-p)(1-R_j^2)}{1-R^2}, \end{aligned} \tag{11.51}$$

since, from (11.50) and (11.48), $(\mathbf{R}_{xx}^{-1})_{jj}=(1-R_j^2)^{-1}$.

The main arguments for scaling seem to be twofold. First, $\mathbf{R}_{xx}$ may be better conditioned than $\tilde{\mathbf{X}}'\tilde{\mathbf{X}}$ as the diagonal elements of $\mathbf{R}_{xx}$ are all equal (Golub [1969: pp. 371, 385]); however, this argument carries little weight since it is better to work with $\tilde{\mathbf{X}}$ directly if conditioning is to be taken into consideration. Second, the correlation coefficients all lie between -1 and $+1$; with numbers in this range the adverse effects of rounding errors are minimized. However, we are no better off unless s_Y and s_j are calculated accurately.

11.8 REGRESSION UPDATING

When data are arriving sequentially, it may be undesirable or impossible to wait for all the data before carrying out a regression analysis. In this situation we need a straightforward algorithm for adding an additional m rows, $(\mathbf{X}_m:\mathbf{Y}_m)$ say, to $(\mathbf{X}:\mathbf{Y})$ after the latter has been reduced to

$$\begin{pmatrix} \mathbf{U} & \mathbf{z} \\ \mathbf{O} & \mathbf{t} \end{pmatrix}$$

by an orthogonal transformation $\mathbf{Q}'$ using Householder or Givens transformations. Now writing

$$\begin{bmatrix} \mathbf{I}_m & \mathbf{O} \\ \mathbf{O} & \mathbf{Q}' \end{bmatrix} \begin{bmatrix} \mathbf{X}_m & \mathbf{Y}_m \\ \mathbf{X} & \mathbf{Y} \end{bmatrix} = \begin{bmatrix} \mathbf{X}_m & \mathbf{Y}_m \\ \mathbf{U} & \mathbf{z} \\ \mathbf{O} & \mathbf{t} \end{bmatrix} \begin{matrix} \}m \\ \}p \\ \}n-p \end{matrix} \tag{11.52}$$

we can apply p Householder transformations of order $(m+p)$ to the first $(m+p)$ rows of the right side of (11.52) and obtain (Golub and Styan [1973])

$$\mathbf{Q}'_H \begin{bmatrix} \mathbf{X}_m & \mathbf{Y}_m \\ \mathbf{X} & \mathbf{Y} \end{bmatrix} = \begin{bmatrix} \mathbf{U}_* & \mathbf{z}_1 \\ \mathbf{O} & \mathbf{z}_2 \\ \mathbf{O} & \mathbf{t} \end{bmatrix} \begin{matrix} \}p \\ \}m \\ \}n-p \end{matrix}$$

where

$$\mathbf{Q}'_H = \begin{pmatrix} \mathbf{P} & \mathbf{O} \\ \mathbf{O} & \mathbf{I}_{n-p} \end{pmatrix} \begin{pmatrix} \mathbf{I}_m & \mathbf{O} \\ \mathbf{O} & \mathbf{Q}' \end{pmatrix}$$

is orthogonal and $\mathbf{P}$ is the product of the p transformations. The new $\hat{\boldsymbol{\beta}}$ for the updated model is the solution of $\mathbf{U}_*\mathbf{x} = \mathbf{z}_1$, and the new residual is [cf.

Equation (11.20) in Section 11.2.4]

$$\mathbf{e}_* = \mathbf{Q}_{H,n+m-p}\begin{pmatrix}\mathbf{z}_2\\ \mathbf{t}\end{pmatrix},$$

where $\mathbf{Q}_{H,n+m-p}$ is the last $n+m-p$ columns of $\mathbf{Q}_H$. The new residual sum of squares is

$$\begin{aligned}\mathbf{e}_*'\mathbf{e}_* &= \mathbf{z}_2'\mathbf{z}_2 + \mathbf{t}'\mathbf{t}\\ &= \mathbf{z}_2'\mathbf{z}_2 + \mathbf{e}'\mathbf{e},\end{aligned} \tag{11.53}$$

so that the "old" residual sum of squares, $\mathbf{e}'\mathbf{e}$, is readily updated.

If Givens transformations are used to carry out the updating then we consider the equations

$$\begin{bmatrix}\mathbf{Q}' & \mathbf{O}\\ \mathbf{O} & \mathbf{I}_m\end{bmatrix}\begin{bmatrix}\mathbf{X} & \mathbf{Y}\\ \mathbf{X}_m & \mathbf{Y}_m\end{bmatrix} = \begin{bmatrix}\mathbf{U} & \mathbf{z}\\ \mathbf{O} & \mathbf{t}\\ \mathbf{X}_m & \mathbf{Y}_m\end{bmatrix} = \mathbf{Z},$$

say. Each row of $\mathbf{X}_m$ can be reduced to zero by transforming it with the first, second,..., pth row of $\mathbf{U}$; that is, there exists an $(n+m)\times(n+m)$ orthogonal matrix $\mathbf{Q}_G$, say, such that

$$\mathbf{Q}'_G\begin{bmatrix}\mathbf{X} & \mathbf{Y}\\ \mathbf{X}_m & \mathbf{Y}_m\end{bmatrix} = \begin{bmatrix}\mathbf{U}_* & \mathbf{z}_1\\ \mathbf{O} & \mathbf{t}\\ \mathbf{O} & \mathbf{z}_0\end{bmatrix}.$$

Here $\mathbf{U}_*$ is the same as for the Householder method as the Cholesky decomposition is unique. Once again we have

$$\mathbf{e}_* = \mathbf{Q}_{G,n+m-p}\begin{pmatrix}\mathbf{t}\\ \mathbf{z}_0\end{pmatrix}$$

and

$$\begin{aligned}\mathbf{e}_*'\mathbf{e}_* &= \mathbf{t}'\mathbf{t} + \mathbf{z}_0'\mathbf{z}_0\\ &= \mathbf{e}'\mathbf{e} + \mathbf{z}_0'\mathbf{z}_0\end{aligned} \tag{11.54}$$

Although $\mathbf{z}_0'\mathbf{z}_0 = \mathbf{z}_2'\mathbf{z}_2$ [cf. (11.53) and (11.54)], it does not necessarily follow that $\mathbf{z}_0 = \mathbf{z}_2$. An Algol program for adding a single row is given by Chambers [1971].

If $\mathbf{t}$ has been reduced to $(\delta,0,\ldots,0)'$ ($\delta > 0$; cf. Section 11.2.3c) then further rotations of δ with what is left of each element in $\mathbf{Y}_m$ reduce the latter to zero and produce a new root residual sum of squares δ_*, say

($=(\mathbf{e}_*'\mathbf{e}_*)^{1/2}$). The last m data rows have now been reduced to zero and can therefore be discarded.

When the rank of $\mathbf{X}$ is less than p, the Householder method of Section 11.5.3 can be modified along the above lines to allow the addition of extra rows. If identifiability constraints are introduced as in Section 11.5.4, the Givens method can be used. Updating procedures also exist for the singular value decomposition method of Section 11.5.5, and the reader is referred to Businger [1970]. The effect of updating on the residuals is studied by Beckman and Trussell [1974].

Sometimes we may wish to remove a row of data from $(\mathbf{X}:\mathbf{Y})$; for example, more accurate observations may become available, or a certain observation becomes suspect. This can be done by reversing the Givens transformations that would have been required to add in the row in question to the reduced $(\mathbf{X}:\mathbf{Y})$; Chambers [1971] gives an Algol proraamme for doing this. The same effect is achieved by adding in the particular row multiplied by $\sqrt{-1}$; no complex arithmetic is actually involved because it amounts to using a weight of -1 (Golub [1969: pp. 378–380], Gentleman [1974a]). Other methods due to Golub are mentioned in Chambers [1971: p. 746] and, in particular, Golub and Styan [1973: p. 264]. However, any method of removing a row is potentially unstable, and must be used with care (Gentleman [1973]).

11.9 ADDING OR REMOVING A SPECIFIED REGRESSOR

11.9.1 Adding a Regressor

Suppose we have fitted a regression model $\mathcal{E}[\mathbf{Y}]=\mathbf{X}\hat{\boldsymbol{\beta}}$, where $\mathbf{X}$ is $n\times p$ of rank p, and obtained $(\mathbf{X}'\mathbf{X})^{-1}$, $\hat{\boldsymbol{\beta}}$, and RSS$=\mathbf{Y}'\mathbf{Y}-\hat{\boldsymbol{\beta}}'\mathbf{X}'\mathbf{Y}$ $(=\mathbf{Y}'\mathbf{R}\mathbf{Y})$. Introducing a further regressor x_k, say, is equivalent to adding a further column to $\mathbf{X}$, namely,

$$G:\mathcal{E}[\mathbf{Y}]=(\mathbf{X},\mathbf{x}_k)\begin{pmatrix}\boldsymbol{\beta}\\ \beta_k\end{pmatrix}.$$

If a subscript G denotes least squares estimation under the enlarged model G, then, from Section 3.7.2, we have

$$\hat{\beta}_{k,G}=\frac{\mathbf{x}_k'\mathbf{R}\mathbf{Y}}{\mathbf{x}_k'\mathbf{R}\mathbf{x}_k},\tag{11.55}$$

$$\hat{\boldsymbol{\beta}}_G=\hat{\boldsymbol{\beta}}-(\mathbf{X}'\mathbf{X})^{-1}\mathbf{X}'\mathbf{x}_k\hat{\beta}_{k,G},\tag{11.56}$$

and

$$\text{RSS}_G = \text{RSS} - \hat{\beta}_{k,G}\mathbf{x}_k'\mathbf{RY} \tag{11.57}$$

where $\mathbf{R} = \mathbf{I}_n - \mathbf{X}(\mathbf{X}'\mathbf{X})^{-1}\mathbf{X}'$. Also, from Equation (3.28) in Section 3.7.1

$$\mathscr{D}\begin{bmatrix} \hat{\boldsymbol{\beta}}_G \\ \hat{\beta}_{k,G} \end{bmatrix} = \sigma^2 \begin{pmatrix} \mathbf{X}'\mathbf{X} & \mathbf{X}'\mathbf{x}_k \\ \mathbf{x}_k'\mathbf{X} & \mathbf{x}_k'\mathbf{x}_k \end{pmatrix}^{-1}$$

$$= \sigma^2 \begin{pmatrix} (\mathbf{X}'\mathbf{X})^{-1} + \mathbf{LL}'m, & -\mathbf{L}m \\ -\mathbf{L}'m, & m \end{pmatrix}$$

where $\mathbf{L} = (\mathbf{X}'\mathbf{X})^{-1}\mathbf{X}'\mathbf{x}_k$ and $m = (\mathbf{x}_k'\mathbf{R}\mathbf{x}_k)^{-1}$.

Once we know $(\mathbf{X}'\mathbf{X})^{-1}$ it is straightforward (theoretically) to update $\hat{\boldsymbol{\beta}}$, RSS, and $(\mathbf{X}'\mathbf{X})^{-1}$ for the augmented model using the above equations. However, for numerical accuracy and stability the following technique is preferred.

Let

$$\mathbf{Q}'(\mathbf{X}, \mathbf{x}_k) = \begin{pmatrix} \mathbf{U} & \mathbf{Q}_p'\mathbf{x}_k \\ \mathbf{O} & \mathbf{Q}_{n-p}'\mathbf{x}_k \end{pmatrix} = \begin{pmatrix} \mathbf{U} & \mathbf{z}_k \\ \mathbf{O} & \mathbf{t}_k \end{pmatrix}, \tag{11.58}$$

say. Therefore, applying the p Householder transformations, represented by $\mathbf{Q}'$, to $\mathbf{x}_k$ we need only one further Householder transformation (Golub and Styan [1973: p. 264]) $\mathbf{H}$, say, of order $n-p$, to annihilate the last $n-p-1$ elements in $\mathbf{Q}'\mathbf{x}_k$ and reduce (11.58) to an upper triangular matrix of order $(p+1)$. Since orthogonal transformations do not change the length of a vector, $\mathbf{HQ}_{n-p}'\mathbf{x}_k = u\boldsymbol{\alpha}_1$, where $\boldsymbol{\alpha}_1' = (1, 0, \ldots, 0)$ and $u = (\mathbf{x}_k'\mathbf{Q}_{n-p}\mathbf{Q}_{n-p}'\mathbf{x}_k)^{1/2} = (\mathbf{t}_k'\mathbf{t}_k)^{1/2}$; and

$$\begin{pmatrix} \mathbf{I}_p & \mathbf{O} \\ \mathbf{O} & \mathbf{H} \end{pmatrix} \mathbf{Q}'(\mathbf{X}, \mathbf{x}_k : \mathbf{Y}) = \begin{pmatrix} \mathbf{U} & \mathbf{Q}_p'\mathbf{x}_k & : & \mathbf{z} \\ \mathbf{O} & u\boldsymbol{\alpha}_1 & : & \mathbf{Ht} \end{pmatrix}.$$

If h_1 is the first element of $\mathbf{H}\boldsymbol{t}$ then

$$\begin{pmatrix} \mathbf{U} & \mathbf{Q}_p'\mathbf{x}_k \\ \mathbf{O} & u \end{pmatrix} \begin{bmatrix} \hat{\boldsymbol{\beta}}_G \\ \hat{\beta}_{k,G} \end{bmatrix} = \begin{pmatrix} \mathbf{z} \\ h_1 \end{pmatrix}, \tag{11.59}$$

and we can solve for $\hat{\boldsymbol{\beta}}_G$ and $\hat{\beta}_{k,G}$ by backward substitution.

We note that

$$\begin{aligned}\text{RSS}_G &= \mathbf{t}'\mathbf{H}'\mathbf{H}\mathbf{t} - h_1^2 \\ &= \mathbf{t}'\mathbf{t} - h_1^2 \\ &= \text{RSS} - h_1^2, \end{aligned} \tag{11.60}$$

where h_1 is the first element of

$$\begin{aligned}\mathbf{H}\mathbf{t} &= (\mathbf{I}_{n-p} - 2\mathbf{v}\mathbf{v}')\mathbf{t} \\ &= \mathbf{t} - 2\mathbf{v}\mathbf{v}'\mathbf{t}.\end{aligned}$$

Here $\mathbf{v}$ is chosen such that $\mathbf{v}'\mathbf{v}=1$ and $\mathbf{H}\mathbf{t}_k = u\boldsymbol{\alpha}_1$, where $\mathbf{t}_k$ $[=(t_{ki})]$ is defined in (11.58). From, for example, Golub and Styan [1973: p. 255] we have

$$2v_1^2 = 1 - \frac{t_{k1}}{r}$$

and

$$2v_1v_i = -\frac{t_{ki}}{r} \qquad (i=2,3,\ldots,n-p),$$

where $r = -\text{sgn}(t_{k1})\cdot\|\mathbf{t}_k\|$, and $\text{sgn}(t_{k1}) = +1$ if $t_{k1} \geqslant 0$ and -1 otherwise. Therefore, substituting for v_1v_i,

$$\begin{aligned}h_1 &= t_1 - 2v_1\mathbf{v}'\mathbf{t} \\ &= t_1 - 2v_1^2t_1 - 2v_1v_2t_2 - \cdots - 2v_1v_{n-p}t_{n-p} \\ &= \frac{t_{k1}t_1 + t_{k2}t_2 + \cdots + t_{k,n-p}t_{n-p}}{r} \\ &= \frac{\mathbf{t}_k'\mathbf{t}}{r}\end{aligned}$$

and

$$\begin{aligned}h_1^2 &= \frac{(\mathbf{t}_k'\mathbf{t})^2}{\mathbf{t}_k'\mathbf{t}_k} \\ &= \frac{(\mathbf{t}_k'\mathbf{t})^2}{u^2}.\end{aligned} \tag{11.61}$$

It is constructive to relate the above theory to Equations (11.55), (11.56), and (11.57). For example, $u^2 = \mathbf{x}_k'\mathbf{Q}_{n-p}\mathbf{Q}_{n-p}'\mathbf{x}_k = \mathbf{x}_k'\mathbf{R}\mathbf{x}_k$ as $\text{RSS} = \mathbf{Y}'\mathbf{R}\mathbf{Y} = \mathbf{t}'\mathbf{t}$

$=\mathbf{Y}'\mathbf{Q}_{n-p}\mathbf{Q}'_{n-p}\mathbf{Y}$; from $u\hat{\beta}_{k,G}=h_1$ and Equation (11.55) we must have

$$h_1=\frac{\mathbf{x}_k'\mathbf{RY}}{u}.$$

Also from

$$\hat{\boldsymbol{\beta}}=\mathbf{U}^{-1}\mathbf{z}=\mathbf{U}^{-1}\mathbf{Q}_p'\mathbf{Y} \qquad \left[=(\mathbf{X}'\mathbf{X})^{-1}\mathbf{X}'\mathbf{Y}\right],$$

and (11.59), we have

$$\begin{aligned}\hat{\boldsymbol{\beta}}_G&=\mathbf{U}^{-1}\mathbf{z}-\mathbf{U}^{-1}\mathbf{Q}_p'\mathbf{x}_k\hat{\beta}_{k,G}\\&=\hat{\boldsymbol{\beta}}-(\mathbf{X}'\mathbf{X})^{-1}\mathbf{X}'\mathbf{x}_k\hat{\beta}_{k,G}.\end{aligned}$$

Finally, from (11.60)

$$\begin{aligned}\mathrm{RSS}_G&=\mathrm{RSS}-h_1^2\\&=\mathrm{RSS}-\frac{(\mathbf{x}_k'\mathbf{RY})^2}{u^2}\\&=\mathrm{RSS}-\mathbf{x}_k'\mathbf{RY}\hat{\beta}_{k,G}.\end{aligned}$$

11.9.2 Removing a Regressor

We now show how Givens transformations can be used for removing a specified regressor from the model; we ignore β_0 (or use $\tilde{\mathbf{X}}$) and assume $p=5$ to simplify the exposition.

To remove x_5 from the model represented by

$$\mathbf{A}=\begin{pmatrix}\mathbf{U} & \mathbf{z}\\ \mathbf{O} & \mathbf{t}\end{pmatrix}$$

we simply omit the last column of $\mathbf{U}$; the resulting increase in residual sum of squares is z_5^2. If we wish to omit x_3, then deleting the third column of $\mathbf{U}$ gives

$$\mathbf{U}^*=\begin{bmatrix}u_{11} & u_{12} & u_{14} & u_{15}\\ 0 & u_{22} & u_{24} & u_{25}\\ 0 & 0 & u_{34} & u_{35}\\ 0 & 0 & u_{44} & u_{45}\\ 0 & 0 & 0 & u_{55}\end{bmatrix}$$

We can now apply Givens transformations to the above matrix (augmented by $\mathbf{z}$ and $\mathbf{t}$) to annihilate the elements u_{44} and u_{55} below the diagonal and thus reduce $\mathbf{U}^*$ to an upper triangular matrix of size 4. This reduction can be accomplished by transforming the third and fourth rows, and the fourth and fifth rows of $\mathbf{A}$ so that $\mathbf{A}$ becomes

$$\begin{bmatrix} u_{11} & u_{12} & u_{13} & u_{14} & u_{15} & z_1 \\ 0 & u_{22} & u_{23} & u_{24} & u_{25} & z_2 \\ 0 & 0 & u'_{33} & u'_{34} & u'_{35} & z'_3 \\ 0 & 0 & u''_{43} & 0 & u''_{45} & z''_4 \\ 0 & 0 & u''_{53} & 0 & 0 & z''_5 \\ & & \mathbf{O} & & & \mathbf{t} \end{bmatrix}.$$

If we now ignore the third column, then $\mathbf{t}$ will "gain" an extra element z''_5 and the increase in the residual sum of squares through removing x_3 will be $(z''_5)^2$.

11.10 HYPOTHESIS TESTING

Suppose that $\mathbf{X}$ is $n\times p$ of rank p and we wish to test $H:\mathbf{A}\boldsymbol{\beta}=\mathbf{c}$, where $\mathbf{A}$ is $q\times p$ of rank q. Then our test statistic is

$$F=\frac{n-p}{q}\cdot\frac{\mathrm{RSS}_H-\mathrm{RSS}}{\mathrm{RSS}}$$

where

$$\mathrm{RSS}_H-\mathrm{RSS}=(\mathbf{A}\hat{\boldsymbol{\beta}}-\mathbf{c})'\left[\mathbf{A}(\mathbf{X}'\mathbf{X})^{-1}\mathbf{A}'\right]^{-1}(\mathbf{A}\hat{\boldsymbol{\beta}}-\mathbf{c}).$$

We now give a method, suggested by Golub and Styan [1973], for computing $\mathrm{RSS}_H-\mathrm{RSS}$. First,

$$\begin{aligned}\mathbf{A}(\mathbf{X}'\mathbf{X})^{-1}\mathbf{A}' &= \mathbf{A}(\mathbf{U}'\mathbf{U})^{-1}\mathbf{A}' \\ &= \mathbf{A}\mathbf{U}^{-1}(\mathbf{U}^{-1})'\mathbf{A}' \\ &= \mathbf{G}'\mathbf{G},\end{aligned}$$

say, where $\mathbf{G}=(\mathbf{U}^{-1})'\mathbf{A}'$ is a $q\times p$ matrix of rank q. Let

$$\mathbf{G}=\mathbf{Q}\begin{pmatrix}\mathbf{V}\\ \mathbf{O}\end{pmatrix} \qquad (11.62\ )$$

be an orthogonal decomposition of $\mathbf{G}$, where $\mathbf{Q}$ is the product of q Householder transformations and $\mathbf{V}$ is an upper triangular $q \times q$ matrix. Then $\mathbf{G}'\mathbf{G} = \mathbf{V}'\mathbf{V}$ and

$$\begin{aligned}\text{RSS}_H - \text{RSS} &= (\mathbf{A}\hat{\boldsymbol{\beta}} - \mathbf{c})'(\mathbf{V}'\mathbf{V})^{-1}(\mathbf{A}\hat{\boldsymbol{\beta}} - \mathbf{c}) \\ &= (\mathbf{A}\hat{\boldsymbol{\beta}} - \mathbf{c})'\mathbf{V}^{-1}(\mathbf{V}^{-1})'(\mathbf{A}\hat{\boldsymbol{\beta}} - \mathbf{c}) \\ &= \mathbf{h}'\mathbf{h},\end{aligned}$$

say, where $\mathbf{h} = (\mathbf{V}^{-1})'(\mathbf{A}\hat{\boldsymbol{\beta}} - \mathbf{c}) = (\mathbf{V}^{-1})'\mathbf{g}$, say. Although not specifically mentioned by Golub and Styan, we see that

$$\begin{aligned}\hat{\boldsymbol{\beta}}_H &= \hat{\boldsymbol{\beta}} + (\mathbf{X}'\mathbf{X})^{-1}\mathbf{A}'\left[\mathbf{A}(\mathbf{X}'\mathbf{X})\mathbf{A}'\right]^{-1}(\mathbf{c} - \mathbf{A}\hat{\boldsymbol{\beta}}) \\ &= \mathbf{U}^{-1}\mathbf{z} - (\mathbf{U}'\mathbf{U})^{-1}\mathbf{A}'(\mathbf{G}'\mathbf{G})^{-1}\mathbf{g} \\ &= \mathbf{U}^{-1}\left(\mathbf{z} - \mathbf{G}(\mathbf{G}'\mathbf{G})^{-1}\mathbf{g}\right).\end{aligned} \tag{11.63}$$

The above algebra suggests the following computational procedure:

(1) Compute $\mathbf{g} = \mathbf{A}\hat{\boldsymbol{\beta}} - \mathbf{c}$, where $\hat{\boldsymbol{\beta}}$ is the solution of $\mathbf{U}\hat{\boldsymbol{\beta}} = \mathbf{z}$.
(2) Compute $\mathbf{G}$ by solving $\mathbf{U}'\mathbf{G} = \mathbf{A}'$, with $\mathbf{U}'$ lower triangular.
(3) Reduce $\mathbf{G}$ to $\mathbf{V}$ by Householder transformations.
(4) Compute $\mathbf{h}$ by solving $\mathbf{V}'\mathbf{h} = \mathbf{g}$, with $\mathbf{V}'$ lower triangular.
(5) Compute

$$F = \frac{(n-p)\mathbf{h}'\mathbf{h}}{q\mathbf{t}'\mathbf{t}},$$

where $\mathbf{t}$ is generated at the same time as $\mathbf{U}$ and $\mathbf{z}$.
(6) If $\hat{\boldsymbol{\beta}}_H$ is required, let $\mathbf{Q} = (\mathbf{Q}_1, \mathbf{Q}_2)$ where $\mathbf{Q}_1$ is $p \times q$. Then $\mathbf{G} = \mathbf{Q}_1\mathbf{V}$ and [Equation (11.63)]

$$\begin{aligned}\mathbf{U}\hat{\boldsymbol{\beta}}_H &= \mathbf{z} - \mathbf{Q}_1\mathbf{V}(\mathbf{V}'\mathbf{V})^{-1}\mathbf{g} \\ &= \mathbf{z} - \mathbf{Q}_1(\mathbf{V}')^{-1}\mathbf{g} \\ &= \mathbf{z} - \mathbf{Q}_1\mathbf{h}\end{aligned}$$

can be solved for $\hat{\boldsymbol{\beta}}_H$.

If $\mathbf{c}=\mathbf{0}$ then

$$\begin{aligned}
\text{RSS}_H-\text{RSS}&=\hat{\boldsymbol{\beta}}'\mathbf{A}'(\mathbf{G}'\mathbf{G})^{-1}\mathbf{A}\hat{\boldsymbol{\beta}}\\
&=(\mathbf{U}^{-1}\mathbf{z})'\mathbf{A}'(\mathbf{G}'\mathbf{G})^{-1}\mathbf{A}(\mathbf{U}^{-1}\mathbf{z})\\
&=\mathbf{z}'\mathbf{G}(\mathbf{G}'\mathbf{G})^{-1}\mathbf{G}'\mathbf{z}\\
&=\mathbf{z}'\mathbf{Q}_1\mathbf{Q}_1'\mathbf{z}\\
&=\mathbf{z}_H'\mathbf{z}_H,
\end{aligned}$$

say. Because

$$\begin{pmatrix}\mathbf{Q}_1'\\ \mathbf{Q}_2'\end{pmatrix}(\mathbf{G}:\mathbf{z})=\begin{pmatrix}\mathbf{V} & \mathbf{z}_H\\ \mathbf{O} & \mathbf{t}_H\end{pmatrix},$$

RSS_H-RSS can be computed by applying the q Householder transformations of (11.62) to $\mathbf{z}$ simultaneously with $\mathbf{G}$, and then summing the squares of the first q elements of the transformed $\mathbf{z}$.

In conclusion we note that the above procedures can be carried out using Givens transformations. When the rank of $\mathbf{X}$ is less than p, Golub and Styan [1973] give two methods for deciding whether a hypothesis is testable or not. They also give a procedure, similar to the one above, for calculating the F-statistic for a testable hypothesis.

11.11 CHECKING OUT COMPUTER PROGRAMS

Longley [1967] gives a number of methods by which a user can check on the accuracy of a given least squares program. Briefly, some of these are: (1) compare means with those calculated by hand; (2) check that the sum of the residuals Σe_i is zero within rounding error; (3) run the program with small numbers in the x- and y-variables and rerun with constant amounts such as 100, 1000, 10,000 added to both sides to determine the limitations of the program; (4) compare the regressions for $\mathbf{Y}^{(1)}$, $\mathbf{Y}^{(2)}$, and $\mathbf{Y}^{(1)}+\mathbf{Y}^{(2)}$ to see if the results are additive; (5) use linear combinations of regressor variables which preserve linear independence (for example, replace x_1 and x_2 by $x_1'=x_1+x_2$ and $x_2'=x_1-x_2$ and, using the same $\mathbf{Y}$, verify that $\hat{\beta}_1=\hat{\beta}_1'+\hat{\beta}_2'$ and $\hat{\beta}_2=\hat{\beta}_1'-\hat{\beta}_2'$); (6) shuffle the columns of $\mathbf{X}$ and rerun the problem; and (7) check a few equations by hand. To this list we might also add, (8) check that $\mathbf{X}'\mathbf{e}=\mathbf{0}$, and (9) use double precision arithmetic (Freund [1963]).

The most obvious way to test the accuracy of a program is to run a set of data for which the regression coefficients and certain test statistics are known accurately; for example, Longley [1967] gives a set of economic data with a number of striking properties. Unfortunately suitable sets of test data are hard to come by so that it is generally more convenient to actually generate the data (cf. Wampler [1970: test data generated from polynomials], Hastings [1972: methods for constructing test data]). However, the accuracy of a particular program is further impaired when the data matrix is ill-conditioned, so that for a user to test a computer program via this method he would need to employ a data matrix that has similar ill-conditioning to the one he desires to run. It seems more appropriate, therefore, to check the accuracy of each problem computed rather than run test data. A useful procedure, given by Mullet and Murray [1971], for examining the effect of rounding error is the following procedure:

(1) Regress Y on $x_1, x_2, \ldots, x_{p-1}$.
(2) Regress $Y + ax_j$ $(a \neq 0)$ on the same set of $p-1$ regressors.
(3) Repeat step (2) with different values of a and different x_j as desired.

It is obvious that the least squares estimates of all the β_i $(i=0,1,\ldots,p-1)$ will be invariant in (*b*) except for $\hat{\beta}_j$, which is increased by a; the residual vector $\mathbf{e}$ and RSS are also invariant. Mullet and Murray applied their method to a set of data from Huang [1970] and obtained the results in Table 11.1. The estimates of β_0, β_2, and β_3 are consistent to four significant figures, and the estimates of β_1 and RSS are consistent to five significant figures. Even though the original data are not particularly ill-conditioned, they do point out the fact that computer printouts involving eight or more "significant" figures are not necessarily giving the apparent accuracy sought.

Table 11.1 The Method of Mullet and Murray [1971] for Checking the Accuracy of a Fitted Regression Applied to Data from Huang [1970]

Dependent Variable	$\hat{\beta}_0$	$\hat{\beta}_1$	$\hat{\beta}_2$	$\hat{\beta}_3$
Y	0.89734	0.67571	0.38889	0.36295
$Y-x_1$	0.89734	-0.32429	0.38890	0.36294
$Y-x_2$	0.89735	0.67571	-0.61109	0.36294
$Y-x_3$	0.89739	0.67571	0.38889	-0.63704

In concluding this section we formally prove the above invariant properties stated.

THEOREM 11.1 Let $Z_i = Y_i + ax_{ij}$, $\mathscr{E}[\mathbf{Z}] = \mathbf{X}\boldsymbol{\gamma}$, and $\hat{\boldsymbol{\gamma}} = (\mathbf{X}'\mathbf{X})^{-1}\mathbf{X}'\mathbf{Z}$. Then

(i) $\hat{\boldsymbol{\gamma}} = \hat{\boldsymbol{\beta}} + a\boldsymbol{\alpha}_j$, where $\boldsymbol{\alpha}_j$ is the unit vector with 1 in the $(j+1)$th position and zeros elsewhere.

(ii) $Z_i - \hat{Z}_i = Y_i - \hat{Y}_i$.

Proof. (i) Now

$$(\mathbf{X}'\mathbf{X})^{-1}\mathbf{X}'\mathbf{X} = \mathbf{I}_p$$

or

$$(\mathbf{X}'\mathbf{X})^{-1}\mathbf{X}'[\mathbf{x}_0, \mathbf{x}_1, \ldots, \mathbf{x}_{p-1}] = [\boldsymbol{\alpha}_0, \boldsymbol{\alpha}_1, \ldots, \boldsymbol{\alpha}_{p-1}].$$

Hence

$$(\mathbf{X}'\mathbf{X})^{-1}\mathbf{X}'\mathbf{x}_j = \boldsymbol{\alpha}_j$$

and

$$\begin{aligned}\hat{\boldsymbol{\gamma}} &= (\mathbf{X}'\mathbf{X})^{-1}\mathbf{X}'(\mathbf{Y} + a\mathbf{x}_j)\\ &= (\mathbf{X}'\mathbf{X})^{-1}\mathbf{X}'\mathbf{Y} + a(\mathbf{X}'\mathbf{X})^{-1}\mathbf{X}'\mathbf{x}_j\\ &= \hat{\boldsymbol{\beta}} + a\boldsymbol{\alpha}_j.\end{aligned}$$

(ii) By part (i)

$$\begin{aligned}Z_j - \hat{Z}_j &= Y_j + ax_{ij} - \hat{Z}_j\\ &= Y_j + ax_{ij} - \left(\hat{Y}_j + ax_{ij}\right)\\ &= Y_j - \hat{Y}_j.\end{aligned}$$

□

MISCELLANEOUS EXERCISES 11

1. Prove that the matrix **K** in Equation (11.3) is unit lower triangular.
2. Prove Equation (11.4).
   [*Hint*: From Exercise 1 show that for suitable **k**

$$\begin{pmatrix}\mathbf{K} & \mathbf{0}\\ \mathbf{k}' & 1\end{pmatrix}\begin{pmatrix}\mathbf{B} & \mathbf{c}\\ \mathbf{c}' & \mathbf{Y}'\mathbf{Y}\end{pmatrix} = \begin{pmatrix}\mathbf{V} & \mathbf{d}\\ \mathbf{0}' & f\end{pmatrix},$$

   where **K** is unit lower triangular.]

3. If $\mathbf{U}'\mathbf{U}$ is the Cholesky decomposition of $\mathbf{X}'\mathbf{X}$, show that

$$|\mathbf{X}'\mathbf{X}| = \prod_{i=1}^{p} u_{ii}^2.$$

4. Suppose that $\mathbf{T}'\mathbf{T}$ is the Cholesky decomposition for the augmented system

$$(\mathbf{X}:\mathbf{Y})'(\mathbf{X}:\mathbf{Y}) = \begin{pmatrix} \mathbf{X}'\mathbf{X} & \mathbf{X}'\mathbf{Y} \\ \mathbf{Y}'\mathbf{X} & \mathbf{Y}'\mathbf{Y} \end{pmatrix}$$

and let

$$\mathbf{T} = \begin{pmatrix} \mathbf{U} & \mathbf{z} \\ \mathbf{0}' & \delta \end{pmatrix},$$

where $\mathbf{U}$ is upper triangular.
(a) Show that $\hat{\boldsymbol{\beta}}$ is the solution of $\mathbf{U}\mathbf{x} = \mathbf{z}$.
(b) Prove that $\delta = \sqrt{\text{RSS}}$.
5. For the case $p = 3$ show how to solve (11.10) for $\mathbf{B}^{-1}$.
6. If the method of Equation (11.23) in Section 11.2.4 is applied to $(\mathbf{X}:\mathbf{Y})$ instead of $\mathbf{X}$, show that the extra element of $\mathbf{D}_1$ is RSS.
7. If we use $(\mathbf{X}:\mathbf{Y})$ instead of $\mathbf{X}$ in Section 11.3, show that the extra element of $\mathbf{D}_2$ is the correct residual sum of squares for the weighted least squares model.
8. Find $\hat{\boldsymbol{\beta}}$ and RSS for the model $\mathbf{Y} = \mathbf{X}\boldsymbol{\beta} + \boldsymbol{\varepsilon}$, where

$$(\mathbf{X}:\mathbf{Y}) = \begin{bmatrix} 1 & 2:1 \\ 1 & 1:2 \\ 1 & 0:3 \\ 1 & 1:4 \end{bmatrix}$$

using
(z) Householder transformations, and
(b) the formulas $\hat{\boldsymbol{\beta}} = (\mathbf{X}'\mathbf{X})^{-1}\mathbf{X}'\mathbf{Y}$ and $\text{RSS} = \|\mathbf{Y} - \mathbf{X}\hat{\boldsymbol{\beta}}\|^2$.
9. Calculate the sample variance of the numbers 3001, 3002, and 3003 using the two formulas
(a) $S^2 = \dfrac{1}{n}\sum_i (x_i - \bar{x})^2$
(b) $S^2 = \dfrac{1}{n}\sum_i x_i^2 - \bar{x}^2$.

What answer would you get using (b) if the calculations were carried out on a computer which worked to seven significant figures only?

10. Given

$$\left(\begin{array}{c|c} \mathbf{U} & \mathbf{z} \\ \hline \mathbf{0}' & \delta \end{array}\right) = \left[\begin{array}{cc|c} 1 & 1 & 1 \\ 0 & 1 & 2 \\ \hline 0 & 0 & 1 \end{array}\right],$$

use Givens transformations to add a further row of data

$$[1 \quad 2:1].$$

11. Let $\mathbf{a}$ be a $p \times 1$ vector and consider the partition

$$\mathbf{a}'\mathbf{\Pi} = (\mathbf{a}_1', \mathbf{a}_2')$$

where $\mathbf{a}_1'$ is $1 \times r$, and $\mathbf{\Pi}$ is the (orthogonal) permutation matrix defined in 11.5.3. Prove that $\mathbf{a}'\boldsymbol{\beta}$ is estimable if and only if

$$\mathbf{a}_1'\mathbf{U}_{11}^{-1}\mathbf{U}_{12} = \mathbf{a}_2'.$$

[*Hint*: prove that $\mathbf{a}'\boldsymbol{\beta}$ is estimable if and only if $\mathbf{a}'\mathbf{X}^*\mathbf{X} = \mathbf{a}'$. The above equation provides a computational method of testing for estimability.]

Golub and Styan [1973: p. 269]

12. Show that the following computing equations can be used instead of the set (11.33) in Section 11.3.4: equations for d', $\bar{s}$, and v_m' are unchanged, the equation for $\bar{c}$ is deleted, $\delta' = d\delta/d'$, and $\tilde{u}_m' = \tilde{u}_m + \bar{s}v_m'$. (This alternative expression for $\tilde{u}_m'$, suggested by Golub, may lead to numerical instability—Gentleman [1973: p. 332]).
13. Let $S_n = \Sigma_{i=1}^n x_i$, $m_n = S_n/n$ and $v_n = \Sigma_{i=1}^n (x_i - m_n)^2$. Obtain the following:
    (a) $m_n = \{(n-1)m_{n-1} + x_n\}/n$,
    $v_n = v_{n-1} + (n-1)(x_n - m_{n-1})^2/n$; and
    (b) $S_n = S_{n-1} + x_n$,
    $v_n = v_{n-1} + (nx_n - S_n)^2/[n(n-1)]$,
    where $m_0 = v_0 = 0$.
    Generalize (b) to give a method for calculating $\Sigma_{i=1}^n (x_i - \bar{x})(y_i - \bar{y})$.

Youngs and Cramer [1971]

14. Verify (11.38).
    [*Hint*: show that $\mathbf{X}^+$ satisfies the four conditions of Section 3.8.1c]

CHAPTER 12

Choosing the "Best" Regression

12.1 INTRODUCTION

A major problem in regression analysis is that of deciding which regressor or predictor variables should be in the model. Suppose that $x_1, x_2, \ldots, x_K$ is the complete set of all possible regressors including any functions such as squares, cross-products, etc., which may seen appropriate. Then there are two conflicting criteria for selecting a subset of regressors. First, the model chosen should include as many of the x's as possible if reliable predictions are to be obtained from the fitted equation. Second, because of the costs involved in obtaining information on a large number of regressors, we would like the equation to include as few x's as possible; also the variance of the predictor increases with the number of regressors (Section 5.4). A suitable compromise between these two extremes is usually called "selecting the best subset" or "selecting the best regression equation." However, the term "best" is subjective; there is no unique statistical procedure for choosing the subset and personal judgment is needed in all the statistical methods described in this chapter. For instance, if two regressors are highly correlated with Y and highly correlated with each other, then it is often sufficient to include just one of the regressors in the regression; once one regressor is in, the *additional* contribution of the other is frequently negligible. The choice of which regressor to include may depend, for example, on which variable is easier or cheaper to measure.

12.2 GENERATING ALL POSSIBLE REGRESSIONS

If we assume, for ease of exposition, that β_0 is always included in the model (though this is not necessary), an obvious approach to this "best subset" problem is to fit every possible regression equation that can be obtained by selecting $0,1,2,\ldots,K$ of the regressors $x_1,x_2,\ldots,x_K$; since there are two possibilities for each regressor, "in" or "out" of the equation, there are 2^K such regressions. For K large (e.g., $2^{10}=1024$), we are faced with comparing a large number of equations so that we need, first, an efficient algorithm for generating all the possibilities and, second, a readily computed measure for comparing the predictive usefulness of the different models. The algorithms described in Section 12.2.2 are generally satisfactory for up to 10, or even possibly 15, regressors.

12.2.1 Order of Generation

A systematic procedure for generating all possible regressions has been given by Garside [1965, 1971] and Schatzoff et al. [1968]. For simplicity of exposition Garside represents each regression by a K-digit binary number; for example, if $K=4$ the binary code 1010 would represent the model $E[Y]=\beta_0+\beta_1x_1+\beta_3x_3$. Since we usually proceed from one model to another by adding or deleting just one regressor at a time, we require a stepwise procedure which, starting with $00\cdots0$ (that is, $E[Y]=\beta_0$) will generate efficiently all 2^K-1 K-digit binary numbers in such a way that at any step only one of the digits is altered. An efficient procedure is one that does not generate the same regression model more than once; for example, when $K=3$ we have the procedure 000–100–110–010–011–111–101–001. Because each binary number can be regarded as the coordinates of a vertex of a K-dimensional hypercube, finding an efficient procedure is equivalent to finding a path along the edges of the hypercube which will pass through each vertex only once (called a Hamiltonian walk). Obviously this path is not unique; for the case $K=3$ we can start with 100, 010, or 001, or alternatively, we can choose a particular path and relabel the regressors. However, the generation of a suitable path can be easily described: using + and − signs to represent adding or removing regressors, we have the following sequences:

K	Sequence (S_K)
1	$+1$
2	$+1\ \ +2\ \ -1$
3	$+1\ \ +2\ \ -1\ \ +3\ \ +1\ \ -2\ \ -1$
4	$\underbrace{+1\ \ +2\ \ -1\ \ +3\ \ +1\ \ -2\ \ -1}_{S_3}\ \ +4\ \ \underbrace{+1\ \ +2\ \ -1\ \ -3\ \ +1\ \ -2\ \ -1}_{T_3}$
t	$S_{t-1},\qquad +t,\qquad T_{t-1}$

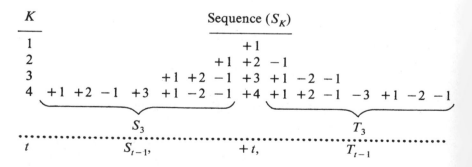

Here $S_K=(S_{K-1},K,T_{K-1})$ ($K=2,3,\ldots,t$; $S_1=1$) where T_K is the sequence S_K, but in reverse order with the sign changed; for $K=4$ see Table 12.4 in Section 12.2.3c.

However, a variety of other orderings are available. For example, Furnival [1971], who concentrates on efficiently generating the residual sum of squares for each regression subset (the RSS being the main statistic used for comparing different regressions, cf. Section 12.2.3), uses the binary order ($K=3$)

001–010–011–100–101–110–111

or

(1)–(2)–(12)–(3)–(13)–(23)–(123).

This is the order used, for example, in defining main effects and interactions for 2^n factorial designs. Furnival and Wilson [1974] also give several other orderings, which they call natural, lexicographic, and familial, that are appropriate to their algorithm.

12.2.2 *Method of Generation*

a. SWEEPING

An $m \times m$ matrix $\mathbf{A}=[(a_{ij})]$ is said to have been swept on the kth row and column (or on the kth pivotal element) when it has been transformed into a matrix $\mathbf{A}^*=[(a_{ij}^*)]$ such that (Beaton [1964])

$$a_{kk}^* = \frac{1}{a_{kk}},$$

$$a_{ik}^* = \frac{-a_{ik}}{a_{kk}} \qquad (i \neq k),$$

$$a_{kj}^* = \frac{a_{kj}}{a_{kk}} \qquad (j \neq k),$$

$$a_{ij}^* = a_{ij} - \frac{a_{ik}a_{kj}}{a_{kk}} \qquad (i,j \neq k).$$

For example, when $k=1$ we have

$$\mathbf{A}^* = \left[\begin{array}{c|c} \dfrac{1}{a_{11}} & \dfrac{a_{12}}{a_{11}}, \ldots, \dfrac{a_{1m}}{a_{11}} \\ \hline \dfrac{-a_{21}}{a_{11}} & \\ \cdots & a_{ij} - \dfrac{a_{i1}a_{1j}}{a_{11}} \\ \dfrac{-a_{m1}}{a_{11}} & \end{array}\right].$$

Schatzoff et al. [1968] point out that the sweep operator possesses the following useful properties:

(1) Sweep is reversible so that sweeping a matrix twice on the same pivotal element is equivalent to not having swept the matrix at all.
(2) Sweep is commutative so that sweeping on the kth and then the lth pivotal element is equivalent to sweeping the matrix in the opposite order.

Using the notation of Section 11.7.1 let $\tilde{\mathbf{X}}$ be the (centered) data matrix for all K regressors. If the augmented matrix

$$\mathbf{A} = \begin{pmatrix} \tilde{\mathbf{X}}'\tilde{\mathbf{X}} & \tilde{\mathbf{X}}'\tilde{\mathbf{Y}} \\ \tilde{\mathbf{Y}}'\tilde{\mathbf{X}} & \tilde{\mathbf{Y}}'\tilde{\mathbf{Y}} \end{pmatrix} \begin{matrix} \}K \\ \}1 \end{matrix} \tag{12.1}$$

is swept on the first K pivotal elements then, provided $(\tilde{\mathbf{X}}'\tilde{\mathbf{X}})^{-1}$ exists, the result is

$$\mathbf{B} = \begin{bmatrix} (\tilde{\mathbf{X}}'\tilde{\mathbf{X}})^{-1} & (\tilde{\mathbf{X}}'\tilde{\mathbf{X}})^{-1}\tilde{\mathbf{X}}'\tilde{\mathbf{Y}} \\ -\tilde{\mathbf{Y}}'\tilde{\mathbf{X}}(\tilde{\mathbf{X}}'\tilde{\mathbf{X}})^{-1} & \tilde{\mathbf{Y}}'\tilde{\mathbf{Y}} - \tilde{\mathbf{Y}}'\tilde{\mathbf{X}}(\tilde{\mathbf{X}}'\tilde{\mathbf{X}})^{-1}\tilde{\mathbf{X}}'\tilde{\mathbf{Y}} \end{bmatrix}$$

$$= \begin{pmatrix} (\tilde{\mathbf{X}}'\tilde{\mathbf{X}})^{-1} & \mathbf{b} \\ -\mathbf{b}' & \text{RSS} \end{pmatrix}$$

where $\mathbf{b}' = (\hat{\beta}_1, \hat{\beta}_2, \ldots, \hat{\beta}_K)$. In general, sweeping $\mathbf{A}$ on any subset of the first K pivotal elements will provide $(\tilde{\mathbf{X}}'\tilde{\mathbf{X}})^{-1}$, $\mathbf{b}$, and RSS corresponding to the regression of $\tilde{Y}$ on that subset of the $\tilde{x}$'s (or equivalently, the regression of Y on that subset of the x's, allowing for a constant term β_0). For example, if the first i pivots are swept ($i \leqslant K$) then we extract the leading $i \times i$ matrix, the first i elements of the last column, and the bottom diagonal element. Also, by the reversibility and commutativity properties of sweep, each application of sweep to a particular pivot of $\mathbf{A}$ either introduces the regressor corresponding to that pivot into the regression model, or removes it if it is already in the model. In this case we can ignore the signs in describing the Hamiltonian walk of the preceding section; thus $S_K = (S_{K-1}, K, S_{K-1})$ $(S_1 = 1)$.

The sweep method can also be applied to

$$\mathbf{A}_1 = \begin{pmatrix} \mathbf{R}_{xx} & \mathbf{r}_{xY} \\ \mathbf{r}'_{xY} & 1 \end{pmatrix} \tag{12.2}$$

where $\mathbf{R}_{xx}$ is the correlation matrix for the K regressors, and $\mathbf{r}_{xY}$ is the correlation vector of the K regressors with Y (see Section 11.7.2 with $p = K+1$). In this case the corresponding $\mathbf{B}$ matrix is

$$\mathbf{B}_1 = \begin{pmatrix} \mathbf{R}_{xx}^{-1} & \mathbf{a} \\ -\mathbf{a}' & 1-R^2 \end{pmatrix}$$

where $\mathbf{a} = \mathbf{R}_{xx}^{-1}\mathbf{r}_{xY}$ and R is the usual multiple correlation coefficient. Once again sweeping $\mathbf{A}_1$ on any subset of the first K pivotal elements produces the appropriate $\mathbf{R}_{xx}^{-1}$, $\mathbf{a}$ and $1-R^2$ corresponding to the regression of $\tilde{Y}$ on that subset. In particular, suppose that the first i pivotal elements of $\mathbf{A}_1$ ($i < K$) are swept, then the $(i+1) \times (i+1)$ leading matrix is of the same form as $\mathbf{B}_1$ except that the role of Y is taken over by x_{i+1}. This means that the $(i+1)$th pivotal element of this leading matrix is $1-R^2$ for the regression of $\tilde{x}_{i+1}$ on $\tilde{x}_1, \tilde{x}_2, \ldots, \tilde{x}_i$, that is, $1 - R^2_{i+1:1,2,\ldots,i}$, say. Similarly, if we consider the $i \times i$ leading matrix augmented by the jth row and column ($i < j \leqslant K$) then, using the same argument, the jth pivotal element becomes $1 - R^2_{j:1,2,\ldots,i}$. Therefore if we wish to bring in a variable x_j which is highly correlated with a linear combination of the i regressors already in the model, then $R^2_{j:1,2,\ldots,i}$ will be near 1 and the jth pivotal element will be small. Since we have to invert this pivotal element we will run into computational difficulties if this element is too small. It is therefore recommended that x_j should not be introduced if the magnitude of the jth pivotal element falls below a certain tolerance.

The sweep method seems to have been first applied to stepwise regression by Efroymson [1960], though in a slightly different format from that given above. Garside [1965] suggested using it in conjunction with his algorithm for generating a Hamiltonian walk. Since, apart from signs, the matrix $\mathbf{A}^*$ is symmetric, we need only work with the upper triangular matrix thereby halving the computations and the storage (Breaux [1968], Schatzoff et al. [1968]); a sweep method that preserves symmetry is described below. Several algorithms for reducing the number of sweeps are given by Schatzoff et al. [1968], Furnival [1971] and Morgan and Tatar [1972]. Schatzoff et al. make use of the fact that not all the matrix **A** needs to be swept each time; sweep is applied to a minimal submatrix only. Furnival reduces the computation still further by storing K additional submatrices in such a way that no pivotal element in a submatrix is swept more than once; as noted in Section 12.2.1 his algorithm uses a binary ordering rather than a Hamiltonian walk to determine the regression sequence. Morgan and Tatar use a method similar in spirit to that given by Schatzoff et al. except that the sweep is modified so that only the residual sum of squares is computed at each step (and not the regression coefficients), and the symmetry of **A** is maintained. The basic operator used in this "symmetric" sweep method is described below.

A different approach to the problem is taken by Newton and Spurrel [1967 a, b] who introduce quantities called *elements* to summarize the set of all 2^K regression sums of squares ($\mathbf{Y}'\mathbf{Y}-\text{RSS}$).

b. SYMMETRIC SWEEP

The sweep method is basically an adaptation of Gauss–Jordan elimination for inverting a matrix in its own space. However, because $\tilde{\mathbf{X}}'\tilde{\mathbf{X}}$ and $\mathbf{R}_{xx}$ are symmetric, it is really necessary to work with only the upper triangle. The symmetry of the matrix **A** can be preserved by changing the sign of the pivot (Stiefel [1963: p. 65], Beale et al. [1967: p. 359], Garside [1971], and Beale [1974]) so that for introducing x_k we have

$$a_{kk}^* = \frac{-1}{a_{kk}},$$

$$a_{ik}^* = a_{ki}^* = a_{ik}a_{kk}^* \qquad (i \neq k)$$

$$a_{ij}^* = a_{ji}^* = a_{ij} + a_{ik}a_{kj}^* \qquad (i,j \neq k),$$

and for removing x_k we have

$$a_{kk}^* = \frac{-1}{a_{kk}}$$

$$a_{ik}^* = a_{ki}^* = -a_{ik}a_{kk}^*$$

$$a_{ij}^* = a_{ji}^* = a_{ij} - a_{ik}a_{kj}^* \qquad (i,j \neq k).$$

If we use this algorithm on the K pivotal elements of $\mathbf{A}_1$ in Equation (12.2) (that is, introduce all K regressors), and work with the upper triangle of $\mathbf{A}_1$ only, we obtain the upper triangular matrix

$$\begin{bmatrix} c_{11} & c_{12} & \cdots & c_{1K} & -a_1 \\ * & c_{22} & \cdots & c_{2K} & -a_2 \\ \cdots & \cdots & \cdots & \cdots & \cdots \\ * & * & \cdots & c_{KK} & -a_K \\ * & * & \cdots & * & 1-R^2 \end{bmatrix},$$

where $\mathbf{C}=[(c_{ij})]=-\mathbf{R}_{xx}^{-1}$ and $\mathbf{a}$ is our usual vector of scaled regression coefficients (Section 11.7.2).

Another method of symmetric sweeping is described by Morgan and Tatar [1972]. They first define a $(K+1)\times 1$ vector $\mathbf{t}=[(t_i)]$ whose elements are initially set equal to $+1$, and also define a cubic matrix $\boldsymbol{\tau}=[(\tau_{ijk})]$ of dimension $(K+1)\times(K+1)\times(K+1)$. The elements of $\boldsymbol{\tau}$ are calculated according to the following rule: $\tau_{ijk}=-1$ if $t_i=t_j=t_k$ and is $+1$ otherwise. (This matrix is not stored explicitly, but rather the rule for computing its elements is stored). Now if x_k is the pivotal variable, that is, x_k is being added to or removed from the regression, we first change the sign of t_k and then perform the following calculations:

$$a_{kk}^* = \frac{1}{a_{kk}},$$

$$a_{ik}^* = a_{ki}^* = t_i a_{ik} a_{kk}^* \qquad (i \neq k),$$

$$a_{ij}^* = a_{ij} - \tau_{ijk} a_{ik} a_{kj} a_{kk}^* \qquad (i,j \neq k).$$

Once again, because of symmetry, we need to work with only the upper triangular portion of $\mathbf{A}$. The vector $\mathbf{t}$ gives us a record of the status of each regressor, whether it is in or out of the regression. A negative value of t_i indicates that x_i is in the regression.

In conclusion we note that both (symmetric) sweep operators described in this section have the same properties of reversibility as the ordinary (nonsymmetric) sweep method of the preceding section.

c. FURNIVAL'S METHOD

A Gaussian elimination method given by Furnival [1971] and Furnival and Wilson [1974] is best described in terms of a "regression tree" (Fig. 12.1). The Gaussian elimination operator is applied to each pivotal element just once in the order given by the binary tree. At the root of the tree (top of

Fig. 12.1) is the full matrix given by Equation (12.1), and at each interior node a submatrix is derived from the parent matrix by a series of pivots (solid lines) and deletions (dotted lines). Thus beginning with the root, the matrix **A** is "split" into two new submatrices; one obtained by pivoting on the first regressor, the other by deleting the row and column associated with that variable. The process is repeated for the submatrices until all the variables have been treated either by pivoting or by deletion; finally each terminal node represents one of the 2^K regressions including the "null" regression $(E[Y]=\beta_0)$. This procedure is readily described by a dot notation similar to that employed for partial correlation coefficients. The integers listed before the dot are the subscripts of the regressors in the submatrix on which pivots have not yet been performed, and the subscripts following the dot correspond to regressors on which pivots have been performed. For example, the submatrix 3.1 has been obtained from **A** by pivoting on x_1 and deleting x_2.

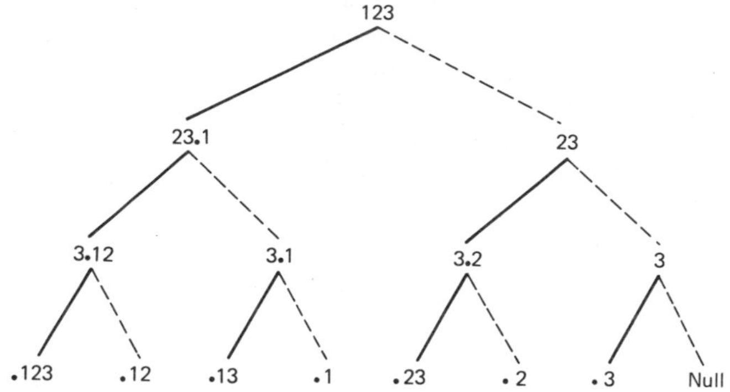

Fig. 12.1 The regression tree.

Furnival's procedure briefly consists of applying Gaussian elimination to the upper half, and to just certain rows and columns, of a submatrix. At the conclusion of a pivot (Gaussian elimination) the bottom diagonal element gives the appropriate RSS.

The regression tree of Fig. 12.1 can be traversed in any "biologically feasible" order; the only restraint is that a father be "born" before his son. By searching the tree using a horizontal, vertical, or hybrid searching technique Furnival obtains a number of regression sequences which he describes as natural, lexicographic, binary, and familial. The reader is referred to his papers for further details.

d. HOUSEHOLDER AND GIVENS TRANSFORMATIONS

The methods given so far require the computation of $\tilde{\mathbf{X}}'\tilde{\mathbf{X}}$ or $\mathbf{R}_{xx}$. However, the formulation of these matrices should generally be avoided because of possible ill-conditioning, and we should work directly with $\tilde{\mathbf{X}}$ (or $\mathbf{W}$, where $\mathbf{W}'\mathbf{W}=\mathbf{R}_{xx}$). This can be done using the methods of Section 11.9; a regressor is brought into the equation using a Householder transformation, and it is eliminated using a Givens transformation. To clarify the procedure consider the following sequence of models: (x_1), (x_1,x_2), (x_2), (x_2,x_3).

Now to bring in x_1 we use a Householder transformation:

$$\mathbf{Q}_1'(\tilde{\mathbf{X}}:\tilde{\mathbf{Y}})=\begin{bmatrix} u_{11}^{(1)} & u_{12}^{(1)} & \cdots & u_{1K}^{(1)} & z_1^{(1)} \\ 0 & u_{22}^{(1)} & \cdots & u_{2K}^{(1)} & t_2^{(1)} \\ \vdots & \vdots & \vdots & \vdots & \vdots \\ 0 & u_{n2}^{(1)} & \cdots & u_{nK}^{(1)} & t_n^{(1)} \end{bmatrix}$$

so that for this model $u_{11}^{(1)}b_1^{(1)}=z_1^{(1)}$ and

$$\begin{aligned} \text{RSS} &= \left(t_2^{(1)}\right)^2+\cdots+\left(t_n^{(1)}\right)^2 \\ &= \tilde{\mathbf{Y}}'\tilde{\mathbf{Y}}-\left(z_1^{(1)}\right)^2. \end{aligned}$$

Again, using a Householder transformation, we can bring in x_2; thus

$$\mathbf{Q}_2'\mathbf{Q}_1'(\tilde{\mathbf{X}}:\tilde{\mathbf{Y}})=\begin{bmatrix} u_{11}^{(2)} & u_{12}^{(2)} & u_{13}^{(2)} & \cdots & u_{1K}^{(2)} & z_1^{(2)} \\ 0 & u_{22}^{(2)} & u_{23}^{(2)} & \cdots & u_{2K}^{(2)} & z_2^{(2)} \\ 0 & 0 & u_{33}^{(2)} & \cdots & u_{3K}^{(2)} & t_3^{(2)} \\ \vdots & \vdots & \vdots & \vdots & \vdots & \vdots \\ 0 & 0 & u_{n3}^{(2)} & \cdots & u_{nK}^{(2)} & t_n^{(2)} \end{bmatrix}$$

(Some of the elements are unchanged, for example, $u_{1j}^{(1)}=u_{1j}^{(2)}$ for $j=1,2,\ldots,K$.) For this model

$$\begin{aligned} u_{11}^{(2)}b_1^{(2)}+u_{12}^{(2)}b_2^{(2)} &= z_1^{(2)} \\ u_{22}^{(2)}b_2^{(2)} &= z_2^{(2)}, \end{aligned}$$

and

$$\text{RSS}=\tilde{\mathbf{Y}}'\tilde{\mathbf{Y}}-\left(z_1^{(2)}\right)^2-\left(z_2^{(2)}\right)^2.$$

To remove x_1 we use a Givens transformation to transform the first and second rows and annihilate $u_{22}^{(2)}$, namely,

$$\mathbf{Q}_3'\mathbf{Q}_2'\mathbf{Q}_1'(\mathbf{X}:\mathbf{Y})=\begin{bmatrix} u_{11}^{(3)} & u_{12}^{(3)} & u_{13}^{(3)} & \cdots & u_{1K}^{(3)} & z_1^{(3)} \\ u_{21}^{(3)} & 0 & u_{23}^{(3)} & \cdots & u_{2K}^{(3)} & t_2^{(3)} \\ 0 & 0 & u_{33}^{(3)} & \cdots & u_{3K}^{(3)} & t_3^{(3)} \\ \vdots & \vdots & \vdots & \vdots & \vdots & \vdots \\ 0 & 0 & u_{n3}^{(3)} & \cdots & u_{nK}^{(3)} & t_n^{(3)} \end{bmatrix}.$$

For this model $u_{12}^{(3)}b_2^{(3)}=z_1^{(3)}$ and (ignoring the first column of the above matrix)

$$\text{RSS}=\tilde{\mathbf{Y}}'\tilde{\mathbf{Y}}-\left(z_1^{(3)}\right)^2.$$

Finally, adding x_3 using a Householder transformation we have

$$\begin{bmatrix} u_{11}^{(4)} & u_{12}^{(4)} & u_{13}^{(4)} & u_{14}^{(4)} & \cdots & u_{1K}^{(4)} & z_1^{(4)} \\ u_{21}^{(4)} & 0 & u_{23}^{(4)} & u_{24}^{(4)} & \cdots & u_{2K}^{(4)} & z_2^{(4)} \\ u_{31}^{(4)} & 0 & 0 & u_{34}^{(4)} & \cdots & u_{3K}^{(4)} & t_3^{(4)} \\ \vdots & \vdots & \vdots & \vdots & \vdots & \vdots & \vdots \\ u_{n1}^{(4)} & 0 & 0 & u_{n4}^{(4)} & \cdots & u_{nK}^{(4)} & t_n^{(4)} \end{bmatrix}.$$

The regression coefficients for this model satisfy

$$\begin{aligned} u_{12}^{(4)}b_2^{(4)}+u_{13}^{(4)}b_3^{(4)}&=z_1^{(4)} \\ u_{23}^{(4)}b_3^{(4)}&=z_2^{(4)}, \end{aligned}$$

and

$$\text{RSS}=\tilde{\mathbf{Y}}'\tilde{\mathbf{Y}}-\left(z_1^{(4)}\right)^2-\left(z_2^{(4)}\right)^2.$$

In general the regression sum of squares when k regressors are in the model is the sum of squares of the first k transformed elements of $\tilde{\mathbf{Y}}$.

12.2.3 Comparing Different Equations

Having calculated all 2^K regression equations, we next choose those equations that are good predictors. We need a method for comparing not only equations with the same number of regressors, but also equations that bear little resemblance to one another both with regard to the number and the choice of regressors. Various comparative measures are in use and we now consider these in detail.

a. COEFFICIENT OF DETERMINATION (R^2)

One measure of goodness of fit of a regression model to data that has been widely used in the past is the coefficient of determination R^2 $[=\Sigma(\hat{Y}_i-\bar{Y})^2/\Sigma(Y_i-\bar{Y})^2]$, the square of the multiple correlation coefficient. Its use is demonstrated, for $K=4$, by Draper and Smith [1966: Chapter 6] using data from Hald [1952: p. 647]. Their procedure is as follows:

(1) Divide the regressions into five classes.
Class A consists of the model $E[Y]=\beta_0$.
Class B consists of the four one-variable models

$$E[Y]=\beta_0+\beta_i x_i.$$

Class C consists of all two-variable models

$$E[Y]=\beta_0+\beta_i x_i+\beta_j x_j.$$

Class D consists of all three-variable models.
Class E consists of the single four-variable model.

(2) Order the models in each class according to the magnitude of R^2.

(3) Examine the leading equations (that is, with the largest R^2) in each class and see if there is a consistent pattern in the way the regressors occur.

The results of the above procedure applied to Hald's data yield the leaders set out in Table 12.1. Draper and Smith argue that the increase in R^2 from class C to class D is small, so that once x_1 and x_2, or x_1 and x_4,

Table 12.1 Regression Subsets with Maximum R^2

Class	Subset	$100R^2$
B	(x_4)	67.5
	(x_2)	66.6
C	(x_1,x_2)	97.9
	(x_1,x_4)	97.2
D	(x_1,x_2,x_4)	98.234
E	(x_1,x_2,x_3,x_4)	98.237

SOURCE. Adapted from Draper and Smith [1966: p. 165].

are in the model there is little to be gained from introducing further regressors. Although (x_1, x_2) has a slightly higher R^2 than (x_1, x_4), it may be considered more appropriate to use the latter model because x_4 provides the best one-variable equation; however, there is little difference between the two models.

From the matrix of sample correlation coefficients

$$\begin{pmatrix} \mathbf{R}_{xx} & \mathbf{r}_{xY} \\ \mathbf{r}'_{xY} & 1 \end{pmatrix}$$

given in Table 12.2 we see that the high pairwise correlations of x_1 and x_2, and particularly x_2 and x_4, would account for the changes in R^2 in Table 12.1.

The above discussion raises the question of when is an R^2 satisfactory? For example, how do we decide whether to choose class C or D in Table 12.1? Perhaps both classes are "satisfactory." Aitkin [1974] has given one solution to this problem by developing a simultaneous test procedure whereby all subsets which have an R^2 not significantly different from R^2 for the full model (R^2_{K+1}, say) can be determined simultaneously. We now briefly outline his method.

Suppose we consider fitting a subset regression model $\mathcal{E}[\mathbf{Y}]=\mathbf{X}_s\boldsymbol{\beta}_s$ which is obtained from the full model $\mathbf{X}\boldsymbol{\beta}$ by setting *any* r elements ($r=1,2,\ldots,K$), except β_0, equal to zero, that is, $\boldsymbol{\beta}_r=\mathbf{0}$. Then assuming that $\mathbf{X}$ is $n\times(K+1)$ of rank $K+1$, and $\mathbf{X}_s$ is $n\times(K+1-r)$ of rank $s=K+1-r$, the F-statistic for testing $\boldsymbol{\beta}_r=\mathbf{0}$ satisfies (Theorem 4.3, Section 4.2)

$$\begin{aligned} rF &= \frac{R^2_{K+1}-R^2_s}{(1-R^2_{K+1})/(n-K-1)} \\ &= U(\mathbf{X}_s), \end{aligned}$$

say. A simultaneous level α test for all hypotheses $\boldsymbol{\beta}_r=\mathbf{0}$, for an arbitrary

Table 12.2 Matrix of Simple Correlation Coefficients for Hald's Data

	x_1	x_2	x_3	x_4	Y
x_1	1.0				
x_2	0.23	1.0			
x_3	−0.82	−0.14	1.0		
x_4	−0.24	−0.97	0.03	1.0	
Y	0.73	0.82	−0.53	−0.82	1.0

SOURCE. Gorman and Toman [1966].

selection $\boldsymbol{\beta}_r$, will be obtained by not rejecting when

$$U(\mathbf{X}_s) < C_{n,K}^{\alpha},$$

where $C_{n,K}^{\alpha}$ is the upper $100\alpha\%$ point of the null distribution of

$$U = \max_{\mathbf{X}_s} U(\mathbf{X}_s),$$

the maximum being taken over all (nonempty) selections $\mathbf{X}_s$. Now the maximum occurs when $r = K$ and $\mathbf{X}_s$ consists of the first column of $\mathbf{X}$ only; in this case R_s^2 is zero. Therefore

$$U = \frac{R_{K+1}^2}{(1 - R_{K+1}^2)/(n-K-1)}$$

and, when all hypotheses $\boldsymbol{\beta}_r = \mathbf{0}$ are *simultaneously* true, $\beta_1 = \beta_2 = \cdots = \beta_K = 0$ and U/K has an $F_{K,n-K-1}$ distribution (Section 4.2 with $p = K+1$). Hence

$$C_{n,K}^{\alpha} = KF_{K,n-K-1}^{\alpha},$$

and the simultaneous test does not reject $\boldsymbol{\beta}_r = \mathbf{0}$, for any arbitrary selection, if

$$\frac{R_{K+1}^2 - R_s^2}{(1 - R_{K+1}^2)/(n-K-1)} < KF_{K,n-K-1}^{\alpha},$$

that is, if

$$R_s^2 > R_0^2 = 1 - (1 - R_{K+1}^2)(1 + d_{n,K}^{\alpha}), \tag{12.3}$$

where

$$d_{n,K}^{\alpha} = \frac{KF_{K,n-K-1}^{\alpha}}{n-K-1}.$$

Aitkin calls any subset of regressor variables (represented by $\mathbf{X}_s$) satisfying (12.3) an *R^2-adequate* (α) *set*. Referring to Hald's data once again we have $n = 13$, $K = 4$, $R_{K+1}^2 = 0.982376$. If we take $\alpha = 0.05$ then

$$R_0^2 = 1 - 0.017624\left(1 + \frac{4F_{4,8}^{0.05}}{8}\right)$$

$$= 0.948538$$

and the subsets which are R^2 adequate are starred in Table 12.3. We see, therefore, that classes C and D in Table 12.1 are both R^2-adequate (0.05), so that we still have not solved our problem as to which class we should choose. However, at least we know that, in terms of the above criterion of "adequacy," they are comparable. The idea of adequacy is also mentioned briefly by Cox and Snell [1974: p. 35] when they refer to "primitive" subsets.

Table 12.3 Values of $100R^2$ for Different Regression Subsets

(1)	(2)	(3)	(4)	(1,2)	(1,3)	(1,4)
53.4	66.6	28.6	67.5	97.9*	54.8	97.2*
(2,3)	(2,4)	(3,4)	(1,2,3)	(1,2,4)	(1,3,4)	(2,3,4)
84.7	68.0	93.5	98.2*	98.2*	98.1*	97.3*

SOURCE. Aitkin [1974: p. 223].
*R^2-adequate (0.05) sets.

Listing the 2^K values of R^2 (including R^2_{K+1}) can be tedious if K is large. Aitkin points out that, in many cases, it will be sufficient to list the R^2 values for *minimal adequate* sets, that is, those R^2-adequate sets which have the property that no subset of the variables in $\mathbf{X}_s$ is also adequate. For example, in Table 12.3 the minimal adequate sets are (1,2), (1,4), and (2,3,4).

As with any simultaneous test procedure, the test for subhypotheses becomes increasingly conservative as the number of regressors in the retained set decreases. Aitkin mentions that the true size of any individual test for a subhypothesis may be determined by interpolation in F-tables or incomplete beta-function tables.

The above procedure also holds for the case of random regressors. By conditioning on the observed values of the regressors, we obtain a test statistic whose null distribution (the F-distribution) does not depend on $\mathbf{X}$. Hence the simultaneous test has size α *unconditionally* as well.

Another multiple comparison technique for comparing regressions has been given by Spjøtvoll [1972c].

b. ADJUSTED COEFFICIENT OF DETERMINATION

Since $1-R^2=\text{RSS}/\Sigma(Y_i-\overline{Y})^2$, maximizing R^2 is equivalent to minimizing the residual sum of squares RSS. In this sense R^2 could therefore be regarded as a measure of goodness of fit. However, as pointed out by Furnival (Barrett [1974]), if RSS is held constant, then with a steeper surface, $\Sigma(Y_i-\overline{Y})^2$ will increase and thus increase R^2. Hence in analyzing two *different* sets of data it is possible that one regression may have a smaller RSS and yet, at the same time, have a smaller R^2 because the particular regression surface is not as steep. However, in our situation we are using the same set of data for different regression models so that the term $\Sigma(Y_i - \bar{Y})^2$ is the same for each regression. This means we can use R^2 as a *relative* measure of goodness of fit, but not as an *absolute* measure.

Even when R^2 is used as a relative measure, there is some difficulty, as we saw in Table 12.3, in comparing regressions with different numbers of regressors. Since introducing an extra regressor increases R^2 (see the comments after Theorem 4.3), it is not just a matter of finding the subset with maximum R^2 (which in any case is the set of all K regressors) but rather that of finding a suitable subset with a high R^2.

To overcome some of the above difficulties a modification of R^2 has been suggested called the "adjusted" or "corrected" R^2 statistic (e.g., Ezekial [1930]) given by

$$\bar{R}_p^2 = 1 - \left[1 - R_p^2\right]\left[\frac{n}{n-p}\right], \tag{12.4}$$

where p is the number of parameters (that is, the number of regressors plus one for β_0) in the equation; we note that if p is large compared to n, then $\bar{R}_p^2$ can be negative. To see the effect on $\bar{R}^2$ of adding extra regressors to the equation, consider the F-statistic for testing the significance of q new additions, namely, (Theorem 4.3, Section 4.2)

$$F = \frac{R_{p+q}^2 - R_p^2}{1 - R_{p+q}^2} \cdot \frac{n-p-q}{q}.$$

Then, using the above equation, we have

$$\begin{aligned}
\bar{R}_{p+q}^2 &= 1 - \left[1 - R_{p+q}^2\right]\left[\frac{n}{n-p-q}\right] \\
&= 1 - \left[1 - R_p^2\right]\left[\frac{n}{qF - q + n - p}\right] \\
&= 1 - \left[1 - R_p^2\right]\left[\frac{n}{n-p}\right]\left[\frac{n-p}{qF - q + n - p}\right] \\
&\geqslant \bar{R}_p^2
\end{aligned}$$

if and only if $F \geqslant 1$. This means that $\bar{R}^2$ will increase with the addition of one or more regressors only if $F > 1$; similar results are proved by Haitovsky [1969] and Edwards [1969]. One criterion, therefore, for selecting the best regression is to choose the regression subset which maximizes $\bar{R}_p^2$ (Haitovsky [1969]). However, the $\bar{R}_p^2$ statistic is related to another well-known statistic that we now discuss.

c. MALLOWS' C_p STATISTIC

As in the above discussion we use the subscript p to denote that a p-parameter model is being considered (that is, a model with β_0 and $p-1$ other β's). Thus $\mathbf{X}_p$ denotes an $n \times p$ data matrix of rank p, and the fitted regression subset at the point $\mathbf{x}' = (1, x_1, x_2, \ldots, x_K)$ will be represented by

$$\begin{aligned}\hat{Y}_p &= \hat{\beta}_0 + \hat{\beta}_1 x_1 + \cdots + \hat{\beta}_{p-1} x_{p-1} \\ &= \mathbf{x}_p' \hat{\boldsymbol{\beta}}_p,\end{aligned}$$

say, where $\mathbf{x}_p' = (1, x_1, \ldots, x_{p-1})$. If $\eta_p = E[\hat{Y}_p]$, then η_p will generally differ from $\mathbf{x}_p'\boldsymbol{\beta}_p$ because of possible bias in the p-parameter model. Therefore, if we use $\hat{Y}_p$ to predict $E[Y]$ $(=\theta)$, where Y represents the (unknown) response at the data point $\mathbf{x}$, then the mean squared error of $\hat{Y}_p$ is

$$\begin{aligned}E\left[\left(\hat{Y}_p - \theta\right)^2\right] &= \operatorname{var}\left[\hat{Y}_p\right] + (\eta_p - \theta)^2 \\ &= \sigma^2 \mathbf{x}_p'(\mathbf{X}_p'\mathbf{X}_p)^{-1}\mathbf{x}_p + (\eta_p - \theta)^2. \qquad (12.5)\end{aligned}$$

This suggests that one criterion for choosing the best regression subset would be to find the subset which minimizes (12.5) for a given future data point $\mathbf{x}$ (e.g., Allen [1971a]). However, if we are interested in more than one future $\mathbf{x}$, it is quite likely that a different regression subset might be recommended for each $\mathbf{x}$. As pointed out by Hocking [1972] it is perhaps more appropriate to use the sum or the average, in some sense, over the future observations of interest. Since it is only safe to predict in the region defined by the initial experiment of n observations, a number of authors recommend some sort of summing or averaging over the rows of $\mathbf{X}$, where $\mathbf{X}$ is the data matrix for the full $K+1$ parameter model. For example, if we define

$$\hat{Y}_{pi} = \hat{\beta}_0 + \hat{\beta}_1 x_{i1} + \cdots + \hat{\beta}_{p-1} x_{i,p-1} \qquad (i = 1, 2, \ldots, n),$$

then a suggested criterion (Mallows [1964, 1966, 1973], Gorman and Toman [1966]) is to minimize the scaled sum of squares

$$\begin{aligned}\Delta_p &= \frac{1}{\sigma^2} E\left[\sum_{i=1}^{n} \left(\hat{Y}_{pi} - \theta_i\right)^2\right] \\ &= \frac{1}{\sigma^2}\left[\sum_{i=1}^{n} \operatorname{var}\left[\hat{Y}_{pi}\right] + \sum_{i=1}^{n} (\eta_{pi} - \theta_i)^2\right]. \qquad (12.6)\end{aligned}$$

If $\hat{\mathbf{Y}}_p = [(\hat{Y}_{pi})]$ and $\mathbf{P}_p = \mathbf{X}_p(\mathbf{X}_p'\mathbf{X}_p)^{-1}\mathbf{X}_p'$, then

$$\begin{aligned}\sum_{i=1}^{n} \text{var}\left[\hat{Y}_{pi}\right] &= \text{tr}\left[\mathcal{D}\left[\hat{\mathbf{Y}}_p\right]\right] \\ &= \text{tr}\left[\mathcal{D}\left[\mathbf{P}_p\mathbf{Y}\right]\right] \\ &= \sigma^2 \,\text{tr}\,\mathbf{P}_p \\ &= \sigma^2 p \qquad \text{(by Theorem 3.1)}\end{aligned} \tag{12.7}$$

and, substituting in (12.6),

$$\Delta_p = p + \frac{\text{SSB}_p}{\sigma^2}. \tag{12.8}$$

Here SSB_p, the "bias" sum of squares, is given by

$$\begin{aligned}\text{SSB}_p &= \sum_{i=1}^{n} (\eta_{pi} - \theta_i)^2 \\ &= (\boldsymbol{\eta}_p - \boldsymbol{\theta})'(\boldsymbol{\eta}_p - \boldsymbol{\theta}),\end{aligned}$$

where $\boldsymbol{\eta}_p = \mathcal{E}[\hat{\mathbf{Y}}_p]$.

What we require now is an unbiased estimate of Δ_p, say, $\hat{\Delta}_p$, so that we can select those regression subsets with small values of $\hat{\Delta}_p$. In addition, when the bias term SSB_p is negligible in (12.8), $\Delta_p \approx p$ so that a plot of $\hat{\Delta}_p$ against p will indicate which regression models have small bias.

Mallows has suggested the estimate

$$C_p = \frac{\text{RSS}_p}{\hat{\sigma}^2} + 2p - n,$$

where $\hat{\sigma}^2$ is a suitable estimate of σ^2. Assuming $\hat{\sigma}^2 \approx \sigma^2$

$$E\left[C_p\right] \approx \frac{1}{\sigma^2} E\left[\text{RSS}_p\right] + 2p - n, \tag{12.9}$$

and, arguing as in Theorem 3.3 (Section 3.3),

$$\begin{aligned}E\left[\text{RSS}_p\right] &= E\left[\left(\mathbf{Y} - \hat{\mathbf{Y}}_p\right)'\left(\mathbf{Y} - \hat{\mathbf{Y}}_p\right)\right] \\ &= E\left[\mathbf{Y}'(\mathbf{I}_n - \mathbf{P}_p)\mathbf{Y}\right] \\ &= \sigma^2 \text{tr}\left[\mathbf{I}_n - \mathbf{P}_p\right] + \boldsymbol{\theta}'(\mathbf{I}_n - \mathbf{P}_p)\boldsymbol{\theta} \\ &= \sigma^2(n-p) + \text{SSB}_p.\end{aligned} \tag{12.10}$$

Hence

$$E[C_p] \approx (n-p) + \frac{\text{SSB}_p}{\sigma^2} + 2p - n$$

$$= p + \frac{\text{SSB}_p}{\sigma^2}$$

$$= \Delta_p,$$

and C_p is approximately an unbiased estimate of Δ_p. The identity $\boldsymbol{\theta}'(\mathbf{I}_n - \mathbf{P}_p)\boldsymbol{\theta} = \text{SSB}_p$ in Equation (12.10) follows from the fact that the second term of $E[\text{RSS}_p]$ can be obtained by replacing each random vector in RSS_p by its expected value; or it can be proved as follows:

$$\boldsymbol{\eta}_p = \mathcal{E}[\hat{\mathbf{Y}}_p]$$

$$= \mathcal{E}[\mathbf{X}_p\hat{\boldsymbol{\beta}}_p]$$

$$= \mathbf{X}_p(\mathbf{X}_p'\mathbf{X}_p)^{-1}\mathbf{X}'_p\mathcal{E}[\mathbf{Y}]$$

$$= \mathbf{P}_p\boldsymbol{\theta}$$

so that

$$\boldsymbol{\theta}'(\mathbf{I}_n - \mathbf{P}_p)\boldsymbol{\theta} = \boldsymbol{\theta}'(\mathbf{I}_n - \mathbf{P}_p)^2\boldsymbol{\theta}$$

$$= (\boldsymbol{\theta} - \mathbf{P}_p\boldsymbol{\theta})'(\boldsymbol{\theta} - \mathbf{P}_p\boldsymbol{\theta})$$

$$= (\boldsymbol{\theta} - \boldsymbol{\eta}_p)'(\boldsymbol{\theta} - \boldsymbol{\eta}_p).$$

(Some authors, e.g., Mallows [1973] and Hocking [1972], assume that $\boldsymbol{\theta} = \mathbf{X}\boldsymbol{\beta}$, though this is not necessary in the above derivation.)

In addition to finding the regression subsets with a small C_p, Mallows [1964] suggested plotting C_p against p for each regression model. The C_p's for those models with small bias, as measured by SSB, tend to cluster about the line $C_p = p$ (Fig. 12.2, point A), and the C_p's for models with substantial bias lie above the line (Fig. 12.2, point B). As pointed out by Gorman and Toman [1966], although point B is above the line $C_p = p$, it is still below point A and therefore represents an equation with a slightly lower total mean square error $\sigma^2\Delta_p$. Therefore, adding q further regressors

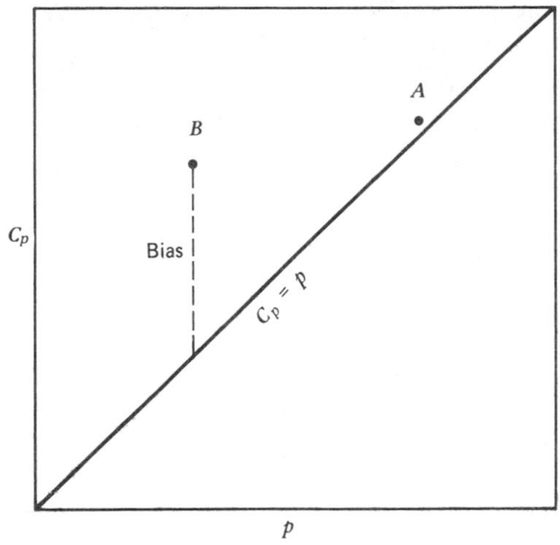

Fig. 12.2 C_p plot.

to a model may reduce the bias term SSB, but at the expense of increasing the variance term from $p\sigma^2$ to $(p+q)\sigma^2$; the tendency for prediction variances to increase was noted in Section 5.4. If the equation is needed for interpolation, it may pay to drop a few regressors and accept some bias in exchange for a smaller Δ_p and a simpler equation.

To calculate C_p we need a suitable estimate $\hat{\sigma}^2$ of σ^2. Often the mean residual sum of squares $\text{RSS}_{K+1}/(n-K-1)$ for the complete equation serves the purpose, but this forces $C_p = p$ for $p = K+1$. Substituting this estimate of σ^2 in the formula for C_p assumes that the complete equation has been carefully chosen so as to give reasonable assurance of negligible bias. When possible this estimate of σ^2 should be checked with previous estimates or with estimates based on "near replicates" (Daniel and Wood [1971: p. 123]). For example, near replicates could be pairs of Y observations taken far apart in time but under nearly the same x-conditions. Even large differences in the levels of a particular regressor can be tolerated provided the regressor has negligible influence on Y.

In applying Mallows' graphical method to the Hald data, Gorman and Toman [1966] obtained Table 12.4. By comparing C_p with p we see that there are four models which could be suitable predictors: (x_1, x_2), (x_1, x_2, x_3), (x_1, x_3, x_4), and (x_1, x_2, x_4). In the absence of technical information about the regressors it may be appropriate to choose the simplest model (x_1, x_2) because it has the smallest C_p value, namely, 2.68. This is the same model as the one recommended by the R^2 criterion above.

Table 12.4 C_p Values for all Regression Subsets (Hald's Data)

Regressors in Equation	p	C_p
None	1	443.2
(1)	2	202.7
(1,2)	3	2.68
(2)	2	142.6
(2,3)	3	62.5
(1,2,3)	4	3.04
(1,3)	3	198.2
(3)	2	315.3
(3,4)	3	22.4
(1,3,4)	4	3.49
(1,2,3,4)	5	5.0
(2,3,4)	4	7.34
(2,4)	3	138.3
(1,2,4)	4	3.03
(1,4)	3	5.51
(4)	2	138.8

SOURCE. Gorman and Toman [1966: Table II], but ordered according to Garside's method.

It is of interest to note that C_p is closely related to the adjusted coefficient of determination $\overline{R}_p^2$ (Kennard [1971]). Noting that [Equation (12.4)]

$$1-\overline{R}_p^2=\left[1-R_p^2\right]\left[\frac{n}{(n-p)}\right]$$

$$=\frac{\text{RSS}_p}{\sum\left(Y_i-\overline{Y}\right)^2}\cdot\frac{n}{(n-p)},$$

and estimating σ^2 by $\hat{\sigma}^2=\text{RSS}_{K+1}/(n-K-1)$, we have

$$C_p=\frac{\text{RSS}_p(n-K-1)}{\text{RSS}_{K+1}}-n+2p$$

or

$$1+\frac{(C_p-p)}{(n-p)}=\frac{\text{RSS}_p(n-K-1)}{\text{RSS}_{K+1}(n-p)}$$

$$=\frac{1-\overline{R}_p^2}{1-\overline{R}_{K+1}^2}.$$

If $n \gg p$ then, noting that $1-\bar{R}_{K+1}^2$ is just a scale factor, we see that $C_p - p$ is roughly equivalent to $1-\bar{R}_p^2$; in fact both quantities give a measure for the size of the bias term SSB_p.

For further examples of the use of C_p statistics the reader is referred to Gorman and Toman [1966], Daniel and Wood [1971], and Mallows [1973].

d. THE MSEP CRITERION

Using the notation at the beginning of the preceding section we now regard $\hat{Y}_p$ as an estimate of the unknown response Y (rather than of $\theta = E[Y]$) at the data point $\mathbf{x}$; its mean square error of prediction is, by (5.21),

$$E\left[\left(\hat{Y}_p - Y\right)^2\right] = \text{var}\left[\hat{Y}_p - Y\right] + \left(E\left[\hat{Y}_p - Y\right]\right)^2$$

$$= \sigma^2 + \text{var}\left[\hat{Y}_p\right] + (\eta_p - \theta)^2$$

$$= \sigma^2\left[1 + \mathbf{x}_p'(\mathbf{X}_p'\mathbf{X}_p)^{-1}\mathbf{x}_p\right] + (\eta_p - \theta)^2, \qquad (12.11)$$

which differs from (12.5) by σ^2. Allen [1971a] suggests that for predicting at a given row vector $\mathbf{x}'$ (of which $\mathbf{x}_p'$ contains the first p elements), the regression subset which minimizes (12.11) should be used. Aitkin [1974] considers the problem of finding the class of regression subsets which are not significantly "worse" than the full equation, and uses the MSEP (and various averages of it over the x-space) as his criterion. For example, averaging over the n data points we have (adding $n\sigma^2$ to $\sigma^2\Delta_p$ in Equation (12.6) and dividing by n),

$$\frac{1}{n}E\left[\sum_{i=1}^{n}\left(\hat{Y}_{pi} - Y_i\right)^2\right] = \frac{\sigma^2}{n}(n + p + \text{SSB}_p),$$

which can be used as a criterion. Assuming that the complete model is unbiased (that is, $\theta = \eta_{K+1}$ and $\boldsymbol{\theta} = \mathbf{X}\boldsymbol{\beta}$), Aitkin uses a simultaneous test procedure when the MSEP (or one of its variants) for a subset differs significantly from the MSEP for the full model; the type of test procedure used is described above [cf. (12.3)]. Both Aitkin [1974] and Narula [1974] investigate the use of the MSEP criterion for selecting a subset when the regressors are random.

e. OTHER MEASURES

Mallows [1967] and Rothman [1963] have suggested another measure for comparing regression subsets, namely,

$$J_p = \frac{n+p}{n-p}\text{RSS}_p.$$

Now when $n \gg p$ we have, from Equation (12.10),

$$\begin{aligned} E[J_p] &= \frac{n+p}{n-p}\{\sigma^2(n-p) + \text{SSB}_p\} \\ &= \sigma^2(n+p) + \left(\frac{n+p}{n-p}\right)\text{SSB}_p \\ &\approx \sigma^2(n+p) + \text{SSB}_p \\ &= \sigma^2 n + \sigma^2 \Delta_p. \end{aligned}$$

This means that J_p has similar properties to C_p when $n \gg p$; otherwise, J_p has rather nebulous theoretical properties.

In addition to J_p, Hocking [1972] mentions another measure, due to Allen [1971b, 1974], called PRESS_p (Prediction Sum of Squares). Unfortunately, in contrast to C_p, PRESS_p is not a simple function of RSS_p and is therefore not so readily calculated; all the methods described in Section 12.2.2 readily generate RSS_p as a by-product of the algorithm used†.

Another criterion, called the Average Estimated Variance (AEV), has been suggested by Helms [1974]. This criterion involves averaging the prediction variance over the whole regression region of interest, rather than for just the data points given, and using a weight function which attaches more weight to the more "important" points in the region. In one very special case (when, in Helms' notation, $\mathbf{M} = (\mathbf{X}'\mathbf{X})/N$)

$$\text{AEV} = \frac{p\,\text{RSS}_p}{n(n-p)}.$$

Helms [1974: p. 269] questions the practice of always including a term β_0; his experience suggests that "intercept terms are frequently primary contributors to variance but their absence often leads to only small contributions to bias."

From the above discussion it is clear that the choice of criterion depends very much on how the chosen regression model will be used. Because further research is obviously needed on the properties of the various measures, it is recommended that several of the measures should always be calculated when comparing the different models.

†See also Hocking [1976], *Biometrics*, **32**, 1–49.

12.3 GENERATING THE BETTER REGRESSIONS ONLY

12.3.1 Searching along Favorable Branches

With the present types of electronic computers it is not unreasonable to investigate all 2^K regressions for $K \leqslant 12$; for example, Daniel and Wood [1971: p. 85] state that, with efficient programming, 2^{12} residual sums of squares (equivalent to fitting 4096 equations) can be computed in less than 10 sec of IBM 360-65 computer time; generally space in the computer core, and not time, tends to be the limiting factor. However, the amount of computation increases exponentially with K and, since regressions with 20 or more variables are not uncommon, we need a suitable method for somehow restricting the search to the more "useful" regressions only. It can be argued that any searching method should at least find, for each value of $k=1,2,\ldots,K$, that k-regressor subset, from among all the $\binom{K}{k}$ possible k-variable models (with β_0 included), which has a minimum RSS_{k+1} (or equivalently a minimum C_{k+1}). Since only a few of the other k subsets would be expected to have an RSS_{k+1} close to this minimum, we would hope that the search for the minimum would also turn up the competitive models as a by-product.

Two similar methods that generally provide most of the more useful models without requiring a complete search through all the subsets have been given by Beale et al. [1967] and Hocking and Leslie [1967]. They basically use a tree-searching technique that avoids searching along unfavorable branches. In the following discussion of Hocking and Leslie's paper is is convenient to talk in terms of the r $(=K-k)$ variables that are not in the model, that is, in terms of the variables which are eliminated from the complete K-regressor model.

Let ϕ_i be the RSS obtained by fitting all K regressors except x_i, and suppose that the regressors are labeled according to the magnitude of the ϕ_i, that is,

$$\phi_1 \leqslant \phi_2 \leqslant \cdots \leqslant \phi_K.$$

After calculating the ϕ_i, the following stages of computation are carried out for $r=2,3,\ldots,K$ $(r=K-k)$.

Stage 1. Compute RSS for the model with regressors labeled $1,2,\ldots,r$ eliminated. If this RSS does not exceed ϕ_{r+1} then the process stops and the regression consisting of the k regressors $x_{r+1}, x_{r+1}, \ldots, x_K$ is the "best" k-regressor subset in the sense of minimum RSS. If, however, the RSS exceeds ϕ_{r+1} then no decision can be made at this point and we proceed to stage 2.

Stage 2. The variable x_{r+1} is now included among the candidates for elimination and the $\binom{r}{1}$ values of RSS obtained by eliminating any set of r regressors selected from the first $r+1$ regressors, but including the $(r+1)$th regressor, are then computed. If the smallest of the $1+\binom{r}{1}$ residual sums of squares computed thus far does not exceed ϕ_{r+1}, the procedure is terminated and the corresponding fitted regression is "best." If not, we proceed to stage 3.

Stage 3. The variable x_{r+2} is now included among the candidates for elimination and the $\binom{r+1}{2}$ values of RSS due to eliminating any set of r regressors selected from the first $r+2$ variables, but including x_{r+2}, are computed. The minimum of the $1+\binom{r}{1}+\binom{r+1}{2}$ values of RSS from the first three stages is now compared with ϕ_{r+3}, and the process is either terminated if the minimum is less than ϕ_{r+3} or continued to the next stage. The general qth stage is as follows:

Stage q. The $\binom{r+q-2}{q-1}$ values of RSS due to eliminating any set of r regressors selected from the first $r+q-1$ variables, but including x_{r+q-1}, are computed. If the minimum of the $\Sigma_{j=1}^{q}\binom{r+j-2}{j-1}$ values of RSS computed up till now does not exceed ϕ_{r+q}, then the corresponding subset is "best"; otherwise we proceed to stage $q+1$ where subsets of size r containing x_{r+q} are considered for elimination.

It is possible that all $k+1$ stages may be completed, and hence all the $\Sigma_{j=1}^{k+1}\binom{r+j-2}{j-1}=\binom{K}{k}$ regressions evaluated. However, it has been observed that this rarely happens except possibly for very small values of k, which, in any case, do not require much computation. In general only a small fraction of the 2^K regressions need to be considered in order to determine the subset of size k $(k=1,2,\ldots,K)$ with minimum RSS.

Within each stage it is convenient to generate the r-subsets in such a way that only one variable changes from one subset to the next. This can be done using the sequence generated in Section 12.2.1; the subsets obtained by Garside's method now refer to the subsets *eliminated*. For example, if $K=4$ then the sequence of regressors to be eliminated is given in Table 12.4. This sequence can now be split into sub-sequences as in Table 12.5.

To show that the method actually arrives at a minimum RSS for a given value of r we note, first of all, that a residual sum of squares either remains the same value or increases when a regressor is eliminated from the regression equation. This means that any fitted regression subset that does

Table 12.5 Order for Generating Regression Subsets of $K-r$ Regressors ($K=4$): Determined by Garside's Method in which Regressors are Changed One at a Time

Number Eliminated (r)	Regressors Eliminated	Stage
1	(x_1)	(ϕ_1)
	(x_2)	(ϕ_2)
	(x_3)	(ϕ_3)
	(x_4)	(ϕ_4)
2	(x_1, x_2)	1
	(x_2, x_3)	2
	(x_1, x_3)	2
	(x_3, x_4)	3
	(x_2, x_4)	3
	(x_1, x_4)	3
3	(x_1, x_2, x_3)	1
	(x_1, x_3, x_4)	2
	(x_2, x_3, x_4)	2
	(x_1, x_2, x_4)	2

not contain x_i cannot have an RSS smaller than ϕ_i. Suppose, therefore, that the RSS obtained by eliminating any set of r regressors, for which the maximum subscript is j, is not greater than ϕ_{j+1}. Then since the value of RSS for this particular model is not greater than ϕ_{j+m} $(m=1,2,\ldots,K-j)$, we cannot reduce the value of RSS by eliminating x_{j+m} and replacing it by one of $x_1, x_2, \ldots, x_j$ not already in the fitted model; otherwise we would be able to find a model with x_{j+m} eliminated which has an RSS less than the allowable minimum of ϕ_{j+m}. When the process stops, say, at the qth stage, we know that variables $x_{r+q}, \ldots, x_K$ must be included in the best regression. It is then a matter of finding the best k-subset which includes these variables by searching through all the values of RSS computed thus far.

In describing the above procedure, Hocking and Leslie [1967] prefer to work with the reduction in the regression sum of squares achieved when variables are eliminated. This reduction is computed directly by inverting a $k \times k$ principal minor of $(\mathbf{X}'\mathbf{X})$ when $k \leqslant r$, and by inverting an $r \times r$ principal minor of $(\mathbf{X}'\mathbf{X})^{-1}$ when $k > r$. Within each stage the sequence of inverses is computed in an efficient manner by using the ordering given in Table 12.5; since the variables come into and go out of the model one at a time, successive submatrices differ from each other only in one row (and column, by symmetry). The authors also note that C_p $(=C_{k+1})$ can be readily expressed in terms of the reduction, and they suggest several methods, based on the values of C_p calculated thus far, for reducing the length of the search. In a later paper, LaMotte and Hocking [1970] give a modification of the above method which reduces the amount of computation considerably; they believe the method to be reasonably efficient for K up to 30, or even possibly 50. The method of Beale et al. [1967], which is similar to the above method, seems to be satisfactory for K not much more than 20 (Beale [1970: p. 913].

If regressors are changed one at a time then the efficient Householder–Givens method of Section 12.2.2 can be used for bringing in one variable and eliminating another.

More recently Furnival and Wilson [1974] have given a different method which seems much faster than those mentioned above.

12.3.2 A t-Directed Search

The test statistic for testing $H:\beta_i=0$ for the full K-regressor model is the t-statistic

$$t_{K,i}=\frac{b_i}{S\sqrt{d_{ii}}}$$

or, equivalently, the F-statistic [cf. (11.47)]

$$\begin{aligned} F_{K,i} &= t_{K,i}^2 \\ &= \frac{b_i^2}{(S^2 d_{ii})} \\ &= \frac{\text{RSS}_{K,i}-\text{RSS}_{K+1}}{S^2} \end{aligned}$$

where $\text{RSS}_{K,i}=\phi_i$ (the residual sum of squares due to fitting all K regressors except x_i) and $S^2=\text{RSS}_{K+1}/(n-K-1)$. Now the preceding method in Section 12.3.1 indicates that those regressors with small ϕ_i will tend to be eliminated early in the proceedings so that the regressors with large ϕ_i, or equivalently with a large $|t_{K,i}|$, will be included in the "best" k-regressor subset for each k (best in the sense of minimum RSS or minimum C_{k+1}). Therefore, suppose that we order the regressors according to the values of $|t_{K,i}|$ $(i=1,2,\ldots,K)$ ranked in decreasing order of magnitude. Then bringing in the regressors one at a time in the given order, we would hope to obtain, in general, the best or one of the near best k-variable subsets for each $k=1,2,\ldots,K$. This so-called t-directed search was proposed by Daniel and Wood [1971], and we demonstrate its use by considering two examples.

EXAMPLE 12.1 ($K=4$: Hald data) The values of $|t_{K,i}|$ for the Hald data are given in Table 12.6. By reference to Table 12.4 we find that the t-directed search produces the best two- and three-variable subsets (1,2) and (1,2,4), respectively; in fact (1,2) is the subset with overall minimum C_p, and (1,2,4) is the subset with the next smallest C_p value. Table 12.6 suggests that x_1 and x_2 form a basic set to be included in any "good" regression model, and we would then confine our search to just those subsets containing x_1 and x_2, namely, (1,2,3) and (1,2,4). [In this example σ^2 is estimated by $\text{RSS}_5/(n-5)$ so that we have the identity $C_5=5$.]

Table 12.6 A t-Directed Search Applied to Hald's Data

Variable (i)	$\|t_{K,i}\|$	Cumulative Subset	p	C_p
1	2.08	(1)	2	202.7
2	0.70	(1,2)	3	2.68*
4	0.20	(1,2,4)	4	3.03
3	0.14	(1,2,3,4)	5	5.00

SOURCE. Daniel and Wood [1971: pp. 89–90].
*Minimum value.

EXAMPLE 12.2 ($K=6$: data from Gorman and Toman [1966]) The t-directed search is set out in Table 12.7 and we find that the method actually provides the best one-, two-, three-, four-, and five-variable subsets. It transpires that the model (1,2,6,4) is the subset with minimum C_p, and (1,2,6) and (1,2,6,4,3) are close competitors. A plot of C_p against p for all the 2^6 models indicates that the subsets with small C_p values are (Gorman and Toman [1966: p. 39]) (1,2,6: $C_4=5.7$), (1,2,4,6: $C_5=4.3$), (1,2,3,4,6: $C_6=5.3$), and (1,2,4,5,6: $C_6=5.5$); the t-directed search produced three of these. In this case our "basic" set of regressors consists of x_1, x_2, and x_6—one less than that suggested by the turning point in the C_p column of Table 12.7 (indicated by *).

Although the above two examples are rather artificial in that we would carry out a t-directed search only when K is large, say, $K>20$, they do indicate broadly what happens in general. When there is a clear winner among the models, with few close competitors, then the turning point (starred in Tables 12.6 and 12.7) will tell us correctly which variables should be in our basic set. In less clear-cut cases, as in Example 12.2, it is

Table 12.7 A t-Directed Search Applied to the Data of Gorman and Toman [1966]

Variable (i)	$\lvert t_{K,i} \rvert$	Cumulative Subset	p	C_p
1	7.0	(1)	2	20.4
2	4.3	(1,2)	3	14.0
6	2.9	(1,2,6)	4	5.7
4	2.1	(1,2,6,4)	5	4.3*
3	0.7	(1,2,6,4,3)	6	5.3
5	0.6	(1,2,6,4,3,5)	7	7.0

SOURCE. Daniel and Wood [1971: p. 96].
*Minimum value.

frequently necessary to include in our basic set one less variable or, more rarely, two less variables. To be conservative Daniel and Wood suggest that the computer search routine be programmed to include in the basic set two less variables than the starred C_p requires. Some idea as to when caution is necessary is given by the relationship of C_p to p. If the smallest C_p in the $t_{K,i}$ listing is less than or equal to p, we can safely include all variables down to and including the turning point. However, if our starred C_p is $C_4=8$ then, since a reduction of p by 1 can produce a reduction of C_p by 2 (since $C_p=\text{RSS}_p/\hat{\sigma}^2+2p-n$), we would take as our basic set all the variables down to two less than the turning point.

Once we have determined our basic set of K_1 variables, we need only search those 2^{K-K_1} models which contain the basic set; this reduction can lead to a considerable saving of computer time and storage. However, if K is very large, 2^{K-K_1} may still be too big for a complete search. In this case we can use the theory of fractional 2^K factorial designs and find the remaining essential variables by looking at certain fractions containing the basic set. Details of the method are given by Gorman and Toman [1966] and Daniel and Wood [1971]; further research, however, is needed into the question of choosing an appropriate fraction.

12.4 STEPWISE REGRESSION

12.4.1 Description of the Method

Stepwise regression is a method whereby, at each step, a single regressor is either added to or eliminated from the current regression model. In the procedure due to Efroymson [1960] we have two F-levels, say, F_{IN} and F_{OUT}. At each step a regressor, x_i, say, is eliminated if its removal would increase the RSS by not more than F_{OUT} times the residual mean square

$[\text{RSS}_p/(n-p)]$; that is, if the F-ratio for testing $\beta_i=0$ in the current regression model is less than or equal to F_{OUT}. If there is a choice of variable to be eliminated, then the one giving the smallest increase in RSS (or, equivalently, the smallest F-ratio) is chosen. If no variable can be eliminated in this way, then a regressor x_j, say, is brought in if its introduction will reduce RSS by at least F_{IN} times the residual mean square calculated after introducing x_j; that is, x_j is introduced if the F-ratio for testing $\beta_j=0$ in the current model plus the x_j term is greater than or equal to F_{IN}. Again, if there is choice of variable to be introduced then the one giving the largest decrease in RSS (or equivalently, the largest F-ratio) is chosen. To begin the procedure we fit β_0 and then endeavor to bring in a regressor. The process ends when a regressor cannot be brought into the model.

Unfortunately the above procedure leads to a single subset and does not suggest alternative good subsets.

12.4.2 Use of Sweeping

The sweep method, as described in Section 12.2.2, can be used for adding or removing a regressor. Here sweeping is applied to the augmented matrix

$$\mathbf{A}_1 = \begin{pmatrix} \mathbf{R}_{xx} & \mathbf{r}_{xy} \\ \mathbf{r}'_{xy} & 1 \end{pmatrix} \begin{matrix} \}K \\ \}1 \end{matrix},$$

where $\mathbf{R}_{xx}$ is the correlation matrix for all K regressors. If we sweep $\mathbf{A}_1$ in any subset of the first K pivotal elements we obtain a new matrix

$$\mathbf{A}_1^* = \begin{pmatrix} \mathbf{B} & \mathbf{c} \\ \mathbf{d}' & f \end{pmatrix}$$

where $\mathbf{B}$, $\mathbf{c}$, and f, respectively, contain the inverse of the correlation matrix, the scaled regression coefficients $\mathbf{a}$, and $1-R^2$ for the regressors that are now in the equation. After each sweep we calculate $V_k = c_k d_k / b_{kk}$ for each $b_{kk} >$ *tolerance*; by controlling the size of b_{kk} we avoid bringing in a regressor which is approximately linearly dependent on the regressors already in the model. Now the elements of $\mathbf{d}$ corresponding to regressors not in the equation are equal to the corresponding elements of $\mathbf{c}$, and the remaining elements of $\mathbf{d}$ and $\mathbf{c}$ are $-\mathbf{a}$ and $\mathbf{a}$, respectively. Hence V_k is negative [being of the form $-a_k^2/b_{kk}$ or, in the notation of Equation (11.51) in Section 11.7.2, $-a_k^2(1-R_k^2)$] if x_k is in the regression equation, and positive $(=c_k^2/b_{kk})$ if x_k is not in the model. To determine which regressor to eliminate, if any, we find the minimum $|V_k|$, V_{MIN}, say, for all $V_k<0$, and eliminate the regressor corresponding to V_{MIN} if [cf. Equation

(11.51) with $\phi = n - p$ and $f = 1 - R^2$]

$$\frac{V_{MIN}\phi}{f} \leqslant F_{OUT}$$

where

$$\phi = n - (\textit{number of regressors in the model} + 1).$$

In a similar manner we determine which regressor to bring in, if any, by finding the maximum V_k, V_{MAX}, say, for all $V_k > 0$. If

$$\frac{V_{MAX}(\phi - 1)}{f'} = \frac{V_{MAX}(\phi - 1)}{f - V_{MAX}} \geqslant F_{IN},$$

then we add the regressor corresponding to V_{MAX}. At each step we do not try to eliminate the variable that was just brought in or add the variable just eliminated. This can be automatically taken care of by choosing $F_{IN} \geqslant F_{OUT}$.

In Efroymson's [1960] original description of the above method he considered sweeping the matrix

$$\mathbf{A}_2 = \begin{bmatrix} \mathbf{R}_{xx} & \mathbf{r}_{xy} & \mathbf{I}_K \\ \mathbf{r}'_{xy} & 1 & \mathbf{0}' \\ -\mathbf{I}_K & \mathbf{0} & \mathbf{O}_K \end{bmatrix}, \tag{12.12}$$

where $\mathbf{O}_K$ is a $K \times K$ matrix of zeros. At any stage of the sweeping the transformed $\mathbf{A}_2$ can be partitioned in the same way, namely,

$$\mathbf{A}_2^* = \begin{bmatrix} \mathbf{B} & \mathbf{c} & \mathbf{F} \\ \mathbf{d}' & f & -\mathbf{g} \\ \mathbf{G} & \mathbf{h} & \mathbf{H} \end{bmatrix} \tag{12.13}$$

where, apart from rounding error, $\mathbf{g} = \mathbf{h}$. In his method the pivotal elements of $\mathbf{B}$ and $\mathbf{H}$ are swept once only; sweeping the ith pivot of $\mathbf{B}$ brings in x_i, and sweeping a nonzero pivot of $\mathbf{H}$, h_{jj}, say, eliminates x_j. At each step the vector $\mathbf{h}$ contains the scaled regression coefficients $\mathbf{a}$ for the model fitted thus far, the remaining elements being zeros; and $\mathbf{H}$ contains zeros and the inverse of the partitioned part of $\mathbf{R}_{xx}$ corresponding to the regressors in the current model. Thus $\mathbf{h}$ and $\mathbf{H}$ simply contain the "useful" parts of $\mathbf{c}$ and $\mathbf{B}$ uncluttered by other nonzero elements.

Since, apart from signs, the transformed $\mathbf{A}_2$ is symmetric, we need work with only the upper triangular matrix, thereby reducing the computation and the storage. Breaux [1968] gives an algorithm for doing this using the symmetric sweep method described in Section 12.2.2. He states that, with this modification, a program can be designed to handle as many as $K=200$ variables with a requirement of less than 21,000 words of storage, whereas the conventional procedure, using $\mathbf{A}_2$, would require over 40,000 words of storage.

12.4.3 Gauss–Jordan Elimination

Instead of sweeping one can use the Jordan elimination method, described toward the end of Section 11.2.1, in which a pivotal element is scaled to unity and the remaining elements in the column are reduced to zero. This method can be applied to $\mathbf{A}_2$ in (12.12) so that a given submatrix of $\mathbf{R}_{xx}$ is reduced to the identity matrix in $\mathbf{B}$ [Equation (12.13)], and the inverse of the submatrix appears in $\mathbf{F}$ (and $\mathbf{H}$). In contrast to the sweep method, which inverts the submatrix in situ, $\mathbf{B}$ no longer contains the nonzero part of $\mathbf{H}$; also $\mathbf{d}$ no longer contains $-\mathbf{a}$, the scaled regression coefficients. This means that V_{MAX} is now the maximum value of c_k^2/b_{kk} for all x_k not in the model, whereas V_{MIN} is the minimum value of c_k^2/h_{kk} for all $h_{kk}>0$ (that is, for all x_k in the model). To bring in a regressor x_i, say, b_{ii} is chosen as the pivotal element to reduce to unity. But to eliminate x_j, say, f_{jj} (and not h_{jj}) is chosen as the pivotal element since we effectively wish to reverse the Jordan process for this variable.

The above modification of Efroymson's technique is described in some detail by Draper and Smith [1966: Section 6.8]. Using the Hald data ($K=4$) they finally arrive at the regression model (x_1,x_2), which agrees with the model obtained by the various methods in Section 12.2.3. The main steps in their computations are as follows:

(1) Test for elimination: this is not appropriate.
Test for adding a new variable: the minimum V_k $(k=1,2,3,4)$ is V_4 and F_{IN} is exceeded so that x_4 is introduced.

(2) Test for elimination: this is not appropriate as there is only one regressor in the model and it has just been brought in.
Test for addition: since the minimum V_k $(k=1,2,3)$ is V_1 and F_{IN} is exceeded, x_1 is introduced.

(3) Test for elimination: the regression subset is now (x_1,x_4) with x_4 as a candidate for elimination; F_{OUT} is exceeded so that x_4 is left in.
Test for addition: the minimum V_k $(k=2,3)$ is V_2; F_{IN} is exceeded and x_2 is brought in.

(4) Test for elimination: the regression subset is (x_1,x_2,x_4) and maximum V_k $(k=1,4)$ is V_4; since F_{OUT} is not exceeded x_4 is eliminated.
Test for addition: the regression subset is (x_1,x_2) and, since x_4 has just been eliminated, only x_3 is a candidate for acceptance; since F_{IN} is not exceeded x_3 is not brought in. Because no variable is added the process stops, and our final regression subset is (x_1,x_2).

12.4.4 Choice of F Values

In using the stepwise regression technique we are faced with the problem of choosing the values of F_{IN} and F_{OUT}. It is common practice to set $F_{IN}=F_{OUT}$ $(=F_0$, say) where F_0 is some arbitrary constant; for example, Efroymson uses $F_0=2.5$ and Draper and Smith set $F_0=3.29$ for the same set of (Hald's) data. Alternatively we can set $F_{IN}=F^{\alpha}_{1,\phi-1}$ and $F_{OUT}=F^{\alpha}_{1,\phi}$, where $\alpha=0.05$ and ϕ is the number of degrees of freedom associated with the current RSS. However, such a choice of F values is not strictly correct for, at each stage, we are looking for either the maximum or the minimum of a set of correlated F variables. For instance, the more variables there are to choose from, the larger V_{MAX}, and therefore the larger the maximum "F to enter," we would expect to observe. Several authors (e.g., Draper et al. [1971] and Pope and Webster [1972]) have investigated this question of ordered dependent F-variables, and little progress has been made. Forsythe et al. [1973] consider this problem for entering variables only and provide a permutation test to replace F_{IN}.

12.4.5 Other One-Step Methods

There are two variations of the stepwise procedure which seem to be popular. One is the so-called forward selection (or step-up) procedure in which variables are not eliminated but are simply brought in one at a time using, say, the F_{IN} criterion. The other is the backward elimination (or step-down) procedure in which the full K-regressor model is fitted and the variables are eliminated one at a time using, say, the F_{OUT} criterion; there are no tests for addition. Unfortunately these two methods do not necessarily arrive at the same subset. For example, Hamaker [1962], using Hald's data, arrived at the subset (x_1,x_2,x_4) using forward selection and (x_1,x_2) using backward elimination.

Some of the pros and cons of the two methods are discussed in detail by Mantel [1970] and Beale [1970]. It appears that most authors prefer backward elimination (e.g., Draper and Smith [1966: p. 177]) and this choice is supported by the analysis given by Kennedy and Bancroft [1971], though in their comparison the order for choosing the regressors is predetermined (as in polynomial regression). Backward elimination is also appropriate for those statisticians who prefer to see all the regressors in the equation once in order "not to miss anything"!

Other variations on the above theme are described in Mantel [1970] and Draper and Smith [1966]. However, it seems that the full stepwise procedure is to be preferred, with one obvious exception—polynomial regression.

In conclusion it would seem that stepwise regression is appropriate when K is very large (a conservative value would be $K > 40$, though this depends on the computer); otherwise the methods of Sections 12.2 and 12.3 should be used.

12.5 OTHER METHODS

12.5.1 Factor and Principal Component Analysis

One method of selecting a regression subset is by factor analysis. After the regressors have been subjected to a factor analysis several factors emerge and the remainder of the variation is ascribed to near multicollinearities which may be ignored. The factors are then rotated to obtain a simple structure in which each factor identifies with some minimal subset of the regressors; a regression of Y on the factors may then suggest suitable subsets of regressors. This technique is used in Massy [1965], Daling and Tamura [1970], and a number of practical studies. However, Hawkins [1973] raises objections to this technique. First, there is no guarantee that Y is dependent on the factors rather than on near multicollinearities which have been ignored; an example is given in Hotelling [1957]. Second, the technique suggests one or more possible regression subsets but, like the stepwise method, gives no explicit information on the number or composition of the alternative good subsets. What is needed, therefore, is a method that shows up multicollinearities and suggests a number of suitable subsets.

Hawkins [1973] feels that principal component analysis can meet these requirements and the reader is referred to his article for details. The idea of using principal component analysis for finding a meaningful regression subset is also suggested by Jeffers [1967] and Cox [1968: p. 272]; the use of component analysis in regression is also discussed briefly by Seber [1966: p. 56] and by Greenberg [1975].

12.5.2 A Bayesian Method

In conclusion some mention should be made of the Bayesian method of Lindley [1968]. If the regression model is to be used for prediction he suggests finding the subset which minimizes

$$(\mathrm{RSS}_{k+1} - \mathrm{RSS}_{K+1})n^{-1} + c_k,$$

where $(\mathrm{RSS}_{k+1} - \mathrm{RSS}_{K+1})$ is the increase in the residual sum of squares through eliminating $K-k$ regressors, and c_k is the cost associated with the particular k-regressor subset, the minimization being taken over all subsets for all values of k. Lindley assumes that

$$E[\mathbf{Y}|\boldsymbol{\beta},\mathbf{x}]=\boldsymbol{\beta}'\mathbf{x},$$

where $\mathbf{x}'=(1,x_1,x_2,\ldots,x_K)$; that is, the full K-regressor model is unbiased.
If all the regressors have the same cost $c=2\sigma^2/n$, and the costs are additive ($c_k=kc$), then Lindley [1968: p. 43] shows that his criterion is equivalent to minimizing Mallows' C_p statistic.

12.6 GENERAL COMMENTS

The whole question of choosing a suitable subset is reviewed briefly by Cox and Snell [1974] and the reader is referred to their article for some helpful practical comments. They recommend that the number of regressors be reduced to about 10 using, for example, techniques like those of Sections 12.3 and 12.5; then all 2^{10} subsets of these 10 regressors can be generated and compared using the methods of Section 12.2. Automatic selection procedures like, for example, the one-step methods of Section 12.4 should be used with caution (if at all); they are more appropriate when the particular variables selected are not of intrinsic interest or when some preliminary reduction in the number of regressors is required.

MISCELLANEOUS EXERCISES 12

1. Show that pivoting on both diagonal elements of the symmetric matrix

$$\mathbf{A}=\begin{bmatrix} a_{11} & a_{12} \\ a_{21} & a_{22} \end{bmatrix}$$

inverts $\mathbf{A}$.

2. Suppose that we wish to omit the regressor x_j from a p-parameter regression model. If F_j is the F-statistic for testing the hypothesis $H:\beta_j=0$, show that

$$C_{p-1}=\frac{F_j \mathrm{RSS}_p}{\hat{\sigma}^2(n-p)}+C_p-2.$$

Gorman and Toman [1966: p. 50]

3. Show that $C_p \leqslant p$ if and only if $F \leqslant 1$ where F is the F-statistic for testing the hypothesis that $K+1-p$ of the regression coefficients β_j are zero.

Hocking [1974]

APPENDIX A

Some Matrix Algebra

A1. TRACE

Provided the matrices are conformable

1. $\text{tr}[\mathbf{A}+\mathbf{B}]=\text{tr}\,\mathbf{A}+\text{tr}\,\mathbf{B}$,
2. $\text{tr}[\mathbf{AC}]=\text{tr}[\mathbf{CA}]$.

   The proofs are straightforward. If $\mathbf{A}$ is a symmetric $n\times n$ matrix with eigenvalues λ_i $(i=1,2,\ldots,n)$, then

3. $\text{tr}\,\mathbf{A}=\sum_{i=1}^{n}\lambda_i$,
4. $\text{tr}[\mathbf{A}^s]=\sum_{i=1}^{n}\lambda_i^s$,
5. $\text{tr}[\mathbf{A}^{-1}]=\sum_{i=1}^{n}\lambda_i^{-1}$, ($\mathbf{A}$ nonsingular).

   Proof. Since $\mathbf{A}$ is symmetric there exists a real orthogonal matrix $\mathbf{T}$ such that $\mathbf{T}'\mathbf{AT}=\text{diag}(\lambda_1,\lambda_2,\ldots,\lambda_n)=\mathbf{\Lambda}$, say. Hence $\Sigma\lambda_i=\text{tr}\,\mathbf{\Lambda}=\text{tr}[\mathbf{T}'\mathbf{AT}]=\text{tr}[\mathbf{ATT}']=\text{tr}\,\mathbf{A}$. Then (4) follows from $\mathbf{\Lambda}^s=(\mathbf{T}'\mathbf{AT})(\mathbf{T}'\mathbf{AT})\ldots(\mathbf{T}'\mathbf{AT})=\mathbf{T}'\mathbf{A}^s\mathbf{T}$ and (5) follows from $\mathbf{\Lambda}^{-1}=(\mathbf{T}'\mathbf{AT})^{-1}=\mathbf{T}'\mathbf{A}^{-1}\mathbf{T}$. [We note that (3) holds for *any* square matrix; from the coefficient of λ^{n-1} in $|\lambda\mathbf{I}_n-\mathbf{A}|=0$ we see that $\text{tr}\,\mathbf{A}$ is the sum of the roots.]

A2. RANK

1. If $\mathbf{A}$ and $\mathbf{B}$ are conformable matrices, then

$$\text{rank}[\mathbf{AB}]\leqslant\text{minimum}(\text{rank}\,\mathbf{A},\text{rank}\,\mathbf{B}).$$

Proof. The rows of $\mathbf{AB}$ are linear combinations of the rows of $\mathbf{B}$ so that the number of linear independent rows of $\mathbf{AB}$ is less than or equal to those of $\mathbf{B}$; thus $\text{rank}[\mathbf{AB}] \leqslant \text{rank}\,\mathbf{B}$. Similarly the columns of $\mathbf{AB}$ are linear combinations of the columns of $\mathbf{A}$ so that $\text{rank}[\mathbf{AB}] \leqslant \text{rank}\,\mathbf{A}$.

2. If $\mathbf{A}$ is any matrix, and $\mathbf{P}$ and $\mathbf{Q}$ are any conformable nonsingular matrices, then $\text{rank}[\mathbf{PAQ}] = \text{rank}[\mathbf{A}]$.

   Proof. $\text{rank}\,\mathbf{A} \leqslant \text{rank}[\mathbf{AQ}] \leqslant [\mathbf{AQQ}^{-1}] = \text{rank}\,\mathbf{A}$, so that $\text{rank}\,\mathbf{A} = \text{rank}[\mathbf{AQ}]$, etc.

3. Let $\mathbf{A}$ be any $m \times n$ matrix such that $r = \text{rank}\,\mathbf{A}$ and $s = \text{nullity}\,\mathbf{A}$ (the dimension of $\mathcal{N}[\mathbf{A}]$, the null space or kernel of $\mathbf{A}$, i.e., dimension $\{\mathbf{x} : \mathbf{Ax} = \mathbf{0}\}$). Then

$$r + s = n.$$

Proof. Let $\boldsymbol{\alpha}_1, \boldsymbol{\alpha}_2, \ldots, \boldsymbol{\alpha}_s$ be a basis for $\mathcal{N}[\mathbf{A}]$. Enlarge this set of vectors to give a basis $\boldsymbol{\alpha}_1, \boldsymbol{\alpha}_2, \ldots, \boldsymbol{\alpha}_s, \boldsymbol{\beta}_1, \boldsymbol{\beta}_2, \ldots, \boldsymbol{\beta}_t$ for E_n, n-dimensional Euclidean space. Every vector in $\mathcal{R}[\mathbf{A}]$, the range space of $\mathbf{A}$, can be expressed in the form

$$\begin{aligned}
\mathbf{Ax} &= \mathbf{A}\left(\sum_{i=1}^{s} a_i \boldsymbol{\alpha}_i + \sum_{j=1}^{t} b_j \boldsymbol{\beta}_j \right) \\
&= \mathbf{A} \sum_{j=1}^{t} b_j \boldsymbol{\beta}_j \\
&= \sum_{j=1}^{t} b_j \mathbf{A} \boldsymbol{\beta}_j \\
&= \sum_{j=1}^{t} b_j \boldsymbol{\gamma}_j,
\end{aligned}$$

say. Now suppose that

$$\sum_{j=1}^{t} c_j \boldsymbol{\gamma}_j = \mathbf{0};$$

then

$$\mathbf{A}\left(\sum_{j=1}^{t} c_j \boldsymbol{\beta}_j \right) = \sum_{j=1}^{t} c_j \boldsymbol{\gamma}_j = \mathbf{0}$$

and $\sum c_j \boldsymbol{\beta}_j \in \mathcal{N}[\mathbf{A}]$. This is possible only if $c_1 = c_2 = \cdots = c_t = 0$, that is, $\boldsymbol{\gamma}_1, \boldsymbol{\gamma}_2, \ldots, \boldsymbol{\gamma}_t$ are linearly independent. Since every vector $\mathbf{Ax}$ in $\mathcal{R}[\mathbf{A}]$ can be expressed in terms of the $\boldsymbol{\gamma}_j$'s, the $\boldsymbol{\gamma}_j$'s form a basis for $\mathcal{R}[\mathbf{A}]$; thus $t = r$. Since $s + t = n$ our proof is complete.

4. $\text{rank}\,\mathbf{A} = \text{rank}\,\mathbf{A}' = \text{rank}[\mathbf{A}'\mathbf{A}] = \text{rank}[\mathbf{A}\mathbf{A}']$.

   Proof. $\mathbf{Ax}=\mathbf{0} \Rightarrow \mathbf{A}'\mathbf{Ax}=\mathbf{0}$ and $\mathbf{A}'\mathbf{Ax}=\mathbf{0} \Rightarrow \mathbf{x}'\mathbf{A}'\mathbf{Ax}=0 \Rightarrow \mathbf{Ax}=\mathbf{0}$. Hence the nullspaces of $\mathbf{A}$ and $\mathbf{A}'\mathbf{A}$ are the same. Since $\mathbf{A}$ and $\mathbf{A}'\mathbf{A}$ have the same number of columns, it follows from *A2.3* that $\text{rank}\,\mathbf{A} = \text{rank}[\mathbf{A}'\mathbf{A}]$. Similarly $\text{rank}\,\mathbf{A}' = \text{rank}[\mathbf{A}\mathbf{A}']$ and the result follows.

5. If $\mathcal{R}[\mathbf{A}]$ is the range space of $\mathbf{A}$ (the space spanned by the columns of $\mathbf{A}$), then $\mathcal{R}[\mathbf{A}'\mathbf{A}] = \mathcal{R}[\mathbf{A}']$.

   Proof. $\mathbf{A}'\mathbf{Aa} = \mathbf{A}'\mathbf{b}$ for $\mathbf{b} = \mathbf{Aa}$ so that $\mathbf{R}[\mathbf{A}'\mathbf{A}] \subset \mathcal{R}[\mathbf{A}']$. However, by *A2.4*, these two spaces must be the same as they have the same dimension.

6. If $\mathbf{A}$ is symmetric then $\text{rank}\,\mathbf{A}$ is equal to the number of nonzero eigenvalues.

   Proof. By *A2.2* $\text{rank}\,\mathbf{A} = \text{rank}[\mathbf{T}'\mathbf{AT}] = \text{rank}\,\mathbf{\Lambda}$.

7. Any $n \times n$ symmetric matrix $\mathbf{A}$ has a set of n orthonormal eigenvectors, and $\mathcal{R}[\mathbf{A}]$ is the space spanned by those eigenvectors corresponding to nonzero eigenvalues.

   Proof. From $\mathbf{T}'\mathbf{AT} = \mathbf{\Lambda}$ we have $\mathbf{AT} = \mathbf{T\Lambda}$ or $\mathbf{At}_i = \lambda_i \mathbf{t}_i$, where $\mathbf{T} = (\mathbf{t}_i, \ldots, \mathbf{t}_n)$; the $\mathbf{t}_i$ are orthonormal as $\mathbf{T}$ is an orthogonal matrix. Suppose $\lambda_i = 0$ $(i = r+1, r+2, \ldots, n)$ and $\mathbf{x} = \Sigma_{i=1}^{n} a_i \mathbf{t}_i$. Then

$$\mathbf{Ax} = \mathbf{A} \sum_{i=1}^{n} a_i \mathbf{t}_i = \sum_{i=1}^{n} a_i \mathbf{At}_i = \sum_{i=1}^{r} a_i \lambda_i \mathbf{t}_i = \sum_{i=1}^{r} b_i \mathbf{t}_i,$$

   and $\mathcal{R}[\mathbf{A}]$ is spanned by $\mathbf{t}_1, \mathbf{t}_2, \ldots, \mathbf{t}_r$.

A3. POSITIVE SEMIDEFINITE MATRICES

A symmetric matrix $\mathbf{A}$ is said to be positive semidefinite† (p.s.d.) if and only if $\mathbf{x}'\mathbf{Ax} \geqslant 0$ for all $\mathbf{x}$.

1. The eigenvalues of a p.s.d. matrix are nonnegative.

   Proof. If $\mathbf{T}'\mathbf{AT} = \mathbf{\Lambda}$, then substituting $\mathbf{x} = \mathbf{Ty}$ we have $\mathbf{x}'\mathbf{Ax} = \mathbf{y}'\mathbf{T}'\mathbf{ATy} = \lambda_1 y_1^2 + \cdots + \lambda_n y_n^2 \geqslant 0$. Setting $y_j = \delta_{ij}$ leads to $0 \leqslant \mathbf{x}'\mathbf{Ax} = \lambda_i$.

2. If $\mathbf{A}$ is p.s.d. then $\text{tr}\,\mathbf{A} \geqslant 0$. This follows from *A3.1* and *A1.3*.
3. $\mathbf{A}$ is p.s.d of rank r if and only if there exists an $n \times n$ matrix $\mathbf{R}$ of rank r such that $\mathbf{A} = \mathbf{RR}'$.

†Some authors use the term nonnegative definite.

Proof. Given $\mathbf{A}$ is p.s.d. of rank r, then, by *A2.6* and *A3.1*, $\boldsymbol{\Lambda}=\mathrm{diag}(\lambda_1,\lambda_2,\ldots,\lambda_r,0,\ldots,0)$ where $\lambda_i>0$ $(i=1,2,\ldots,r)$. Let $\boldsymbol{\Lambda}^{1/2}=\mathrm{diag}(\lambda_1^{1/2},\lambda_2^{1/2},\ldots,\lambda_r^{1/2},0,\ldots,0)$; then $\mathbf{T'AT}=\boldsymbol{\Lambda}$ implies that $\mathbf{A}=\mathbf{T}\boldsymbol{\Lambda}^{1/2}\boldsymbol{\Lambda}^{1/2}\mathbf{T}'=\mathbf{RR}'$, where $\mathrm{rank}\,\mathbf{R}=\mathrm{rank}\,\boldsymbol{\Lambda}^{1/2}=r$. Conversely, if $\mathbf{A}=\mathbf{RR}'$ then $\mathrm{rank}\,\mathbf{A}=\mathrm{rank}\,\mathbf{R}=r$ (*A2.4*) and $\mathbf{x'Ax}=\mathbf{x'RR'x}=\mathbf{y'y}\geqslant 0$, where $\mathbf{y}=\mathbf{R'x}$.

4. If $\mathbf{A}$ is an $n\times n$ p.s.d. matrix of rank r, then there exists an $n\times r$ matrix $\mathbf{S}$ of rank r such that $\mathbf{S'AS}=\mathbf{I}_r$.

   Proof. From

$$\mathbf{T'AT}=\begin{pmatrix}\boldsymbol{\Lambda}_r & \mathbf{O}\\ \mathbf{O} & \mathbf{O}\end{pmatrix}$$

   we have $\mathbf{T}_1'\mathbf{AT}_1=\boldsymbol{\Lambda}_r$, where $\mathbf{T}_1$ consists of the first r columns of $\mathbf{T}$. Setting $\mathbf{S}=\mathbf{T}_1\boldsymbol{\Lambda}_r^{1/2}$ leads to the required result.

5. If $\mathbf{A}$ is p.s.d. then $\mathbf{X'AX}=\mathbf{O}\Rightarrow\mathbf{AX}=\mathbf{O}$.

   Proof. From *A3.3*, $\mathbf{O}=\mathbf{X'AX}=\mathbf{X'RR'X}=\mathbf{B'B}$ $(\mathbf{B}=\mathbf{R'X})$, which implies that $\mathbf{b}_i'\mathbf{b}_i=0$, that is, $\mathbf{b}_i=\mathbf{0}$ for every column $\mathbf{b}_i$ of $\mathbf{B}$. Hence $\mathbf{AX}=\mathbf{RB}=\mathbf{O}$.

A4. POSITIVE DEFINITE MATRICES

A symmetric matrix $\mathbf{A}$ is said to be positive definite (p.d.) if $\mathbf{x'Ax}>0$ for all $\mathbf{x}$, $\mathbf{x}\neq\mathbf{0}$. We note that a p.d. matrix is also p.s.d.

1. The eigenvalues of a p.d. matrix $\mathbf{A}$ are all positive (proof is similar to *A3.1*); thus $\mathbf{A}$ is also nonsingular (*A2.6*).
2. $\mathbf{A}$ is p.d. if and only if there exists a nonsingular $\mathbf{R}$ such that $\mathbf{A}=\mathbf{RR}'$.

   Proof. This follows from *A3.3* with $r=n$.

3. If $\mathbf{A}$ is p.d. then so is $\mathbf{A}^{-1}$.

   Proof. $\mathbf{A}^{-1}=(\mathbf{RR}')^{-1}=(\mathbf{R}')^{-1}\mathbf{R}^{-1}=(\mathbf{R}^{-1})'\mathbf{R}^{-1}=\mathbf{SS}'$, where $\mathbf{S}$ is nonsingular. The result then follows from *A4.2* above.

4. If $\mathbf{A}$ is p.d. then $\mathrm{rank}\,[\mathbf{CAC}']=\mathrm{rank}\,\mathbf{C}$.

   Proof.
$$\begin{aligned}\mathrm{rank}[\mathbf{CAC}'] &= \mathrm{rank}[\mathbf{CRR'C}']\\ &= \mathrm{rank}[\mathbf{CR}] \qquad (\text{by } A2.4)\\ &= \mathrm{rank}\,\mathbf{C} \qquad (\text{by } A2.2).\end{aligned}$$

5. If $\mathbf{A}$ is an $n\times n$ p.d. matrix and $\mathbf{C}$ is $p\times n$ of rank p, then $\mathbf{CAC}'$ is p.d.

Proof. $\mathbf{x}'\mathbf{CAC}'\mathbf{x}=\mathbf{y}'\mathbf{Ay}\geqslant 0$ with equality $\Leftrightarrow \mathbf{y}=\mathbf{0}\Leftrightarrow \mathbf{C}'\mathbf{x}=\mathbf{0}\Leftrightarrow \mathbf{x}=\mathbf{0}$ (since the columns of $\mathbf{C}'$ are linearly independent). Hence $\mathbf{x}'\mathbf{CAC}'\mathbf{x}>0$ all $\mathbf{x}, \mathbf{x}\neq\mathbf{0}$.

6. If $\mathbf{X}$ is $n\times p$ of rank p then $\mathbf{X}'\mathbf{X}$ is p.d.

Proof. $\mathbf{x}'\mathbf{X}'\mathbf{X}\mathbf{x}=\mathbf{y}'\mathbf{y}\geqslant 0$ with equality $\Leftrightarrow \mathbf{X}\mathbf{x}=\mathbf{0}\Leftrightarrow \mathbf{x}=\mathbf{0}$ (since the columns of $\mathbf{X}$ are linearly independent).

7. $\mathbf{A}$ is p.d. if and only if all the leading minor determinants of $\mathbf{A}$ (including $|A|$ itself) are positive.

Proof. Given $\mathbf{x}'\mathbf{A}\mathbf{x}$ is p.d., then

$$|\mathbf{A}|=|\mathbf{T}'\mathbf{T}||\mathbf{A}|=|\mathbf{T}'\mathbf{AT}|=|\mathbf{\Lambda}|=\prod_i \lambda_i>0 \qquad \text{(by } A4.1\text{)}.$$

Let

$$\mathbf{A}_r=\begin{pmatrix} a_{11} & \cdots & a_{1r} \\ \cdots & \cdots & \cdots \\ a_{r1} & \cdots & a_{rr}\end{pmatrix} \quad \text{and} \quad \mathbf{x}_r=\begin{pmatrix} x_1 \\ \cdots \\ x_r\end{pmatrix};$$

then

$$\mathbf{x}_r'\mathbf{A}_r\mathbf{x}_r=(\mathbf{x}_r',\mathbf{0}')\mathbf{A}\begin{pmatrix}\mathbf{x}_r \\ \mathbf{0}\end{pmatrix}>0 \text{ for } \mathbf{x}_r\neq\mathbf{0}$$

and $\mathbf{A}_r$ is positive definite. Hence if $\mathbf{A}$ is $n\times n$, it follows from the above argument that $|\mathbf{A}_r|>0$ $(r=1,2,\ldots,n)$. Conversely, suppose that all the leading minor determinants of $\mathbf{A}$ are positive; then we wish to show that $\mathbf{A}$ is p.d. Let

$$\mathbf{A}=\begin{pmatrix}\mathbf{A}_{n-1}, & \mathbf{c} \\ \mathbf{c}', & a_{nn}\end{pmatrix} \quad \text{and} \quad \mathbf{R}=\begin{pmatrix}\mathbf{I}_{n-1}, & \boldsymbol{\alpha} \\ \mathbf{0}', & -1\end{pmatrix},$$

where $\boldsymbol{\alpha}=\mathbf{A}_{n-1}^{-1}\mathbf{c}$. Then

$$\mathbf{R}'\mathbf{AR}=\begin{pmatrix}\mathbf{A}_{n-1}, & \mathbf{0} \\ \mathbf{0}', & k\end{pmatrix},$$

where

$$k=|\mathbf{R}'\mathbf{AR}|/|\mathbf{A}_{n-1}|=|\mathbf{R}|^2|\mathbf{A}|/|\mathbf{A}_{n-1}|>0 \qquad \text{(since } \mathbf{R} \text{ is nonsingular)}.$$

We now proceed by induction. The result is trivially true for $n=1$; assume it is true for matrices of orders up to $n-1$. If we set $\mathbf{y}=\mathbf{R}^{-1}\mathbf{x}$ $(\mathbf{x}\neq\mathbf{0})$, $\mathbf{x}'\mathbf{A}\mathbf{x}=\mathbf{y}'\mathbf{R}'\mathbf{AR}\mathbf{y}=\mathbf{y}_{n-1}'\mathbf{A}_{n-1}\mathbf{y}_{n-1}+ky_n^2>0$, since $\mathbf{A}_{n-1}$ is p.d. by the inductive hypothesis, and $\mathbf{y}\neq\mathbf{0}$. Hence the result is true for matrices of order n.

8. The diagonal elements of a p.d. matrix are all positive.

   Proof. Setting $x_j = \delta_{ij}$ $(j=1,2,\ldots,n)$ we have $0 < \mathbf{x}'\mathbf{A}\mathbf{x} = a_{ii}$.

9. If $\mathbf{A}$ is an $n \times n$ p.d. matrix and $\mathbf{B}$ is an $n \times n$ symmetric matrix, then $\mathbf{A} - t\mathbf{B}$ is p.d. for $|t|$ sufficiently small.

   Proof. The ith leading minor determinant of $\mathbf{A} - t\mathbf{B}$ is a function of t which is positive when $t=0$ (by *A*4.7 above). Since this function is continuous it will be positive for $|t| < \delta_i$ for δ_i sufficiently small. Let $\delta = \text{minimum } (\delta_1, \delta_2, \ldots, \delta_n)$, then all the leading minor determinants will be positive for $|t| < \delta$ and the result follows from *A*4.7.

10. (Cholesky decomposition). If $\mathbf{A}$ is p.d. there exists a unique upper triangular matrix $\mathbf{U}$ with positive diagonal elements such that $\mathbf{A} = \mathbf{U}'\mathbf{U}$.

    Proof. We proceed by induction and assume that the unique factorisation holds for matrices of orders up to $n-1$. Thus

$$\mathbf{A} = \begin{pmatrix} \mathbf{A}_{n-1}, & \mathbf{c} \\ \mathbf{c}', & a_{nn} \end{pmatrix}$$

$$= \begin{pmatrix} \mathbf{U}'_{n-1}\mathbf{U}_{n-1}, & \mathbf{c} \\ \mathbf{c}', & a_{nn} \end{pmatrix},$$

where $\mathbf{U}_{n-1}$ is a unique upper triangular matrix of order $n-1$ with positive diagonal elements. Since the determinant of a triangular matrix is the product of its diagonal elements, $\mathbf{U}_{n-1}$ is nonsingular and we can define

$$\mathbf{U} = \begin{pmatrix} \mathbf{U}_{n-1}, & \mathbf{d} \\ \mathbf{0}', & \sqrt{k} \end{pmatrix},$$

where $\mathbf{d} = (\mathbf{U}'_{n-1})^{-1}\mathbf{c}$ and $k = a_{nn} - \mathbf{d}'\mathbf{d}$. Since $\mathbf{U}$ is unique and $\mathbf{A} = \mathbf{U}'\mathbf{U}$, we have the required decomposition of $\mathbf{A}$ provided $k > 0$. We take determinants

$$|\mathbf{A}| = |\mathbf{U}'\mathbf{U}| = |\mathbf{U}|^2 = |\mathbf{U}_{n-1}|^2 k,$$

so that k is positive as $|\mathbf{A}| > 0$ (*a*4.7) and $|\mathbf{U}_{n-1}| \neq 0$. Thus the factorization also holds for positive definite matrices of order n.

11. If $\mathbf{L}$ is positive definite, then for any $\mathbf{b}$

$$\underset{\mathbf{h}:\mathbf{h}\neq\mathbf{0}}{\text{supremum}} \left\{ \frac{(\mathbf{h}'\mathbf{b})^2}{\mathbf{h}'\mathbf{L}\mathbf{h}} \right\} = \mathbf{b}'\mathbf{L}^{-1}\mathbf{b}.$$

Proof. For all a

$$\begin{aligned}0 &\leqslant \|(\mathbf{v}-a\mathbf{u})\|^2\\ &= a^2\|\mathbf{u}\|^2 - 2a\mathbf{u}'\mathbf{v} + \|\mathbf{v}\|^2\\ &= \left(a\|\mathbf{u}\| - \frac{\mathbf{u}'\mathbf{v}}{\|\mathbf{u}\|}\right)^2 + \|\mathbf{v}\|^2 - \frac{(\mathbf{u}'\mathbf{v})^2}{\|\mathbf{u}\|^2}.\end{aligned}$$

Hence, given $\mathbf{u}\neq\mathbf{0}$, we have the Cauchy–Schwartz inequality $\|\mathbf{v}\|^2\|\mathbf{u}\|^2 \geqslant (\mathbf{u}'\mathbf{v})^2$ with equality if and only if $\mathbf{v}=a\mathbf{u}$ for some a; thus

$$\underset{\mathbf{v}:\mathbf{v}\neq\mathbf{0}}{\text{supremum}} \left\{\frac{(\mathbf{u}'\mathbf{v})^2}{\mathbf{v}'\mathbf{v}}\right\} = \mathbf{u}'\mathbf{u}.$$

Because $\mathbf{L}$ is positive definite there exists a nonsingular matrix $\mathbf{R}$ such that $\mathbf{L}=\mathbf{R}\mathbf{R}'$ (*A4.2*). Setting $\mathbf{v}=\mathbf{R}'\mathbf{h}$ and $\mathbf{u}=\mathbf{R}^{-1}\mathbf{b}$ leads to the required result.

A5. IDEMPOTENT MATRICES

A matrix $\mathbf{P}$ is idempotent if $\mathbf{P}^2=\mathbf{P}$. A symmetric idempotent matrix is called a projection matrix.

1. If $\mathbf{P}$ is symmetric, then $\mathbf{P}$ is idempotent and of rank r if and only if it has r eigenvalues equal to unity and $n-r$ eigenvalues equal to zero.

   Proof. Given $\mathbf{P}^2=\mathbf{P}$, then $\mathbf{P}\mathbf{x}=\lambda\mathbf{x}$ $(\mathbf{x}\neq\mathbf{0})$ implies that $\lambda\mathbf{x}'\mathbf{x}=\mathbf{x}'\mathbf{P}\mathbf{x}=\mathbf{x}'\mathbf{P}^2\mathbf{x}=(\mathbf{P}\mathbf{x})'(\mathbf{P}\mathbf{x})=\lambda^2\mathbf{x}'\mathbf{x}$, and $\lambda(\lambda-1)=0$. Hence the eigenvalues are 0 or 1 and, by *A2.6*, $\mathbf{P}$ has r eigenvalues equal to unity and $n-r$ eigenvalues equal to zero. Conversely, if the eigenvalues are 0 or 1, then we can assume without loss of generality that the first r eigenvalues are unity. Hence there exists an orthogonal matrix $\mathbf{T}$ such that

   $$\mathbf{T}'\mathbf{P}\mathbf{T} = \begin{pmatrix}\mathbf{I}_r & \mathbf{O}\\ \mathbf{O} & \mathbf{O}\end{pmatrix} = \mathbf{\Lambda}, \text{ or } \mathbf{P}=\mathbf{T}\mathbf{\Lambda}\mathbf{T}'.$$

   Therefore $\mathbf{P}^2=\mathbf{T}\mathbf{\Lambda}\mathbf{T}'\mathbf{T}\mathbf{\Lambda}\mathbf{T}'=\mathbf{T}\mathbf{\Lambda}^2\mathbf{T}'=\mathbf{T}\mathbf{\Lambda}\mathbf{T}'=\mathbf{P}$, and rank $\mathbf{P}=r$ (*A2.2*).

2. If $\mathbf{P}$ is a projection matrix then $\operatorname{tr}\mathbf{P}=\operatorname{rank}\mathbf{P}$.

   Proof. If $\operatorname{rank}\mathbf{P}=r$ then, by *A5.1* above, $\mathbf{P}$ has r unit eigenvalues and $n-r$ zero eigenvalues. Hence $\operatorname{tr}\mathbf{P}=r$ (*A1.3*).

3. If $\mathbf{P}$ is idempotent, so is $\mathbf{I}-\mathbf{P}$.

   Proof. $(\mathbf{I}-\mathbf{P})^2=\mathbf{I}-2\mathbf{P}+\mathbf{P}^2=\mathbf{I}-2\mathbf{P}+\mathbf{P}=\mathbf{I}-\mathbf{P}$.

4. Projections matrices are positive semidefinite.

   Proof. $\mathbf{x}'\mathbf{P}\mathbf{x}=\mathbf{x}'\mathbf{P}^2\mathbf{x}=(\mathbf{P}\mathbf{x})'(\mathbf{P}\mathbf{x})\geqslant 0.$

5. If $\mathbf{P}_i$ $(i=1,2)$ is a projection matrix and $\mathbf{P}_1-\mathbf{P}_2$ is p.s.d., then
   (a) $\mathbf{P}_1\mathbf{P}_2=\mathbf{P}_2\mathbf{P}_1=\mathbf{P}_2$,
   (b) $\mathbf{P}_1-\mathbf{P}_2$ is a projection matrix.

   Proof. (a) Given $\mathbf{P}_1\mathbf{x}=\mathbf{0}$, then $0\leqslant\mathbf{x}'(\mathbf{P}_1-\mathbf{P}_2)\mathbf{x}=-\mathbf{x}'\mathbf{P}_2\mathbf{x}$. Since $\mathbf{P}_2$ is positive semidefinite (*A5.4*), $\mathbf{x}'\mathbf{P}_2\mathbf{x}=0$ and $\mathbf{P}_2\mathbf{x}=\mathbf{0}$. Hence for any $\mathbf{y}$, $\mathbf{P}_2(\mathbf{I}-\mathbf{P}_1)\mathbf{y}=\mathbf{0}$ as $\mathbf{P}_1(\mathbf{I}-\mathbf{P}_1)\mathbf{y}=\mathbf{0}$. Thus $\mathbf{P}_2\mathbf{P}_1\mathbf{y}=\mathbf{P}_2\mathbf{y}$, which implies that $\mathbf{P}_2\mathbf{P}_1=\mathbf{P}_2$ (*A9.1*). Taking transposes leads to $\mathbf{P}_1\mathbf{P}_2=\mathbf{P}_2$, and (a) is proved.
   (b) $(\mathbf{P}_1-\mathbf{P}_2)^2=\mathbf{P}_1^2-\mathbf{P}_1\mathbf{P}_2-\mathbf{P}_2\mathbf{P}_1+\mathbf{P}_2^2=\mathbf{P}_1-\mathbf{P}_2-\mathbf{P}_2+\mathbf{P}_2=\mathbf{P}_1-\mathbf{P}_2$.

A6. VECTOR DIFFERENTIATION

If

$$\frac{d}{d\boldsymbol{\beta}}=\left[\left(\frac{d}{d\beta_i}\right)\right]$$

then

1. $$\frac{d(\boldsymbol{\beta}'\mathbf{a})}{d\boldsymbol{\beta}}=\mathbf{a}.$$
2. $$\frac{d(\boldsymbol{\beta}'\mathbf{A}\boldsymbol{\beta})}{d\boldsymbol{\beta}}=2\mathbf{A}\boldsymbol{\beta}\qquad(\mathbf{A}\text{ symmetric}).$$

   Proof. (1) is trivial. Also

$$\begin{aligned}\frac{d(\boldsymbol{\beta}'\mathbf{A}\boldsymbol{\beta})}{d\beta_i}&=\frac{d}{d\beta_i}\left(\sum_i\sum_j a_{ij}\beta_i\beta_j\right)\\&=2a_{ii}\beta_i+2\sum_{j\neq i}a_{ij}\beta_j\\&=2\sum_j a_{ij}\beta_j\\&=2(\mathbf{A}\boldsymbol{\beta})_i.\end{aligned}$$

A7. PARTITIONED MATRICES

If $\mathbf{A}$ and $\mathbf{D}$ are symmetric, and all inverses exist,

$$\begin{pmatrix}\mathbf{A}&\mathbf{B}\\\mathbf{B}'&\mathbf{D}\end{pmatrix}^{-1}=\begin{pmatrix}\mathbf{A}^{-1}+\mathbf{F}\mathbf{E}^{-1}\mathbf{F}'&-\mathbf{F}\mathbf{E}^{-1}\\-\mathbf{E}^{-1}\mathbf{F}'&\mathbf{E}^{-1}\end{pmatrix}$$

where $\mathbf{E}=\mathbf{D}-\mathbf{B}'\mathbf{A}^{-1}\mathbf{B}$ and $\mathbf{F}=\mathbf{A}^{-1}\mathbf{B}$.

Proof. As the inverse of a matrix is unique we only have to check that the matrix times its inverse is the identity matrix.

A8. SOLUTION OF LINEAR EQUATIONS

Every solution of the consistent equations $\mathbf{Bx}=\mathbf{c}$ can be expressed in the form $\mathbf{B}^{-}\mathbf{c}$ for some $\mathbf{B}^{-}$, a generalized inverse of $\mathbf{B}$.

Proof. We first show that all solutions of $\mathbf{Bx}=\mathbf{c}$ are, for any specific $\mathbf{B}^{-}$ $(=\mathbf{C}$, say), generated by

$$\tilde{\mathbf{x}}=\mathbf{Cc}+(\mathbf{CB}-\mathbf{I})\mathbf{z}, \tag{1}$$

where $\mathbf{z}$ is arbitrary.

Now

$$\begin{aligned}\mathbf{B}\tilde{\mathbf{x}}&=\mathbf{BCc}+(\mathbf{BCB}-\mathbf{B})\mathbf{z}\\&=\mathbf{BCc}\\&=\mathbf{c}\ \left[\text{by Equation (3.45), Section 3.8.1c}\right]\end{aligned}$$

so that $\tilde{\mathbf{x}}$ is a solution. Conversely if $\hat{\mathbf{x}}$ is any solution, then setting $\mathbf{z}=-\hat{\mathbf{x}}$, we have

$$\begin{aligned}\tilde{\mathbf{x}}&=\mathbf{Cc}-(\mathbf{CB}-\mathbf{I})\hat{\mathbf{x}}\\&=\mathbf{Cc}-\mathbf{Cc}+\hat{\mathbf{x}}\\&=\hat{\mathbf{x}}\end{aligned}$$

and $\hat{\mathbf{x}}$ is a member of (1) above. This establishes the equivalence of the two sets of solutions.

If $\hat{\mathbf{x}}$ is a solution, then by the above argument, $\hat{\mathbf{x}}$ is given by (1) for some $\mathbf{z}$. If we choose $\mathbf{M}$ such that $\mathbf{z}=-\mathbf{Mc}$ (select $c_k\neq 0$ and let $\mathbf{M}=[(m_{ij})]$, where $m_{ij}=-\delta_{jk}z_i c_k^{-1}$), then

$$\begin{aligned}\hat{\mathbf{x}}&=\mathbf{Cc}-(\mathbf{CB}-\mathbf{I})\mathbf{Mc}\\&=(\mathbf{C}-\mathbf{CBM}+\mathbf{M})\mathbf{c}\\&=\mathbf{Dc},\end{aligned}$$

say, where

$$\begin{aligned}\mathbf{BDB}&=\mathbf{BCB}-\mathbf{BCBMB}+\mathbf{BMB}\\&=\mathbf{B}-\mathbf{BMB}+\mathbf{BMB}\\&=\mathbf{B}.\end{aligned}$$

Thus **D** is a generalized inverse of **B**, and $\hat{\mathbf{x}}$ is of the form $\mathbf{B}^{-}\mathbf{c}$. (This proof is based on Searle [1971: Chapter 1]).

A9. TWO EQUATIONS

1. If $\mathbf{Ax}=\mathbf{0}$ for all $\mathbf{x}$, then $\mathbf{A}=\mathbf{O}$.

   Proof. Setting $x_k=\delta_{ik}$ $(k=1,2,\ldots,n)$, we have $\mathbf{Ax}=\mathbf{a}_i=\mathbf{0}$ where $\mathbf{a}_i$ is the ith column of $\mathbf{A}$.

2. If $\mathbf{A}$ is symmetric and $\mathbf{x}'\mathbf{Ax}=0$ for all $\mathbf{x}$, then $\mathbf{A}=\mathbf{O}$.

   Proof. Setting $x_k=\delta_{ik}$ $(k=1,2,\ldots,n)$ then $a_{ii}=0$. If we set $x_k=\delta_{ik}+\delta_{jk}$ $(k=1,2,\ldots,n)$, then $\mathbf{x}'\mathbf{Ax}=0 \Rightarrow a_{ii}+2a_{ij}=0 \Rightarrow a_{ij}=0$.

A10. SINGULAR VALUE DECOMPOSITION

Let $\mathbf{X}$ be an $n\times p$ matrix. Then $\mathbf{X}$ can be expressed in the form

$$\mathbf{X}=\mathbf{P}\boldsymbol{\Sigma}\mathbf{Q}',$$

where $\mathbf{P}$ is an $n\times p$ matrix consisting of p orthonormalized eigenvectors associated with the p largest eigenvalues of $\mathbf{XX}'$, $\mathbf{Q}$ is a $p\times p$ orthogonal matrix consisting of the orthonormalized eigenvectors of $\mathbf{X}'\mathbf{X}$, and $\boldsymbol{\Sigma}=\operatorname{diag}(\sigma_1,\sigma_2,\ldots,\sigma_p)$ is a $p\times p$ diagonal matrix. Here $\sigma_1\geqslant\sigma_2\geqslant\cdots\geqslant\sigma_p\geqslant 0$, called the singular values of $\mathbf{X}$, are the square roots of the (nonnegative) eigenvalues of $\mathbf{X}'\mathbf{X}$.

Proof. Suppose $\operatorname{rank}[\mathbf{X}'\mathbf{X}]=\operatorname{rank}\mathbf{X}=r$ (*A2.5*). Then there exists a $p\times p$ orthogonal matrix $\mathbf{T}$ such that

$$\mathbf{X}'\mathbf{XT}=\mathbf{T}\boldsymbol{\Lambda}$$

where $\boldsymbol{\Lambda}=\operatorname{diag}(\sigma_1^2,\sigma_2^2,\ldots,\sigma_r^2,0,\ldots,0), \sigma_i^2>0$. Let

$$\mathbf{s}_i=\sigma_i^{-1}\mathbf{Xt}_i \qquad (i=1,2,\ldots,r);$$

then $\mathbf{X}'\mathbf{s}_i=\sigma_i^{-1}\mathbf{X}'\mathbf{Xt}_i=\sigma_i\mathbf{t}_i$ and $\mathbf{XX}'\mathbf{s}_i=\sigma_i\mathbf{Xt}_i=\sigma_i^2\mathbf{s}_i$. Thus the $\mathbf{s}_i$ $(i=1,2,\ldots,r)$ are eigenvectors of $\mathbf{XX}'$ corresponding to the eigenvalues σ_i^2 $(i=1,2,\ldots,r)$. Now $\mathbf{s}_i'\mathbf{s}_i=1$ and, since the eigenvectors corresponding to different eigenvectors of a symmetric matrix are orthogonal, the $\mathbf{s}_i$ are orthonormal. By *A2.3* and *A2.4* there exists an orthonormal set $\{\mathbf{s}_{r+1},\mathbf{s}_{r+2},\ldots,\mathbf{s}_n\}$ spanning $\mathcal{N}[\mathbf{XX}']$ $(=\mathcal{N}[\mathbf{X}'])$. But $\mathcal{N}[\mathbf{X}']\perp\mathcal{R}[\mathbf{X}]$ and $\mathbf{s}_i\in\mathcal{R}[\mathbf{X}]$ $(i=1,2,\ldots,r)$ so that $\mathbf{S}=(\mathbf{s}_1,\mathbf{s}_2,\ldots,\mathbf{s}_n)$ is an $n\times n$ orthogonal matrix. Hence

$$\begin{aligned}(\mathbf{S}'\mathbf{XT})_{ij}=\mathbf{s}_i'(\mathbf{Xt}_j)&=\sigma_i\mathbf{s}_i'\mathbf{s}_j, \qquad i=1,2,\ldots,r\\ &=0, \qquad i=r+1,\ldots,n\end{aligned}$$

and $\mathbf{S}'\mathbf{X}\mathbf{T}=\begin{pmatrix}\mathbf{\Sigma}\\ \mathbf{O}\end{pmatrix}$. Finally

$$\mathbf{X}=\mathbf{S}\begin{pmatrix}\mathbf{\Sigma}\\ \mathbf{O}\end{pmatrix}\mathbf{T}'=\mathbf{P}\mathbf{\Sigma}\mathbf{Q}'$$

where $\mathbf{P}$ is the first p columns of $\mathbf{S}$ and $\mathbf{Q}=\mathbf{T}$.

A11. SOME MISCELLANEOUS STATISTICAL RESULTS

1. For any random variable $X, \gamma_2 \geqslant -2$.

Proof. Let $\mu = E[X]$; then

$$\begin{aligned}0 &\leqslant \operatorname{var}\left[(X-\mu)^2\right]\\ &= E\left[(X-\mu)^4\right]-\left\{E\left[(X-\mu)^2\right]\right\}^2\\ &= \mu_4-\mu_2^2\\ &= \mu_2^2\left(\frac{\mu_4}{\mu_2^2}-3+2\right)\\ &= \mu_2^2(\gamma_2+2)\end{aligned}$$

and $\gamma_2+2 \geqslant 0$.

2. Let X be a nonnegative nondegenerate (that is not identically equal to a constant) random variable. If the expectations exist then

$$E[X^{-1}]>(E[X])^{-1}.$$

Proof. Let $f(x)=x^{-1}$ and let $\mu=E[X]$ (>0, since X is not identically zero). Taking a Taylor expansion we have

$$f(X)=f(\mu)+(X-\mu)f'(\mu)+\tfrac{1}{2}(X-\mu)^2 f''(X_0),$$

where X_0 lies between X and μ. Now $f''(X_0)=2X_0^{-3}>0$ so that $E[(X-\mu)^2 f''(X_0)]>0$. Hence

$$E[X^{-1}]=E[f(X)]>f(\mu)=(E[X])^{-1}.$$

APPENDIX B

Orthogonal Projections

B1. ORTHOGONAL DECOMPOSITION OF VECTORS

1. Given Ω, a vector subspace of E_n (n-dimensional Euclidean space), every $n\times 1$ vector $\mathbf{y}$ can be expressed uniquely in the form $\mathbf{y}=\mathbf{u}+\mathbf{v}$, where $\mathbf{u}\in\Omega$ and $\mathbf{v}\in\Omega^{\perp}$.

   Proof. Suppose there are two such decompositions $\mathbf{y}=\mathbf{u}_i+\mathbf{v}_i$ $(i=1,2)$; then $(\mathbf{u}_1-\mathbf{u}_2)+(\mathbf{v}_1-\mathbf{v}_2)=\mathbf{0}$. Because $(\mathbf{u}_1-\mathbf{u}_2)\in\Omega$ and $(\mathbf{v}_1-\mathbf{v}_2)\in\Omega^{\perp}$ we must have $\mathbf{u}_1=\mathbf{u}_2$ and $\mathbf{v}_1=\mathbf{v}_2$.

2. If $\mathbf{u}=\mathbf{P}_\Omega\mathbf{y}$ then $\mathbf{P}_\Omega$ is unique.

   Proof. Given two such matrices $\mathbf{P}_i$ $(i=1,2)$, then, since $\mathbf{u}$ is unique for *every* $\mathbf{y}$, $(\mathbf{P}_1-\mathbf{P}_2)\mathbf{y}=\mathbf{0}$ for all $\mathbf{y}$; hence $\mathbf{P}_1-\mathbf{P}_2=\mathbf{O}$ (*A9.1*).

3. The matrix $\mathbf{P}_\Omega$ can be expressed in the form $\mathbf{P}_\Omega=\mathbf{T}\mathbf{T}'$ where the columns of $\mathbf{T}$ form an orthonormal basis for Ω.

   Proof. Let $\mathbf{T}=(\boldsymbol{\alpha}_1,\boldsymbol{\alpha}_2,\ldots,\boldsymbol{\alpha}_r)$, where r is the dimension of Ω. Expand the set of $\boldsymbol{\alpha}_i$ to give an orthonormal basis for E_n, namely, $\boldsymbol{\alpha}_1,\ldots,\boldsymbol{\alpha}_r,\boldsymbol{\alpha}_{r+1},\ldots,\boldsymbol{\alpha}_n$. Then

$$\mathbf{y}=\sum_{i=1}^{n} c_i\boldsymbol{\alpha}_i=\sum_{i=1}^{r} c_i\boldsymbol{\alpha}_i+\sum_{i=r+1}^{n} c_i\boldsymbol{\alpha}_i=\mathbf{u}+\mathbf{v},$$

   where $\mathbf{u}\in\Omega$ and $\mathbf{v}\in\Omega^{\perp}$. But $\boldsymbol{\alpha}_i'\boldsymbol{\alpha}_j=\delta_{ij}$ so that $\boldsymbol{\alpha}_i'\mathbf{y}=\mathbf{c}_i$. Hence

$$\mathbf{u}=(\boldsymbol{\alpha}_1,\ldots,\boldsymbol{\alpha}_r)\begin{bmatrix}\boldsymbol{\alpha}_1'\mathbf{y}\\ \cdots\\ \boldsymbol{\alpha}_r'\mathbf{y}\end{bmatrix}=\mathbf{T}\mathbf{T}'\mathbf{y}.$$

   By (2), $\mathbf{P}_\Omega=\mathbf{T}\mathbf{T}'$.

4. $\mathbf{P}_\Omega$ is symmetric and idempotent.

Proof. $\mathbf{P}_\Omega = \mathbf{T}\mathbf{T}'$, which is obviously symmetric, and

$$\mathbf{P}_\Omega^2 = \mathbf{T}\mathbf{T}'\mathbf{T}\mathbf{T}' = \mathbf{T}\mathbf{I}_r\mathbf{T}' = \mathbf{T}\mathbf{T}' = \mathbf{P}_\Omega.$$

5. $\mathcal{R}[\mathbf{P}_\Omega] = \Omega$.

Proof. Clearly $\mathcal{R}[\mathbf{P}_\Omega] \subset \Omega$ since $\mathbf{P}_\Omega$ projects onto Ω. Conversely, if $\mathbf{x} \in \Omega$ then $\mathbf{x} = \mathbf{P}_\Omega \mathbf{x} \in \mathcal{R}[\mathbf{P}]$. Thus the two spaces are the same.

6. $\mathbf{I}_n - \mathbf{P}_\Omega$ represents an orthogonal projection on $\Omega^\perp$.

Proof. From the identity $\mathbf{y} = \mathbf{P}_\Omega \mathbf{y} + (\mathbf{I}_n - \mathbf{P}_\Omega)\mathbf{y}$ we have that $\mathbf{v} = (\mathbf{I}_n - \mathbf{P}_\Omega)\mathbf{y}$. The above results then apply by interchanging the roles of Ω and $\Omega^\perp$.

7. If $\mathbf{P}$ is a symmetric idempotent $n \times n$ matrix, then $\mathbf{P}$ represents an orthogonal projection onto $\mathcal{R}[\mathbf{P}]$.

Proof. Let $\mathbf{y} = \mathbf{P}\mathbf{y} + (\mathbf{I}_n - \mathbf{P})\mathbf{y}$. Then $(\mathbf{P}\mathbf{y})'(\mathbf{I}_n - \mathbf{P})\mathbf{y} = \mathbf{y}'(\mathbf{P} - \mathbf{P}^2)\mathbf{y} = 0$ so that this decomposition gives orthogonal components of $\mathbf{y}$. The result then follows from (5).

8. If $\Omega = \mathcal{R}[\mathbf{X}]$ then $\mathbf{P}_\Omega = \mathbf{X}(\mathbf{X}'\mathbf{X})^{-}\mathbf{X}'$, where $(\mathbf{X}'\mathbf{X})^{-}$ is any generalized inverse of $\mathbf{X}'\mathbf{X}$ (i.e., if $\mathbf{B} = \mathbf{X}'\mathbf{X}$ then $\mathbf{B}\mathbf{B}^{-}\mathbf{B} = \mathbf{B}$).

Proof. Let $\mathbf{c} = \mathbf{X}'\mathbf{Y}$. Then $\mathbf{B}(\mathbf{B}^{-}\mathbf{c}) = \mathbf{B}\mathbf{B}^{-}\mathbf{B}\boldsymbol{\beta} = \mathbf{B}\boldsymbol{\beta}$ and $\hat{\boldsymbol{\beta}} = \mathbf{B}^{-}\mathbf{c}$ is a solution of $\mathbf{B}\boldsymbol{\beta} = \mathbf{c}$, that is, of $\mathbf{X}'\mathbf{X}\boldsymbol{\beta} = \mathbf{X}'\mathbf{Y}$. Hence writing $\hat{\boldsymbol{\theta}} = \mathbf{X}\hat{\boldsymbol{\beta}}$, we have $\mathbf{Y} = \hat{\boldsymbol{\theta}} + (\mathbf{Y} - \hat{\boldsymbol{\theta}})$ where

$$\begin{aligned}\hat{\boldsymbol{\theta}}'(\mathbf{Y} - \hat{\boldsymbol{\theta}}) &= \hat{\boldsymbol{\beta}}'\mathbf{X}'(\mathbf{Y} - \mathbf{X}'\mathbf{X}\hat{\boldsymbol{\beta}}) \\ &= \hat{\boldsymbol{\beta}}'(\mathbf{X}'\mathbf{Y} - \mathbf{X}'\mathbf{X}\hat{\boldsymbol{\beta}}) \\ &= 0.\end{aligned}$$

Thus we have an orthogonal decomposition of $\mathbf{Y}$ such that $\hat{\boldsymbol{\theta}} \in \mathcal{R}[\mathbf{X}]$ and $(\mathbf{Y} - \hat{\boldsymbol{\theta}}) \perp \mathcal{R}[\mathbf{X}]$. Since $\hat{\boldsymbol{\theta}} = \mathbf{X}\hat{\boldsymbol{\beta}} = \mathbf{X}(\mathbf{X}'\mathbf{X})^{-}\mathbf{X}'\mathbf{Y}$ we have that $\mathbf{P}_\Omega = \mathbf{X}(\mathbf{X}'\mathbf{X})^{-}\mathbf{X}'$ [by (2)].

9. When the columns of $\mathbf{X}$ are linearly independent in (8) then $\mathbf{P}_\Omega = \mathbf{X}(\mathbf{X}'\mathbf{X})^{-1}\mathbf{X}'$.

Proof. Although (9) follows from (8) the result can be proved directly since $\mathbf{X} = \mathbf{T}\mathbf{C}$ for nonsingular $\mathbf{C}$ [by (3) above] and

$$\mathbf{P}_\Omega = \mathbf{X}\mathbf{C}^{-1}(\mathbf{C}^{-1})'\mathbf{X}' = \mathbf{X}(\mathbf{C}'\mathbf{C})^{-1}\mathbf{X}' = \mathbf{X}(\mathbf{X}'\mathbf{X})^{-1}\mathbf{X}'.$$

B2. ORTHOGONAL COMPLEMENTS

1. If $\mathcal{N}[\mathbf{C}]$ is the null space (kernel) of the matrix $\mathbf{C}$, then $\mathcal{N}[\mathbf{C}] = \{\mathcal{R}[\mathbf{C}']\}^{\perp}$.

   Proof. If $\mathbf{x} \in \mathcal{N}[\mathbf{C}]$ then $\mathbf{Cx} = \mathbf{0}$ and $\mathbf{x}$ is orthogonal to each row of $\mathbf{C}$. Hence $\mathbf{x} \perp \mathcal{R}[\mathbf{C}']$. Conversely, if $\mathbf{x} \perp \mathcal{R}[\mathbf{C}']$ then $\mathbf{Cx} = \mathbf{0}$ and $\mathbf{x} \in \mathcal{N}[\mathbf{C}]$.

2. $(\Omega_1 \cap \Omega_2)^{\perp} = \Omega_1^{\perp} + \Omega_2^{\perp}$.

   Proof. Let $\mathbf{C}_i$ be such that $\Omega_i = \mathcal{N}[\mathbf{C}_i]$ $(i = 1, 2)$. Then

$$
\begin{aligned}
(\Omega_1 \cap \Omega_2)^{\perp} &= \left\{ \mathcal{N}\begin{bmatrix} \mathbf{C}_1 \\ \mathbf{C}_2 \end{bmatrix} \right\}^{\perp} \\
&= \mathcal{R}[\mathbf{C}_1', \mathbf{C}_2'] \qquad \text{(by } B2.1\text{)} \\
&= \mathcal{R}[\mathbf{C}_1'] + \mathcal{R}[\mathbf{C}_2'] \\
&= \Omega_1^{\perp} + \Omega_2^{\perp}
\end{aligned}
$$

B3. PROJECTIONS ON SUBSPACES

1. Given $\omega \subset \Omega$ then $\mathbf{P}_{\Omega}\mathbf{P}_{\omega} = \mathbf{P}_{\omega}\mathbf{P}_{\Omega} = \mathbf{P}_{\omega}$.

   Proof. Since $\omega \subset \Omega$ and $\omega = \mathcal{R}[\mathbf{P}_{\omega}]$ (by *B1.5*) we have $\mathbf{P}_{\Omega}\mathbf{P}_{\omega} = \mathbf{P}_{\omega}$. The result then follows by the symmetry of $\mathbf{P}_{\omega}$ and $\mathbf{P}_{\Omega}$.

2. $\mathbf{P}_{\Omega} - \mathbf{P}_{\omega} = \mathbf{P}_{\omega^{\perp} \cap \Omega}$.

   Proof. Consider $\mathbf{P}_{\Omega}\mathbf{y} = \mathbf{P}_{\omega}\mathbf{y} + (\mathbf{P}_{\Omega} - \mathbf{P}_{\omega})\mathbf{y}$. Now $\mathbf{P}_{\Omega}\mathbf{y}$ and $\mathbf{P}_{\omega}\mathbf{y}$ belong to Ω so that $(\mathbf{P}_{\Omega} - \mathbf{P}_{\omega})\mathbf{y} \in \Omega$. Hence the preceding equation represents an orthogonal decomposition of Ω into ω and $\omega^{\perp} \cap \Omega$ since $\mathbf{P}_{\omega}(\mathbf{P}_{\Omega} - \mathbf{P}_{\omega}) = \mathbf{O}$ (by *B3.1*).

3. If $\mathbf{A}_1$ is any matrix such that $\omega = \mathcal{N}[\mathbf{A}_1] \cap \Omega$, then $\omega^{\perp} \cap \Omega = \mathcal{R}[\mathbf{P}_{\Omega}\mathbf{A}_1']$.

   Proof.

$$
\begin{aligned}
\omega^{\perp} \cap \Omega &= \{\Omega \cap \mathcal{N}[\mathbf{A}_1]\}^{\perp} \cap \Omega \\
&= \{\Omega^{\perp} + \mathcal{R}[\mathbf{A}_1']\} \cap \Omega \qquad \text{(by } B2.1 \text{ and } B2.2\text{)}.
\end{aligned}
$$

   If $\mathbf{x}$ belongs to the right side then

$$
\mathbf{x} = \mathbf{P}_{\Omega}\mathbf{x} = \mathbf{P}_{\Omega}\{(\mathbf{I}_n - \mathbf{P}_{\Omega})\boldsymbol{\alpha} + \mathbf{A}_1'\boldsymbol{\beta}\} = \mathbf{P}_{\Omega}\mathbf{A}_1'\boldsymbol{\beta} \in \mathcal{R}[\mathbf{P}_{\Omega}\mathbf{A}_1'].
$$

   Conversely, if $\mathbf{x} \in \mathcal{R}[\mathbf{P}_{\Omega}\mathbf{A}_1']$ then $\mathbf{x} \in \mathcal{R}[\mathbf{P}_{\Omega}] = \Omega$. Also if $\mathbf{z} \in \omega$, then $\mathbf{x}'\mathbf{z} = \boldsymbol{\beta}'\mathbf{A}_1\mathbf{P}_{\Omega}\mathbf{z} = \boldsymbol{\beta}'\mathbf{A}_1\mathbf{z} = 0$, that is, $\mathbf{x} \in \omega^{\perp}$. Thus $\mathbf{x} \in \omega^{\perp} \cap \Omega$.

4. If $\mathbf{A}_1$ is a $q \times n$ matrix of rank q, then $\text{rank}[\mathbf{P}_\Omega \mathbf{A}_1'] = q$ if and only if $\mathcal{R}[\mathbf{A}_1'] \cap \Omega^\perp = \mathbf{0}$.

   Proof. $\text{rank}[\mathbf{P}_\Omega \mathbf{A}_1'] \leqslant \text{rank}\, \mathbf{A}_1$ (by *A2.1*). Let the rows of $\mathbf{A}_1$ be $\mathbf{a}_i'$ ($i = 1, 2, \ldots, q$) and suppose that $\text{rank}[\mathbf{P}_\Omega \mathbf{A}_1'] < q$. Then the columns of $\mathbf{P}_\Omega \mathbf{A}_1'$ are linearly dependent so that $\Sigma_{i=1}^q c_i \mathbf{P}_\Omega \mathbf{a}_i = \mathbf{0}$, that is, there exists a vector $\Sigma_i c_i \mathbf{a}_i \in \mathcal{R}[\mathbf{A}_1']$ which is perpendicular to Ω. Hence $\mathcal{R}[\mathbf{A}_1'] \cap \Omega^\perp \neq \mathbf{0}$, which is a contradiction. [By selecting the linearly independent rows of $\mathbf{A}_1$ we find that the above result is true if $\mathbf{A}_1$ is $k \times n$ ($k \geqslant q$).]

APPENDIX C

Normal Probability Paper

If $Z \sim N(0, 1)$, then the graph of the distribution function $y = \Phi(z)$ $(= \text{pr}(Z \leqslant z))$ is the S-shaped curve in Fig. 1. However, this curve can be transformed into a straight line by applying a nonlinear transformation Φ to the vertical scale. The form of this transformation can be visualized by thinking of the curve as plotted on an elastic sheet which is now stretched out to make the curve a straight line; the direction and amount of stretching is shown in Fig. 1. There is a special graph paper (see Fig. 2) which incorporates this nonlinear vertical scale (expressed as a percentage) called normal-probability paper or simply *probability paper*. Thus in Fig. 2 the graph of $y[= 100\Phi(z)]$ against z is now the straight line $y = z$.

If $Z \sim N(\theta, \sigma^2)$ the corresponding graph in Fig. 2 is $y = (z - \theta)/\sigma$, which is still a straight line but with different slope and location.

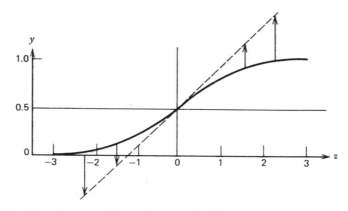

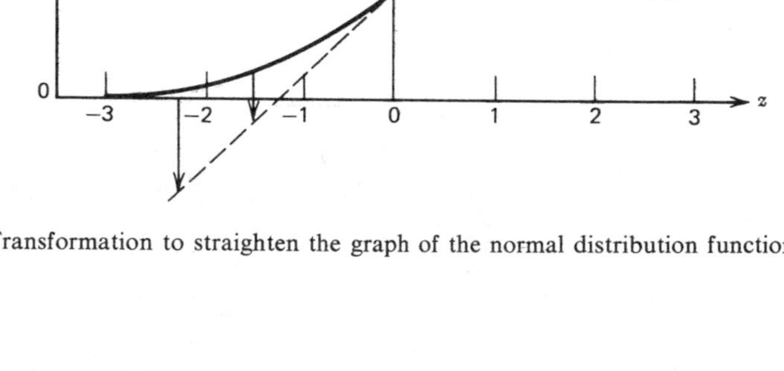

Fig. 1 Transformation to straighten the graph of the normal distribution function. (Appendix C)

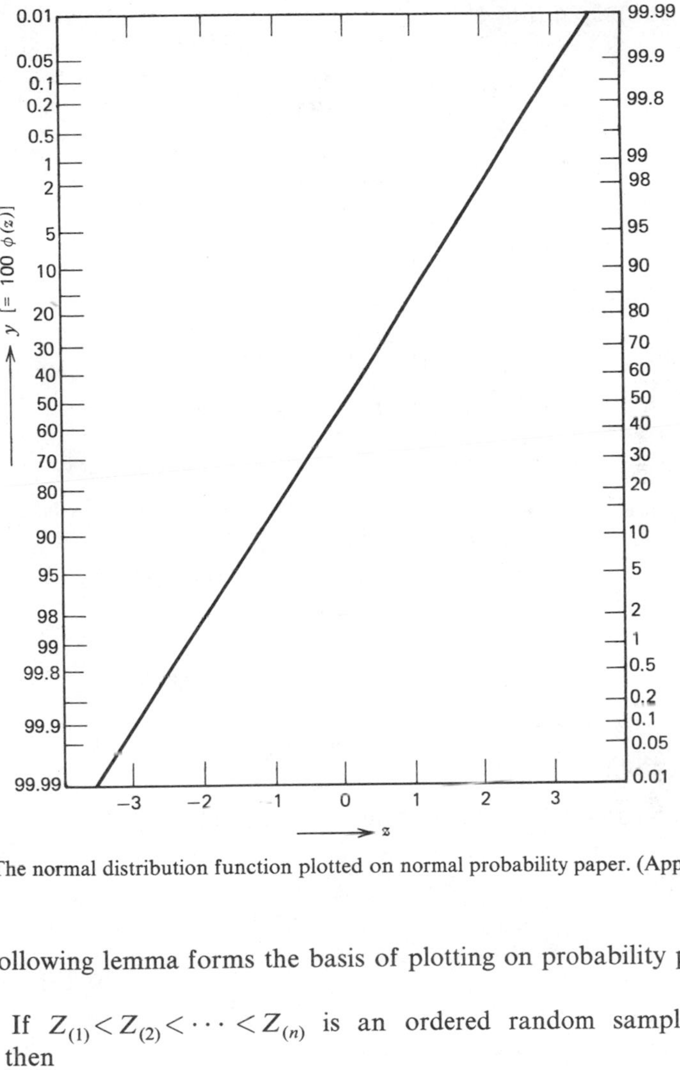

Fig. 2 The normal distribution function plotted on normal probability paper. (Appendix C)

The following lemma forms the basis of plotting on probability paper.

LEMMA If $Z_{(1)} < Z_{(2)} < \cdots < Z_{(n)}$ is an ordered random sample from $N(0,1)$, then

$$E\left[\Phi(Z_{(i)})\right] = \frac{i}{n+1} \qquad (i = 1,2,\ldots,n).$$

Proof. (outline only). Consider the transformation $U = \Phi(Z)$. Since Φ is a monotonic increasing function, the inverse function exists so that $Z = \Phi^{-1}(U) = Z(U)$, say. Now if ϕ is the probability density function (p.d.f.)

of Z, then the p.d.f. of U is

$$\begin{aligned} g(u) &= \phi(z(u))\left|\frac{dz}{du}\right| \qquad 0 \leqslant u \leqslant 1 \\ &= \phi(z(u))\left|\frac{du}{dz}\right|^{-1} \\ &= \frac{\phi(z(u))}{\Phi'(z)} \\ &= \frac{\phi(z(u))}{\phi(z(u))} \\ &= 1, \end{aligned}$$

which is the uniform distribution. Therefore, if we set $U_{(i)} = \Phi(Z_{(i)})$, $U_{(1)} < U_{(2)} < \cdots < U_{(n)}$ is an ordered random sample from the uniform (rectangular) distribution. From symmetry we would expect the $U_{(i)}$ to divide the interval $[0,1]$ into roughly $n+1$ equal parts; thus

$$E[\Phi(Z_{(i)})] = E[U_{(i)}] = \frac{i}{n+1}. \tag{1}$$

(More rigorously, it can be shown that $U_{(i)}$ has a beta distribution; e.g., David [1970].) □

It can also be shown that (David [1970: pp. 64–67, 161–163])

$$E[Z_{(i)}] \cong \Phi^{-1}\left(\frac{i-\frac{1}{2}}{n}\right) \tag{2}$$

or

$$\Phi(E[Z_{(i)}]) \cong \frac{i-\frac{1}{2}}{n}. \tag{3}$$

Thus from Equation (2) we see that the assumption $Z \sim N(0,1)$ can be checked by comparing $E[Z_{(i)}]$ with $Z_{(i)}$. Alternatively, using Equations (3) and (1), we can compare either $\Phi(E[Z_{(i)}]$ or $E[\Phi(Z_{(i)})]$ with $\Phi(Z_{(i)})$. If we use probability paper then we plot $(i-\frac{1}{2})/n$ or $i/(n+1)$ against $Z_{(i)}$; for n sufficiently large (so that observations are close to their expected values), the plot should be approximately linear if Z is normal.

For mainly empirical reasons $(i-\frac{1}{2})/n$ is preferred to $i/(n+1)$. Also it is common practice to orient the probability paper so that the nonlinear scale is on the x-axis; $Z_{(i)}$ is plotted against $(i-\frac{1}{2})/n$. Several examples of such plots with sample sizes ranging from 8 to 384 are given in Daniel and Wood [1971: pp. 34–43] and should be referred to. The authors conclude that samples of size 8 tell us almost nothing about normality; sets of 16 show shocking wobbles; sets of 32 are visibly better behaved; sets of 64 nearly always appear straight in their central regions but fluctuate at their ends; sets of 384 seem very stable except for their few lowest and highest points. It is recommended that n be at least 20 and preferably greater than 50. Further details and references regarding probability plotting are given in Wilk and Gnanadesikan [1968]. Some interesting nonnormal plots are graphed in Hahn and Shapiro [1967: Chapter 8].

If plotting is done automatically by computer then we plot $Z_{(i)}$ against $\Phi^{-1}[(i-\frac{1}{2})/n)]$ [cf. Equation (2)]; a number of efficient numerical approximations for Φ^{-1} are available. Andrews and Tukey [1973] describe a crude but useful method of plotting using a six-line printout from a teletypewriter.

In addition to its use in examining residuals (Section 6.6), probability plotting has been used for examining various hypotheses in some analysis of variance situations. For example, Daniel [1959] used so-called "half-normal" plotting in the analysis of 2^n factorial experiments. On the null assumption of no real treatment effects for any of the factors, the absolute values of the various contrasts (main effects, first-order interactions, second-order interactions, etc.) would behave like a random sample from a half-normal distribution. The ordered absolute contrasts are then plotted against representative values from the standard half-normal distribution and, in the absence of real effects or interactions, the plot is expected to be linear through the origin. Real effects and interactions would show up as large deviations of the corresponding contrasts from the linear configuration.

It is, however, more difficult to apply probability plotting to a general analysis of variance because even under the null model of no effects, interactions, etc., the mean squares (SS/df) in the usual analysis of variance table do not have the same distribution owing to their different degrees of freedom. One method of tackling this problem is given by Gnanadesikan and Wilk [1970].

APPENDIX D

Percentage Points of the Bonferroni t-Statistic

Tabulation of $t_\nu^{\alpha/(2k)}$ for Different Values of α, k, and ν, Where

$$\text{pr}\left[T \geqslant t_\nu^{\alpha/(2k)}\right] = \frac{\alpha}{2k}$$

and T has the t-Distribution with ν Degrees of Freedom (see Section 5.1.1a in the text)

					$\alpha=0.05$							
$k\backslash\nu$	5	7	10	12	15	20	24	30	40	60	120	∞
2	3.17	2.84	2.64	2.56	2.49	2.42	2.39	2.36	2.33	2.30	2.27	2.24
3	3.54	3.13	2.87	2.78	2.69	2.61	2.58	2.54	2.50	2.47	2.43	2.39
4	3.81	3.34	3.04	2.94	2.84	2.75	2.70	2.66	2.62	2.58	2.54	2.50
5	4.04	3.50	3.17	3.06	2.95	2.85	2.80	2.75	2.71	2.66	2.62	2.58
6	4.22	3.64	3.28	3.15	3.04	2.93	2.88	2.83	2.78	2.73	2.68	2.64
7	4.38	3.76	3.37	3.24	3.11	3.00	2.94	2.89	2.84	2.79	2.74	2.69
8	4.53	3.86	3.45	3.31	3.18	3.06	3.00	2.94	2.89	2.84	2.79	2.74
9	4.66	3.95	3.52	3.37	3.24	3.11	3.05	2.99	2.93	2.88	2.83	2.77
10	4.78	4.03	3.58	3.43	3.29	3.16	3.09	3.03	2.97	2.92	2.86	2.81
15	5.25	4.36	3.83	3.65	3.48	3.33	3.26	3.19	3.12	3.06	2.99	2.94
20	5.60	4.59	4.01	3.80	3.62	3.46	3.38	3.30	3.23	3.16	3.09	3.02
25	5.89	4.78	4.15	3.93	3.74	3.55	3.47	3.39	3.31	3.24	3.16	3.09
30	6.15	4.95	4.27	4.04	3.82	3.63	3.54	3.46	3.38	3.30	3.22	3.15
35	6.36	5.09	4.37	4.13	3.90	3.70	3.61	3.52	3.43	3.34	3.27	3.19
40	6.56	5.21	4.45	4.20	3.97	3.76	3.66	3.57	3.48	3.39	3.31	3.23
45	6.70	5.31	4.53	4.26	4.02	3.80	3.70	3.61	3.51	3.42	3.34	3.26
50	6.86	5.40	4.59	4.32	4.07	3.85	3.74	3.65	3.55	3.46	3.37	3.29
100	8.00	6.08	5.06	4.73	4.42	4.15	4.04	3.90	3.79	3.69	3.58	3.48
250	9.68	7.06	5.70	5.27	4.90	4.56	4.4*	4.2*	4.1*	3.97	3.83	3.72

$\alpha = 0.01$

$k \backslash \nu$	5	7	10	12	15	20	24	30	40	60	120	∞
2	4.78	4.03	3.58	3.43	3.29	3.16	3.09	3.03	2.97	2.92	2.86	2.81
3	5.25	4.36	3.83	3.65	3.48	3.33	3.26	3.19	3.12	3.06	2.99	2.94
4	5.60	4.59	4.01	3.80	3.62	3.46	3.38	3.30	3.23	3.16	3.09	3.02
5	5.89	4.78	4.15	3.93	3.74	3.55	3.47	3.39	3.31	3.24	3.16	3.09
6	6.15	4.95	4.27	4.04	3.82	3.63	3.54	3.46	3.38	3.30	3.22	3.15
7	6.36	5.09	4.37	4.13	3.90	3.70	3.61	3.52	3.43	3.34	3.27	3.19
8	6.56	5.21	4.45	4.20	3.97	3.76	3.66	3.57	3.48	3.39	3.31	3.23
9	6.70	5.31	4.53	4.26	4.02	3.80	3.70	3.61	3.51	3.42	3.34	3.26
10	6.86	5.40	4.59	4.32	4.07	3.85	3.74	3.65	3.55	3.46	3.37	3.29
15	7.51	5.79	4.86	4.56	4.29	4.03	3.91	3.80	3.70	3.59	3.50	3.40
20	8.00	6.08	5.06	4.73	4.42	4.15	4.04	3.90	3.79	3.69	3.58	3.48
25	8.37	6.30	5.20	4.86	4.53	4.25	4.1*	3.98	3.88	3.76	3.64	3.54
30	8.68	6.49	5.33	4.95	4.61	4.33	4.2*	4.13	3.93	3.81	3.69	3.59
35	8.95	6.67	5.44	5.04	4.71	4.39	4.3*	4.26	3.97	3.84	3.73	3.63
40	9.19	6.83	5.52	5.12	4.78	4.46	4.3*	4.1*	4.01	3.89	3.77	3.66
45	9.41	6.93	5.60	5.20	4.84	4.52	4.3*	4.2*	4.1*	3.93	3.80	3.69
50	9.68	7.06	5.70	5.27	4.90	4.56	4.4*	4.2*	4.1*	3.97	3.83	3.72
100	11.04	7.80	6.20	5.70	5.20	4.80	4.7*	4.4*	4.5*		4.00	3.89
250	13.26	8.83	6.9*	6.3*	5.8*	5.2*	5.0*	4.9*	4.8*			4.11

SOURCE. Dunn [1961: Tables 1 and 2].
*Obtained by graphical interpolation.

APPENDIX E

Distribution of the Largest Absolute Value of k Student t Variates

Tabulation of $u^{\alpha}_{k,\nu,\rho}$ for Different Values of ρ, α, k, and ν, where

$$\text{pr}\left[U \geqslant u^{\alpha}_{k,\nu,\rho}\right] = \alpha$$

and U is the Maximum Absolute Value of k Student t-Variables, Each Based on ν Degrees of Freedom and Having a Common Pairwise Correlation ρ (see Section 5.1.1b in the text)

	$\rho=0.0$										
$\nu\backslash k$	1	2	3	4	5	6	8	10	12	15	20
					$\alpha=0.10$						
3	2.353	2.989	3.369	3.637	3.844	4.011	4.272	4.471	4.631	4.823	5.066
4	2.132	2.662	2.976	3.197	3.368	3.506	3.722	3.887	4.020	4.180	4.383
5	2.015	2.491	2.769	2.965	3.116	3.239	3.430	3.576	3.694	3.837	4.018
6	1.943	2.385	2.642	2.822	2.961	3.074	3.249	3.384	3.493	3.624	3.790
7	1.895	2.314	2.556	2.725	2.856	2.962	3.127	3.253	3.355	3.478	3.635
8	1.860	2.262	2.494	2.656	2.780	2.881	3.038	3.158	3.255	3.373	3.522
9	1.833	2.224	2.447	2.603	2.723	2.819	2.970	3.086	3.179	3.292	3.436
10	1.813	2.193	2.410	2.562	2.678	2.771	2.918	3.029	3.120	3.229	3.368
11	1.796	2.169	2.381	2.529	2.642	2.733	2.875	2.984	3.072	3.178	3.313
12	1.782	2.149	2.357	2.501	2.612	2.701	2.840	2.946	3.032	3.136	3.268
15	1.753	2.107	2.305	2.443	2.548	2.633	2.765	2.865	2.947	3.045	3.170
20	1.725	2.065	2.255	2.386	2.486	2.567	2.691	2.786	2.863	2.956	3.073
25	1.708	2.041	2.226	2.353	2.450	2.528	2.648	2.740	2.814	2.903	3.016
30	1.697	2.025	2.207	2.331	2.426	2.502	2.620	2.709	2.781	2.868	2.978
40	1.684	2.006	2.183	2.305	2.397	2.470	2.585	2.671	2.741	2.825	2.931
60	1.671	1.986	2.160	2.278	2.368	2.439	2.550	2.634	2.701	2.782	2.884

APPENDIX E (continued)

$\rho = 0.0$

$\nu \backslash k$	1	2	3	4	5	6	8	10	12	15	20
						$\alpha = 0.05$					
3	3.183	3.960	4.430	4.764	5.023	5.233	5.562	5.812	6.015	6.259	6.567
4	2.777	3.382	3.745	4.003	4.203	4.366	4.621	4.817	4.975	5.166	5.409
5	2.571	3.091	3.399	3.619	3.789	3.928	4.145	4.312	4.447	4.611	4.819
6	2.447	2.916	3.193	3.389	3.541	3.664	3.858	4.008	4.129	4.275	4.462
7	2.365	2.800	3.056	3.236	3.376	3.489	3.668	3.805	3.916	4.051	4.223
8	2.306	2.718	2.958	3.128	3.258	3.365	3.532	3.660	3.764	3.891	4.052
9	2.262	2.657	2.885	3.046	3.171	3.272	3.430	3.552	3.651	3.770	3.923
10	2.228	2.609	2.829	2.984	3.103	3.199	3.351	3.468	3.562	3.677	3.823
11	2.201	2.571	2.784	2.933	3.048	3.142	3.288	3.400	3.491	3.602	3.743
12	2.179	2.540	2.747	2.892	3.004	3.095	3.236	3.345	3.433	3.541	3.677
15	2.132	2.474	2.669	2.805	2.910	2.994	3.126	3.227	3.309	3.409	3.536
20	2.086	2.411	2.594	2.722	2.819	2.898	3.020	3.114	3.190	3.282	3.399
25	2.060	2.374	2.551	2.673	2.766	2.842	2.959	3.048	3.121	3.208	3.320
30	2.042	2.350	2.522	2.641	2.732	2.805	2.918	3.005	3.075	3.160	3.267
40	2.021	2.321	2.488	2.603	2.690	2.760	2.869	2.952	3.019	3.100	3.203
60	2.000	2.292	2.454	2.564	2.649	2.716	2.821	2.900	2.964	3.041	3.139
						$\alpha = 0.01$					
3	5.841	7.127	7.914	8.479	8.919	9.277	9.838	10.269	10.616	11.034	11.559
4	4.604	5.462	5.985	6.362	6.656	6.897	7.274	7.565	7.801	8.087	8.451
5	4.032	4.700	5.106	5.398	5.625	5.812	6.106	6.333	6.519	6.744	7.050
6	3.707	4.271	4.611	4.855	5.046	5.202	5.449	5.640	5.796	5.985	6.250
7	3.500	3.998	4.296	4.510	4.677	4.814	5.031	5.198	5.335	5.502	5.716
8	3.355	3.809	4.080	4.273	4.424	4.547	4.742	4.894	5.017	5.168	5.361
9	3.250	3.672	3.922	4.100	4.239	4.353	4.532	4.672	4.785	4.924	5.103
10	3.169	3.567	3.801	3.969	4.098	4.205	4.373	4.503	4.609	4.739	4.905
11	3.106	3.485	3.707	3.865	3.988	4.087	4.247	4.370	4.470	4.593	4.750
12	3.055	3.418	3.631	3.782	3.899	3.995	4.146	4.263	4.359	4.475	4.625
15	2.947	3.279	3.472	3.608	3.714	3.800	3.935	4.040	4.125	4.229	4.363
20	2.845	3.149	3.323	3.446	3.541	3.617	3.738	3.831	3.907	3.999	4.117
25	2.788	3.075	3.239	3.354	3.442	3.514	3.626	3.713	3.783	3.869	3.978
30	2.750	3.027	3.185	3.295	3.379	3.448	3.555	3.637	3.704	3.785	3.889
40	2.705	2.969	3.119	3.223	3.303	3.367	3.468	3.545	3.607	3.683	3.780
60	2.660	2.913	3.055	3.154	3.229	3.290	3.384	3.456	3.515	3.586	3.676

APPENDIX E (continued)

$\nu \backslash k$	1	2	3	4	5	6	8	10	12	15	20
					$\rho=0.2$						
					$\alpha=0.10$						
3	2.353	2.978	3.347	3.607	3.806	3.967	4.216	4.405	4.557	4.739	4.969
4	2.132	2.653	2.958	3.172	3.337	3.470	3.676	3.833	3.960	4.112	4.303
5	2.015	2.482	2.753	2.943	3.089	3.207	3.390	3.530	3.642	3.778	3.948
6	1.943	2.377	2.627	2.802	2.937	3.045	3.213	3.342	3.446	3.570	3.728
7	1.895	2.306	2.542	2.707	2.833	2.935	3.093	3.214	3.312	3.429	3.577
8	1.860	2.255	2.481	2.638	2.759	2.856	3.007	3.122	3.214	3.326	3.468
9	1.833	2.217	2.435	2.586	2.702	2.796	2.941	3.052	3.141	3.248	3.384
10	1.813	2.187	2.399	2.546	2.658	2.749	2.889	2.997	3.083	3.187	3.319
11	1.796	2.163	2.370	2.513	2.623	2.711	2.848	2.952	3.036	3.138	3.266
12	1.782	2.143	2.346	2.487	2.594	2.680	2.814	2.916	2.998	3.097	3.222
15	1.753	2.101	2.295	2.429	2.531	2.613	2.741	2.837	2.915	3.009	3.128
20	1.725	2.060	2.245	2.373	2.470	2.548	2.669	2.761	2.835	2.923	3.036
25	1.708	2.036	2.217	2.341	2.435	2.510	2.627	2.716	2.787	2.873	2.981
30	1.697	2.020	2.198	2.319	2.412	2.485	2.600	2.686	2.756	2.839	2.945
40	1.684	2.000	2.174	2.293	2.383	2.455	2.566	2.649	2.717	2.798	2.900
60	1.671	1.981	2.151	2.267	2.354	2.424	2.532	2.613	2.679	2.757	2.856
					$\alpha=0.05$						
3	3.183	3.946	4.403	4.727	4.976	5.178	5.492	5.731	5.923	6.154	6.445
4	2.777	3.371	3.725	3.975	4.168	4.325	4.569	4.755	4.906	5.087	5.316
5	2.571	3.082	3.383	3.596	3.760	3.893	4.102	4.261	4.390	4.545	4.742
6	2.447	2.908	3.178	3.369	3.516	3.635	3.821	3.964	4.079	4.218	4.395
7	2.365	2.793	3.042	3.218	3.353	3.463	3.634	3.766	3.872	4.000	4.163
8	2.306	2.711	2.946	3.111	3.238	3.340	3.501	3.624	3.724	3.844	3.997
9	2.262	2.650	2.874	3.031	3.151	3.249	3.402	3.518	3.613	3.727	3.873
10	2.228	2.603	2.818	2.969	3.084	3.178	3.324	3.436	3.527	3.637	3.776
11	2.201	2.565	2.774	2.919	3.031	3.122	3.263	3.371	3.458	3.564	3.698
12	2.179	2.535	2.738	2.879	2.988	3.075	3.212	3.317	3.402	3.504	3.635
15	2.132	2.469	2.660	2.793	2.895	2.977	3.105	3.203	3.282	3.377	3.499
20	2.086	2.406	2.586	2.711	2.806	2.883	3.002	3.093	3.166	3.255	3.367
25	2.060	2.370	2.543	2.663	2.754	2.828	2.942	3.029	3.099	3.183	3.291
30	2.042	2.346	2.515	2.632	2.721	2.792	2.903	2.987	3.055	3.137	3.241
40	2.021	2.317	2.481	2.594	2.679	2.748	2.855	2.936	3.001	3.097	3.179
60	2.000	2.288	2.447	2.556	2.639	2.705	2.808	2.886	2.948	3.023	3.119

APPENDIX E (continued)

$\alpha=0.01$

3	5.841	7.104	7.871	8.418	8.841	9.184	9.721	10.132	10.462	10.860	11.360
4	4.604	5.447	5.958	6.323	6.607	6.838	7.200	7.477	7.702	7.973	8.316
5	4.032	4.690	5.085	5.369	5.589	5.769	6.051	6.268	6.444	6.658	6.930
6	3.707	4.263	4.595	4.832	5.017	5.168	5.405	5.588	5.736	5.917	6.147

$\rho=0.2$

$\nu\backslash k$	1	2	3	4	5	6	8	10	12	15	20
					$\alpha=0.01$						
7	3.500	3.991	4.283	4.491	4.653	4.786	4.994	5.155	5.286	5.445	5.648
8	3.355	3.803	4.068	4.257	4.403	4.523	4.711	4.857	4.975	5.119	5.303
9	3.250	3.666	3.911	4.086	4.221	4.331	4.505	4.639	4.748	4.881	5.051
10	3.169	3.562	3.792	3.956	4.082	4.186	4.348	4.474	4.576	4.700	4.859
11	3.106	3.480	3.699	3.854	3.974	4.071	4.225	4.344	4.440	4.558	4.708
12	3.055	3.414	3.623	3.771	3.886	3.979	4.126	4.239	4.331	4.443	4.587
15	2.947	3.276	3.466	3.599	3.703	3.787	3.919	4.020	4.103	4.204	4.332
20	2.845	3.146	3.318	3.439	3.532	3.607	3.725	3.816	3.890	3.980	4.094
25	2.788	3.072	3.235	3.348	3.435	3.506	3.616	3.701	3.769	3.853	3.959
30	2.750	3.025	3.181	3.289	3.373	3.440	3.545	3.626	3.692	3.771	3.872
40	2.705	2.967	3.115	3.218	3.297	3.361	3.460	3.536	3.598	3.672	3.767
60	2.660	2.911	3.052	3.150	3.224	3.285	3.378	3.449	3.507	3.577	3.666

$\rho=0.4$

$\nu\backslash k$	1	2	3	4	5	6	8	10	12	15	20
					$\alpha=0.10$						
3	2.353	2.941	3.282	3.519	3.700	3.845	4.069	4.237	4.373	4.534	4.737
4	2.132	2.623	2.905	3.101	3.250	3.370	3.556	3.696	3.809	3.943	4.113
5	2.015	2.455	2.706	2.880	3.013	3.120	3.284	3.410	3.510	3.630	3.781
6	1.943	2.352	2.584	2.745	2.867	2.965	3.117	3.233	3.325	3.436	3.575
7	1.895	2.283	2.502	2.653	2.768	2.861	3.004	3.112	3.199	3.304	3.435
8	1.860	2.233	2.442	2.587	2.697	2.786	2.922	3.026	3.109	3.208	3.334
9	1.833	2.195	2.398	2.538	2.644	2.729	2.860	2.960	3.040	3.136	3.257
10	1.813	2.166	2.363	2.499	2.602	2.684	2.812	2.909	2.986	3.079	3.196
11	1.796	2.142	2.335	2.468	2.568	2.649	2.773	2.867	2.943	3.034	3.148
12	1.782	2.123	2.312	2.442	2.541	2.620	2.742	2.834	2.908	2.996	3.108
15	1.753	2.081	2.263	2.387	2.481	2.556	2.673	2.760	2.831	2.916	3.022
20	1.725	2.041	2.216	2.334	2.424	2.496	2.606	2.690	2.757	2.837	2.938
25	1.708	2.018	2.188	2.303	2.390	2.460	2.567	2.649	2.713	2.791	2.888
30	1.697	2.003	2.169	2.283	2.368	2.437	2.542	2.621	2.684	2.760	2.856
40	1.684	1.984	2.146	2.257	2.341	2.408	2.510	2.587	2.650	2.723	2.816
60	1.671	1.965	2.124	2.233	2.315	2.379	2.479	2.554	2.615	2.686	2.776

APPENDIX E (continued)

$\alpha=0.05$

3	3.183	3.902	4.324	4.620	4.846	5.028	5.309	5.522	5.693	5.898	6.155
4	2.777	3.337	3.665	3.894	4.069	4.210	4.430	4.596	4.730	4.891	5.093
5	2.571	3.053	3.333	3.528	3.677	3.798	3.986	4.128	4.243	4.381	4.555

$\rho=0.4$

$\nu \backslash k$	1	2	3	4	5	6	8	10	12	15	20
					$\alpha=0.05$						
6	2.447	2.883	3.134	3.309	3.443	3.552	3.719	3.847	3.950	4.074	4.230
7	2.365	2.770	3.002	3.164	3.288	3.388	3.543	3.661	3.756	3.870	4.014
8	2.306	2.690	2.909	3.061	3.177	3.271	3.417	3.528	3.617	3.725	3.860
9	2.262	2.630	2.839	2.984	3.095	3.184	3.323	3.429	3.513	3.616	3.745
10	2.228	2.584	2.785	2.925	3.032	3.117	3.250	3.352	3.433	3.531	3.655
11	2.201	2.547	2.742	2.877	2.980	3.063	3.192	3.290	3.369	3.464	3.583
12	2.179	2.517	2.707	2.838	2.939	3.020	3.145	3.240	3.317	3.409	3.525
15	2.132	2.452	2.632	2.756	2.850	2.927	3.043	3.133	3.205	3.291	3.400
20	2.086	2.391	2.560	2.677	2.766	2.837	2.947	3.031	3.098	3.178	3.280
25	2.060	2.355	2.520	2.631	2.718	2.786	2.891	2.971	3.036	3.113	3.211
30	2.042	2.332	2.492	2.602	2.685	2.751	2.854	2.933	2.995	3.070	3.165
40	2.021	2.304	2.459	2.565	2.646	2.711	2.810	2.885	2.945	3.018	3.110
60	2.000	2.275	2.426	2.530	2.608	2.670	2.766	2.838	2.897	2.966	3.054
					$\alpha=0.01$						
3	5.841	7.033	7.740	8.240	8.623	8.932	9.414	9.780	10.074	10.428	10.874
4	4.604	5.401	5.874	6.209	6.467	6.675	7.000	7.249	7.448	7.688	7.991
5	4.032	4.655	5.024	5.284	5.485	5.648	5.902	6.096	6.253	6.442	6.682
6	3.707	4.235	4.545	4.764	4.934	5.071	5.285	5.449	5.582	5.742	5.946
7	3.500	3.967	4.241	4.435	4.583	4.704	4.893	5.038	5.155	5.297	5.477
8	3.355	3.783	4.031	4.207	4.343	4.452	4.624	4.755	4.861	4.990	5.154
9	3.250	3.648	3.879	4.041	4.167	4.268	4.427	4.549	4.647	4.766	4.918
10	3.169	3.545	3.763	3.916	4.034	4.129	4.277	4.392	4.484	4.596	4.739
11	3.106	3.464	3.671	3.817	3.929	4.019	4.160	4.269	4.357	4.463	4.598
12	3.055	3.400	3.598	3.737	3.844	3.931	4.066	4.170	4.254	4.356	4.484
15	2.947	3.263	3.444	3.571	3.668	3.746	3.869	3.962	4.039	4.131	4.247
20	2.845	3.135	3.301	3.415	3.504	3.574	3.685	3.769	3.837	3.921	4.026
25	2.788	3.063	3.219	3.327	3.410	3.477	3.581	3.660	3.725	3.802	3.900
30	2.750	3.016	3.166	3.270	3.349	3.415	3.514	3.590	3.650	3.726	3.820
40	2.705	2.959	3.103	3.202	3.277	3.337	3.432	3.505	3.562	3.632	3.722
60	2.660	2.904	3.040	3.134	3.207	3.264	3.353	3.421	3.477	3.542	3.628

APPENDIX E (continued)

$\nu \backslash k$	1	2	3	4	5	6	8	10	12	15	20
						$\rho=0.5$					
						$\alpha=0.10$					
3	2.353	2.912	3.232	3.453	3.621	3.755	3.962	4.117	4.242	4.390	4.576
4	2.132	2.598	2.863	3.046	3.185	3.296	3.468	3.597	3.701	3.825	3.980
5	2.015	2.434	2.669	2.832	2.956	3.055	3.207	3.323	3.415	3.525	3.664

$\nu \backslash k$	1	2	3	4	5	6	8	10	12	15	20
						$\rho=0.5$					
						$\alpha=0.05$					
6	1.943	2.332	2.551	2.701	2.815	2.906	3.047	3.153	3.238	3.340	3.469
7	1.895	2.264	2.471	2.612	2.720	2.806	2.938	3.038	3.119	3.215	3.336
8	1.860	2.215	2.413	2.548	2.651	2.733	2.860	2.956	3.032	3.124	3.239
9	1.833	2.178	2.369	2.500	2.599	2.679	2.801	2.893	2.967	3.055	3.167
10	1.813	2.149	2.335	2.463	2.559	2.636	2.755	2.844	2.916	3.002	3.110
11	1.796	2.126	2.308	2.433	2.527	2.602	2.718	2.805	2.875	2.959	3.064
12	1.782	2.107	2.286	2.408	2.500	2.574	2.687	2.773	2.841	2.923	3.026
15	1.753	2.066	2.238	2.355	2.443	2.514	2.622	2.704	2.769	2.847	2.945
20	1.725	2.027	2.192	2.304	2.388	2.455	2.559	2.637	2.699	2.773	2.867
25	1.708	2.004	2.165	2.274	2.356	2.421	2.522	2.597	2.658	2.730	2.820
30	1.697	1.989	2.147	2.254	2.335	2.399	2.498	2.572	2.631	2.701	2.790
40	1.687	1.970	2.125	2.230	2.309	2.372	2.468	2.540	2.598	2.667	2.753
60	1.671	1.952	2.104	2.207	2.284	2.345	2.439	2.509	2.565	2.632	2.716
						$\alpha=0.05$					
3	3.183	3.867	4.263	4.538	4.748	4.916	5.176	5.372	5.529	5.718	5.953
4	2.777	3.310	3.618	3.832	3.995	4.126	4.328	4.482	4.605	4.752	4.938
5	2.57	3.03	3.29	3.48	3.62	3.73	3.90	4.03	4.14	4.26	4.42
6	2.45	2.86	3.10	3.26	3.39	3.49	3.64	3.76	3.86	3.97	4.11
7	2.36	2.75	2.97	3.12	3.24	3.33	3.47	3.58	3.67	3.78	3.91
8	2.31	2.67	2.88	3.02	3.13	3.22	3.35	3.46	3.54	3.64	3.76
9	2.26	2.61	2.81	2.95	3.05	3.14	3.26	3.36	3.44	3.53	3.65
10	2.23	2.57	2.76	2.89	2.99	3.07	3.19	3.29	3.36	3.45	3.57
11	2.20	2.53	2.72	2.84	2.94	3.02	3.14	3.23	3.30	3.39	3.50
12	2.18	2.50	2.68	2.81	2.90	2.98	3.09	3.18	3.25	3.34	3.45
15	2.13	2.44	2.61	2.73	2.82	2.89	3.00	3.08	3.15	3.23	3.33
20	2.09	2.38	2.54	2.65	2.73	2.80	2.90	2.98	3.05	3.12	3.22
25	2.060	2.344	2.500	2.607	2.688	2.752	2.852	2.927	2.987	3.059	3.150
30	2.04	2.32	2.47	2.58	2.66	2.72	2.82	2.89	2.95	3.02	3.11
40	2.02	2.29	2.44	2.54	2.62	2.68	2.77	2.85	2.90	2.97	3.06
60	2.00	2.27	2.41	2.51	2.58	2.64	2.73	2.80	2.86	2.92	3.00

APPENDIX E (continued)

$\alpha = 0.01$

3	5.841	6.974	7.639	8.104	8.459	8.746	9.189	9.527	9.797	10.123	10.532
4	4.604	5.364	5.809	6.121	6.361	6.554	6.855	7.083	7.267	7.488	7.766
5	4.03	4.63	4.98	5.22	5.41	5.56	5.80	5.98	6.12	6.30	6.52
6	3.71	4.21	4.51	4.71	4.87	5.00	5.20	5.35	5.47	5.62	5.81
7	3.50	3.95	4.21	4.39	4.53	4.64	4.82	4.95	5.06	5.19	5.36
8	3.36	3.77	4.00	4.17	4.29	4.40	4.56	4.68	4.78	4.90	5.05
9	3.25	3.63	3.85	4.01	4.12	4.22	4.37	4.48	4.57	4.68	4.82

$\rho = 0.5$

$\nu \backslash k$	1	2	3	4	5	6	8	10	12	15	20
10	3.17	3.53	3.74	3.88	3.99	4.08	4.22	4.33	4.42	4.52	4.65
11	3.11	3.45	3.65	3.79	3.89	3.98	4.11	4.21	4.29	4.39	4.52
12	3.05	3.39	3.58	3.71	3.81	3.89	4.02	4.12	4.19	4.29	4.41
15	2.95	3.25	3.43	3.55	3.64	3.71	3.83	3.92	3.99	4.07	4.18
20	2.85	3.13	3.29	3.40	3.48	3.55	3.65	3.73	3.80	3.87	3.97
25	2.788	3.055	3.205	3.309	3.388	3.452	3.551	3.626	3.687	3.759	3.852
30	2.75	3.01	3.15	3.25	3.33	3.39	3.49	3.56	3.62	3.69	3.78
40	2.70	2.95	3.09	3.19	3.26	3.32	3.41	3.48	3.53	3.60	3.68
60	2.66	2.90	3.03	3.12	3.19	3.25	3.33	3.40	3.45	3.51	3.59

SOURCE. Hahn and Hendrickson [1971].

APPENDIX F

Working–Hotelling Confidence Bands for Finite Intervals

The Working–Hotelling confidence band for a straight line over the interval $a \leqslant x \leqslant b$ is the region between the two curves given by Equation (7.7) in Section 7.2.3.

Tabulation of λ for Different Values of $n-2$ and c, Where c is Given by Equation (7.9)

	$\alpha=0.01$									
	$c \backslash n-2$	5	10	15	20	30	40	60	120	∞
One point	0.0	4.03	3.16	2.95	2.84	2.75	2.70	2.66	2.62	2.58
	0.05	4.10	3.22	2.99	2.88	2.79	2.74	2.69	2.65	2.61
	0.1	4.18	3.27	3.03	2.93	2.83	2.78	2.73	2.69	2.64
	0.15	4.26	3.32	3.07	2.96	2.86	2.81	2.76	2.72	2.67
	0.2	4.33	3.36	3.11	3.00	2.89	2.84	2.80	2.75	2.70
	0.3	4.45	3.44	3.18	3.06	2.95	2.90	2.85	2.80	2.75
	0.4	4.56	3.50	3.24	3.11	3.00	2.95	2.89	2.84	2.79
	0.6	4.73	3.61	3.32	3.20	3.07	3.02	2.96	2.91	2.86
	0.8	4.85	3.68	3.39	3.25	3.13	3.07	3.01	2.95	2.90
	1.0	4.94	3.74	3.43	3.30	3.17	3.11	3.05	2.99	2.94
	1.5	5.05	3.81	3.50	3.36	3.22	3.16	3.10	3.04	2.98
	2.0	5.10	3.85	3.53	3.38	3.25	3.19	3.15	3.06	3.01
	∞	5.15	3.89	3.57	3.42	3.28	3.22	3.15	3.10	3.04

$\alpha = 0.05$

	$c \backslash n-2$	5	10	15	20	30	40	60	120	∞
One point	0.0	2.57	2.23	2.13	2.08	2.04	2.02	2.00	1.98	1.96
	0.05	2.62	2.27	2.17	2.12	2.08	2.06	2.03	2.02	1.99
	0.1	2.68	2.31	2.21	2.16	2.12	2.10	2.07	2.05	2.03
	0.15	2.74	2.36	2.25	2.20	2.15	2.13	2.11	2.08	2.06
	0.2	2.79	2.40	2.29	2.23	2.18	2.16	2.14	2.11	2.09
	0.3	2.88	2.47	2.35	2.30	2.42	2.22	2.19	2.17	2.15
	0.4	2.97	2.53	2.41	2.35	2.29	2.27	2.24	2.21	2.19
	0.6	3.10	2.62	2.49	2.43	2.37	2.34	2.31	2.29	2.26
	0.8	3.19	2.69	2.55	2.49	2.43	2.38	2.37	2.34	2.31
	1.0	3.25	2.74	2.60	2.53	2.47	2.44	2.41	2.38	2.35
	1.5	3.33	2.81	2.67	2.59	2.52	2.49	2.46	2.43	2.40
	2.0	3.36	2.83	2.68	2.61	2.55	2.51	2.48	2.45	2.42
	∞	3.40	2.86	2.71	2.64	2.58	2.54	2.51	2.48	2.45

$\alpha = 0.10$

	$c \backslash n-2$	5	10	15	20	30	40	60	120	∞
One point	0.0	2.01	1.81	1.75	1.72	1.68	1.68	1.67	1.66	1.65
	0.05	2.06	1.85	1.79	1.76	1.73	1.72	1.70	1.69	1.68
	0.1	2.11	1.89	1.83	1.80	1.77	1.85	1.74	1.73	1.71
	0.15	2.16	1.93	1.87	1.84	1.81	1.79	1.77	1.76	1.75
	0.2	2.21	1.97	1.90	1.87	1.84	1.82	1.81	1.79	1.78
	0.3	2.30	2.04	1.97	1.93	1.90	1.88	1.87	1.85	1.84
	0.4	2.37	2.10	2.02	1.99	1.95	1.93	1.92	1.90	1.88
	0.6	2.49	2.19	2.12	2.07	2.03	2.01	1.99	1.98	1.96
	0.8	2.57	2.26	2.17	2.13	2.09	2.07	2.05	2.03	2.01
	1.0	2.62	2.31	2.22	2.17	2.13	2.11	2.09	2.07	2.05
	1.5	2.69	2.37	2.27	2.23	2.18	2.16	2.15	2.12	2.10
	2.0	2.72	2.39	2.29	2.25	2.20	2.18	2.16	2.14	2.12
	∞	2.75	2.42	2.32	2.27	2.23	2.21	2.19	2.17	2.14

SOURCE. Wynn and Bloomfield [1971: Appendix A].

Outline Solutions to Exercises

EXERCISES 1a

4. $\text{cov}[X_i, X_j] = \sigma^2 \rho^{|i-j|}$.

EXERCISES 1b

1. $E[\mathbf{X}'\mathbf{A}\mathbf{X}] = E[\text{tr}[\mathbf{X}'\mathbf{A}\mathbf{X}]] = E[\text{tr}[\mathbf{A}\mathbf{X}\mathbf{X}']] = \text{tr}[\mathbf{A}\mathscr{E}[\mathbf{X}\mathbf{X}']]$. Also $\mathscr{E}[\mathbf{X}\mathbf{X}'] = \mathscr{D}[\mathbf{X}] + \boldsymbol{\theta}\boldsymbol{\theta}'$.
3. (a) $\text{var}[\bar{X}] = \frac{\sigma^2}{n}\{1 + (n-1)\rho\} \geqslant 0$.

   (b) $E[Q] = \sigma^2\{an + bn(1 + (n-1)\rho\} + \theta^2(an + bn^2) \equiv \sigma^2 + 0$. Hence $b = -1/\{n(n-1)(1-\rho)\}$, $a = -bn$.

4. $\exp(\theta t + \frac{1}{2}\sigma^2 t^2)$.
5. Use Corollary 1 with $\mathbf{A}\boldsymbol{\theta} = \theta\mathbf{A}\mathbf{1}_n = \mathbf{0}$.
   (a) $\text{tr}\,\mathbf{A}^2 = \text{tr}\,\mathbf{A} = n - 1$.
   (c) $\text{tr}\,\mathbf{A}^2 = \Sigma\Sigma a_{ij}^2 = 6n - 8$. Hence $\text{var}[Q] = 2\sigma^4(6n-8)/4(n-1)^2$.

EXERCISES 1c

3. $\text{cov}[X, Y] = p_{11} - p_{\cdot 1}p_{\cdot 1} = 0$ implies that $p_{ij} = p_{i\cdot}p_{\cdot j}$ for all i, j (e.g., $p_{10} = p_{1\cdot} - p_{1\cdot}p_{\cdot 1} = p_{1\cdot}(1 - p_{\cdot 1}) = p_{1\cdot}p_{\cdot 0}$).

4. $E[X^{2r+1}]=0$ because $f(x)$ is an even function. Hence

$$\text{cov}[X,X^2]=E[X^3]-E[X]E[X^2]=0.$$

5. $f(x,y)=\frac{1}{4}=f_1(x)f_2(y)$.

MISCELLANEOUS EXERCISES 1

3. (a) 17; (b) $\begin{pmatrix} 12 & 15 \\ 15 & 20 \end{pmatrix}$.
4. $2\frac{2}{5}$, obtained by substituting in Theorem 1.8 with $\boldsymbol{\theta}=\mathbf{0}$, $\text{tr}\,\mathbf{A}^2=\Sigma\Sigma a_{ij}^2=18$, $\mathbf{a}'\mathbf{a}=12$, $\mu_4=\frac{1}{5}$, and $\mu_2=\frac{1}{3}$.
5. Argue as in Theorems 1.8 and 1.7; thus

$$\begin{aligned}\text{cov}[\mathbf{X}'\mathbf{A}\mathbf{X},\mathbf{X}'\mathbf{B}\mathbf{X}] &= E[\mathbf{X}'\mathbf{A}\mathbf{X}\mathbf{X}'\mathbf{B}\mathbf{X}]-(E[\mathbf{X}'\mathbf{A}\mathbf{X}]E[\mathbf{X}'\mathbf{B}\mathbf{X}]) \\ &= \mu_4\sum_i a_{ii}b_{ii}+\mu_2^2\Big(\sum_{i\neq k}\sum a_{ii}b_{kk}+2\sum_{i\neq k}\sum a_{ik}b_{ik}\Big)-\mu_2^2\,\text{tr}\,\mathbf{A}\,\text{tr}\,\mathbf{B}, \\ &= (\mu_4-3\mu_2^2)\mathbf{a}'\mathbf{b}+2\mu_2^2\,\text{tr}[\mathbf{AB}].\end{aligned}$$

Or else consider var[$\mathbf{X}'(\mathbf{A}+\mathbf{B})\mathbf{X}$].

EXERCISES 2a

1. $2\pi|\boldsymbol{\Sigma}^{-1}|^{1/2}=\pi/\sqrt{3}$ [by Equation (2.6)].
3. Let $\sigma_{ii}=\sigma_i^2$, $\sigma_{12}=\sigma_1\sigma_2\rho$. Expanding the expression gives

$$\frac{1}{(1-\rho^2)}\left(\frac{\rho Y_1}{\sigma_1}-\frac{Y_2}{\sigma_2}\right)^2=\frac{1}{(1-\rho^2)}Y_3^2,$$

where $Y_3 \sim N(0,1-\rho^2)$.

EXERCISES 2b

1. $\exp[\theta_1 t_1+\theta_2 t_2+\frac{1}{2}(\sigma_1^2 t_1^2+2\rho\sigma_1\sigma_2 t_1 t_2+\sigma_2^2 t_2^2)]$.
3. Multivariate normal with mean $\begin{pmatrix} 5 \\ 1 \end{pmatrix}$ and dispersion matrix $\begin{pmatrix} 10 & 0 \\ 0 & 3 \end{pmatrix}$.
4. Let $\mathbf{Z}=\mathbf{CY}$, where $\mathbf{C}=\begin{pmatrix} \mathbf{a}' \\ \mathbf{b}' \end{pmatrix}$ has independent rows. Hence $\mathbf{Z}$ is bivariate normal with diagonal dispersion matrix $\mathbf{CC}'$.

5.

$$E\left[\exp(tQ)\right]=k^{-1}\int\cdots\int\exp\left\{(\mathbf{y}-\boldsymbol{\theta})'\left[(1-2t)^{-1}\boldsymbol{\Sigma}\right]^{-1}(\mathbf{y}-\boldsymbol{\theta})\right\}d\mathbf{y}$$

$$=(2\pi)^{(1/2)n}(1-2t)^{-(1/2)n}|\Sigma|^{1/2}/k=(1-2t)^{-(1/2)n}.$$

7. $N_2\left(\boldsymbol{\theta},\frac{1}{n}\boldsymbol{\Sigma}\right)$. Use moment generating functions, that is,

$$E\left[\exp\{t_1\bar{X}+t_2\bar{Y}\}\right]=\left\{E\left[\exp\left(\frac{1}{n}t_1X+\frac{1}{n}t_2Y\right)\right]\right\}^n \quad \text{etc.}$$

10. $$E\left[\exp\left(s\bar{Y}+\sum_{i=1}^{n}t_i(Y_i-\bar{Y})\right)\right]=\prod_{i=1}^{n}E\left[\exp\left\{\left(t_i-\bar{t}+\frac{s}{n}\right)Y_i\right\}\right]$$

$$=\prod_{i=1}^{n}\exp\left[\tfrac{1}{2}\left(t_i-\bar{t}+\frac{s}{n}\right)^2\right]$$

$$=\exp\left(\tfrac{1}{2}\Sigma(t_i-\bar{t})^2+\frac{s^2}{2n}\right),$$

which factorizes.

EXERCISES 2c

2. Use the transformation

$$\begin{pmatrix}\mathbf{W}\\ \mathbf{X}_2\end{pmatrix}=\begin{pmatrix}\mathbf{I}_p, & -\boldsymbol{\Sigma}_{12}\boldsymbol{\Sigma}_{22}^{-1}\\ \mathbf{0}, & \mathbf{I}_{n-p}\end{pmatrix}\begin{pmatrix}\mathbf{X}_1\\ \mathbf{X}_2\end{pmatrix}$$

or $\mathbf{Y}=\mathbf{C}\mathbf{X}$, which is multivariate normal. Then

$$h(\mathbf{x}_1|\mathbf{x}_2)=\frac{f(\mathbf{x}_1,\mathbf{x}_2)}{f_2(\mathbf{x}_2)}$$

$$=\frac{g(\mathbf{w},\mathbf{x}_2)|\mathbf{C}|}{f_2(\mathbf{x}_2)}$$

$$=\frac{g_1(\mathbf{w})f_2(\mathbf{x}_2)}{f_2(\mathbf{x}_2)} \qquad (|\mathbf{C}|=1)$$

$$=g_1(\mathbf{w}).$$

Since $\mathfrak{D}[\mathbf{W}]=\boldsymbol{\Sigma}_{11}-\boldsymbol{\Sigma}_{12}\boldsymbol{\Sigma}_{22}^{-1}\boldsymbol{\Sigma}_{21}=\boldsymbol{\Sigma}_{11.2}$, $\mathbf{W}\sim N_p(\boldsymbol{\alpha}_1-\boldsymbol{\Sigma}_{12}\boldsymbol{\Sigma}_{22}^{-1}\boldsymbol{\alpha}_2,\boldsymbol{\Sigma}_{11.2})$. The result then follows by writing out the density function of $\mathbf{W}$ in terms of $\mathbf{x}_1$ and $\mathbf{x}_2$.

3. $N[\theta_1+(\rho\sigma_1/\sigma_2)(y_2-\theta_2),\sigma_1^2(1-\rho^2)]$.
4. Use Theorem 2.7, Corollary 2 and Example 2.4.
5. $\rho=-\frac{1}{2}$ (use Theorem 2.7).

EXERCISES 2d

1. Use (2.5) and *A4.9* to find $M(t)$. Then choose orthogonal $\mathbf{T}$ such that $|\mathbf{I}_n-2t\mathbf{A}|=|\mathbf{T}'\mathbf{T}|\ |\mathbf{I}_n-2t\mathbf{A}|=|\mathbf{I}_n-2t\mathbf{T}'\mathbf{A}\mathbf{T}|$ where $\mathbf{T}'\mathbf{A}\mathbf{T}=\text{diag}(\mathbf{1}_r',0,\ldots,0)$ (since $\mathbf{A}$ is idempotent with r unit eigenvalues and $n-r$ zero eigenvalues).
2. $M(t_1,t_2)=|\mathbf{I}_n-2t_1\mathbf{A}-2t_2\mathbf{B}|^{-1/2}$. When $\mathbf{AB}=\mathbf{0}$, $M(t_1,t_2)=M(t_1,0)M(0,t_2)$, and Q_1 and Q_2 are independent. Conversely, from $\text{var}[Q_1+Q_2]=\text{var}[Q_1]+\text{var}[Q_2]$, we obtain (Theorem 1.8, Corollary 2) $0=\text{tr}[\mathbf{AB}]=\text{tr}[\mathbf{RR}'\mathbf{SS}']=\text{tr}[\mathbf{S}'\mathbf{RR}'\mathbf{S}]=\text{tr}[\mathbf{K}'\mathbf{K}]=\Sigma\Sigma_{ij}k_{ij}^2$ which implies $\mathbf{K}=\mathbf{0}$ and $\mathbf{AB}=\mathbf{0}$. (The condition of positive definiteness can be dropped, e.g., Lancaster [1969]. Thus Q_1 and Q_2 are independent if and only if $\mathbf{AB}=\mathbf{0}$.)
3. $\mathbf{A}_i\mathbf{A}_j=\mathbf{0}$ for all $i\neq j$ and hence

$$M(t_1,t_2,\ldots,t_n)=M(t_1,0,\ldots,0)M(0,t_2,0,\ldots,0)M(0,0,\ldots,0,t_n).$$

4. The moment generating function of $\mathbf{Z}=\mathbf{Y}-\boldsymbol{\theta}$ is $\exp(\frac{1}{2}\mathbf{t}'\boldsymbol{\Sigma}\mathbf{t})$ which, expanded in powers of the t_i, gives zero coefficients for t_i^3, $t_i^2t_j$, and $t_it_jt_k$. Then

$$\begin{aligned}\mathcal{C}[\mathbf{Y},\mathbf{Y}'\mathbf{A}\mathbf{Y}]&=\mathcal{E}\left[(\mathbf{Y}-\boldsymbol{\theta})\mathbf{Y}'\mathbf{A}\mathbf{Y}\right]\\&=\mathcal{E}\left[(\mathbf{Y}-\boldsymbol{\theta})\{(\mathbf{Y}-\boldsymbol{\theta})'\mathbf{A}(\mathbf{Y}-\boldsymbol{\theta})+2\boldsymbol{\theta}'\mathbf{A}(\mathbf{Y}-\boldsymbol{\theta})+\boldsymbol{\theta}'\mathbf{A}\boldsymbol{\theta}\}\right]\\&=\mathcal{E}\left[\mathbf{ZZ}'\mathbf{AZ}+2\mathbf{Z}\boldsymbol{\theta}'\mathbf{AZ}\right]\\&=2\mathcal{E}\left[\mathbf{ZZ}'\mathbf{A}\boldsymbol{\theta}\right]=2\boldsymbol{\Sigma}\mathbf{A}\boldsymbol{\theta},\end{aligned}$$

since

$$E\left[\sum_r\sum_s a_{rs}Z_rZ_sZ_i\right]=\sum_r\sum_s a_{rs}E[Z_rZ_sZ_i]=0.$$

5. $\boldsymbol{\Sigma}=\mathbf{RR}'$ and $\mathbf{Z}=\mathbf{R}^{-1}\mathbf{Y}\sim N(\mathbf{0},\mathbf{I}_n)$. Hence $\mathbf{Y}'\mathbf{AY}=\mathbf{Z}'\mathbf{R}'\mathbf{ARZ}\sim\chi_r^2$ if $\mathbf{R}'\mathbf{ARR}'\mathbf{AR}=\mathbf{R}'\mathbf{AR}$, or $\mathbf{A}\boldsymbol{\Sigma}\mathbf{A}=\mathbf{A}$.

6. Let $\mathbf{A}=\mathbf{A}_1+\mathbf{A}_2+\mathbf{A}_3=\mathbf{A}_1+\mathbf{B}_1$, say. Since $Q_2\sim\chi^2_{r_2}$, $\mathbf{A}_2\geqslant\mathbf{0}$ (that is, positive semidefinite) and hence $\mathbf{B}_1\geqslant\mathbf{0}$. Therefore, by Theorem 2.9, $\mathbf{A}_1\mathbf{B}_1=\mathbf{0}$ and $\mathbf{B}_1^2=\mathbf{B}_1$. Similarly $\mathbf{B}_1=\mathbf{A}_2+\mathbf{A}_3$ so that $\mathbf{A}_3\geqslant 0$ implies that $\mathbf{A}_2\mathbf{A}_3=\mathbf{0}$ and $\mathbf{A}_3^2=\mathbf{A}_3$. If we now write $\mathbf{A}=\mathbf{A}_2+(\mathbf{A}_1+\mathbf{A}_3)$ we obtain $\mathbf{A}_1\mathbf{A}_3=\mathbf{0}$, and hence $\mathbf{A}_1\mathbf{A}_2=\mathbf{0}$. Thus the Q_i are pairwise independent and therefore mutually independent (by Exercise 3).

MISCELLANEOUS EXERCISES 2

1. $\pi\,\mathrm{tr}[\mathbf{A\Sigma}]=2\pi$.
2. Let $\mathbf{X}=\mathbf{TY}$ where $\mathbf{T}$ is Helmert's transformation. Then $\mathrm{var}[X_1]=(1+(n-1)\rho)\sigma^2$, $\mathrm{var}[X_i]=(1-\rho)\sigma^2$ $(i=2,3,\ldots,n)$, and $\mathrm{cov}[X_i,X_j]=0$ $(i\neq j)$. Hence $\mathscr{D}[\mathbf{X}]$ is diagonal and the X_i are independent. Using the same argument given in Example 2.3 we have
$$Q=\frac{\sum_{i=2}^{n}X_i^2}{\sigma^2(1-\rho)}\sim\chi^2_{n-1}$$
3. Use moment generating functions.
4. (a) Using the moment generating function, $\overline{\mathbf{Y}}\sim N_n\left(\boldsymbol{\theta},\frac{1}{n}\boldsymbol{\Sigma}\right)$.

   (b) $\mathscr{C}[\overline{\mathbf{Y}},\mathbf{Y}_i-\overline{\mathbf{Y}}]=\mathscr{C}[\overline{\mathbf{Y}},\mathbf{Y}_i]-\mathscr{D}[\overline{\mathbf{Y}}]=\frac{1}{n}\mathscr{D}[\mathbf{Y}_i]-\frac{1}{n}\boldsymbol{\Sigma}=\mathbf{0}$.

   (c)
$$\mathscr{E}\left[\sum_i(\mathbf{Y}_i-\overline{\mathbf{Y}})(\mathbf{Y}_i-\overline{\mathbf{Y}})'\right]=\sum_i\mathscr{E}\left[\mathbf{Y}_i-\boldsymbol{\theta}-(\overline{\mathbf{Y}}-\boldsymbol{\theta})(\mathbf{Y}_i-\boldsymbol{\theta}-(\overline{\mathbf{Y}}-\boldsymbol{\theta}))'\right]$$
$$=\sum_i\left\{\mathscr{D}[\mathbf{Y}_i]-2\mathscr{C}[\overline{\mathbf{Y}},\mathbf{Y}_i]+\mathscr{D}[\overline{\mathbf{Y}}]\right\},\quad\text{etc.}$$
5. Use Theorem 2.7 and $\mathrm{cov}[\overline{Y},Y_i-Y_{i+1}]=0$.
6. $f(y)$ is an odd function of each y_i so that the square bracket vanishes when y_i is integrated out.
7. From Theorem 1.8, $\mathrm{var}[\mathbf{Y}'\mathbf{AY}]=2\,\mathrm{tr}\,\mathbf{A}^2=2\Sigma\Sigma a_{ij}^2=12n-16$.
8. $\mathbf{W}=\mathbf{U}+\mathbf{V}=(\mathbf{B}+\mathbf{C})\mathbf{Y}=\mathbf{DY}$. $\mathscr{C}[\mathbf{X},\mathbf{W}]=\mathbf{0}$. Let $\mathbf{D}_1$ consist of the linearly independent rows of $\mathbf{D}$. Then, if we argue as in Theorem 2.7, $\mathbf{X}$ is independent of $\mathbf{W}_1=\mathbf{D}_1\mathbf{Y}$ and therefore of $\mathbf{W}$ (which is a function of $\mathbf{W}_1$).
9. (a) Q can be negative. (b) $|\mathbf{I}_4-2t\mathbf{A}|^{-1/2}=(1-t^2)^{-1}$.

10. Choose orthogonal $\mathbf{T}$ with first row $\frac{1}{\|\mathbf{a}\|}\mathbf{a}'$. Let $\mathbf{Z}=\mathbf{TY}$. Then $Z_1=0$ and $\mathbf{Y}'\mathbf{Y}=\mathbf{Z}'\mathbf{Z}=Z_2^2+\cdots+Z_n^2$. Since the Z_i are i.i.d. $N(0,1)$, the conditional distribution of the Z_i $(i\neq 1)$ is the same as the unconditional distribution.
11. Let $X_i=Y_i-\theta$, then $\mathbf{X}\sim N_n(\mathbf{0},\mathbf{\Sigma})$ and $Q=\mathbf{X}'\mathbf{A}\mathbf{X}/(1-\rho)$. Now $M(t)=|\mathbf{I}_n-2t\mathbf{A}\mathbf{\Sigma}/(1-\rho)|^{-1/2}=|\mathbf{I}_n-2t\mathbf{A}|^{-1/2}$ [by *A4.9*, (2.6), and (1.7)] which does not depend on ρ. Hence Q has the same distribution as for $\rho=0$.

EXERCISES 3a

2. $\partial\|\mathbf{Y}-\mathbf{X}\boldsymbol{\beta}\|^2/\partial\beta_0=0\Rightarrow\Sigma(Y-\hat{\beta}_0-\hat{\beta}_1x_{i1}-\cdots-\hat{\beta}_{p-1}x_{i,p-1})=0.$
3. $\hat{\theta}=\frac{1}{6}(Y_1+2Y_2+Y_3),\quad \hat{\phi}=\frac{1}{5}(2Y_3-Y_2).$
4. $\hat{\beta}_0=\overline{Y},\quad \hat{\beta}_1=\frac{1}{2}(Y_3-Y_1),\quad \hat{\beta}_2=\frac{1}{6}(Y_1-2Y_2+Y_3).$
5. Let $x=\sin\theta$; then $\hat{w}=\Sigma_iT_ix_i/\Sigma_ix_i^2$.
6. $\mathbf{P}\boldsymbol{\alpha}=\mathbf{X}(\mathbf{X}'\mathbf{X})^{-1}\mathbf{X}'\boldsymbol{\alpha}=\mathbf{X}\boldsymbol{\beta}$ so that $\mathcal{R}[\mathbf{P}]\subset\mathcal{R}[\mathbf{X}]$. Conversely, if $\mathbf{y}=\mathbf{X}\boldsymbol{\gamma}$ then $\mathbf{P}\mathbf{y}=\mathbf{y}$ and $\mathcal{R}[\mathbf{X}]\subset\mathcal{R}[\mathbf{P}]$.
7. $\hat{\mathbf{Y}}'(\mathbf{Y}-\hat{\mathbf{Y}})=\mathbf{Y}'\mathbf{P}(\mathbf{I}_n-\mathbf{P})\mathbf{Y}=\mathbf{Y}'(\mathbf{P}-\mathbf{P}^2)\mathbf{Y}=0.$
8. See Example 4.4 in Section 4.2 ($k=1$).

EXERCISES 3b

1. $\hat{\beta}_0=\overline{Y}-\hat{\beta}_1\bar{x}$, $\hat{\beta}_1=\Sigma_iY_i(x_i-\bar{x})/\Sigma(x_i-\bar{x})^2$. From $(\mathbf{X}'\mathbf{X})^{-1}$, $\text{cov}[\hat{\beta}_0,\hat{\beta}_1]=-\bar{x}/\Sigma(x_i-\bar{x})^2$.
2. It is helpful to express the model in the form

$$\begin{bmatrix}\mathbf{U}\\ \mathbf{V}\\ \mathbf{W}\end{bmatrix}=\begin{bmatrix}\mathbf{1}_m & \mathbf{0}\\ \mathbf{1}_m & \mathbf{1}_m\\ \mathbf{1}_n & -2\mathbf{1}_n\end{bmatrix}\begin{bmatrix}\theta\\ \phi\end{bmatrix}+\boldsymbol{\varepsilon}.$$

Then

$$\hat{\theta}=\frac{1}{m(m+13n)}\left\{(m+4n)\sum_i U_i+6n\sum_j V_j+3m\sum_k W_k\right\},$$

$$\hat{\phi}=\frac{1}{m(m+13n)}\left\{(2n-m)\sum_i U_i+(m+3n)\sum_j V_j-5m\sum_k W_k\right\}.$$

3. $\mathbf{Y}=\theta\mathbf{1}_n+\boldsymbol{\varepsilon}$. BLUE of θ is $\overline{Y}$.
4. Let

$$\mathbf{A}=\begin{bmatrix} \Sigma(x_{i1}-\bar{x}_1)^2 & \Sigma(x_{i1}-\bar{x}_1)(x_{i2}-\bar{x}_2) \\ \Sigma(x_{i1}-\bar{x}_1)(x_{i2}-\bar{x}_2) & \Sigma(x_{i2}-x_2)^2 \end{bmatrix}.$$

The variance–covariance matrix of $\hat{\beta}_1$ and $\hat{\beta}_2$ is $\sigma^2\mathbf{A}^{-1}$. Hence $\text{var}[\hat{\beta}_1] = \sigma^2\Sigma(x_{i2}-\bar{x}_2)^2/|\mathbf{A}|$.

EXERCISES 3c

1. (a) From Theorem 3.4 $\text{var}[S^2]=\text{var}[\mathbf{Y}'\mathbf{R}\mathbf{Y}]/(n-p)^2=2\sigma^4/(n-p)$.
   (b) $E[(\mathbf{Y}'\mathbf{R}\mathbf{Y}/(n-p+2)-\sigma^2)^2]=2\sigma^4/(n-p+2)$.
2. $\Sigma_i(Y_i-\overline{Y})^2/(n-1)$, since the diagonal elements of the associated matrix $\mathbf{A}$ are all equal.

EXERCISES 3d

1. $Y_i=\theta+\varepsilon_i$, which is a regression model with $\beta_0=\theta$.
2. $\mathbf{U}=(\mathbf{Y}-\mathbf{X}\hat{\boldsymbol{\beta}})=(\mathbf{I}_n-\mathbf{P})\mathbf{Y}=(\mathbf{I}_n-\mathbf{P})(\mathbf{Y}-\mathbf{X}\boldsymbol{\beta})=(\mathbf{I}_n-\mathbf{P})\mathbf{Z}$, say. $\mathbf{V}=\mathbf{X}(\hat{\boldsymbol{\beta}}-\boldsymbol{\beta})=\mathbf{X}\{(\mathbf{X}'\mathbf{X})^{-1}\mathbf{X}'\mathbf{Y}-(\mathbf{X}'\mathbf{X})^{-1}\mathbf{X}'\mathbf{X}\boldsymbol{\beta})\}=\mathbf{P}(\mathbf{Y}-\mathbf{X}\boldsymbol{\beta})=\mathbf{P}\mathbf{Z}$.

EXERCISES 3e

2. $\phi(x)=x^2-\frac{2}{3}$.
3. (a) $|\mathbf{W}'\mathbf{W}|/|\mathbf{X}'\mathbf{X}|$ is the bottom diagonal element of $(\mathbf{X}'\mathbf{X})^{-1}$.
   (b) $\mathbf{x}_k'\mathbf{x}_k-\mathbf{x}_k'\mathbf{P}_W\mathbf{x}_k \leqslant \mathbf{x}_k'\mathbf{x}_k$ with equality if and only if $\mathbf{P}_W\mathbf{x}_k=\mathbf{0}$ or $\mathbf{W}'\mathbf{x}_k=\mathbf{0}$.
4. Omit the condition $\Sigma_i x_{ij}=0$.
5. Let $\mathbf{Y}=\mathbf{X}\boldsymbol{\beta}+\boldsymbol{\varepsilon}$. Apply the Gram–Schmidt process to $\mathbf{X}$ (see Equation 11.15) and obtain an orthonormal basis $\mathbf{z}_1,\ldots,\mathbf{z}_p$ for $\mathcal{R}[\mathbf{X}]$. Thus $\mathbf{X}=\mathbf{Z}\mathbf{U}$, where $\mathbf{U}$ is upper triangular and $\mathbf{Z}=(\mathbf{z}_1,\mathbf{z}_2,\ldots,\mathbf{z}_p)$. Also $\boldsymbol{\gamma}=\mathbf{U}\boldsymbol{\beta}$ so that $\gamma_r=u_{rr}\beta_r+u_{r,r+1}\beta_{r+1}+\cdots+u_{rp}\beta_p$.
6. (a) From Exercise 3 the minimum variance is achieved when the columns of $\mathbf{X}$ are mutually orthogonal. This minimum variance $\sigma^2/(\mathbf{x}_k'\mathbf{x}_k))$ is least when every element of $\mathbf{x}_k$ is nonzero.
   (b) For the optimum design $\text{var}[\hat{\beta}_k]=\sigma^2/n$.

EXERCISES 3f

1. $\frac{1}{3}(2Y_1 - Y_2)$, $\frac{2}{3}\sigma^2$.
2. $\Sigma_i w_i Y_i / \Sigma_i w_i$, $\sigma^2 / \Sigma_i w_i$.
3. $\theta^* = \dfrac{1}{n}\Sigma_i (Y_i / i)$.
4. $\mathbf{V}^{-1}\mathbf{X} = \mathbf{V}^{-1}\mathbf{1}_n = c\mathbf{1}_n = c\mathbf{X}$, etc.
5. Reduce the model as in Section 3.6.
6. Use Lagrange multipliers to show that $\boldsymbol{\theta}^* = (\mathbf{I}_n - \mathbf{VA}'(\mathbf{AVA}')^{-1}\mathbf{A})\mathbf{Y}$.

EXERCISES 3g

1. $\mathbf{Y}'\mathbf{RY} - \mathbf{Y}'\mathbf{R}_G\mathbf{Y} = \hat{\gamma}_G'\mathbf{Z}'\mathbf{RY} = \hat{\gamma}_G'\mathbf{Z}'\mathbf{RZ}\hat{\gamma}_G$.
2. By Theorem 3.7 (v), $\text{var}[\hat{\beta}_{G,i}] - \text{var}[\hat{\beta}_i] = \sigma^2(\mathbf{LML}')_{ii} \geqslant 0$ since $\mathbf{LML}'$ is positive definite (or zero when $\mathbf{X}'\mathbf{Z} = \mathbf{0}$).
3. $\hat{\theta} = \overline{Y} - \hat{\gamma}\overline{x}$, $\hat{\gamma} = \Sigma_i Y_i(x_i - \overline{x}) / \Sigma_i (x_i - \overline{x})^2$.

EXERCISES 3h

2. $\mathbf{CX} = \mathbf{I}$ implies that $\mathbf{X}$ has full rank.
3. $\text{rank}\begin{bmatrix}\mathbf{X}\\ \mathbf{L}\end{bmatrix} = \text{rank}\,\mathbf{L}$ if and only if the rows of $\mathbf{X}$ are linearly dependent on the rows of $\mathbf{L}$, that is, $\mathbf{X}' = \mathbf{L}'\mathbf{K}'$. $\mathbf{K}$ has rank r only if $\text{rank}\,\mathbf{X} = \text{rank}\,\mathbf{L}$.
4. (b) From $\mathbf{P} = \mathbf{BB}^-$, $\text{rank}\,\mathbf{P} \leqslant \text{rank}\,\mathbf{B}$; from $\mathbf{PB} = \mathbf{B}$, $\text{rank}\,\mathbf{B} \leqslant \text{rank}\,\mathbf{P}$.
5. (a) $0 = \text{tr}[\mathbf{B}'\mathbf{B}] = \Sigma\Sigma b_{ij}^2$.
6. (a) Transpose $\mathbf{X}'\mathbf{XCX}'\mathbf{X} = \mathbf{X}'\mathbf{X}$. (b) Use $\frac{1}{2}(\mathbf{C} + \mathbf{C}')$.
   (c) Applying Exercise 5 (b) to the above equation gives $\mathbf{XCX}'\mathbf{X} = \mathbf{X}$.
   (d) $\mathbf{X}'\mathbf{XC}_1\mathbf{X}'\mathbf{X} = \mathbf{X}'\mathbf{XC}_2\mathbf{X}'\mathbf{X} \Rightarrow \mathbf{XC}_1\mathbf{X}' = \mathbf{XC}_2\mathbf{X}'$ (apply Exercise 5 twice).
   (e) By (d) and (a) $\mathbf{XCX}' = \mathbf{XC}'\mathbf{X}' = (\mathbf{XCX}')'$. By (c) $\mathbf{XCX}' = \mathbf{XX}^-$ which is idempotent by Exercise 4.
   (f) Use hint and $\mathbf{XCX}'\boldsymbol{\alpha} = \mathbf{X}\boldsymbol{\beta}$ (i.e., $\mathcal{R}[\mathbf{XCX}'] \subset \mathcal{R}[\mathbf{X}]$).
7. (a) $\mathbf{X}'\mathbf{X}(\mathbf{G}'\mathbf{G})^{-1}\mathbf{H}'\mathbf{y} + \mathbf{H}'\mathbf{H}(\mathbf{G}'\mathbf{G})^{-1}\mathbf{H}'\mathbf{y} = \mathbf{H}'\mathbf{y}$ for all $\mathbf{y}$. Hence $\mathbf{X}'\mathbf{X}(\mathbf{G}'\mathbf{G})^{-1}\mathbf{H}'\mathbf{y} = \mathbf{H}'(\mathbf{I}_{p-r} - \mathbf{H}(\mathbf{G}'\mathbf{G})^{-1}\mathbf{H}')\mathbf{y}$ or $\mathbf{X}'\boldsymbol{\alpha} = \mathbf{H}'\boldsymbol{\beta}$ ($=\mathbf{0}$ for identifiability).
   (1) $\mathbf{X}'\mathbf{X}(\mathbf{G}'\mathbf{G})^{-1}\mathbf{H}'\mathbf{y} = \mathbf{0}$ all $\mathbf{y} \Rightarrow \mathcal{R}[\mathbf{X}] \supset \mathcal{R}[\mathbf{X}(\mathbf{G}'\mathbf{G})^{-1}\mathbf{H}'] \perp \mathcal{R}[\mathbf{X}]$. Hence $\mathbf{X}(\mathbf{G}'\mathbf{G})^{-1}\mathbf{H}' = \mathbf{0}$.
   (2) $\mathbf{H}'\boldsymbol{\beta} = \mathbf{0} \Rightarrow \boldsymbol{\beta} = \mathbf{0} \Rightarrow \mathbf{H}(\mathbf{G}'\mathbf{G})^{-1}\mathbf{H}' = \mathbf{I}_{p-r}$.
   (b) $\mathbf{X}'\mathbf{X}(\mathbf{G}'\mathbf{G})^{-1}\mathbf{X}'\mathbf{X} = (\mathbf{X}'\mathbf{X} + \mathbf{H}'\mathbf{H})(\mathbf{G}'\mathbf{G})^{-1}\mathbf{X}'\mathbf{X} = \mathbf{X}'\mathbf{X}$.
8. $\mathbf{A}\boldsymbol{\alpha} = \begin{pmatrix}\mathbf{X}'\mathbf{X} & \mathbf{H}'\\ \mathbf{H} & \mathbf{0}\end{pmatrix}\begin{pmatrix}\boldsymbol{\alpha}_1\\ \boldsymbol{\alpha}_2\end{pmatrix} = \mathbf{0} \Rightarrow \mathbf{X}'\mathbf{X}\boldsymbol{\alpha}_1 + \mathbf{H}'\boldsymbol{\alpha}_2 = \mathbf{0}$ and $\mathbf{H}\boldsymbol{\alpha}_1 = \mathbf{0} \Rightarrow \mathbf{X}'\mathbf{X}\boldsymbol{\alpha}_1 = \mathbf{0}$,

$\mathbf{H}'\boldsymbol{\alpha}_2=\mathbf{0}$ and $\mathbf{H}\boldsymbol{\alpha}_1=\mathbf{0}\Rightarrow\boldsymbol{\alpha}_2=\mathbf{0}$ and $\mathbf{G}'\mathbf{G}\boldsymbol{\alpha}_1=\mathbf{0}$ (i.e., $\boldsymbol{\alpha}_1=\mathbf{0}$). Thus the columns of $\mathbf{A}$ are linearly independent. From $\mathbf{A}\mathbf{A}^{-1}=\mathbf{I}$ we obtain $\mathbf{X}'\mathbf{X}\mathbf{C}_{12}'+\mathbf{H}'\mathbf{C}_{22}=\mathbf{0}$ which implies that $\mathbf{H}'\mathbf{C}_{22}=\mathbf{0}$, i.e., $\mathbf{C}_{22}=\mathbf{0}$. Then $\mathbf{X}'\mathbf{X}\mathbf{C}_{11}+\mathbf{H}'\mathbf{C}_{22}=\mathbf{I}_p\Rightarrow\mathbf{X}'\mathbf{X}\mathbf{C}_{11}=\mathbf{I}_p\Rightarrow\mathbf{X}'\mathbf{X}\mathbf{C}_{11}\mathbf{X}'\mathbf{X}=\mathbf{X}'\mathbf{X}$.

EXERCISES 3i

2. $\mathbf{a}'\mathcal{E}[\hat{\boldsymbol{\beta}}]=\mathbf{a}'(\mathbf{X}'\mathbf{X})^-\mathbf{X}'\mathbf{X}\boldsymbol{\beta}=\mathbf{c}'\boldsymbol{\beta}$, where $\mathbf{c}=\mathbf{X}'\mathbf{X}(\mathbf{X}'\mathbf{X})^{-\prime}\mathbf{a}\in\mathcal{R}[\mathbf{X}']$.
4. Let $\hat{\boldsymbol{\beta}}_i=\mathbf{b}_i+\mathbf{c}_i$ $(i=1,2)$ be two solutions. Then $\mathbf{b}_1-\mathbf{b}_2\in\mathcal{R}[\mathbf{X}']$. Hence $\mathbf{X}'\mathbf{X}\hat{\boldsymbol{\beta}}_1=\mathbf{X}'\mathbf{X}\hat{\boldsymbol{\beta}}_2$ $(=\mathbf{X}'\mathbf{Y})\Rightarrow\mathbf{X}'\mathbf{X}(\mathbf{b}_1-\mathbf{b}_2)=\mathbf{0}\Rightarrow(\mathbf{b}_1-\mathbf{b}_2)\perp\mathcal{R}[\mathbf{X}'\mathbf{X}]$ $(=\mathcal{R}[\mathbf{X}'])$ $\Rightarrow\mathbf{b}_1-\mathbf{b}_2=\mathbf{0}$. $\hat{\mathbf{Y}}=\mathbf{X}\mathbf{b}$. Part (i) follows from $\mathbf{a}'\hat{\boldsymbol{\beta}}=\boldsymbol{\alpha}'\mathbf{X}'\mathbf{X}\hat{\boldsymbol{\beta}}=\boldsymbol{\alpha}'\mathbf{X}'\mathbf{X}\mathbf{b}$.
5. Using Exercise 4, if $\mathbf{X}'\mathbf{X}\mathbf{b}=\mathbf{X}'\mathbf{Y}$ then $\mathbf{b}+\mathbf{c}$ is a solution for all $\mathbf{c}\perp\mathcal{R}[\mathbf{X}']$. Thus $\mathbf{a}'\mathbf{c}$ is invariant for all such $\mathbf{c}$ including $\mathbf{c}=\mathbf{0}$. Hence $\mathbf{a}'\mathbf{c}=\mathbf{0}$ and $\mathbf{a}\in\mathcal{R}[\mathbf{X}']$.
6. If $\mathbf{a}'=\boldsymbol{\alpha}'\mathbf{X}'\mathbf{X}$ the result follows. Conversely, given the result,

$$E\left[\mathbf{a}'\hat{\boldsymbol{\beta}}\right]=E\left[\mathbf{a}'(\mathbf{X}'\mathbf{X})^-\mathbf{X}'\mathbf{Y}\right]=\mathbf{a}'(\mathbf{X}'\mathbf{X})^-\mathbf{X}'\mathbf{X}\boldsymbol{\beta}=\mathbf{a}'\boldsymbol{\beta}$$

   and $\mathbf{a}'\boldsymbol{\beta}$ is estimable.
7. Use $\hat{\boldsymbol{\beta}}=(\mathbf{X}'\mathbf{X})^-\mathbf{X}'\mathbf{Y}$ and Exercise 6.
8. $\mathbf{X}_*'\mathbf{X}_*\boldsymbol{\alpha}=(\mathbf{X}'\mathbf{X}+\mathbf{x}_{n+1}\mathbf{x}_{n+1}')\boldsymbol{\alpha}=\lambda\boldsymbol{\alpha}+c^2\boldsymbol{\alpha}\boldsymbol{\alpha}'\boldsymbol{\alpha}$.

EXERCISES 3j

1. Given $(\mathbf{I}_n-\mathbf{P})\mathbf{Z}\boldsymbol{\alpha}=\mathbf{0}$ then $\mathbf{Z}\boldsymbol{\alpha}\in\mathcal{R}[\mathbf{P}]=\mathcal{R}[\mathbf{X}]$. But $\mathcal{R}[\mathbf{X}]\cap\mathcal{R}[\mathbf{Z}]=\mathbf{0}$ so that $\boldsymbol{\alpha}=\mathbf{0}$. Hence $(\mathbf{I}_n-\mathbf{P})\mathbf{Z}$ has full rank. Then

$$\mathbf{Z}'(\mathbf{I}_n-\mathbf{P})'(\mathbf{I}_n-\mathbf{P})\mathbf{Z}=\mathbf{Z}'(\mathbf{I}_n-\mathbf{P})\mathbf{Z}.$$

2. Clearly the rows of $(\mathbf{H},\mathbf{0})$ are linearly independent of the rows of $(\mathbf{X},\mathbf{Z})$. Since $\mathcal{R}[\mathbf{X}]\cap\mathcal{R}[\mathbf{Z}]=\mathbf{0}$ and the columns of $\mathbf{Z}$ are linearly independent, the columns of $\begin{pmatrix}\mathbf{X} & \mathbf{Z}\\ \mathbf{H} & \mathbf{0}\end{pmatrix}$ are linearly independent. Hence $\mathbf{X}'\mathbf{X}\hat{\boldsymbol{\beta}}_G=\mathbf{X}'(\mathbf{Y}-\mathbf{Z}\hat{\boldsymbol{\gamma}}_G)$ and $\mathbf{H}'\mathbf{H}\hat{\boldsymbol{\beta}}_G=\mathbf{0}$.

EXERCISES 3k

1. $\mathbf{Y}'(\mathbf{I}_n-\mathbf{P})\mathbf{X}(\hat{\boldsymbol{\beta}}-\hat{\boldsymbol{\beta}}_H)=\mathbf{0}$ by Theorem 3.1 (iii).
2. The second expression in $\mathcal{D}[\hat{\boldsymbol{\beta}}_H]$ is positive semidefinite so that the diagonal elements of the underlying matrix are nonnegative.

3. $\hat{\mathbf{Y}} - \hat{\mathbf{Y}}_H = \mathbf{X}\hat{\boldsymbol{\beta}} - \mathbf{X}\hat{\boldsymbol{\beta}}_H = \frac{1}{2}\mathbf{X}(\mathbf{X}'\mathbf{X})^{-1}\mathbf{A}'\hat{\boldsymbol{\lambda}}_H$. $\|\mathbf{Y} - \hat{\mathbf{Y}}_H\|^2 = \frac{1}{4}\hat{\boldsymbol{\lambda}}_H'\mathbf{A}(\mathbf{X}'\mathbf{X})^{-1}\mathbf{A}'\hat{\boldsymbol{\lambda}}_H$, etc.
4. Suppose $\mathcal{R}[\mathbf{M}] \cap \Omega^{\perp} \neq \mathbf{0}$. Then there exists $\boldsymbol{\alpha} \neq \mathbf{0}$ such that $\mathbf{M}'\boldsymbol{\alpha} = (\mathbf{I}_n - \mathbf{P})\boldsymbol{\beta}$, i.e., $\mathbf{A}'\boldsymbol{\alpha} = \mathbf{X}'\mathbf{M}'\boldsymbol{\alpha} = \mathbf{0}$ which implies $\boldsymbol{\alpha} = \mathbf{0}$.
5. $\mathbf{XB}\boldsymbol{\alpha} = \mathbf{0} \Rightarrow \mathbf{B}\boldsymbol{\alpha} = \mathbf{0} \Rightarrow \boldsymbol{\alpha} = \mathbf{0}$, that is, columns of $\mathbf{XB}$ are linearly independent.

MISCELLANEOUS EXERCISES 3

1. $\Sigma_i a_i b_i = 0$.
2. Use Lagrange multipliers or show that $\mathbf{I}_n - \mathbf{P}_{\Omega} = \mathbf{A}'(\mathbf{AA}')^{-1}\mathbf{A}'$.
3. Use Section 3.7.1 with $\mathbf{X} = \mathbf{X}_1$ and $\mathbf{Z} = \mathbf{X}_2$.

$$\mathcal{D}\left[\hat{\boldsymbol{\beta}}_2\right] = \sigma^2 \mathbf{X}_2'\left(\mathbf{I}_n - \mathbf{X}_1(\mathbf{X}_1'\mathbf{X}_1)^{-1}\mathbf{X}_1'\right)\mathbf{X}_2.$$

4. Let $\mathbf{c}'\mathbf{Y} = (\mathbf{a}+\mathbf{b})'\mathbf{Y}$ be any other linear unbiased estimate of $\mathbf{a}'\mathbf{X}\boldsymbol{\beta}$ i.e., $\mathbf{b}'\mathbf{X} = \mathbf{0}'$. Then $\mathrm{var}[\mathbf{a}'\mathbf{Y}] + \mathrm{var}[\mathbf{b}'\mathbf{Y}] + 2\,\mathrm{cov}[\mathbf{a}'\mathbf{Y}, \mathbf{b}'\mathbf{Y}] \geqslant \mathrm{var}[\mathbf{a}'\mathbf{Y}]$ with equality if and only if $\mathbf{b} = \mathbf{0}$.
5. $\mathrm{tr}\,\mathcal{D}[\hat{\mathbf{Y}}] = \mathrm{tr}\,\mathcal{D}[\mathbf{PY}] = \sigma^2\,\mathrm{tr}\,\mathbf{P} = \sigma^2\,\mathrm{rank}\,\mathbf{P} = \sigma^2 p$.
6. 9.95, 5.0, 4.15, 1.1.
7. $\frac{1}{8}(3Y_{1..} - Y_{.1.} - Y_{..1})$, $\frac{1}{8}(-Y_{1..} + 3Y_{.1.} - Y_{..1})$, $\frac{1}{8}(-Y_{1..} - Y_{.1.} + 3Y_{..1})$ where $Y_{1..} = \Sigma_j \Sigma_k Y_{1jk}$, etc.
8. (a) $(\Sigma_i Y_i x_i)/(\Sigma_i x_i^2)$; (b) $(\Sigma_i Y_i)/(\Sigma_i x_i)$; (c) $\dfrac{1}{n}\Sigma_i (Y_i/x_i)$.
9. $\mathbf{X}\boldsymbol{\beta} = \mathbf{KL}\boldsymbol{\beta} = \mathbf{K}\boldsymbol{\alpha}$ and $\boldsymbol{\alpha}$ is estimated by $(\mathbf{K}'\mathbf{K})^{-1}\mathbf{K}'\mathbf{Y}$.
11. $\mathbf{C}_1'\mathbf{X}'\mathbf{X} + \mathbf{C}_2'\mathbf{H} = \mathbf{I}$. Transposing and multiplying on the right by $\mathbf{X}'\mathbf{y}$ leads to $\mathbf{X}'\mathbf{a} + \mathbf{H}'\mathbf{b} = \mathbf{0}$, that is, $\mathbf{H}'\mathbf{b} = \mathbf{0}$ and $\mathbf{C}_2\mathbf{X}' = \mathbf{0}$ (since $\mathcal{R}[\mathbf{X}'] \cap \mathcal{R}[\mathbf{H}'] = \mathbf{0}$). Hence $\mathbf{X}'\mathbf{X}\mathbf{C}_1'\mathbf{X}'\mathbf{X} = \mathbf{X}'\mathbf{X}$.
12. Use the identity $\mathbf{Y} - \mathbf{X}\boldsymbol{\beta} = \mathbf{Y} - \mathbf{X}\boldsymbol{\beta}^* + \mathbf{X}(\boldsymbol{\beta}^* - \boldsymbol{\beta})$. Then

$$(\mathbf{X}\boldsymbol{\beta}^* - \mathbf{X}\boldsymbol{\beta})'\mathbf{V}^{-1}(\mathbf{Y} - \mathbf{X}\boldsymbol{\beta}^*) = (\boldsymbol{\beta}^* - \boldsymbol{\beta})'(\mathbf{X}'\mathbf{V}^{-1}\mathbf{Y} - \mathbf{X}'\mathbf{V}^{-1}\mathbf{X}\boldsymbol{\beta}^*) = \mathbf{0}.$$

13. $\hat{\theta}_1 = \frac{1}{3}(Y_1 + Y_3)$, $\hat{\theta}_2 = \frac{1}{3}(Y_1 - Y_2)$.
$\hat{\theta}_1 = \frac{1}{3}(-Y_1 - 2Y_2 + 3Y_3)$, $\hat{\theta}_2 = \frac{1}{3}(Y_1 - Y_2)$, $\hat{\theta}_3 = Y_1 + Y_2 - Y_3$.
14. $\bar{Y} = \dfrac{1}{n}\mathbf{1}_n'\mathbf{Y}$ and $\mathbf{Y} - \hat{\mathbf{Y}} = (\mathbf{I}_n - \mathbf{P})\mathbf{Y}$. Now $\mathcal{C}[\mathbf{1}_n'\mathbf{Y}, (\mathbf{I}_n - \mathbf{P})\mathbf{Y}] = \mathbf{0}$ since $(\mathbf{I}_n - \mathbf{P})\mathbf{1}_n = \mathbf{0}$ ($\mathbf{1}_n \in \mathcal{R}[\mathbf{X}]$). Result follows from Theorem 3.7.

15.

$$\mathfrak{D}[\mathbf{u}]=\sigma^2\begin{bmatrix} 1 & \rho & \rho^2 & \cdots & \rho^{n-1} \\ \rho & 1 & \rho & \cdots & \rho^{n-2} \\ \cdots & \cdots & \cdots & \cdots & \cdots \\ \rho^{n-1} & \rho^{n-2} & \cdots & \cdots & 1 \end{bmatrix}=\sigma^2\mathbf{V}, \quad \text{say.}$$

$$\begin{aligned}\text{var}[\hat{\boldsymbol{\beta}}] &= \sigma^2(\mathbf{x}'\mathbf{x})^{-1}\mathbf{x}'\mathbf{V}\mathbf{x}(\mathbf{x}'\mathbf{x})^{-1} \\ &= \frac{\sigma^2}{(\mathbf{x}'\mathbf{x})^2}(\mathbf{x}'\mathbf{x}+f(\rho)) > \frac{\sigma^2}{(\mathbf{x}'\mathbf{x})}.\end{aligned}$$

16. $\mathbf{X}'\mathbf{X}$ is diagonal. Hence $\hat{\beta}_0=\overline{Y}$, $\hat{\beta}_1=(2/n)\Sigma_{t=1}^n Y_t\cos(2\pi k_1 t/n)$, $\hat{\beta}_2=(2/n)\Sigma_{t=1}^n Y_t\sin(2\pi k_2 t/n)$.

EXERCISES 4a

1. Proceed as in Theorem 4.1 (iv) except that $\mathbf{A}\hat{\boldsymbol{\beta}}=\mathbf{A}(\mathbf{X}'\mathbf{X})^{-1}\mathbf{X}'\mathbf{Y}$ is replaced by $\mathbf{A}\hat{\boldsymbol{\beta}}-\mathbf{c}=\mathbf{A}(\mathbf{X}'\mathbf{X})^{-1}\mathbf{X}'(\mathbf{Y}-\mathbf{X}\boldsymbol{\beta})$ when $\mathbf{A}\boldsymbol{\beta}=\mathbf{c}$.
2. $\hat{\boldsymbol{\lambda}}_H=2[\mathbf{A}(\mathbf{X}'\mathbf{X})^{-1}\mathbf{A}']^{-1}(\mathbf{A}\hat{\boldsymbol{\beta}}-\mathbf{c})$, etc.

EXERCISES 4b

1. Since $\mathbf{1}_n\in\mathfrak{R}[\mathbf{X}]$, $(\mathbf{I}_n-\mathbf{P})\mathbf{1}_n=\mathbf{0}$ so that $(\mathbf{Y}-c\mathbf{1}_n)'(\mathbf{I}_n-\mathbf{P})(\mathbf{Y}-c\mathbf{1}_n)=\mathbf{Y}'(\mathbf{I}_n-\mathbf{P})\mathbf{Y}$. The same is true for H.
2. $H:(1,0)\boldsymbol{\beta}=0$. Using the general matrix theory $F=\hat{\beta}_0^2\big/(\Sigma x_i^2 S^2/n\Sigma(x_i-\bar{x})^2)$ where $\hat{\beta}_0=\overline{Y}-\hat{\beta}_1\bar{x}$, etc.
3. $$\frac{(\hat{\beta}_1-\overline{Y})^2}{S^2\left\{(1/n)+\left[1/\sum(x_i-\bar{x})^2\right]\right\}}.$$
4. $F=(\hat{\theta}_1-2\hat{\theta}_2)^2/(\frac{7}{6}S^2)$ where $\hat{\theta}_1=\frac{1}{2}(Y_1-Y_3)$, $\hat{\theta}_2=\frac{1}{6}(Y_1+2Y_2+Y_3)$, and $S^2=Y_1^2+Y_2^2+Y_3^2-2\hat{\theta}_1^2-6\hat{\theta}_2^2$.

EXERCISES 4c

1. $\text{RSS}_H-\text{RSS}=\Sigma_i(\hat{Y}_i-\hat{Y}_{iH})^2=\Sigma_i(\hat{Y}_i-\overline{Y})^2$, etc.
2. From Exercise 1 F is distributed as $F_{p-1,n-p}$; hence show that the distribution of R^2 is beta.

EXERCISES 4d

1. $\mathbf{X}_A = (1 \;\; 1 \;\; 3)'$; $\mathbf{X}_A = \mathbf{1}_n$ $(n = n_1 + n_2)$.
2. $\mathbf{X} = \mathbf{T}_p\mathbf{C}$, where $\mathbf{T}_p = (\boldsymbol{\alpha}_1, \boldsymbol{\alpha}_2, \ldots, \boldsymbol{\alpha}_p)$ and $\mathbf{C}$ is nonsingular. Then $\mathbf{T}'\mathbf{X}\boldsymbol{\beta} = \mathbf{T}'\mathbf{T}_p\mathbf{C}\boldsymbol{\beta} = \begin{pmatrix} \mathbf{I}_p \\ \mathbf{0} \end{pmatrix}\boldsymbol{\mu}$, where $\boldsymbol{\mu} = \mathbf{C}\boldsymbol{\beta}$, etc.
3. $$F_{1,n-1} = \frac{(x_{n+1} - \bar{x})^2 n(n-1)}{\sum_{i=1}^{n} (x_i - \bar{x})^2 (n+1)}.$$

EXERCISES 4e

1. $\mathbf{A}\hat{\boldsymbol{\beta}} = \mathbf{MX}(\mathbf{X}'\mathbf{X})^{-}\mathbf{X}'\mathbf{Y} = \mathbf{MP}_\Omega\mathbf{Y}$ and $\mathscr{D}[\mathbf{A}\hat{\boldsymbol{\beta}}] = \sigma^2\mathbf{MP}_\Omega\mathbf{M}' = \sigma^2\mathbf{MP}_\Omega\mathbf{M}' = \sigma^2\mathbf{A}(\mathbf{X}'\mathbf{X})^{-}\mathbf{A}'$. From Theorem 4.6 $\mathbf{MP}_\Omega\mathbf{M}'$ has rank q.
2. $$E[\text{RSS}_H - \text{RSS}] = E[\tilde{\mathbf{Y}}(\mathbf{P}_\Omega - \mathbf{P}_\omega)\tilde{\mathbf{Y}}]$$
$$= \sigma^2 \operatorname{tr}[\mathbf{P}_\Omega - \mathbf{P}_\omega] + (\mathbf{X}\boldsymbol{\beta} - \mathbf{X}\boldsymbol{\beta}_0)'(\mathbf{P}_\Omega - \mathbf{P}_\omega)(\mathbf{X}\boldsymbol{\beta} - \mathbf{X}\boldsymbol{\beta}_0)$$
and then proceed as in Theorem 4.6 (ii).

MISCELLANEOUS EXERCISES 4

1. Minimizing $\|\mathbf{Y} - \boldsymbol{\theta}\|^2$ subject to $\mathbf{1}_4'\boldsymbol{\theta} = 2\pi$ using a Lagrange multiplier gives us $\hat{\theta}_i = Y_i - \bar{Y} + \frac{1}{2}\pi$. Then $\text{RSS} = \Sigma(Y_i - \hat{\theta}_i)^2 = 4(\bar{Y} - \frac{1}{2}\pi)^2$. Under H, $\theta_1 = \theta_3 = \phi_1$, $\theta_2 = \theta_4 = \pi - \phi_1$, and $\hat{\phi}_1 = \frac{1}{4}(Y_1 - Y_2 + Y_3 - Y_4 + 2\pi)$. Hence $\text{RSS}_H = (Y_1 - \hat{\phi}_1)^2 + (Y_2 - \pi + \hat{\phi}_1)^2 + (Y_3 - \hat{\phi}_1)^2 + (Y_1 - \pi + \hat{\phi}_1)^2$. Finally $F = \frac{1}{2}(\text{RSS}_H - \text{RSS})/\text{RSS}$.
2. From Theorem 4.1 (iv) with $q = p_1$ and $\mathbf{P}_H = \mathbf{P}_2$ we have $E[\text{RSS}_H - \text{RSS}] = \sigma^2 p_1 + \boldsymbol{\beta}'\mathbf{X}'(\mathbf{P} - \mathbf{P}_2)\mathbf{X}\boldsymbol{\beta}$. Also $(\mathbf{P} - \mathbf{P}_2)\mathbf{X}\boldsymbol{\beta} = (\mathbf{I}_n - \mathbf{P}_2)\mathbf{X}\boldsymbol{\beta} = (\mathbf{I}_n - \mathbf{P}_2)(\mathbf{X}_1\boldsymbol{\beta}_1 + \mathbf{X}_2\boldsymbol{\beta}_2) = (\mathbf{I}_n - \mathbf{P}_2)\mathbf{X}_1\boldsymbol{\beta}_1$.
3. Let $\phi_i = \beta_i - \beta_q$ $(i = 1, 2, \ldots, q-1)$. Then H is true $\Leftrightarrow$ each $\phi_i = 0 \Leftrightarrow \Sigma h_i\phi_i = 0$ for all $\{h_i\} \Leftrightarrow \Sigma_{i=1}^{q} c_i\beta_i = 0$ (see Example 5.2 in Section 5.1.3). Also use the fact that a linear combination of estimable functions is estimable.

4. (a) The distribution of $\mathbf{A}\hat{\boldsymbol{\beta}}$ is multivariate normal as $\mathbf{A}\hat{\boldsymbol{\beta}}=\mathbf{A}(\mathbf{X}'\mathbf{X})^{-}\mathbf{X}'\mathbf{Y}$ $=\mathbf{C}\mathbf{Y}$ and has a nonsingular variance–covariance matrix (Exercise 1 of Exercises 4e).
   (b) Follows from (a) and Theorem 2.1 (iii).
   (c) $\mathbf{P}=\mathbf{X}(\mathbf{X}'\mathbf{X})^{-}\mathbf{X}'$ and $\mathbf{P}\mathbf{X}=\mathbf{X}$. Hence $\mathcal{C}[\mathbf{A}\hat{\boldsymbol{\beta}},(\mathbf{I}_n-\mathbf{P})\mathbf{Y}]=\mathbf{0}$.
5. Use $H:(1,-1)\boldsymbol{\beta}=0$ and $F=\mathbf{A}\hat{\boldsymbol{\beta}}[\mathbf{A}(\mathbf{X}'\mathbf{X})^{-1}\mathbf{A}']^{-1}\mathbf{A}\hat{\boldsymbol{\beta}}/qS^2$.

MISCELLANEOUS EXERCISES 5

1. $\hat{\alpha}_0=\overline{Y}$ and the least squares estimates of the β_j are the same. Let $\hat{\boldsymbol{\beta}}'=(\hat{\alpha}_0,\hat{\beta}_1,\ldots,\hat{\beta}_{p-1})$ and $\mathbf{v}'=(x_1-\bar{x}_{\cdot 1},\ldots,x_{p-1}-\bar{x}_{\cdot p-1})$. Then

$$\mathfrak{D}[\hat{\boldsymbol{\beta}}]=\sigma^2\begin{bmatrix}\frac{1}{n} & \mathbf{0}'\\ \mathbf{0} & \mathbf{C}\end{bmatrix}$$

and

$$\begin{aligned}\operatorname{var}[\hat{Y}]&=(1,\mathbf{v}')\mathfrak{D}[\hat{\boldsymbol{\beta}}](1,\mathbf{v}')'\\ &=\sigma^2\left(\frac{1}{n}+\mathbf{v}'\mathbf{C}\mathbf{v}\right)\geqslant\frac{\sigma^2}{n}\end{aligned}$$

since $\mathbf{C}$ is positive definite. Equality occurs when $\mathbf{v}=\mathbf{0}$.
2. Let $\hat{Y}_{*G}=(\mathbf{x}_*',\mathbf{z}_*')\hat{\boldsymbol{\delta}}_G$, then using an identical argument we find that

$$\begin{aligned}\operatorname{var}\left[\hat{Y}_{*G}\right]&=\sigma^2\mathbf{x}_*'(\mathbf{X}'\mathbf{X})^{-1}\mathbf{x}_*+\sigma^2(\mathbf{L}'\mathbf{x}_*-\mathbf{z}_*)'\mathbf{M}(\mathbf{L}'\mathbf{x}_*-\mathbf{z}_*)\\ &\geqslant\sigma^2\mathbf{x}_*'(\mathbf{X}'\mathbf{X})^{-1}\mathbf{x}_*,\end{aligned}$$

because $\mathbf{M}$ is positive definite.
3. By (5.13), $a_0\hat{\beta}_0+a_1\hat{\beta}_1\pm(2F^{\alpha}_{2,n-2})^{1/2}\hat{v}^{1/2}$, where

$$\hat{v}=\frac{S^2\left\{a_0^2(\Sigma x_i^2/n)-2a_0a_1\bar{x}+a_1^2\right\}}{\Sigma(x_i-\bar{x})^2}.$$

MISCELLANEOUS EXERCISES 6

1. $E[\hat{\beta}_0]=\beta_3+4\beta_2$, $E[\hat{\beta}_1]=\beta_1+7\beta_3$.
3. Use Exercise 1 at the end of Section 4.1.1.

5. From (6.22) $E[Z] \sim \frac{1}{2}(f_2^{-1} - f_1^{-1})(1 + \frac{1}{2}\gamma_2 A)$. Here $f_1 = k$, $f_2 = n - k - 1$, $\mathbf{P}_1 = \mathbf{P} - \frac{1}{n}\mathbf{1}_n\mathbf{1}_n'$ [cf. (4.26) in Section 4.2] and $\mathbf{P}_2 = \mathbf{I}_n - \mathbf{P}$, where $\mathbf{P} = \mathbf{X}(\mathbf{X}'\mathbf{X})^{-1}\mathbf{X}' = [(p_{ij})]$. Hence

$$A = \frac{k^2 \Sigma (1 - p_{ii})^2 - (n - k - 1)^2 \Sigma \left(\frac{1}{n} - p_{ii} \right)^2}{k(n - k - 1)(2k - n + 1)}.$$

6. From Example 4.2 (Section 4.1.3), $\text{RSS}_H - \text{RSS} = n_1 n_2 (\overline{U} - \overline{V})^2 / (n_1 + n_2) = \mathbf{Y}'\mathbf{P}_1\mathbf{Y}$ and $\text{RSS} = \Sigma (U_i - \overline{U})^2 + \Sigma (V_j - \overline{V})^2 = \mathbf{Y}'\mathbf{P}_2\mathbf{Y}$. We then use (6.21) and (6.22) with $f_1 = 1$, $f_2 = n_1 + n_2 - 2$,

$$\mathbf{p}_1' = \frac{n_1 n_2}{n_1 + n_2} \left(\frac{1}{n_1^2} \mathbf{1}_{n_1}', \frac{1}{n_2^2} \mathbf{1}_{n_2}' \right)$$

and

$$\mathbf{p}_2' = \left(\left(1 - \frac{1}{n_1} \right) \mathbf{1}_{n_1}', \left(1 - \frac{1}{n_2} \right) \mathbf{1}_{n_2}' \right).$$

When $n_1 = n_2$, F is quadratically balanced.

MISCELLANEOUS EXERCISES 7

1. Follows from $Y_* = \overline{Y} + \hat{\beta}_1 (x_* - \overline{x})$ and $(\mathbf{X}'\mathbf{X})^{-1}$.
2. The log likelihood function is

$$L(\phi, \beta_1) = -\frac{n}{2} \log \sigma^2 - \frac{1}{2\sigma^2} \sum_i (Y_i - \phi\beta_1 + \beta_1 x_i)2.$$

$$\frac{\partial \log L}{\partial \phi} = 0 \Rightarrow \overline{Y} - \hat{\phi}\hat{\beta}_1 + \hat{\beta}_1 \overline{x} = 0.$$

$$\frac{\partial \log L}{\partial \beta_1} = 0 \Rightarrow \Sigma x_i \left(Y_i - \overline{Y} - \hat{\beta}_1 (x_i - \overline{x}) \right) = 0, \text{ etc.}$$

3. Apply the method of Section 7.2.2 to $U = \mathbf{a}_1' \hat{\boldsymbol{\beta}} - \phi \mathbf{a}_2' \hat{\boldsymbol{\beta}}$, where $\phi = \mathbf{a}_1' \boldsymbol{\beta} / \mathbf{a}_2' \boldsymbol{\beta}$. Let $\sigma_U^2 = (\mathbf{a}_1 - \phi \mathbf{a}_2)' \mathscr{D}[\hat{\boldsymbol{\beta}}](\mathbf{a}_1 - \phi \mathbf{a}_2)$ and show that

$$T = \frac{U / \sigma_U}{\sqrt{S^2 / \sigma^2}} \sim t_{n-p}.$$

Then consider $T^2 = F^{\alpha}_{1,n-p}$ as a quadratic function of ϕ.

4. $\dot{x}_*/\hat{x}_* = \dot{\beta}_1 \hat{\beta}_1 = r^2$.
5. (a) $\beta_1^* = \frac{1}{n}\Sigma\left(\frac{Y_i}{x_i}\right)$. (b) $\beta_1^* = \frac{\Sigma Y_i}{\Sigma x_i}$.
6. Same as Equation (7.18) with Σx_i^2 replaced by $\Sigma w_i x_i^2$, and $\tilde{\beta}$ and S^2replaced by β^* and S_W^2 of (7.26) and (7.27), respectively.
8. Under $H: E[Y_{ki}] = b + \beta_k(x_{ki} - a)$. Obtain RSS_H by minimizing $\Sigma_k\Sigma_i[Y_{ki} - b - \beta_k(x_{ki} - a)^2]$ with respect to β_1 and β_2. Also equivalent to shifting the origin to (a,b) and testing the hypothesis that both lines go through the origin.
9. An estimate of the distance δ is (in the notation of Section 7.5.2) $d = (\tilde{\alpha}_2 - \tilde{\alpha}_1)/\tilde{\beta}$. Use the method of Section 7.2.2 and consider $U = (\tilde{\alpha}_2 - \tilde{\alpha}_1) - \delta\tilde{\beta} = (\overline{Y}_{2\cdot} - \overline{Y}_{1\cdot}) + \tilde{\beta}(\bar{x}_{1\cdot} - \bar{x}_{2\cdot} - \delta)$. Then $E[U] = 0$ and, since $\text{cov}[\overline{Y}_{k\cdot}, Y_{ki} - \overline{Y}_{k\cdot}] = 0$,

$$\sigma_U^2 = \text{var}[U] = \sigma^2\left\{\frac{1}{n_2} + \frac{1}{n_1} + \frac{(\bar{x}_{1\cdot} - \bar{x}_{2\cdot} - \delta)^2}{\Sigma\Sigma(x_{ki} - \bar{x}_{k\cdot})^2}\right\}.$$

Let $S^2 = \text{RSS}_{H_1}/(n_1 + n_2 - 3)$; then the confidence limits are the roots of the quadratic in δ, $T^2 = F^{\alpha}_{1,n_1+n_2-3}$, where $T = (U/\sigma_U)/\sqrt{S^2/\sigma^2}$.

10. $\frac{1}{y} = \frac{1}{\alpha} + \sin^2\theta\left(\frac{1}{\beta} - \frac{1}{\alpha}\right)$. Let $x = \sin^2\theta$, etc.

MISCELLANEOUS EXERCISES 8

1. $n = 5$, $x = -2, -1, 0, 1, 2$.

$$\hat{\beta}_1 = \frac{1}{10}(-2Y_1 - Y_2 + Y_4 + 2Y_5) = 1.65,$$

$$\hat{\beta}_2 = \frac{1}{14}(2Y_1 - Y_2 - 2Y_3 - Y_4 + 2Y_5) = -0.064,$$

$$\hat{\beta}_3 = \frac{1}{10}(-Y_1 + 2Y_2 - 2Y_4 + Y_5) = -0.167.$$

$$\hat{Y} = \overline{Y} + \hat{\beta}_1 x + \hat{\beta}_2(x^2 - 2) + \hat{\beta}_3 \frac{5}{6}\left(x^3 - \frac{68}{20}x\right)$$

$$= 13.02 + 1.65x - 0.064(x^2 - 2) - 0.167(x^3 - 3.4x).$$

$$\text{RSS} = \Sigma(Y_i - \overline{Y})^2 - 10\hat{\beta}_1^2 - 14\hat{\beta}_2^2 - 10\hat{\beta}_3^3 = 0.00514.$$

$F_{1,1} = \frac{10\hat{\beta}_3^2}{\text{RSS}} = 78$. $H_0: \beta_3 = 0$ is not rejected.

3. Bias is zero as the extra column is orthogonal to the original **X** (cf. Section 6.1.1). $\hat{\beta}_{12} = \frac{1}{4}(Y_1 - Y_a - Y_b + Y_{ab})$.

4. Differentiating we have $x_m = -\beta_1/(2\beta_2)$. Let $\hat{x}_m = -\hat{\beta}_1/(2\hat{\beta}_2)$ and consider $U = \hat{\beta}_1 + x_m 2\hat{\beta}_2$. Then $E[U]=0$,

$$\sigma_U^2 = \text{var}[U] = \text{var}[\hat{\beta}_1] + 4x_m \text{cov}[\hat{\beta}_1, \hat{\beta}_2] + 4x_m^2 \text{var}[\hat{\beta}_2],$$

where $\mathcal{D}[\hat{\boldsymbol{\beta}}] = \sigma^2(\mathbf{X}'\mathbf{X})^{-1}$, and $T = (U/\sigma_U)/\sqrt{S^2/\sigma^2} \sim t_{n-3}$. The confidence limits are the roots of $T^2 = F_{1,n-3}^{\alpha}$, a quadratic in x_m.

MISCELLANEOUS EXERCISES 9

1. Show that $\text{cov}[\bar{\varepsilon}_{r\cdot} - \bar{\varepsilon}_{\cdot\cdot}, \varepsilon_{ij} - \bar{\varepsilon}_{i\cdot} - \bar{\varepsilon}_{\cdot j} + \bar{\varepsilon}_{\cdot\cdot}] = 0$ and evoke Theorem 2.7 in Section 2.3.
2. Differentiating $\Sigma\Sigma\varepsilon_{ij}^2 + \lambda\Sigma_i d_i \alpha_i$ with respect to μ and α_i gives us

$$\Sigma\Sigma(Y_{ij} - \mu - \alpha_i) = 0 \qquad \text{and} \qquad -2\sum_j (Y_{ij} - \mu - \alpha_i) + \lambda d_i = 0.$$

Summing the second equation over i leads to $\lambda \Sigma d_i = 0$ or $\lambda = 0$. Then $\hat{\mu} = \Sigma d_i \bar{Y}_{i\cdot}/\Sigma d_i$ and $\hat{\alpha}_i = \bar{Y}_{i\cdot} - \hat{\mu}$.

3. (a) $\varepsilon_{ijk} = \bar{\varepsilon}_{\cdots} + (\bar{\varepsilon}_{i\cdot\cdot} - \bar{\varepsilon}_{\cdots}) + (\bar{\varepsilon}_{ij\cdot} - \bar{\varepsilon}_{i\cdot\cdot}) + \varepsilon_{ijk} - \bar{\varepsilon}_{ij\cdot}$. Squaring and summing on i, j, k the cross-product terms vanish. Hence, substituting for ε_{ijk}, we obtain $\hat{\mu} = \bar{Y}_{\cdots}$, $\hat{\alpha}_i = \bar{Y}_{i\cdot\cdot} - \bar{Y}_{\cdots}$, $\hat{\beta}_{ij} = \bar{Y}_{ij\cdot} - \bar{Y}_{i\cdot\cdot}$. Zero covariance and Theorem 2.7 imply independence.

   (b) Test for H_1 is

$$F = \frac{\Sigma\Sigma\Sigma\left(\bar{Y}_{ij\cdot} - \bar{Y}_{i\cdot\cdot}\right)^2 / I(J-1)}{\Sigma\Sigma\Sigma\left(Y_{ijk} - \bar{Y}_{ij\cdot}\right)^2 / IJ(K-1)}.$$

4. (a) $H: \begin{pmatrix} 1 & -2 & 0 & 0 \\ 0 & 2 & -3 & 0 \end{pmatrix} \boldsymbol{\mu} = \mathbf{0}$. $F = [(Q_H - Q)/2]/[Q/(4J-4)]$ where $Q = \Sigma\Sigma(Y_{ij} - \bar{Y}_{i\cdot})^2$, $Q_H = J\{(\bar{Y}_{1\cdot} - 3\hat{\mu}_{3H})^2 + (\bar{Y}_{2\cdot} - \frac{3}{2}\hat{\mu}_{3H})^2 + (\bar{Y}_{3\cdot} - \hat{\mu}_{3H})^2\}$, and $\hat{\mu}_{3H} = (1/49)(12\bar{Y}_{1\cdot} + 6\bar{Y}_{2\cdot} + 4\bar{Y}_{3\cdot})$.

   (b) Show that $\Sigma_{i=1}^2 \Sigma_{j=1}^J (\bar{Y}_{i\cdot} - \bar{Y}_{\cdot\cdot})^2 = (\bar{Y}_{1\cdot} - \bar{Y}_{2\cdot})^2/(2/J)$.
5. (a) $\mu = \bar{\mu}_{\cdots}$, $\alpha_i = \bar{\mu}_{i\cdot\cdot} - \bar{\mu}_{\cdots}$, $\beta_j = \bar{\mu}_{\cdot j\cdot} - \bar{\mu}_{\cdots}$, $\gamma_k = \bar{\mu}_{\cdot\cdot k} - \bar{\mu}_{\cdots}$.

   (b) Use the decomposition

$$\varepsilon_{ijk} = \bar{\varepsilon}_{\cdots} + (\bar{\varepsilon}_{i\cdot\cdot} - \bar{\varepsilon}_{\cdots}) + (\bar{\varepsilon}_{\cdot j\cdot} - \bar{\varepsilon}_{\cdots}) + (\bar{\varepsilon}_{\cdot\cdot k} - \bar{\varepsilon}_{\cdots})$$
$$+ (\varepsilon_{ijk} - \bar{\varepsilon}_{i\cdot\cdot} - \bar{\varepsilon}_{\cdot j\cdot} - \bar{\varepsilon}_{\cdot\cdot k} + 2\bar{\varepsilon}_{\cdots}).$$

Hence find RSS and RSS_H by inspection and obtain

$$F = \frac{\Sigma\Sigma\Sigma\left(\bar{Y}_{i..} - \bar{Y}_{...}\right)^2/(I-1)}{\Sigma\Sigma\Sigma\left(Y_{ijk} - \bar{Y}_{i..} - \bar{Y}_{.j.} - \bar{Y}_{..k} + 2\bar{Y}_{...}\right)^2/(IJK - I - J - K + 2)}.$$

(c)

$$\Sigma\Sigma\Sigma\left(\bar{\varepsilon}_{ij.} - \bar{\varepsilon}_{i..}\right)^2 = \Sigma\Sigma\Sigma\left\{\left(\bar{\varepsilon}_{ij.} - \bar{\varepsilon}_{i..} - \bar{\varepsilon}_{.j.} + \bar{\varepsilon}_{...}\right) + \left(\bar{\varepsilon}_{.j.} - \bar{\varepsilon}_{...}\right)\right\}^2$$

$$= \Sigma\Sigma\Sigma\left(\bar{\varepsilon}_{ij.} - \bar{\varepsilon}_{i..} - \bar{\varepsilon}_{.j.} + \bar{\varepsilon}_{...}\right)^2 + \Sigma\Sigma\Sigma\left(\bar{\varepsilon}_{.j.} - \bar{\varepsilon}_{...}\right)^2$$

or $Q_1 = (Q_1 - Q_2) + Q_2$. Now $Q_1/\sigma^2 \sim \chi^2_{IJ-I}$, $Q_2/\sigma^2 \sim \chi^2_{J-1}$, $Q_1 - Q_2 \geqslant 0$ so that by Theorem 2.9 $(Q_1 - Q_2)/\sigma^2 \sim \chi^2_{IJ-I-J+1}$.

6. See Scheffé [1959: p. 93, Theorem 1].

7. Using the weights given in the hint we find that the decomposition (9.29) in Section 6.2.2 is still orthogonal provided we define $\bar{\varepsilon}_{i..} = \Sigma_j v_j \bar{\varepsilon}_{ij.} = \Sigma_j K_{.j}\bar{\varepsilon}_{ij.}/K_{..} = \Sigma_j K_{ij}\bar{\varepsilon}_{ij}/K_{i.} = \Sigma_j\Sigma_k \varepsilon_{ijk}/K_{i.}$ and $\bar{\varepsilon}_{...} = \Sigma_i\Sigma_j\Sigma_k u_i v_j \bar{\varepsilon}_{ij.} = \Sigma_i\Sigma_j\Sigma_k K_{ij}\bar{\varepsilon}_{ij.}/K_{..} = \Sigma_i\Sigma_j\Sigma_k \varepsilon_{ijk}/K_{..}$, etc.
Least squares estimates are $\hat{\mu} = \bar{Y}_{...}$, $\hat{\alpha}_i = \bar{Y}_{i..} - \bar{Y}_{...}$, $\hat{\beta}_j = \bar{Y}_{.j.} - \bar{Y}_{...}$, and $\widehat{(\alpha\beta)}_{ij} = \bar{Y}_{ij.} - \bar{Y}_{i..} - \bar{Y}_{.j.} + \bar{Y}_{...}$. The F-statistic is

$$F = \frac{\sum_i \sum_j K_{ij}\widehat{(\alpha\beta)}^2_{ij}/(I-1)(J-1)}{\Sigma\Sigma\Sigma\left(Y_{ijk} - \bar{Y}_{ij.}\right)^2/(K_{..} - IJ)}$$

9. (a) Split up ε_{ijk} in the same way as μ_{ijk} and obtain an orthogonal decomposition. Hence $\hat{\mu} = \bar{Y}_{...}$, $\hat{\alpha}_i = \bar{Y}_{i..} - \bar{Y}_{...}$, $\hat{\beta}_{ij} = \bar{Y}_{ij.} - \bar{Y}_{i..}$, $\hat{\gamma}_k = \bar{Y}_{..k} - \bar{Y}_{...}$.
(b)

$$F = \frac{\sum_i \sum_j \sum_k \left(\bar{Y}_{i..} - \bar{Y}_{...}\right)^2/(I-1)}{\sum_i \sum_j \sum_k \left(Y_{ijk} - \bar{Y}_{ij.} - \bar{Y}_{..k} + \bar{Y}_{...}\right)^2/(IJK - IJ - K + 1)}.$$

10. (a) $n_1 n_2 n_3 \equiv C(n_1, C(n_2, n_3))$
$= 1 + \nu_1 + \nu_2 + \nu_3 + \nu_1\nu_2 + \nu_2\nu_3 + \nu_1\nu_3 + \nu_1\nu_2\nu_3.$
(b) $n_1 n_2 n_3 \equiv C(N(n_1, n_2), n_3)$
$= 1 + \nu_1 + \nu_3 + n_1\nu_2 + \nu_1\nu_3 + n_1\nu_2\nu_3.$

MISCELLANEOUS EXERCISES 10

1. $R_{yz} = \Sigma\Sigma Y_{ij}(z_{ij} - \bar{z}_{i\cdot} - \bar{z}_{\cdot j} + \bar{z}_{\cdot\cdot})$.
2. (a) $\hat{\gamma}_1 = (R_{ww}R_{yz} - R_{zw}R_{yw})/(R_{zz}R_{ww} - R_{yz}^2)$.
   (b) $\sigma^2 \begin{pmatrix} R_{zz} & R_{zw} \\ R_{zw} & R_{ww} \end{pmatrix}^{-1}$. (c) $R_{zw} = 0$.
3. (a) $\hat{\gamma}_{ij} = R_{yzij}/R_{zzij}$, where $R_{yzij} = \Sigma_k (Y_{ijk} - \overline{Y}_{ij\cdot})(z_{ijk} - \bar{z}_{ij\cdot})$.

$$\text{RSS} = R_{yy} - \sum_i \sum_j \left(R_{yzij}^2 / R_{zzij}\right), \quad \text{RSS}_H = R_{yy} - \left(R_{yz}^2 / R_{zz}\right).$$

$$F = \frac{(\text{RSS}_H - \text{RSS})/(IJ-1)}{\text{RSS}/(IJK - 2IJ)}.$$

   (b) $\hat{\gamma} = R_{yz}/R_{zz}$, $\text{var}[\hat{\gamma}] = \sigma^2/R_{zz}$. Use the result:

$$(\hat{\gamma} - \gamma)\left(\frac{R_{zz}}{\text{RSS}_H/(IJK - IJ - 1)}\right)^{1/2} \sim t_{IJK-IJ-1}.$$

4. (a) $\text{RSS} = R_{yy} - (R_{yz}^2/R_{zz})$ where $R_{yz} = \Sigma\Sigma\Sigma(Y_{ijk} - \overline{Y}_{ij\cdot})(z_{ijk} - \bar{z}_{ij\cdot})$.

$$F = \frac{R_{yz}^2/R_{zz}}{\left(R_{yy} - R_{yz}^2/R_{zz}\right)/(IJK - IJ - 1)}.$$

   (b) $\text{RSS} = R_{yy} - (R_{yz}^2/R_{zz})$, $\text{RSS}_H = T_{yy} - (T_{yz}^2/T_{zz})$ where

$$T_{yz} = \Sigma\Sigma\Sigma\left(Y_{ijk} - \overline{Y}_{ij\cdot} + \overline{Y}_{i\cdot\cdot} - \overline{Y}_{\cdots}\right)\left(z_{ijk} - \bar{z}_{ij\cdot} + \bar{z}_{i\cdot\cdot} - \bar{z}_{\cdots}\right).$$

$$F = \frac{(\text{RSS}_H - \text{RSS})/(I-1)}{\text{RSS}/(IJK - IJ - 1)}.$$

6. $u = Y_{1\bullet}/(J-1)$, $v = Y_{\bullet\bullet}/(IJ-1)$, and $b = (u-v)^2\left(1 - \frac{1}{IJ}\right)$.
7. From Exercise 2 of Miscellaneous Exercises 9 we have that $\hat{\mu} = \overline{Y}_{\cdot\cdot}$ and $\hat{\alpha}_i = \overline{Y}_{i\cdot} - \overline{Y}_{\cdot\cdot}$, where $\overline{Y}_{\cdot\cdot} = \Sigma_i\Sigma_j Y_{ij}/\Sigma_i J_i$. Then, following Example 10.4 (Section 10.1.2), we have $\hat{\psi}_G = \Sigma_i c_i \overline{Y}_{i\cdot} - \hat{\gamma}_G \Sigma_i c_i \bar{z}_{i\cdot}$, where $R_{yz} = \Sigma\Sigma(Y_{ij} - \overline{Y}_{i\cdot})(z_{ij} - \bar{z}_{i\cdot})$ and $\hat{\gamma}_G = R_{yz}/R_{zz}$. Hence $\text{var}[\hat{\psi}_G] = \sigma^2\{\Sigma c_i^2/J_i + (\Sigma c_i \bar{z}_{i\cdot})^2/R_{zz}\}$. Finally $1 - \alpha = \text{pr}[\hat{\psi}_G \in \psi_G \pm ((I-1)F^{\alpha}_{I-1, \Sigma J_i - I} S^2 v)^{1/2}$, all contrasts].

10. Using the "star" notation of Section 10.2.2 we have:

$$\text{Exercise 5:}\quad \hat{Y}_{123}=\frac{IY_{1**}+JY_{*2*}+KY_{**3}-2Y_{***}}{IJK-I-J-K+2}.$$

$$\text{Exercise 9:}\quad \hat{Y}_{123}=\frac{IJY_{12*}+KY_{**3}-Y_{***}}{IJK-IJ-K+1}.$$

MISCELLANEOUS EXERCISES 11

1. Show that each transformation is unit lower triangular, and the product of such transformations is also unit triangular.
2. $\mathbf{k}'=\mathbf{c}'\mathbf{B}^{-1}$ and $\mathbf{K}=\mathbf{V}\mathbf{B}^{-1}$, that is, the same set of transformations on $\mathbf{V}$ and $\mathbf{c}'$.
3. $|\mathbf{X}'\mathbf{X}|=|\mathbf{U}'\mathbf{U}|=|\mathbf{U}|^2$.
4. $\mathbf{T}'\mathbf{T}=\begin{pmatrix}\mathbf{U}'\mathbf{U} & \mathbf{U}'\mathbf{z}\\ \mathbf{z}'\mathbf{U} & \mathbf{z}'\mathbf{z}+\delta^2\end{pmatrix}=\begin{pmatrix}\mathbf{B} & \mathbf{c}\\ \mathbf{c}' & \mathbf{Y}'\mathbf{Y}\end{pmatrix}$.
   (a) $\mathbf{U}'\mathbf{U}=\mathbf{B}$ and $\mathbf{U}'\mathbf{z}=\mathbf{c}\Rightarrow\mathbf{U}\mathbf{x}=\mathbf{z}$.
   (b) $\delta^2=\mathbf{Y}'\mathbf{Y}-\mathbf{z}'\mathbf{z}=\text{RSS}$ (cf. Equation (11.8) and following comments).
5. Let $\mathbf{B}^{-1}=\mathbf{C}$ and $(\mathbf{U}')^{-1}=\mathbf{V}$. We solve

$$\begin{bmatrix}u_{11} & u_{12} & u_{13}\\ 0 & u_{22} & u_{23}\\ 0 & 0 & u_{33}\end{bmatrix}\begin{bmatrix}c_{11} & c_{12} & c_{13}\\ c_{21} & c_{22} & c_{23}\\ c_{31} & c_{32} & c_{33}\end{bmatrix}=\begin{bmatrix}u_{11}^{-1} & 0 & 0\\ v_{21} & u_{22}^{-1} & 0\\ v_{31} & v_{32} & u_{33}^{-1}\end{bmatrix}$$

   by solving $\mathbf{U}\mathbf{c}_i=\mathbf{v}_i$ for $i=3,2,1$.
6. From Exercise 4 above $(\mathbf{X}:\mathbf{Y})=\mathbf{R}_{p+1}\mathbf{T}$, say, so that by the uniqueness of the Cholesky decomposition the extra element of $\mathbf{D}_1$ $(=\delta^2)$ is RSS.
7. Use Exercise 4 and replace $\mathbf{X}$ and $\mathbf{Y}$ by $\mathbf{W}^{1/2}\mathbf{X}$ and $\mathbf{W}^{1/2}\mathbf{Y}$. Thus

$$\mathbf{T}=\begin{pmatrix}\mathbf{W}^{1/2}\mathbf{U} & \mathbf{z}_*\\ \mathbf{0}' & \delta_*\end{pmatrix}$$

   and $\delta_*^2=\mathbf{Y}'\mathbf{W}\mathbf{Y}-\mathbf{Y}'\mathbf{W}\mathbf{X}(\mathbf{X}'\mathbf{W}\mathbf{X})^{-1}\mathbf{X}'\mathbf{W}\mathbf{Y}=(\mathbf{Y}-\mathbf{X}\boldsymbol{\beta}_*)'\mathbf{V}^{-1}(\mathbf{Y}-\mathbf{X}\boldsymbol{\beta}_*)$.
8. (a) Let $\mathbf{X}=(\mathbf{x}_1,\mathbf{x}_2)$, then $u_{11}=\|\mathbf{x}_1\|=2$. We use the theory of Section 11.9.1 with $\mathbf{t}_k=\mathbf{x}_1$ and $\mathbf{t}=\mathbf{x}_2$ or $\mathbf{Y}$. Thus $2v_1^2=1+|x_{11}|/\|\mathbf{x}_1\|$ and $2v_1v_i=x_{1i}/\|\mathbf{x}_1\|$, so that $v_1=\sqrt{3}/2$ and $v_2=v_3=v_4=1/2\sqrt{3}$. Then $\mathbf{H}\mathbf{x}_2=\mathbf{x}_2-2\mathbf{v}\mathbf{v}'\mathbf{x}_2$, etc.
   (b) $\hat{\boldsymbol{\beta}}'=(\frac{7}{2},-1)$, RSS$=3$.

9. 1.

10.

$$\begin{bmatrix} \sqrt{2} & 3/\sqrt{2} & \sqrt{2} \\ 0 & \sqrt{(3/2)} & 2\sqrt{(2/3)} \\ 0 & 0 & \sqrt{(7/3)} \end{bmatrix}.$$

11. $\mathbf{a}'\boldsymbol{\beta}$ estimable$\Rightarrow \mathbf{a}' = \boldsymbol{\alpha}'\mathbf{X}'\mathbf{X}$ and $\mathbf{a}'\mathbf{X}^*\mathbf{X} = \boldsymbol{\alpha}'\mathbf{X}'\mathbf{X}\mathbf{X}^*\mathbf{X} = \boldsymbol{\alpha}'\mathbf{X}'\mathbf{X} = \mathbf{a}$. The condition $\Rightarrow E[\mathbf{a}'\mathbf{X}^*\mathbf{Y}] = \mathbf{a}'\mathbf{X}^*\mathbf{X}\boldsymbol{\beta} = \mathbf{a}'\boldsymbol{\beta} \Rightarrow \mathbf{a}'\boldsymbol{\beta}$ is estimable.

$$\mathbf{a}'\boldsymbol{\Pi} = \mathbf{a}'\mathbf{X}^*\mathbf{X}\boldsymbol{\Pi} = \mathbf{a}'\boldsymbol{\Pi}\begin{pmatrix} \mathbf{I}_r & \mathbf{U}_{11}^{-1}\mathbf{U}_{12} \\ \mathbf{0} & \mathbf{0} \end{pmatrix}, \quad \text{etc.}$$

13. $v_n = v_{n-1} + (nx_n - S_{xn})(ny_n - S_{yn})/[n(n-1)]$.

MISCELLANEOUS EXERCISES 12

3.

$$C_p = \frac{(\text{RSS}_p - \text{RSS}_{K+1})(n-K-1)}{\text{RSS}_{K+1}} + 2p - n + n - K - 1$$

$$= (K+1-p)F + p + (p-K-1), \quad \text{so that}, \quad C_p \leqslant p \Leftrightarrow F \leqslant 1.$$

References

Abrahamse, A. P. J. and Koerts, J. (1971). New estimates of disturbances in regression analysis. *J. Am. Stat. Assoc.*, **66**, 71–74.

Abrahamse, A. P. J., *see also* Koertts, J. and ... (1969)

Aitkin, M. A. (1974). Simultaneous inference and the choice of variable subsets. *Technome trics*, **16**, 221–227.

Albert, A. (1972). *Regression and the Moore-Penrose Pseudoinverse*. Academic Press: New York.

Allen, D. M. (1971a). Mean square error of prediction as a criterion for selecting variables. *Technometrics*, **13**, 469–475.

Allen, D. M. (1971b). The prediction sum of squares as a criterion for selecting prediction variables. *Tech. Rep. 23*. Department of Statistics, University of Kentucky.

Allen, D. M. (1974). The relationship between variable selection and data augmentation and a method for prediction. *Technometrics*, **16**, 125–127.

Anderson, D. A. (1972). Overall confidence levels of the least significant difference procedure. *Am. Stat.*, **26** (4), 30–32.

Anderson, M. R. (1971). A characterization of the multivariate normal distribution. *Ann. Math. Stat.*, **42**, 824–827.

Anderson, R. L. and Bancroft, T. A. (1952). *Statistical Theory in Research*. McGraw-Hill: New York.

Anderson, T. W. (1958). *An Introduction to Multivariate Statistical Analysis*. Wiley: New York.

Anderson, T. W. (1962). The choice of the degree of a polynomial regression as a multiple decision problem. *Ann. Math. Stat.*, **33**, 255–265.

Anderson, T. W. (1971). *The Statistical Analysis of Time Series*. Wiley: New York.

Andrews, D. F. (1971a). Significance tests based on residuals. *Biometrika*, **58**, 139–148.

Andrews, D. F. (1971b). A note on the selection of data transformations. *Biometrika*, **58**, 249–254.

Andrews, D. F. (1974). A robust method for multiple linear regression. *Technometrics*, **16**, 523–531.

Andrews, D. F., Bickel, P. J., Hampel, F. R., Huber, P. J., Rogers, W. H., and Tukey, J. W. (1972). *Robust Estimates of Location: Survey and Advances*. Princeton University Press: Princeton, N. J.

Andrews, D. F. and Tukey, J. W. (1973). Teletypewriter plots for data analysis can be fast: 6-line plots, including probability plots. *Appl. Stat..*, **22**, 192–202.

Anscombe, F. J. (1961) Examination of residuals. *Proc. Fourth Berkeley Symp. Math. Stat. Probab.*, **1**, 1–36.

Anscombe, F. J. (1967). Topics in the investigation of linear relations fitted by the method of least squares. *J. R. Stat. Soc. B*, **29**, 1–52.

Anscombe, F. J. (1973). Graphs in statistical analysis. *Am. Stat.*, **27** (2), 17–21.

Anscombe, F. J. and Tukey, J. W. (1963). The examination and analysis of residuals. *Technometrics*, **5**, 141–160.

Atiqullah, M. (1962). The estimation of residual variance in quadratically balanced least squares problems and the robustness of the *F*-test. *Biometrika*, **49**, 83–91.

Atiqullah, M. (1964). The robustness of the analysis of covariance analysis of a one-way classification. *Biometrika*, **51**, 365–372.

Atkinson, A. C. (1972). Planning experiments to detect inadequate regression models. *Biometrika*, **59**, 275–293.

Atkinson, A. C. (1973). Testing transformations to normality. *J. R. Stat. Soc. **B***, **35**, 473–479.

Atkinson, A. C. and Cox, D. R. (1974). Planning experiments for discriminating between models. *J. R. Stat. Soc. B*, **36**, 321–334.

Atwood, C. L. (1971). Robust procedures for estimating polynomial regression. *J. Am. Stat. Assoc.*, **66**, 855–860.

Bancroft, T. A., *see also* Anderson, R. L. and ... (1952).

Bancroft, T. A., *see also* Kennedy, W. J. and ... (1971).

Banerjee, K. S. and Carr, R. N. (1971). A comment on ridge regression. Biased estimation for non-orthogonal problems. *Technometrics*, **13**, 895–898.

Barrett, J. P. (1974). The coefficient of determination—some limitations. *Am. Stat.*, **28** (1), 19–20.

Bartlett, M. S. (1937a). Properties of sufficiency and statistical tests. *Proc. R. Soc., A*, **160**, 268–282.

Bartlett, M. S. (1937b). Some examples of statistical methods of research in agriculture and applied biology. *J. R. Stat. Soc. Suppl.*, **4**, 137–170.

Bauer, F. L. (1965). Elimination with weighted row combinations for solving linear equations and least squares problems. *Numer. Math.*, **7**, 338–352.

Beale, E. M. L. (1970). Note on procedures for variable selection in multiple regression. *Technometrics*, **12**, 909–914.

Beale, E. M. L. (1974). The scope of Jordan elimination in statistical computing. *J. Inst. Math. Appl.*, **10**, 138–140.

Beale, E. M. L., Kendall, M. G., and Mann, D. W. (1967). The discarding of variables in multivariate analysis. *Biometrika*, **54**, 357–366.

Beaton, A. E. (1964). *The Use of Special Matrix Operators in Statistical Calculus*. Research Bulletin RB-64-51. Education Testing Service: Princeton, New Jersey.

Beaton, A. E. and Tukey, J. W. (1974). The fitting of power series, meaning polynomials, illustrated on band-spectroscopic data. *Technometrics*, **16**, 147–185.

Beckman, R. J. and Trussell, H. J. (1974). The distribution of an arbitrary studentized residual and the effects of updating in multiple regression. *J. Am. Stat. Assoc.*, **69**, 199–201.

Beckman, R. J., *see also* Tietjen, G. L. and ... (1972).

Beckman, R. J., *see also* Tietjen, G. L., Moore, R. H., and ... (1973).

Behnken, D. W. and Draper, N. R. (1972). Residuals and their variance patterns. *Technometrics*, **14**, 101–111.

Bellman, R. and Roth, R. (1969). Curve fitting by segmented straight lines. *J. Am. Stat. Assoc.*, **64**, 1079–1084.

Berkson, J. (1950). Are there two regressions? *J. Am. Stat. Assoc.*, **45**, 164–180.

Berry, G., *see also* Farebrother, R. W. and ... (1974).

Bhargava, R. P. and Srivastava, M. S. (1973). On Tukey's confidence intervals for the contrasts in the means of the intraclass correlation model. *J. R. Stat. Soc. B*, **35**, 147–152.

Bickel, P. J. (1975). One-step Huber estimates in the linear model. *J. Am. Stat. Assoc.*, **70**, 428–434.

Björck, A. (1967a). Solving linear least squares problems by Gram-Schmidt orthonormalization. *BIT* (Nord. Tidskr. Informations-Behandl.), **7**, 1–21.

Björck, A. (1967b). Iterative refinement of linear least square solutions I. *BIT* (Nord. Tidskr. Informations-Behandl.), **7**, 257–278.

Björck, A. (1968). Iterative refinement of linear least squares solutions II. *BIT* (Nord. Tidskr. Informations-Behandl.), **8**, 8–30.

Björck, A. and Golub, G. H. (1967). Iterative refinement of linear least square solutions by Householder transformation. *BIT* (Nord. Tidskr. Informations-Behandl.), **7**, 322–337.

Bloomfield, P. and Watson, G. S. (1975). The inefficiency of least squares. *Biometrika*, **62**, 121–128.

Bloomfield, P., *see also* Wynn, H. P. and ... (1971).

Bock, R. D. (1963). Programming univariate and multivariate analysis of variance. *Technometrics*, **5**, 95–117.

Bock, R. D. (1965). A computer program for univariate and multivariate analysis of variance. In *Proceedings of the IBM Scientific Computing Symposium on Statistics, October 21–23, 1963*, 69–111. IBM Data Processing Division: White Plains, N. Y.

Boddy, R., *see also* Goldsmith, P. L. and ... (1973).

Bohrer, R. (1973). An optimality property of Scheffé bounds. *Ann. Stat.*, **1**, 766–772.

Bohrer, R. and Francis, G. K. (1972). Sharp one-sided confidence bounds for linear regression over intervals. *Biometrika*, **59**, 99–107.

Bolch, B. W., *see also* Huang, C. J. and ... (1974).

Boswell, M. T., *see also* Patil, G. P. and ... (1970).

Boullion, T. L. and Odell, P. L. (1971). *Generalized Inverse Matrices*. Wiley: New York.

Bowden, D. C. (1970). Simultaneous confidence bands for linear regression models. *J. Am. Stat. Assoc.*, **65**, 413–421.

Bowden, D. C. and Graybill, F. A. (1966). Confidence bands of uniform and proportional width for linear models. *J. Am. Stat. Assoc.*, **61**, 182–198.

Bowden, D. C., *see also* Graybill, F. A. and ... (1967).

Bower, D. R., *see also* Swindel, B. F. and ... (1972).

Box, G. E. P. (1953). Non-normality and tests on variances. *Biometrika*, **40**, 318–335.

Box, G. E. P. (1957). Evolutionary operation: a method for increasing industrial productivity. *Appl. Stat.*, **6**, 3–23.

Box, G. E. P. (1966). Use and abuse of regression. *Technometrics*, **8**, 625–629.

Box, G. E. P. and Cox, D. R. (1964). An analysis of transformations. *J. R. Stat. Soc. B*, **26**, 211–252.

Box, G. E. P. and Draper, N. R. (1959). A basis for the selection of a response surface design. *J. Am. Stat. Assoc.*, **54**, 622–654.

Box, G. E. P. and Draper, N. R. (1963). The choice of a second order rotatable design. *Biometrika*, **50**, 335–352.

Box, G. E. P. and Draper, N. R. (1969). *Evolutionary Operation*. Wiley: New York.

Box, G. E. P. and Draper, N. R. (1975). Robust designs. *Biometrika*, **62**, 347–352.

Box, G. E. P. and Hill, W. J. (1974). Correcting inhomogeneity of variance with power transformation weighting. *Technometrics*, **16**, 385–389.

Box, G. E. P. and Tidwell, P. W. (1962). Transformation of the independent variables. *Technometrics*, **4**, 531–550.

Box, G. E. P. and Watson, G. S. (1962). Robustness to non-normality of regression tests. *Biometrika*, **49**, 93–106.

Box, M. J. and Draper, N. R. (1971). Factorial designs, the $|X'X|$ criterion, and some related matters. *Technometrics*, **13**, 731–742.

Bradu, D. and Gabriel, K. R. (1974). Simultaneous statistical inference on interactions in two-way analysis of variance. *J. Am. Stat. Assoc.*, **69**, 428–439.

Breaux, H. J. (1968). A modification of Efromyson's technique for stepwise regression analysis. *Commun. Assoc. Comp. Mach.*, **8**, 556–557.

Brown, M. B. and Forsythe, A. B. (1974). Robust tests for the equality of variances. *J. Am. Stat. Assoc.*, **69**, 364–367.

Brown, R. L., Durbin, J., and Evans, J. M. (1975). Techniques for testing the constancy of regression relationships over time. *J. R. Stat. Soc. B*, **37**, 149–163.

Brunk, H. D. (1965). *An Introduction to Mathematical Statistics*, 2nd ed. Blaisdell: Waltham, Mass.

Businger, P. A. (1970). Updating a singular value decomposition. *BIT* (Nord. Tidskr. Informations-Behandl.), **10**, 376–385.

Businger, P. and Golub, G. H. (1965). Linear least squares solutions by Householder transformations. *Numer. Math.*, **7**, 269–276.

Cadwell, J. H. and Williams, D. E. (1961). Some orthogonal methods of curve and surface fitting. *Comput. J.*, **4**, 260–264.

Caliński, T., *see also* Pearce, S. C., ..., and Marshall, T. F. de C. (1974).

Canner, P. L. (1969). Some curious results using minimum variance linear unbiased estimators. *Am. Stat.*, **23** (5), 39–40.

Carleton, W. T., *see also* McGee, V. E. and ... (1970).

Carmer, S. G. and Swanson, M. R. (1973). Evaluation of ten pairwise multiple comparison procedures by Monte Carlo methods. *J. Am. Stat. Assoc.*, **68**, 66–74.

Carr, R. N., *see also* Banerjee, K. S. and ... (1971).

Chambers, J. M. (1971). Regression updating. *J. Am. Stat. Assoc.*, **66**, 744–748.

Chen, E. H. and Dixon, W. J. (1972). Estimates of parameters of a censored regression sample. *J. Am. Stat. Assoc.*, **67**, 664–675.

Chen, H. J., *see also* Shapiro, S. S., Wilk, M. B., and ... (1968).

Christensen, L. R. (1973). Simultaneous statistical inference in the normal multiple linear regression model. *J. Am. Stat. Assoc.*, **68**, 457–461.

Clayton, D. G. (1971). Algorithm AS 46: Gram-Schmidt orthogonalisation. *Appl. Stat.*, **20**, 335–338.

Clenshaw, C. W. (1955). A note on the summation of Chebyshev series. *Math. Tables Aids Comput.*, **9**, 118.

Clenshaw, C. W. (1960). Curve fitting with a digital computer. *Comput. J.*, **2**, 170.

Clenshaw, C. W. and Hayes, J. G. (1965). Curve and surface fitting. *J. Inst. Math. Appl.*, **1**, 164–183.

Cochran, W. G. (1934). The distribution of quadratic forms in a normal system, with applications to the analysis of covariance. *Proc. Camb. Phil. Soc.*, **30**, 178–191.

Cochran, W. G. (1938). The omission or addition of an independent variate in multiple linear regression. *J. R. Stat. Soc. Suppl.*, **5**, 171–176.

Cochran, W. G. (1941). The distribution of the largest of a set of estimated variances as a fraction of their total. *Ann. Eugenics Lond.*, **11**, 47–52.

Cochran, W. G. (1957). Analysis of covariance: its nature and uses. *Biometrics*, **13**, 261–281.

Cochran, W. G. (1969). The use of covariance in observational studies. *Appl. Stat.*, **18**, 270–275.

Cochran, W. G. and Cox, G. M. (1957). *Experimental Designs*, 2nd . Wiley: New York.

Cooper, B. E. (1968). The use of orthogonal polynomials: Algorithm AS 10. *Appl. Stat.*, **17**, 283–287.

Cooper, B. E. (1971a). The use of orthogonal polynomials with equal x-values: Algorithm AS 42. *Appl. Stat.*, **20**, 208–213.

Cooper, B. E. (1971b). A remark on algorithm AS 10. *Appl. Stat.*, **20**, 216.

Cote, R., Manson, A. R., and Hader, R. J. (1973). Minimum bias approximation of a general regression model. *J. Am. Stat. Assoc.*, **68**, 633–638.

Cox, C. P. (1971). Interval estimating for X-predictions from linear Y—on—X regression lines through the origin. *J. Am. Stat. Assoc.*, **66**, 749–751.

Cox, D. R. (1961). Tests of separate families of hypotheses. *Proc. 4th Berkeley Symp.*, **1**, 105–123.

Cox, D. R. (1962). Further results on tests of separate families of hypotheses. *J. R. Stat. Soc. B*, **24**, 406–424.

Cox, D. R. (1968). Notes on some aspects of regression analysis. *J. R. Stat. Soc. B*, **30**, 265–279.

Cox, D. R. and Hinkley, D. V. (1968). A note on the efficiency of least squares estimates. *J. R. Stat. Soc. B*, **30**, 284–289.

Cox, D. R. and Snell, E. J. (1968). A general definition of residuals. *J. R. Stat. Soc. B*, **30**, 248–275.

Cox, D. R. and Snell, E. J. (1974). The choice of variables in observational studies. *Appl. Stat.*, **23**, 51–59.

Cox, D. R., *see also* Atkinson, A. C. and ... (1974).

Cox, D. R., *see also* Box, G. E. P. and ... (1964).

Cox, D. R., *see also* Draper, N. R. and ... (1969).

Cox, D. R., *see also* Herzberg, A. M. and ... (1972).

Cox, G. M., *see also* Cochran, W. G. and ... (1957).

Craig, A. T., *see also* Hogg, R. V. and ... (1958).

Craig, A. T., *see also* Hogg, R. V. and ... (1970).

Cramer, E. M. (1972). Missing values in experimental design models. *Am. Stat.*, **26** (4), 58.

Cramer, E. M., *see also* Youngs, E. A. and ... (1971).

Csorgo, M., Seshadri, V., and Yalovsky, M. (1973). Some exact tests for normality in the presence of unknown parameters. *J. R. Stat. Soc. B*, **35**, 507–522.

Curnow, R. N. (1973). A smooth population response curve based on an abrupt threshold and plateau model for individuals. *Biometrics*, **29**, 1–10.

Curry, H. B. and Schoenberg, I. J. (1966). On Pólya frequency functions IV: the fundamental spline functions and their limits. *J. Anal. Math.*, **17**, 71–107.

Daling, J. R. and Tamura, H. (1970). Use of orthogonal factors for selection of variables in a regression equation—an illustration. *Appl. Stat.*, **19**, 260–268.

Daniel, C. (1959). Use of half-normal plots in interpreting factorial two-level experiments. *Technometrics*, **1**, 311–342.

Daniel, C. and Wood, F. S. (1971). *Fitting Equations to Data*. Wiley-Interscience: New York.

David, H. A. (1952). Upper 5 and 1% points of the maximum *F* ratio. *Biometrika*, **39**, 422–424.

David, H. A. (1956). The ranking of variances in normal populations. *J. Am. Stat. Assoc.*, **51**, 621–626.

David, H. A. (1970). *Order Statistics*. Wiley: New York.

Davies, M. (1967). Linear approximation using the criterion of least total deviations. *J. R. Stat. Soc. B*, **29**, 101–109.

Davies, O. L. (Ed.), (1960). *The design and analysis of industrial experiments*, 2nd ed. Oliver and Boyd: London.

Davies, R. B. and Hutton, B. (1975). The effects of errors in the independent variables in linear regression. *Biometrika*, **62**, 383–391.

Dayton, C. M. and Schafer, W. D. (1973). Extended tables of *t* and chi-square for Bonferroni tests with unequal error allocation. *J. Am. Stat. Assoc.*, **68**, 78–83.

DeGracie, J. S. and Fuller, W. A. (1972). Estimation of the slope and analysis of covariance when the concomitant variable is measured with error. *J. Am. Stat. Assoc.*, **67**, 930–937.

De La Garza, A. (1954). Spacing of information in poflynomial regression. *Ann. Math. Stat.*, **25**, 123–130.

Dershowitz, A. F., *see also* Hahn, G. J. and ... (1974).

Dixon, W. J. and Massey, F. J., Jr. (1969). *Introduction to Statistical Analysis*, 3rd ed. McGraw-Hill: New York.

Dixon, W. J., *see also* Chen, E. H. and ... (1972).

Doolittle, M. H. (1878). Method employed in the solution of normal equations and the adjustment of a triangulation. *U.S. Coast Geod. Surv. Rep.*, 115–120.

Draper, N. R. and Cox, D. R. (1969). On distributions and their transformation to normality. *J. R. Stat. Soc. B*, **31**, 472–476.

Draper, N. R., Guttman, I., and Kanemasu, H. (1971). The distribution of certain regression statistics. *Biometrika*, **58**, 295–298.

Draper, N. R. and Hunter, W. G. (1969). Transformations: some examples revisited. *Technometrics*, **11**, 23–40.

Draper, N. and Smith, H. (1966). *Applied Regression Analysis*. Wiley: New York.

Draper, N. R., *see also* Behnken, D. W. and ... (1972).

Draper, N. R., *see also* Box, G. E. P. and ... (1959).

Draper, N. R., *see also* Box, G. E. P. and ... (1963).

Draper, N. R., *see also* Box, G. E. P. and ... (1969).

Draper, N. R., *see also* Box, G. E. P. and ... (1975).

Draper, N. R., *see also* Box, M. J. and ... (1971).

Draper, N. R., *see also* St. John, R. C. and ... (1975).

Drygas, H. (1970). *The coordinate-free approach to Gauss-Markov estimation*. Lecture Notes in Operations Research and Mathematical Systems No. 40. Springer-Verlag: New York.

Duncan, D. B., *see also* Waller, R. A. and ... (1969).

Dunn, O. J. (1959). Confidence intervals for the means of dependent, normally distributed variables. *J. Am. Stat. Assoc.*, **54**, 613–621.

Dunn, O. J. (1961). Multiple comparisons among means. *J. Am. Stat. Assoc.*, **56**, 52–64.

Dunn, O. J. (1968). A note on confidence bands for a regression line over finite range. *J. Am. Stat. Assoc.*, **63**, 1028–1033.

Durbin, J. (1969). Tests for serial correlation in regression analysis based on the periodogram of least-squares residuals. *Biometrika*, **56**, 1–15.

Durbin, J. and Watson, G. S. (1950). Testing for serial correlation in least squares regression. I. *Biometrika*, **37**, 409–428.

Durbin, J. and Watson, G. S. (1951). Testing for serial correlation in least squares regression. II. *Biometrika*, **38**, 159–178.

Durbin, J. and Watson, G. S. (1971). Testing for serial correlation in least squares regression. III. *Biometrika*, **58**, 1–19.

Durbin, J., *see also* Borwn, R. L., ..., and Evans, J. M. (1975).

Dwyer, P. S. (1941). The Doolittle technique. *Ann. Math. Stat.*, **12**, 449–458.

Dwyer, P. S. (1944). A matrix presentation of least squares and correlation theory with matrix justification of improved methods of solution. *Ann. Math. Stat.*, **15**, 82–89.

Dwyer, P. S. (1945). The square root method and its use in correlation and regression. *J. Am. Stat. Assoc.*, **40**, 493–503.

Dyer, A. R. (1974). Comparison of tests for normality with a cautionary note. *Biometrika*, **61**, 185–189.

Dykstra, R. L., Hewett, J. E., and Thompson, W. A., Jr. (1973). Events which are almost independent. *Ann. Stat.*, **1**, 674–681.

Dykstra, R. L., *see also* Pierce, D. A. and ... (1969).

Edwards, J. B. (1969). The relation between the *F*-test and $\bar{R}^2$. *Am. Stat.*, **23** (5), 28.

Efroymson, M. A. (1960). Multiple regression analysis. *In* A. Ralston and H. S. Wilf (Eds.), *Mathematical Methods for Digital Computers*, Vol. 1, pp. 191–203.

Eicker, F. (1963). Asymptotic normality and consistency of the least squares estimators for families of linear regressions. *Ann. Math. Stat.*, **34**, 447–456.

Elfving, G. (1959). Design of linear experiments. *In* Ulf Grenander (Ed.), *Probability and statistics*, Harold Cramér Volume, pp. 58–74. Wiley: New York.

Ellenberg, J. H. (1973). The joint distribution of the standardized least squares residuals from a general linear regression. *J. Am. Stat. Assoc.*, **68**, 941–943.

Engleman, L., *see also* Forsythe, A. B.,..., Jennrich, R., and May, P. R. A. (1973).

Evans, J., *see also* Brown, R. L., Durbin, J. and ... (1975).

Ezekiel, M. (1930). *Methods of Correlation Analysis*. Wiley: New York.

Ezekiel, M. and Fox, K. A. (1959). *Methods of Correlation and Regression Analysis*, 3rd ed. Wiley: New York.

Farebrother, R. W. (1974). Algorithm AS 79: Gram-Schmidt regression. *Appl. Stat.*, **23**, 470–476.

Farebrother, R. W. and Berry, G. (1974). Remark AS R12. A remark on algorithm AS 6: Triangular decomposition of a symmetric matrix. *Appl. Stat.*, **23**, 477–478.

Farley, J. U. and Hinich, M. J. (1970). A test for a shifting slope coefficient in a linear model. *J. Am. Stat. Assoc.*, **65**, 1320–1329.

Feder, P. I. (1974). Graphical techniques in statistical data analysis—tools for extracting information from data. *Technometrics*, **16**, 287–299.

Feder, P. I. (1975). On asymptotic distribution theory in segmented regression problems—identified case. *Ann. Stat.*, **3**, 49–83.

Feller, W. (1968). *An introduction to Probability Theory and its Applications*, 3rd ed. Wiley: New York.

Fieller, E. C. (1940). The biological standardization of insulin. *J. R. Stat. Soc. Suppl.*, **7**, 1–64.

Fienberg, S., *see also* Schatzoff, M., Tsau, R., and ... (1968).

Fisher, R. A. and Yates, F. (1957). *Statistical Tables for Biological, Agricultural and Medical Research*, 5th ed. Oliver and Boyd: London and Edinburgh.

Fletcher, R. H. (1975). On the iterative refinement of least squares solutions. *J. Am. Stat. Assoc.*, **70**, 109–112.

Forsythe, A. B., Engleman, L., Jennrich, R., and May, P. R. A. (1973). A stopping rule for variable selection in multiple regression. *J. Am. Stat. Assoc.*, **68**, 75–77.

Forsythe, A. B., *see also* Brown, M. B. and ... (1974).

Forsythe, G. E. (1957). Generation and use of orthogonal polynomials for data-fitting with a digital computer. *J. Soc. Indust. Appl. Math., 5, 74–87.*

Fowlkes, E. B. (1969). Some operators for ANOVA calculations. *Technometrics*, **11**, 511–526.

Fox, K. A., *see also* Ezekiel, M. and ... (1959).

Fox, L. (1964). *An Introduction to Numerical Linear Algebra*. Oxford University Press: London.

Fox, L. and Hayes, J. G. (1951). Practical methods for the inversion of matrices. *J. R. Stat. Soc. B*, **13**, 83–91.

Francia, R. S., *see also* Shapiro, S. S. and ... (1972).

Francis, G. K., *see also* Bohrer, R. and ... (1972).

Francis, I. (1973). A comparison of several analysis of variance programs. *J. Am. Stat. Assoc.*, **68**, 860–865.

Freeman, G. H. and Jeffers, J. N. R. (1962). Estimation of means and standard errors in the analysis of non-orthogonal experiments by electronic computers. *J. R. Stat. Soc. B*, **24**, 435–446.

Freund, R. J. (1963). A warning of roundoff errors in regression. *Am. Stat.*, **17**, 13–15.

Fuller, W. A., *see also*, DeGracie, J. S. and ... (1972).

Fuller, W. A., *see also* Gallant, A. R. and ... (1973).

Furnival, G. M. (1971). All possible regressions with less computation. *Technometrics*, **13**, 403–408.

Furnival, G. M. and Wilson, R. W. M. Jr. (1974). Regressions by leaps and bounds. *Technometrics*, **16**, 499–511.

Gabriel, K. R., Putter, J., and Wax, Y. (1973). Simultaneous confidence intervals for product-type interaction contrasts. *J. R. Stat. Soc. B*, **35**, 234–244.

Gabriel, K. R., *see also* Bradu, D. and ... (1974).

Gafarian, A. V. (1964). Confidence bands in straight line regression. *J. Am. Stat. Assoc.*, **59**, 182–213.

Gallant, A. R. and Fuller, W. A. (1973). Fitting segmented polynomial regression models whose joins points have to be estimated. *J. Am. Stat. Assoc.*, **68**, 144–147.

Garside, M. J. (1965). The best sub-set in multiple regression analysis. *Appl. Stat.*, **14**, 196–200.

Garside, M. J. (1971). Some computational procedures for the best subset problem. *Appl. Stat.*, **20**, 8–15.

Gartside, P. S. (1972). A study of methods for comparing several variances. *J. Am. Stat. Assoc.*, **67**, 342–346.

Gaylor, D. W. and Sweeny, H. C. (1965). Design for optimal prediction in simple linear regression. *J. Am. Stat. Assoc.*, **60**, 205–216.

Gaylor, D. W., *see also* Haseman, J. K. and ... (1973).*wGentleman, W. M.* (1973). *Least squares computations by Givens transformations without square roots. J. Inst. Math. Appl.*, **12**, 329–336.

Gentleman, W. M. (1974a). Algorithm AS 75: Basic procedures for large, sparse or weighted linear least squares problems. *Appl. Stat.*, **23**, 448–454.

Gentleman, W. M. (1947b). Regression problems and the QR decomposition. *J. Inst. Math. and Appl.*, **10**, 195–197.

Ghosh, M. N. and Sharma, D. (1963). Power of Tukey's tests for non-additivity. *J. R. Stat. Soc. B*, **25**, 213–219.

Gnanadesikan, R. and Wilk, M. B. (1970). A probability plotting procedure for general analysis of variance. *J. R. Stat. Soc. B*, **32**, 88–101.

Gnanadesikan, R., *see also* Wilk, M. B. and ... (1968).

Goldberger, A. S. (1964). *Econometric Theory*. Wiley: New York.

Goldsmith, P. L. and Boddy, R. (1973). Critical analysis of factorial experiments and orthogonal fractions. *Appl. Stat.*, **22**, 141–160.

Goldstein, M. Smith, A. F. M. (1974). Ridge-type estimators for regression analysis. *J. R. Stat. Soc. B*, **36**, 284–291.

Golub, G. H. (1965). Numerical methods for solving linear least squares problems. *Numer. Math.*, **7**, 206–216.

Golub, G. H. (1969). Matrix decompositions and statistical calculations. *In* R. C. Milton and J. A. Nelder (Eds.), *Statistical Computation*, pp. 365–397. Academic Press: New York.

Golub, G. H. Reinsch, C. (1970). Singular value decomposition and least squares solutions. *Numer. Math.*, **14**, 403–420.

Golub, G. H. and Styan, G. P. (1973). Numerical computations for univariate linear models. *J. Stat. Comput. Simul.*, **2**, 253–274.

Golub, G. H. and Styan, G. P. (1974). Some aspects of numerical computations for linear models. *Interface*—Proceedings of Computer Science and Statistics, 7th Annual Symposium on the Interface (August 1973), pp. 189–192. Statistical Computing Laboratory: Iowa State University.

Golub, G. H. and Wilkinson, J. H. (1966). Note on the iterative refinement of least squares solutions. *Numer. Math., 9, 139–148.*

Golub, G. H., *see also* Björck, A. and ... (1967).

Golub, G. H., *see also* Businger, P. and ... (1965).

Good, I. J. (1963). On the independence of quadratic expressions. *J. R. Stat. Soc. B*, **25**, 377–382.

Gorman, J. W. and Toman, R. J. (1966). Selection of variables for fitting equations to data. *Technometrics*, **8**, 27–51.

Gower, J. C., *see also* Preece, D. A. and ... (1974).

Graybill, F. A. (1961). *An Introduction to Linear Statistical Models*, Vol. I. McGraw-Hill: New York.

Graybill, F. A. (1969). *Introduction to Matrices with Applications in Statistics*. Wadsworth: Belmont, California.

Graybill, F. A. and Bowden, D. C. (1967). Linear segment confidence bands for simple linear models. *J. Am. Stat. Assoc.*, **62**, 403–408.

Graybill, F. A., *see also* Bowden, D. C. and ... (1966).

Graybill, F. A., *see also* Johnson, D. E. and ... (1972a).

Graybill, F. A., *see also* Johnson, D. E. and ... (1972b).

Graybill, F. A., *see also* Kingman, A. and ... (1970).

Greenberg, E. (1975). Minimum variance properties of principal component regression. *J. Am. Stat. Assoc.*, **70**, 194–197.

Grether, D. M. (1972). Missing values in experimental design models. *Am. Stat.*, **26** (4), 57–58.

Grossman, S. I. and Styan, G. P. H. (1972). Optimal Properties of Theil's BLUS residuals. *J. Am. Stat. Assoc.*, **67**, 672–673.

Guest, P. G. (1958). The spacing of observations in polynomial regression. *Ann. Math. Stat.*, **29**, 294–299.

Gujarati, D. (1970). Use of dummy variables in testing for equality between sets of coefficients in linear regressions: a generalization. *Am. Stat.*, **24** (5), 18–22.

Gunst, R. F., *see also* Webster, J. T.,..., and Mason, R. L. (1974).

Gurian, J., *see also* Halperin, M. and... (1968).

Gurian, J., *see also* Halperin, M. and... (1971).

Guttman, I., Wilks, S. S., and Hunter, J. S. (1971). *Introductory Engineering Statistics*, 2nd ed. Wiley: New York.

Guttman, I., *see also* Draper, N. R.,..., and Kanemasu, H. (1971).

Haberman, S. J. (1975). How much do Guss–Markov and least square estimates differ? A coordinate-free approach. *Ann. Stat.*, **3**, 982–990.

Hader, R. J., *see also* Cote, R., Manson, A. R., and... (1973).

Hader, R. J., *see also* Karson, M. J., Manson, A. R. and... (1969).

Hahn, G. J. (1972). Simultaneous prediction intervals for a regression model. *Technometrics*, **14**, 203–214.

Hahn, G. J. and Dershowitz, A. F. (1974). Evolutionary operation today—some survey results and observations. *Appl. Stat.*, **23**, 214–218.

Hahn, G. J. and Hendrickson, R. W. (1971). A table of precentage points of the distribution of the largest absolute value of k Student t variates and its applications. *Biometrika*, **58**, 323–332.

Hahn, G. J. and Shapiro, S. S. (1967). *Statistical Models in Engineering*. Wiley: New York.

Hahn, G. J., *see also* Nelson, W. and... (1972).

Hahn, G. J., *see also* Nelson W. and... (1973).

Haitovsky, Y. (1969). A note on the maximization of $\bar{R}^2$. *Am. Stat.*, **23** (1), 20–21.

Hald, A. (1952), *Statistical Theory with Engineering Applications*. Wiley: New York.

Halperin, M. (1970). On inverse estimation in linear regression. *Technometrics*, **12**, 727–736.

Halperin, M. and Gurian, J. (1968). Confidence bands in linear regression with constraints on the independent variables. *J. Am. Stat. Assoc.*, **63**, 1020–1027.

Halperin, M. and Gurian, J. (1971). A note on estimation in straight line regression when both variables are subject to error. *J. Am. Stat. Assoc.*, **66**, 587–589.

Halperin, M., Rastogi, S. C., Ho, I., and Yang, Y. Y. (1967). Shorter confidence bands in linear regression. *J. Am. Stat. Assoc.*, **62**, 1050–1067.

Halperin, E. F. (1973). Polynomial regression from a Bayesian approach. *J. Am. Stat. Assoc.*, **68**, 137–143.

Hamaker, H. C. (1962). On multiple regression analysis. *Stat. Neerl.*, **16**, 31–56.

Hamilton, M. A., *see also* Lieberman, G. J., Miller, R. G., Jr., and... (1967).

Han, C. P. (1968). Testing the homogeneity of a set of correlated variances. *Biometrika*, **55**, 317–326.

Han, C. P. (1969). Testing the homogeneity of variances in a two-way classification. *Biometrics*, **25**, 153–158.

Harter, H. L. (1970). Multiple comparison procedures for interactions. *Am. Stat.*, **24** (5), 30–32.

Hartley, H. O. (1950). The maximum F-ratio as a short cut test for heterogeneity of variance. *Biometrika*, **37**, 308–312.

Hartley, H. O. (1956). Programming analysis of variance for general purpose computers. *Biometrics*, **12**, 110–122.

Hartley, H. O. and Jayatillake, K. S. E. (1973). Estimation for linear models with unequal variances. *J. Am. Stat. Assoc.*, **68**, 189–192.

Hartley, H. O., *see also* Pearson, E. S. and... (1970).

Hartley, H. O., *see also* Sielken, R. L., Jr. and... (1973).

Harvey, A. C., *see also* Phillips, G. D. A. and... (1974).

Harvey, J. R., *see also* Rohde, C. A. and... (1965).

Haseman, J. K. and Gaylor, D. W. (1973). An algorithm for non-iterative estimation of multiple missing values for non-iterative estimation of multiple missing values for crossed classifications. *Technometrics*, **15**, 631–636.

Hastings, W. K. (1972). Test data for statistical algorithms: least squares and ANOVA. *J. Am. Stat. Assoc.*, **67**, 874–879.

Hawkins, D. M. (1973). On the investigations of alternative regressions by principal component analysis. *Appl. Stat.*, **22**, 275–286.

Hayes, D. G. (1969). A method of storing the orthogonal polynomials used for curve and surface fitting. *Comput. J.*, **12**, 148–150.

Hayes, J. G. (1970a). Curve fitting by polynomials in one variable. *In* J. G. Hayes (Ed.), *Numerical Approximation to Functions and Data*, pp. 43–64. Athlone Press: London.

Hayes, J. G. (Ed.), (1970b). *Numerical Approximation to Functions and Data*. Athlone Press: London.

Hayes, J. G. (1974). Numerical methods for curve and surface fitting. *J. Inst. Math. Appl.*, **10**, 144–152.

Hayes, J. G., *see also* Clenshaw, C. W. and... (1965).

Hayes, J. G., *see also* Fox, L. and... (1951).

Healey, M. J. R. (1968a). Multiple regression with a singular matrix. *Appl. Stat.*, **17**, 110–117.

Healey, M. J. R. (1968b). Algorithm AS6: Triangular decomposition of a symmetric matrix; Algorithm AS7: Inversion of a positive semi-definite symmetric matrix. *Appl. Stat.*, **17**, 195–199.

Healey, M. J. R. and Westmacott, M. (1956). Missing values in experiments analyzed on automatic computers. *Appl. Stat.*, **5**, 203–206.

Hedayat, A. and Robson, D. S. (1970). Independent stepwise residuals for testing homoscedasticity. *J. Am. Stat. Assoc.*, **65**, 1573–1581.

Helms, R. W. (1974). The average estimated variance criterion for the selection-of-varibles problem in general linear models. *Technometrics*, **16**, 261–273.

Hemmerle, W. J. (1974). Nonorthogonal analysis of variance using iterative improvement and balanced residuals. *J. Am. Stat. Assoc.*, **69**, 772–778.

Hendrickson, R. W., *see also* Hahn, G. J. and... (1971).

Herzberg, A. M. and Cox, D. R. (1972). Some optimal designs for interpolation and extrapolation. *Biometrika*, **59**, k51–561.

Hewett, J. E., *see also* Dykstra, R. L.,..., and Thompson, W. A., Jr. (1973).

Hill, W. J. and Hunter, W. G. (1966). A review of response surface methodology: a literature survey. *Technometrics*, **8**, 571–590.

Hill, W. J., *see also* Box, G. E. F. and... (1974).

Hinich, M. J., *see also* Farley, J. U. and... (1970).

Hinkley, D. V. (1969a). On the ratio of two correlated normal random variables. *Biometrika*, **56**, 635–639.

Hinkley, D. V. (1969b). Inference about the intersection in two-phase regression. *Biometrika*, **56**, 495–504.

Hinkley, D. V. (1971). Inference in two-phase regression. *J. Am. Stat. Assoc.*, **66**, 736–743.

Hinkley, D. V., *see also* Cox, D. R. and... (1968).

Ho, I., *see also* Halperin, M., Rastogi, S. C., and... (1967).

Hoadley, B. (1970). A bayesian look at inverse linear regression. *J. Am. Stat. Assoc.*, **65**, 356–369.

Hocking, R. R. (1972). Criteria for selection of a subset regression: which one should be used? *Technometrics*, **14**, 967–970.

Hocking, R. R. (1974). Misspecification in regression. *Am. Stat.*, **28** (1), 39–40.

Hocking, R. R. and Leslie, R. N. (1967). Selection of the best subset in regression analysis. *Technometrics*, **9**, 531–540.

Hocking, R. R., *see also* La Motte, L. R. and... (1970).

Hodges, S. D. and Moore, P. G. (1972). Data uncertainties and least squares regression. *Appl. Stat.*, **21**, 185–195.

Hoel, P. G. (1958). Efficiency problems in polynomial estimation. *Ann. Math. Stat.*, **29**, 1134–1145.

Hoel, P. G. (1968). On testing for the degree of a polynomial. *Technometrics*, **10**, 757–767.

Hoel, P. G. and Levine, A. (1964). Optimal spacing and weighting in polynomial prediction. *Ann. Math. Stat.*, **35**, 1553–1560.

Hoerl, A. E. and Kennard, R. W. (1970a). Ridge regression. Biased estimation for non-orthogonal problems. *Technometrics*, **12**, 55–67.

Hoerl, A. E. and Kennard, R. W. (1970b). Ridge regression. Applications to non-orthogonal problems. *Technometrics*, **12**, 69–82.

Hogg, R. V. (1974). Adaptive robust procedures: a partial review and some suggestions for future applications and theory. *J. Am. Stat. Assoc.*, **69**, 909–925.

Hogg, R. V. and Craig, A. T. (1958). On the decomposition of certain chi-square variables. *Ann. Math. Stat.*, **29**, 608–610.

Hogg, R. V. and Craig, A. T. (1970). *Introduction to Mathematical Statistics*, 3rd ed. Macmillan: New York.

Hotelling, H. (1943). Some new methods in matrix calculation. *Ann. Math. Stat.*, **14**, 1–34.

Hotelling, H. (1957). The relations of the newer multivariate statistical methods to factor analysis. *Brit. J. Stat. Psychol.*, **10**, 69–79.

Hotelling, H., *see also* Working, H. and... (1929).

Hsu, P. L. (1938). On the best unbiased quadratic estimate of the variance. *Stat. Res. Mem.*, **2**, 91–104.

Huang, C. J. and Bolch, B. W. (1974). On the testing of regression disturbances for normality. *J. Am. Stat. Assoc.*, **69**, 330–335.

Huang, D. S. (1970). *Regression and Econometric Methods*. Wiley: New York.

Hudson, D. J. (1966). Fitting segmented curves whose join points have to be estimated. *J. Am. Stat. Assoc.*, **61**, 1097–1129.

Hudson, D. J. (1969). Least squares fitting of a polynomial constrained to be either non-negative, non-decreasing or convex. *J. R. Stat. Soc. B*, **31**, 113–118.

Hunter, J. S., *see also* Guttman, I., Wilks, S. S., and... (1971).

Hunter, W. G., *see also* Draper, N. R. and... (1969).

Hunter, W. G., *see also* Hill, W. J. and... (1966).

Hutton, B., *see also* Davies, R. B. and... (1975).

I.B.M. (1968). *System* 360 *Scientific Sub-routines Package*, 360*A-CM*-03*X Version III.*

Jaech, J. L. (1966). An alternative approach to missing value estimation. *Am. Stat.*, **20** (5), 27–29.

James, A. T. and Wilkinson, G. N. (1971). Factorization of the residual operator and canonical decomposition of non-orthogonal factors in analysis of variance. *Biometrika*, **58**, 279–294.

James, W. and Stein, C. (1961). Estimation with quadratic loss. *Proc. Fourth Berkeley Symp. Math. Stat. Probab.*, **1**, 361–379.

Jayatillake, K. S. E., *see also* Hartley, H. O. and... (1973).

Jeffers, J. N. R. (1967). Two case studies in the application of principal component analysis. *Appl. Stat.*, **16**, 225–236.

Jeffers, J. N. R., *see also* Freeman, G. H. and... (1962).

Jeffers, J. R. N., *see also* Pearce, S. C. and... (1971).

Jennrich, R. I. and Sampson, P. I. (1971). A remark on algorithm AS 10. *Appl. Stat.*, **20**, 117–118.

Jennrich, R., *see also* Forsythe, A. B., Engleman, L.,..., and May, P. R. A. (1973).

John, J. A. and Smith, T. M. F. (1974). Sum of squares in non-full rank general linear hypothesis. *J. R. Stat. Soc. B*, **36**, 107–109.

John, P. W. M. (1971). *Statistical Design and Analysis of Experiments*. Macmillan: New York.

Johnson, A. F. (1971). Linear combinations in designing experiments. *Technometrics*, **13**, 575–587.

Johnson, D. E. and Graybill, F. A. (1972a). Estimation of σ^2 in a two-way classification model with interaction. *J. Am. Stat. Assoc.*, **67**, 388–394.

Johnson, D. E. and Graybill, F. A. (1972b). An analysis of a two-way model with interaction and no replication. *J. Am. Stat. Assoc.*, **67**, 862–868.

Jordan, T. L. (1968). Experiments on error growth associated with some linear least squares procedures. *Math. Comp.*, **22**, 579–588.

Joshi, S. W. (1970). Construction of certain bivariate distributions. *Am. Stat.*, **24** (2), 32.

Kalotay, A. J. (1971). Structural solution to the linear calibration problem. *Technometrics*, **13**, 761–769.

Kanemasu, H., *see also* Draper, N. R., Guttman, I., and... (1971).

Karlin, S. and Studden, W. J. (1966). Optimal experimental designs. *Ann. Math. Stat.*, **37**, 783–815.

Karson, M. J., Manson, A. R., and Hader, R. J. (1969). Minimum bias estimation and experimental design for response surfaces. *Technometrics*, **11**, 461–475.

Kelley, T. L. (1948). *The Kelley Statistical Tables*. (Revised 1948). Harvard University Press: Cambridge, Mass.

Kendall, M. G. and Stuart, A. (1968). *The Advanced Theory of Statistics*, **3**. Griffin: London.

Kendall, M. G., *see also* Beale, E. M. L.,..., and Mann, D. W. (1967).

Kennard, R. W. (1971). A note on the C_p statistic. *Technometrics*, **13**, 899–900.

Kennard, R. W., *see also* Hoerl, A. E. and... (1970a).

Kennard, R. W., *see also* Hoerl, A. E. and... (1970b).

Kennedy, W. J. and Bancroft, T. A. (1971). Model building for prediction in regression based upon repeated significance tests. *Ann. Math. Stat.*, **42**, 1273–1284.

Kiefer, J. (1959). Optimum experimental designs. *J. R. Stat. Soc. B*, **21**, 273–319.

Kiefer, J. and Wolfowitz, J. (1959). Optimum designs in regression problems. *Ann. Math. Stat.*, **30**, 271–294.

Kiefer, J. and Wolfowitz, J. (1960). The equivalence of two extremum problems. *Can. J. Math.*, **12**, 363–366.

Kiefer, J. and Wolfowitz, J. (1965). On a theorem of Hoel and Levine on extrapolation designs. *Ann. Math. Stat.*, **36**, 1627–1655.

Kingman, A. and Graybill, F. A. (1970). A non-linear characterization of the normal distribution. *Ann. Math. Stat.*, **41**, 1889–1895.

Kiountouzis, E. A. (1973). Linear programming techniques in regression analysis. *Appl. Stat.*, **22**, 69–73.

Knott, M., *see also* Scott, A. J. and... (1974).

Koerts, J. (1967). Some further notes on disturbance estimates in regression analysis. *J. Am. Stat. Assoc.*, **62**, 169–183.

Koerts, J. and Abrahamse, A. P. J. (1969). *On the Theory and Application of the General Linear Model*. Rotterdam University Press: Rotterdam.

Koerts, J., *see also* Abrahamse, A. J. P. and... (1971).

Kowalski, C. (1970). The performance of some rough tests for bivariate normality before and after coordinate transformations to normality. *Technometrics*, **12**, 517–544.

Kowalski, C. J. (1973). Non-normal bivariate distributions with normal marginals. *Am. Stat.*, **27** (3), 103–106.

Kruskal, W. (1960). The coordinate free approach to Gauss-Markov estimation and its application to missing and extra observations. *Proc. Fourth Berkeley Symp. Math. Stat. Probab.*, **1**, 435–451.

Kruskal, W. (1968). When are Gauss-Markov and least squares estimators identical? A coordinate-free approach. *Ann. Math. Stat.*, **39**, 70–75.

Kruskal, W. (1975). The geometry of generalized inverses. *J. R. Stat. Soc. B*, **37**, 272–283.

Krutchoff, R. G. (1967). Classical and inverse regression methods of calibration. *Technomet rics*, **9**, 425–439.

Krutchoff, R. G. (1969). Classical and inverse regression methods of calibration in extrapolation. *Technometrics*, **11**, 605–608.

Kshirsagar, A. M. (1971). Bias due to missing plots. *Am. Stat.*, **25** (1), 47–50.

Kuiper, N. H. (1960). Tests concerning random points on a circle. *Proc. K. Ned. Acad. Wet., Ser. A*, **63**, 38–47.

Kupper, L. L. (1972). Letter to the editor. *Am. Stat.*, **26** (1), 52.

Kupper, L. L. (1973). A note on the admissibility of a response surface. *J. R. Stat. Soc. B*, **35**, 28–32.

Kupper, L. L. and Meydrech, E. F. (1973). A new approach to mean squared error estimation of response surfaces. *Biometrika*, **60**, 573–579.

Kussmaul, K. (1966). Protection against assuming the wrong degree in polynomial regression. *Technometrics*, **11**, 677–682.

Laha, R. G. (1957). On a characterization of the normal distribution from properties of suitable linear statistics. *Ann. Math. Stat.*, **28**, 126–139.

La Motte, L. R. and Hocking, R. R. (1970). Computational efficiency in the selection of regression variables. *Technometrics*, **12**, 83–93.

Lancaster, H. O. (1954). Traces and cumulants of quadratic forms in normal variables. *J. R. Stat. Soc. B*, **16**, 247–254.

Lancaster, H. O. (1969). *The Chi-Squared Distribution*. Wiley: New York.

Larsen, W. A. and McCleary, S. J. (1972). The use of partial residual plots in regression analysis. *Technometrics*, **14**, 781–790.

Layard, M. W. J. (1973). Robust large-sample tests for homgeneity of variances. *J. Am. Stat. Assoc.*, **68**, 195–198.

Leone, F. C., *see also* Moussa-Hamouda, E. A. and... (1974).

Leslie, R. N., *see also* Hocking, R. R. and... (1967).

Levene, H. (1960). Robust tests for equality of variances. *In* I. Olkin (Ed.), *Contributions to Probability and Statistics*, pp. 278–292. Stanford University Press: Palo Alto, Calif.

Levine, A., *see also* Hoel, P. G. and... (1964).

Lieberman, G. J. (1961). Prediction region for several predictions from a single regression line. *Technometrics*, **3**, 21–27.

Lieberman, G. J. and Miller, R. G., Jr. (1963). Simultaneous tolerance intervals in regression. *Biometrika*, **50**, 155–168.

Lieberman, G. J., Miller, R. G., Jr., and Hamilton, M. A. (1967). Unlimited simultaneous discrimination intervals in regression. *Biometrika*, **54**, 133–145.

Lindley, D. V. (1968). The choice of variables in multiple regression. *J. R. Stat. Soc. B*, **30**, 31–53.

Lindley, D. V. and Smith, A. F. M. (1972). Bayes estimates for the linear model. *J. R. Stat. Soc. B*, **34**, 1–18.

Ling, R. F. (1974). Comparison of several algorithms for computing sample means and variances. *J. Am. Stat. Assoc.*, **69**, 859–866.

Longley, J. W. (1967). An appraisal of least squares programs for the electronic computer from the point of view of use. *J. Am. Stat. Assoc.*, **62**, 819–841.

Lowe, C. W. (1974). Evolutionary operation in action. *Appl. Stat.*, **23**, 218–226.

Lowerre, J. M. (1974). On the mean square error of parameter estimates for some biased estimators. *Technometrics*, **16**, 461–464.

Lukacs, E. (1959). Characterization of populations by properties of suitable statistics. *Proc. 3rd Berkeley Symp.*, **2**, 195–214.

Lukacs, E., *see also* Savage, I. R. and... (1954).

McCleary, S. J., *see also* Larsen, W. A. and... (1972).

McElroy, F. W. (1967). A necessary and sufficient condition that ordinary least-squares estimators be best linear unbiased. *J. Am. Stat. Assoc.*, **62**, 1302–1304.

McGee, V. E. and Carleton, W. T. (1970). Piecewise regression. *J. Am. Stat. Assoc.*, **65**, 1109–1124.

Marcus, M. (1964). *Basic Theorem in Matrix Theory*. National Bureau of Standards Applied Mathematics Series, No. 57. U.S. Government Printing Office: Washington, D.C.

Malinvaud, E. (1970). *Statistical Methods of Econometrics* (translated by A. Silvey). Amsterdam.

Mallows, C. L. (1964). *Choosing Variables in a Linear Regression: a Graphical Aid*. Presented at the Cental Regional Meeting of the Institute of Mathematical Statistics, Manhattan, Kansas.

Mallows, C. L. (1966). *Choosing a Subset Regression*. Presented at the Joint Statistical Meeting, Los Angeles, Calif.

Mallows, C. L. (1967 approx.). *Choosing a Subset Regression*. Unpublished report. Bell Telephone Laboratories.

Mallows, C. L. (1973). Some comments on C_p. *Technometrics*, **15**, 661–675.

Mann, D. W., *see also* Beale, E. M. L., Kendall, M. G., and... (1967).

Manson, A. R., *see also* Cote, R.,..., and Hader, R. J. (1973).

Manson, A. R., *see also* Karson, M. J.,..., and Hader, R. J. (1969).

Mantel, N. (1970). Why stepdown procedures in variable selection. *Technometrics*, **12**, 621–625.

Marcus, M. (1964). *Basic Theorems in Matrix Theory*. National Bureau of Standards. Applied Mathematics Series, No. 57. U.S. Government Printing Office: Washington, D.C.

Marsaglia, G. (1964). Conditional means and covariances of normal variables with singular covariance matrix. *J. Am. Stat. Assoc.*, **59**, 1203–1204.

Marshall, T. F. de C., *see also* Pearce, S. C., Caliński, T., and... (1974).

Martin, R. S., Peters, G., and Wilkinson, J. H. (1965). Symmetric decomposition of a positive definite matrix. *Numer. Math.*, **7**, 362–383.

Mason, R. L., *see also* Webster, J. T., Gunst, R. F., and... (1974).

Massey, F. J., Jr., *see also* Dixon, W. J. and... (1969).

Massy, W. F. (1965). Principal component regression in exploratory statistical research. *J. Am. Stat. Assoc.*, **60**, 234–256.

May, P. R. A., *see also* Forsythe, A. B., Engelman, L., Jennrich, R., and... (1973).

Mayer, L. S. and Willke, T. A. (1973). On biased estimation in linear models. *Technometrics*, **15**, 497–508.

Mendenhall, W., *see also* Ott, L. and... (1972).

Meydrech, E. F., *see also* Kupper, L. L. and... (1973).

Miller, K. S. (1975). *Multivariate Distributions*. R. E. Krieger: New York.

Miller, R. G., Jr. (1966). *Simultaneous Statistical Inference*. McGraw-Hill: New York.

Miller, R. G., Jr., *see also* Lieberman, G. J. and... (1963).

Miller, R. G., Jr., *see also* Lieberman, G. J.,..., and Hamilton, M. A. (1967).

Mitchell, T. J. (1974a). An algorithm for the construction of "D-optimal" experimental designs. *Technometrics*, **16**, 203–210.

Mitchell, T. J. (1974b). Computer construction of "D-optimal" first-order designs. *Technometrics*, **16**, 211–220.

Mitra, S. K., *see also* Rao, C. R. and... (1971a).

Mitra, S. K., *see also* Rao, C. R. and... (1971b).

Moler, C. E. (1967). Iterative refinement in floating point. *J. Assoc. Comput. Mach.*, **14**, 316–321.

Moore, P. G., *see also* Hodges, S. D. and... (1972).

Moore, R. H., *see also* Tietjen, G. L.,..., and Beckman, R. J. (1973).

Moran, P. A. P. (1970). Fitting a straight line when both variables are subject to error. *In* R. S. Anderssen and M. R. Osborne (Eds.), *Data Presentation*, pp. 25–28. University of Queensland Press.

Morgan, J. A. and Tatar, J. F. (1972). Calculation of the residual sum of squares for all possible regressions. *Technometrics*, **14**, 317–325.

Moussa-Hamouda, E. A. and Leone, F. C. (1974). The O-BLUE estimators for complete and censored samples. *Technometrics*, **16**, 441–446.

Muller, M. E. (1970). Computers as an instrument for data analysis. *Technometrics*, **12**, 259–294.

Mullet, G. M. and Murray, T. W. (1971). A new method for examining rounding error in least squares regression computer programs. *J. Am. Stat. Assoc.*, **66**, 496–498.

Murray, T. W., *see also* Mullet, G. M. and... (1971).

Murty, V. N. (1971). Minimax designs. *J. Am. Stat. Assoc.*, **66**, 319–320.

Murty, V. N. and Studden, W. J. (1972). Optimal designs for estimating the slope of a polynomial regression. *J. Am. Stat. Assoc.*, **67**, 869–873.

Myers, R. H. (1971). *Response Surface Methodology*. Allyn and Bacon: Boston.

Myers, R. H., *see also* Ott, R. L. and... (1968).

Narula, S. C. (1974). Predictive mean square error and stochastic regressor variables. *Appl. Stats.*, **23**, 11–16.

Nelder, J. A. (1965a). The analysis of randomized experiments with orthogonal block structure. I. Block structure and the null analysis of variance. *Proc. R. Soc., A*, **283**, 147–162.

Nelder, J. A. (1965b). The analysis of randomized experiments with orthogonal block structure. II. Treatment structure and the general analysis of variance. *Proc. R. Soc. A*, **283**, 163–178.

Nelder, J. A. (1968). Regression, Model-building and invariance. *J. R. Stat. Soc. A*, **131**, 303–315.

Nelder, J. A. (1972). Discussion of a paper by D. V. Lindley and A. F. M. Smith. *J. R. Stat. Soc. B*, **34**, 18–20.

Nelder, J. A. (1974). Analysis of variance programs, least squares and two-way tables. Appl. Stat., **23**, 232.

Nelson, W. (1973). Analysis of residuals from censored data. *Technometrics*, **15**, 697–715.

Nelson, W. and Hahn, G. J. (1972), Linear estimation of a regression relationship from censored data—Part I. Simple methods and their application. *Technometrics*, **14**, 247–269.

Nelson, W. and Hahn, G. J. (1973). Linear estimation of a regression relationship from censored data—Part II. Best linear unbiased estimation and theory. *Technometrics*, **15**, 133–150.

Newton, R. G. and Spurrell, D. J. (1967a). A development of multiple regression for the analysis of routine data. *Appl. Stat.*, **16**, 51–64.

Newton, R. G. and Spurrell, D. J. (1967b). Examples of the use of elements for clarifying regression analysis. *Appl. Stats.*, **16**, 165–172.

Odell, P. L., *see also* Boullion, T. L. and... (1971).

Odén, A. (1973). Simultaneous confidence intervals in inverse linear regression. *Biometrika*, **60**, 339–343.

Olshen, R. A. (1973). The conditional level of the *F*-test. *J. Am. Stat. Assoc.*, **68**, 692–698.

O'Neill, R. and Wetherill, G. B. (1971). The present state of multiple comparison methods. *J. R. Stat. Soc. B*, **33**, 218–250.

Ott, L. and Mendenhall, W. (1972). Designs for estimating the slope of a second order linear model. *Technometrics*, **14**, 341–353.

Ott, R. L. and Myers, R. H. (1968). Optimal experimental designs for estimating the independent variable in regression. *Technometrics*, **10**, 811–823.

Patil, G. P. and Boswell, M. T. (1970). A characteristic property of the multivariate normal density function and some of its applications. *Ann. Math. Stat.*, **41**, 1970–1977.

Payne, J. A. (1970). An automatic curve-fitting package. *In* J. G. Hayes (Ed.), *Numerical Approximation of Functions and Data*, pp. 98–106. Athlone Press: London.

Pearce, S. C. (1965). *Biological Statistics: an Introduction*. McGraw-Hill: New York.

Pearce, S. C., Caliński, T., and Marshall, T. F. de C. (1974). The basic contrasts of an experimental design with special reference to the analysis of data. *Biometrika*, **61**, 449–460.

Pearce, S. C. and Jeffers, J. R. N. (1971). Block designs and missing data. *J. R. Stat. Soc. B*, **33**, 131–136.

Pearson, E. S. and Hartley, H. O. (1970). *Biometrika tables for statisticians*, Vol. 1, 3rd ed. Cambridge University Press.

Perng, S. K. and Tong, Y. L. (1974). A sequential solution to the inverse linear regression problem. *Ann. Stat.*, **2**, 535–539.

Peters, G. and Wilkinson, J. H. (1970). The least squares problem and pseudoinverses. *Comput. J.*, **13**, 309–316.

Peters, G., *see also* Martin, R. S.,..., and Wilkinson, J. H. (1965).

Phillips, G. D. A. and Harvey, A. C. (1974). A simple test for serial correlation in regression analysis. *J. Am. Stat. Assoc.*, **69**, 935–939.

Pierce, D. A. and Dykstra, R. L. (1969). Independence and the normal distribution. *Am. Stat.*, **23** (4), 39.

Plackett, R. L. (1950). Some theorems in least squares. *Biometrika*, **37**, 149–157.

Plackett, R. L. (1960). *Regression Analysis*. Clarendon Press: Oxford.

Poirier, D. J. (1973). Piecewise regression using cubic splines. *J. Am. Stat. Assoc.*, **68**, 515–524.

Pope, P. T. and Webster, J. T. (1972). The use of an *F*-statistic in stepwise regression procedures. *Technometrics*, **14**, 327–340.

Preece, D. A. (1971). Iterative procedures for missing values in experiments. *Technometrics*, **13**, 743–753.

Preece, D. A. and Gower, J. C. (1974). An iterative computer procedure for mixed-up values in experiments. *Appl. Stats.*, **23**, 73–74.

Prentice, R. L. (1974). Degrees-of-freedom modifications for *F* tests based on nonnormal errors. *Biometrika*, **61**, 559–563.

Pringle, R. M. and Rayner, A. A. (1971). *Generalized Inverse Matrices with Applications to Statistics*. Griffin: London.

Putter, J. (1967). Orthonormal bases of error spaces and their use for investigating the normality and variance of residuals. *J. Am. Stat. Assoc.*, **62**, 1022–1036.

Putter, J., *see also* Gabriel, K. R.,..., and Wax, Y. (1973).

Quenouille, M. H. (1950). An application of least squares to family diet surveys. *Econometrica*, **18**, 27–44.

Rahman, N. A. (1967). *Exercises in Probability and Statistics*. Griffin: London.

Ramsey, J. B. (1969). Tests for specification errors in classical linear least-squares regression analysis. *J. R. Stat. Soc. B*, **31**, 350–371.

Rao, C. R. (1952). Some theorems on minimum variance estimation. *Sankhyā*, **12**, 27–42.

Rao, C. R. (1969). Some characterizations of the multivariate normal distribution. *In* P. R. Krishnaiah (Ed.), *Multivariate Analysis* Vol. II, pp. 321–328. Academic Press: New York.

Rao, C. R. (1970). Estimation of heteroscedastic variances in linear models. *J. Am. Stat. Assoc.*, **65**, 161–172.

Rao, C. R. (1972a). Recent trends of research work in multivariate analysis. *Biometrics*, **28**, 3–22.

Rao, C. R. (1972b). Estimation of variance and covariance components in linear models. *J. Am. Stat. Assoc.*, **67**, 112–115.

Rao, C. R. (1973). *Linear Statistical Inference and its Applications*, 2nd ed. Wiley: New York.

Rao, C. R. (1974). Projectors, generalized inverses and the BLUE's. *J. R. Stat. Soc. B*, **36**, 442–448.

Rao, C. R. and Mitra, S. K. (1971a). *Generalized Inverse of Matrices and its Applications*. Wiley: New York.

Rao, C. R. and Mitra, S. K. (1971b). Further contributions to the theory of generalized inverse of matrices and its applications. *Sankhyā, Series A*, **33**, 289–300.

Rastogi, S. C., *see also* Halperin, M., Ho, I., and... (1967).

Rayner, A. A., *see also* Pringle, R. M. and... (1971).

Reinsch, C., *see also* Golub, G. H. and... (1970).

Reinsch, J., *see also* Wilkinson, J. H. and... (1971).

Rice, J. R. (1966). Experiments on Gram-Schmidt orthonalization. *Math. Comp.*, **20**, 325–328.

Richardson, D. H. and Wu, De-Min (1970). Alternative estimators in the error in variables model. *J. Am. Stat. Assoc.*, **65**, 724–748.

Robison, D. E. (1964). Estimates for the points of intersection of two polynomial regressions. *J. Am. Stat. Assoc.*, **59**, 214–224.

Robson, D. S., *see also* Hedayat, A. and... (1970).

Rogers, C. E. and Wilkinson, G. N. (1974). Regression, curve fitting and smoothing numerical problems in recursive analysis of variance algorithms. *J. Inst. Math. Appl.*, **10**, 141–143.

Rogers, C. E., *see also* Wilkinson, G. N. and... (1973).

Rohde, C. A. and Harvey, J. R. (1965). Unified least squares analysis. *J. Am. Stat. Assoc.*, **60**, 523–527.

Roth, R., *see also* Bellman, R. and... (1969).

Rothman, D. (1968). Letter to the editor. *Technometrics*, **10**, 432.

Rubin, D. B. (1972). A non-iterative algorithm for least squares estimation of missing values in any analysis of variance design. *Appl. Stats.*, **21**, 136–141.

Sadovski, A. N. (1974). L1-norm fit of a straight line: algorithm AS 74. *Appl. Stats.*, **23**, 244–248.

St. John, R. C. and Draper, N. R. (1975). *D*-optimality for regression designs: a review. *Technometrics*, **17**, 15–23.

Sampson, P. I., *see also* Jennrich, R. I. and... (1971).

Savage, I. R. and Lukacs, E. (1954). Tables of inverses of finite segments of the Hilbert matrix. *In* O. Taussky (Ed.), *Contributions to the Solution of Systems of Linear Equations and the Determination of Eigenvalues*. National Bureau of Standards Applied Mathematics Series 39, pp. 105–108. U.S. Govt. Printing Office: Washington, D.C.

Saw, J. G. (1966). A conservative test for the concurrence of several regression lines and related problems. *Biometrika*, **53**, 272–275.

Schafer, W. D., *see also* Dayton, C. M. and... (1973).

Schatzoff, M., Tsao, R., and Fienberg, S. (1968). Efficient calculation of all possible regressions. *Technometrics*, **10**, 769–779.

Scheffé, H. (1953). A method of judging all contrasts in the analysis of variance. *Ann. Math. Stat.*, **40**, 87–104.

Scheffé, H. (1959). *The Analysis of Variance*. Wiley: New York.

Schlesselman, J. (1971). Power families: a note on the Box and Cox transformation. *J. R. Stat. Soc. B*, **33**, 307–311.

Schlossmacher, E. J. (1973). An iterative technique for absolute deviations curve fitting. *J. Am. Stat. Assoc.*, **68**, 857–859.

Schoenberg, I. J. (1946). Contributions to the problem of approximation of equidistant data by analytic functions. *Q. J. Appl. Math.*, **4**, 45–99; 112–141.

Schoenberg, I. J., *see also* Curry, H. B. and... (1966).

Schweitzer, A., *see also* Theil, H. and... (1961).

Sclove, S. L. (1968). Improved estimators for coefficients in linear regression. *J. Am. Stat. Assoc.*, **63**, 597–606.

Sclove, S. L. (1972). (Y vs. X) or ($\log Y$ vs. X)? *Technometrics*, **14**, 391–403.

Scott, A. J. and Knott, M. (1974). A cluster analysis method for grouping means in the analysis of variance. *Biometrics*, **30**, 507–512.

Scott, A. and Smith, T. M. F. (1970). A note on Moran's approximation to Student's t. *Biometrika*, **57**, 681–682.

Scott, A. J. and Smith, T. M. F. (1971). Interval estimates for linear combinations of means. *Appl. Stat.*, **20**, 276–285.

Searle, S. R. (1971). *Linear Models*. Wiley: New York.

Seber, G. A. F. (1966). *The Linear Hypothesis: a General Theory*. Griffin's Statistical Monographs No. 19. Griffin: London.

Seber, G. A. F. (1973). *The Estimation of Animal Abundance and Related Parameters*. Griffin: London.

Seely, J. and Zyskind, G. (1971). Linear spaces and minimum variance unbiased estimation. *Ann. Math. Stat.*, **42**, 691–703.

Seshadri, V., *see also* Csorgo, M.,..., and Yalovsky, M. (1973).

Shapiro, S. S. and Francia, R. S. (1972). An approximate analysis of variance test for normality. *J. Am. Stat. Assoc.*, **67**, 215–216.

Shapiro, S. S. and Wilk, M. B. (1965). An analysis-of-variance test for normality (complete samples). *Biometrika*, **52**, 591–611.

Shapiro, S. S., Wilk, M. B., and Chen, H. J. (1968). A comparative study of various tests for normality. *J. Am. Stat. Assoc.*, **63**, 1343–1372.

Shapiro, S. S., *see also* Hahn, G. J. (1967).

Sharma, D., *see also* Ghosh, M. N. and... (1963).

Shearer, P. R. (1973). Missing data in quantitative designs. *Appl. Stats.*, **22**, 135–140.

Sidak, Z. (1968). On multivariate normal probabilities of rectangles. *Ann. Math. Stat.*, **39**, 1425–1434.

Sielken, R. L., Jr. and Hartley, H. O. (1973). Two linear programming algorithms for unbiased estimation of linear models. *J. Am. Stat. Assoc.*, **68**, 639–641.

Silvey, S. D. (1969). Multicollinearity and imprecise estimation. *J. R. Stat. Soc. B*, **31**, 539–552.

Silvey, S. D. (1970). *Statistical Inference*. Penguin Books.

Silvey, S. D. and Titterington, D. M. (1974). A Lagrangian approach to optimal design. *Biometrika*, **61**, 299–302.

Smith, A. M. F., *see also* Goldstein, M. and... (1974).

Smith, A. F. M., *see also* Lindley, D. V. and... (1972).

Smith, H., *see also* Draper, N. R. and... (1966).

Smith, J. H. (1972). Families of transformations for use in regression analysis. *Am. Stat.*, **26** (3), 59–61.

Smith, T. M. F., *see also* John, J. A. and... (1974).

Smith, T. M. F., *see also* Scott, A. and... (1970).

Smith, T. M. F., *see also* Scott, A. J. and... (1971).

Snee, R. D. (1971). A note on the use of residuals for examining the assumptions of covariance analysis. *Technometrics*, **13**, 430–437.

Snell, E. J., *see also* Cox, D. R. and... (1968).

Snell, E. J., *see also* Cox, D. R. and... (1974).

Spjøtvoll, E. (1972a). On the optimality of some multiple comparison procedures. *Ann. Math. Stat.*, **43**, 398–411.

Spjøtvoll, E. (1972b). Joint confidence intervals for all linear functions of means in the one way layout with unknown group variances. *Biometrika*, **59**, 683–685.

Spjøtvoll, E. (1972c). Multiple comparison of regression functions. *Ann. Math. Stat.*, **43**, 1076–1088.

Spjøtvoll, E. and Stoline, M. R. (1973). An extension of the *T* method of multiple comparison to include the cases with unequal sample sizes. *J. Am. Stat. Assoc.*, **68**, 975–978.

Sprent, P. (1961). Some hypotheses concerning two phase regression lines. *Biometrics*, **17**, 634–645.

Sprent, P. (1969). *Models in Regression and Related Topics*. Methuen: London.

Sprent, P. (1971). Parallelism and concurrence in linear regression. *Biometrics*, **27**, 440–444.

Spurrell, D. J., *see also* Newton, R. G. and... (1967a).

Spurrell, D. J., *see also* Newton, R. G. and... (1967b).

Srivastava, M. S., *see also* Bhargava, R. P. and... (1973).

Stefansky, W. (1971). Rejecting outliers by maximum normed residual. *Ann. Math. Stat.*, **42**, 35–45.

Stefansky, W. (1972). Rejecting outliers in factorial designs. *Technometrics*, **14**, 469–479.

Stein, C. (1960). Multiple regression. In *Contributions to Probability and Statistics*, "Essays in honor of Harold Hotelling," pp. 424–443. Stanford University Press: Palo Alto, Calif.

Stein, C., *see also* James, W. and... (1961).

Stiefel, E. L. (1963). *An introduction to Numerical Mathematics*. Academic Press: New York and London.

Stigler, S. M. (1971). Optimal experimental design for polynomial regression. *J. Am. Stat. Assoc.*, **66**, 311–318.

Stoline, M. R., *see also* Spjøtvoll, E. and... (1973).

Stuart, A., *see also* Kendall, M. G. and... (1968).

Studden, W. J. (1968). Optimal designs on Tchebycheff points. *Ann. Math. Stat.*, **5**, 1435–1447.

Studden, W. J., *see also* Karlin, S. and... (1966).

Studden, W. J., *see also* Murty, V. N. and... (1972).

Styan, G. P. (1970). Notes on the distribution of quadratic forms in singular normal variables. *Biometrika*, **57**, 567–572.

Styan, G. P., *see also* Golub, G. H. and... (1973).

Styan, G. P., *see also* Golub, G. H. and... (1974).

Styan, G. P. H., *see also* Grossman, S. I. and... (1972).

Swanson, M. R., *see also* Carmer, S. G. and... (1973).

Sweeny, H. C., *see also* Gaylor, D. W. and... (1965).

Swindel, B. F. (1968). On the bias of some least-squares estimators of variance in a general linear model. *Biometrika*, **55**, 313–316.

Swindel, B. F. and Bower, D. R. (1972). Rounding errors in the independent variables in a general linear model. *Technometrics*, **14**, 215–218.

Tamura, H., *see also* Daling, J. R. and... (1970).

Tatar, J. F., *see also* Morgan, J. A. and... (1972).

Theil, H. (1965). The analysis of disturbances in regression analysis. *J. Am. Stat. Assoc.*, **60**, 1067–1079.

Theil, H. (1968). A simplification of the BLUS procedure for analyzing regression disturbances. *J. Am. Stat. Assoc.*, **63**, 242–251.

Theil, H. and Schweitzer, A. (1961). The best quadratic estimator of the residual variance in regression analysis. *Stat. Neerl.*, **15**, 19–23.

Theobald, C. M. (1974). Generalizations of mean square error applied to ridge regression. *J. R. Stat. Soc. B*, **36**, 103–106.

Thompson, J. R. (1968). Some shrinkage techniques for estimating the mean. *J. Am. Stat. Assoc.*, **63**, 113–122.

Thompson, W. A., Jr., *see also* Dykstra, R. L., Hewett, J. E., and... (1973).

Thompson, W. O. (1973). Secondary criteria in the selection of minimum bias designs in two variables. *Technometrics*, **15**, 319–328.

Tidwell, P. W., *see also* Box, G. E. P. and... (1962).

Tietjen, G. L. and Beckman, R. J. (1972). Tables for use of the maximum *F*-ratio in multiple comparison procedures. *J. Am. Stat. Assoc.*, **67**, 581–583.

Tietjen, G. L., Moore, R. H., and Beckman, R. J. (1973). Testing for a single outlier in simple linear regression. *Technometrics*, **15**, 717–721.

Titterington, D. M., *see also* Silvey, S. D. and... (1974).

Tocher, K. D. (1952). The design and analysis of block experiments. *J. R. Stat. Soc. B*, **14**, 45–91.

Todd, J. (1954). The condition of the finite segments of the Hilbert matrix. *In* O. Taussley (Ed.), *Contributions to the Solution of Systems of Linear Equations and the Determination of Eigenvalues*, National Bureau of Standards Applied Mathematics Series 39, pp. 109–116. U.S. Govt. Printing Office: Washington, D.C.

Todd, J. (1961). Computational problems concerning the Hilbert matrix. *J. Res. Nat. Bur. St.*, **65**, 19–22.

Toman, R. J., *see also* Gorman, J. W. and... (1966).

Tong, Y. L. (1970). Some probability inequalities of multivariate normal and multivariate *t*. *J. Am. Stat. Assoc.*, **65**, 1243–1247.

Tong, Y. L., *see also* Perng, S. K. and... (1974).

Trussell, H. J., *see also* Beckman, R. J. and... (1974).

Tsao, R., *see also* Schatzoff, M.,..., and Fienberg, S. (1968).

Tukey, J. W. (1957). On the comparative anatomy of transformations. *Ann. Math. Stat.*, **28**, 602–632.

Tukey, J. W. (1949). One degree of freedom for non-additivity. *Biometrics*, **5**, 232–242.

Tukey, J. W. (1954). Causation, regression and path analysis. *In* O. Kempthorne (Ed.), *Statistics and Mathematics in Biology*, pp. 35–66. Iowa State College Press: Ames.

Tukey, J. W., *see also* Andrews, D. F. and... (1973).

Tukey, D. W., *see also* Anscombe, F. J. and... (1963).

Tukey, J. W., *see also* Beaton, A. E. and... (1974).

Turner, M. E. (1960). Straight line regression through the origin. *Biometrics*, **16**, 483–485.

Waller, R. A. and Duncan, D. B. (1969). A Bayes rule for the symmetric multiple comparisons problem. *J. Am. Stat. Assoc.*, **64**, 1484–1503.

Walls, R. C. and Weeks, D. L. (1969). A note on the variance of a predicted response in regression. *Am. Stat.*, **23** (3), 24–25.

Wampler, R. H. (1970). A report on the accuracy of some widely least squares computer programs. *J. Am. Stat. Assoc.*, **65**, 549–565.

Warren, W. G. (1971). Correlation or regression: bias or precision. *Appl. Stats.*, **20**, 148–164.

Waterman, M. S. (1974). A restricted least squares problem. *Technometrics*, **16**, 135–136.

Watson, G. S. (1955). Serial correlation in regression analysis. I. *Biometrika*, **42**, 327–341.

Watson, G. S. (1967). Linear least squares regression. *Ann. Math. Stat.* **38**, 1679–1699.

Watson, G. S. (1972). Prediction and efficiency of least squares. *Biometrika*, **59**, 91–98.

Watson, G. S., *see also* Bloomfield, P. and... (1975).

Watson, G. S., *see also* Box, G. E. P. and... (1962).

Watson, G. S., *see also* Durbin, J. and... (1950).

Watson, G. S., *see also* Durbin, J. and... (1951).

Watson, G. S., *see also* Durbin, J. and... (1971).

Wax, Y., *see also* Gabriel, K. R., Putter, J., and... (1973).

Webster, J. T., Gunst, R. F., and Mason, R. L. (1974). Latent root regression analysis. *Technometrics*, **16**, 513–522.

Webster, J. T., *see also* Pope, P. T. and... (1972).

Wedderburn, R. W. M. (1974). Generalized linear models specified in terms of constraints. *J. R. Stat. Soc. B*, **36**, 449–454.

Weeks, D. L., *see also* Walls, R. C. and... (1969).

Westmacott, M., *see also* Healy, M. J. R. and... (1956).

Wetherill, G. B., *see also* O'Neill, R. and... (1971).

Whittle, P. (1973). Some general points in the theory of optimal experimental design. *J. R. Stat. Soc. B*, **35**, 123–130.

Wilk, M. B. and Gnanadesikan, R. (1968). Probability plotting methods for the analysis of data. *Biometrika*, **55**, 1–17.

Wilk, M. B., *see also* Gnanadesikan, R. and... (1970).

Wilk, M. B., *see also* Shapiro, S. S. and... (1965).

Wilk, M. B., *see also* Shapiro, S. S.,..., and Chen, H. J. (1968).

Wilkinson, G. N. (1957). The analysis of covariance with incomplete data. *Biometrics*, **13**, 363–372.

Wilkinson, G. N. (1958a). Estimation of missing values for the analysis of incomplete data. *Biometrics*, **14**, 257–286.

Wilkinson, G. N. (1958b). The analysis of variance and derivation of standard errors for incomplete data. *Biometrics*, **14**, 360–384.

Wilkinson, G. N. (1960). Comparison of missing value procedures. *Aust. J. Stat.*, **2**, 53–65.

Wilkinson, G. N. (1970). A general recursive procedure for analysis of variance. *Biometrika*, **57**, 19–46.

Wilkinson, G. N. and Rogers, C. E. (1973). Symbolic description of factorial models for analysis of variance. *Appl. Stat.*, **22**, 392–399.

Wilkinson, G. N., *see also* James, A. T. and... (1971).

Wilkinson, G. N., *see also* Rogers, C. E. and... (1974).

Wilkinson, J. H. (1965). *The Algebraic Eigenvalue Problem*. Oxford University Press: London.

Wilkinson, J. H. (1967). The solution of ill-conditioned linear equations. *In* A. Ralston and H. S. Wilf (Eds.), *Mathematical Methods for Digital Computers*, Vol. 2, pp. 65–93.

Wilkinson, J. H. (1974). The classical error analyses for the solution of linear systems. *J. Inst. Math. Appl.*, **10**, 175–180.

Wilkinson, J. H. and Reinsch, J. (1971). *Handbook for Automatic Computation*. Vol. III, *Linear Algebra*. Springer-Verlag: Berlin.

Wilkinson, J. H., *see also* Golub, G. H. and... (1966).

Wilkinson, J. H., *see also* Martin, R. S., Peters, G., and... (1965).

Wilkinson, J. H., *see also* Peters, G. and... (1970).

Wilks, S. S., *see also* Guttman, I.,..., and Hunter, J. S. (1971).

Williams, D. A. (1970). Discrimination between regression models to determine the pattern of enzyme synthesis in synchronous cell cultures. *Biometrics*, **26**, 23–32.

Williams, D. A. (1973). Letter to the editors. *Appl. Stat.*, **22**, 407–408.

Williams, D. E., *see also* Cadwell, J. H. and... (1961).

Williams, E. J. (1959). *Regression Analysis*. Wiley: New York.

Williams, E. J. (1969). A note on regression methods in calibration. *Technometrics*, **11**, 189–192.

Wilke, T. A., *see also* Mayer, L. S. and... (1973).

Wilson, R. W., Jr., *see also* Furnival, G. M. and... (1974).

Wold, S., (1974). Spline functions in data analysis. *Technometrics*, **16**, 1–11.

Wolfowitz, J., *see also* Kiefer, J. and... (1959).

Wolfowitz, J., *see also* Kiefer, J. and... (1960).

Wolfowitz, J., *see also* Kiefer, J. and... (1965).

Wood, F. S. (1973). The use of individual effects and residuals in fitting equations to data. *Technometrics*, **15**, 677–695.

Wood, F. S., *see also* Daniel, C. and... (1971).

Wood, J. T. (1974). An extension of the analysis of transformations of Box and Cox. *Appl. Stat.*, **23**, 278–283.

Working, H. and Hotelling, H. (1929). Application of the theory of error to the interpretation of trends. *J. Am. Stat. Assoc., Suppl.* (*Proc.*), **24**, 73–85.

Wu, De-Min, *see also* Richardson, D. H. and... (1970).

Wynn, H. P. and Bloomfield, P. (1971). Simultaneous confidence bands in regression analysis. *J. R. Stat. Soc. B*, **33**, 202–217.

Yalovski, M., *see also* Csorgo, M., Seshadri, V., and... (1973).

Yates, F. (1933). The analysis of replicated experiments when the field results are incomplete. *Emp. J. Exp. Agr.*, **1**, 129–142.

Yates, F. (1972), A Monte-Carlo trial on the behavior of the non-additivity test with non normal data. *Biometrika*, **59**, 253–261.

Yates, F., *see also* Fisher, R. A. and... (1957).

Youngs, E. A. and Cramer, E. M. (1971). Some results relevant to choice of sum and sum-of-product algorithms. *Technometrics*, **13**, 657–665.

Zyskind, G., *see also* Seely, J. and... (1971).

Index

Applied Probability and Statistics (Continued)

JAGERS · Branching Processes with Biological Applications

JOHNSON and KOTZ · Distributions in Statistics
Discrete Distributions
Continuous Univariate Distributions-1
Continuous Univariate Distributions-2
Continuous Multivariate Distributions

JOHNSON and KOTZ · Urn Models and Their Application

JOHNSON and LEONE · Statistics and Experimental Design: In Engineering and the Physical Sciences, Volumes I and II, *Second Edition*

KEENEY and RAIFFA · Decisions with Multiple Objectives

LANCASTER · The Chi Squared Distribution

LANCASTER · An Introduction to Medical Statistics

LEWIS · Stochastic Point Processes

MANN, SCHAFER and SINGPURWALLA · Methods for Statistical Analysis of Reliability and Life Data

MEYER · Data Analysis for Scientists and Engineers

OTNES and ENOCHSON · Digital Time Series Analysis

PRENTER · Splines and Variational Methods

RAO and MITRA · Generalized Inverse of Matrices and Its Applications

SARD and WEINTRAUB · A Book of Splines

SEAL · Stochastic Theory of a Risk Business

SEARLE · Linear Models

THOMAS · An Introduction to Applied Probability and Random Processes

WHITTLE · Optimization under Constraints

WONNACOTT and WONNACOTT · Econometrics

WONNACOTT and WONNACOTT · Introductory Statistics for Business and Economics, *Second Edition*

YOUDEN · Statistical Methods for Chemists

ZELLNER · An Introduction to Bayesian Inference in Econometrics

Tracts on Probability and Statistics

BHATTACHARYA and RAO · Normal Approximation and Asymptotic Expansions

BILLINGSLEY · Convergence of Probability Measures

CRAMER and LEADBETTER · Stationary and Related Stochastic Processes

JARDINE and SIBSON · Mathematical Taxonomy

RIORDAN · Combinatorial Identities